ECOLOGY, IMPACT ASSESSMENT, AND ENVIRONMENTAL PLANNING

WALTER E. WESTMAN

University of California, Los Angeles

A Wiley-Interscience Publication

JOHN WILEY & SONS

New York   Chichester   Brisbane   Toronto   Singapore

Library of Congress Cataloging in Publication Data:

Westman, Walter E., 1945–
    Ecology, impact assessment, and environmental planning.
    (Environmental science and technology, ISSN 0194-0287)

    "A Wiley-Interscience publication."
    Includes bibliographical references and index.
    1. Ecology. 2. Environmental impact analysis. 3. Land
use—Planning. I. Title. II. Series.

QH541.W43   1984   333.7   84-11867
ISBN 0-471-89621-7
ISBN 0-471-80895-4 (pbk.)

Printed in the United States of America

10 9 8 7 6 5

FOREWORD

Walter Westman brings an extraordinary range of interest and experience to bear on the awkward, almost impossible problem of measuring the value of resources and predicting changes due to disturbance. Economic gradients seem all-powerful in determining details of management of forests, fisheries, air, water, and land. Emphasis on immediate profits generates overwhelming pressures to discount the future at virtually any cost. How can we see that appropriate values are assigned to clean water, clean air, and the biotic systems of land and sea that maintain a stable biosphere?

The first step is measurement. What are the dimensions of the renewable resources of the earth? What tools do we have, how accurately can the measurements be made, what are the values gained and lost through management? And what units are best understood by those who join in such important decisions?

Westman addresses these and related issues from the standpoint of a biologist experienced in ecology. But he is also experienced in law, in government, in politics, in economics, in geography, and planning. About these subjects, all obviously close to his heart, he has written fully and well.

We need this book. We need it for the synthesis it brings to the challenge of measuring and interpreting the values of resources that are so commonly discounted or ignored. We need it to enable anticipation of effects of intensified human activity. We need it too for the discussion and argument it will engender, for the stimulation of analysis and the potential that it shows for development of still better methods. The book will be vital to anyone interested in developing his or her own objective appraisal of the limits of the earth for support of humankind.

G. M. Woodwell

Woods Hole, Massachusetts
July 1984

Director, The Ecosystems Center
Marine Biological Laboratory

SERIES PREFACE

Environmental Science and Technology

The Environmental Science and Technology Series of Monographs, Textbooks, and Advances is devoted to the study of the quality of the environment and to the technology of its conservation. Environmental science therefore relates to the chemical, physical, and biological changes in the environment through contamination or modification, to the physical nature and biological behavior of air, water, soil, food, and waste as they are affected by agricultural, industrial, and social activities, and to the application of science and technology to the control and improvement of environmental quality.

The deterioration of environmental quality, which began when people first collected into villages and utilized fire, has existed as a serious problem under the ever-increasing impacts of exponentially increasing population and of industrializing society. Environmental contamination of air, water, soil and food has become a threat to the continued existence of many plant and animal communities of the ecosystem and may ultimately threaten the very survival of the human race.

It seems clear that if we are to preserve for future generations some semblance of the biological order of the world of the past and hope to improve on the deteriorating standards of urban public health, environmental science and technology must quickly come to play a dominant role in designing our social and industrial structure for tomorrow. Scientifically rigorous criteria of environmental quality must be developed. Based in part on these criteria, realistic standards must be established, and our technological progress must be tailored to meet them. It is obvious that civilization will continue to require increasing amounts of fuel, transportation, industrial chemicals, fertilizers, pesticides, and countless other products and that it will continue to produce waste products of all descriptions. What is urgently needed is a total systems approach to modern civilization through which the pooled talents of scientists and engineers, in cooperation with social scientists and the medical profession, can be focused on the development of order and equilibrium in the presently disparate segments of the human environment. Most of the skills and tools that are needed are already in existence. We surely have a right to hope that a technology that has created such manifold environmental problems is also capable of solving them. It is our hope that this Series in Environmental Sciences and Technology will not only serve to make this challenge more explicit to established professionals but that it also will help to stimulate students to pursue career opportunities in this vital area.

Robert L. Metcalf
Werner Stumm

PREFACE

Elucidating the interrelations between the environment and human activities is a central goal of much research in both the natural and social sciences. Yet the task of *predicting* the impact of human actions on ecosystems is in many ways a nascent field, replete with challenges to the student and professional. I discuss here some concepts and methods for assessing the effects of human activities on natural ecosystems. By the "assessment" process I refer to the collection and analysis of data, the prediction of effects, and the evaluation of the human significance of the results. I also discuss the legal and planning context within which assessment takes place. I have written to introduce readers to issues and literature in this complex field.

I have aimed to draw together the perspectives of three subdisciplines which have evolved within distinct traditions: applied ecology, environmental planning, and ecological impact assessment. The approaches used to prepare environmental impact statements are subsumed within this larger context. My focus is on biological and physical, rather than socioeconomic, components of environmental systems.

By drawing from the research literature, I have given this book a deliberate theoretical focus. The last five chapters constitute a review of modern concepts in applied ecology of relevance to environmental planning and analysis and provide an extensive biological core for the book. The earlier chapters, which deal with topics in environmental planning and impact assessment, are written from an ecological perspective.

In addition to reviewing commonly used methods of assessment, I also describe some that are newborn and teething. While some techniques will be of aid to those involved in rapid survey tasks, other techniques are appropriate only for longer-term studies. Most of the approaches have been selected to highlight important conceptual issues in a rapidly evolving field.

I stress the interdisciplinary nature of the assessment task, drawing on perspectives offered by planning, law, decision theory, economics, physical geography, ecology, toxicology, and other fields. The broader social science context (law and planning, evaluation and decision making) is discussed in the first five chapters, followed by predominantly natural science approaches to the prediction of impacts to ecosystem components—land, air, water (Chapters 6,7), and biota (Chapters 8–12). Quantitative approaches and probabilistic ways of thinking are emphasized, and applications of the computer are noted throughout.

G. Evelyn Hutchinson has written (1975, p. vii), "The ecologist is continually having to look at aspects of nature with which he is unfamiliar and perforce must be an amateur for much of his working time."* This experience is shared by environmental analysts of all stripes. Because of the nature of environmental assessment, the range of topics in this book is unabashedly large. My aim has been to present a spectrum of concepts and methods to illustrate both the reach of the subject area and its commonalities. Although I have stressed the conceptual basis for the topics addressed, and used numerous examples, discussion of many issues is condensed. Citations to approximately 1200 recent books and articles provide a source for readers seeking more detailed treatment of a topic.

The book is appropriate as a text in an upper division undergraduate, graduate, or professional course in impact assessment, applied ecology, environmental planning, resource management, or environmental design. The book should also be useful as a reference for environmental consultants, resource managers, planners, and decision makers. Some background in the environmental sciences is assumed, including an elementary knowledge of ecology, economics, and statistics.

I write from the perspective of an academic with primarily ecological training. I have worked in the legislative branch of government and in university teaching and research in departments of botany, urban planning, and geography, primarily in the United States and Australia. I have sought to broaden the viewpoint of this book by surveying techniques used in professional and academic circles worldwide and have written for an international audience. Nevertheless, the United States serves as the most frequent point of reference.

I wish to thank the Department of Forestry and Resource Management at the University of California, Berkeley, for hosting me during a sabbatical leave when much of this book was written, and students and colleagues at the Department of Geography, University of California, Los Angeles, for their support. I am also grateful to Dr. George M. Woodwell, Director of the Ecosystems Center, Marine Biological Laboratory, Woods Hole, for writing the Foreword, and to the following people for their numerous helpful comments on the outline and drafts of particular chapters: S. Beatty (UCLA), J. Cairns, Jr. and D. Conn (Virginia Polytechnic Institute and State University), R. Corwin (Marin County Planning Department), J. T. Gray (Dames and Moore), E. Hobbs (Colgate University), D. Liverman (University of Wisconsin), T. Meredith (McGill University), C. Salter (UCLA), and G. Wandesforde-Smith (University of California, Davis).

WALTER E. WESTMAN

Los Angeles, California
July 1984

* Hutchinson, G. E. (1975). *A Treatise on Limnology.* Vol. 3. Wiley-Interscience, New York.

CONTENTS

PART V.   PREDICTING IMPACTS: THE BIOTA

PART ONE

INTRODUCTION

1

ECOLOGICAL IMPACT ASSESSMENT AS A DISCIPLINE

The demand for impact analysis has propelled ecologists, geographers, and planners into what has at times seemed a murky world of futurology. Imprecision in predicting the response of ecosystems to human activities has stemmed from at least two sources: the difficulty of extending a largely descriptive ecological literature into a predictive mode, and the complexity and interconnectedness of ecosystems themselves. Yet within the recent literature in a range of disciplines lie concepts and methods that provide the framework for a quantitative science of ecological impact analysis.

Ecological impact assessment involves not only analysis but also evaluation of the significance of the predicted ecological alterations to human society. The evaluation process contains its own imprecisions in striving to apply human values in an explicit, quantitative fashion. In this book the broad themes of ecological impact analysis and evaluation are reviewed.

The prediction of impacts may form part of tasks in environmental planning and design, resource management, and applied ecology. To design an urban development project, a park reserve, a rehabilitation scheme, or a forest management plan requires understanding how the natural environment will respond to the proposed manipulation. This calls upon a knowledge of ecology, principles of environmental design, and a predictive capability. New policies, laws, or technological innovations may also be evaluated using principles of impact assessment.

There is not yet a universally acknowledged body of theory and methodology that can be applied to the analysis and evaluation of future impacts on the natural environment. Most impact analysts began their task by drawing on concepts and methods in a range of traditional disciplines, and the practice of impact assessment remains interdisciplinary. Nevertheless, new approaches of particular application to ecological impact assessment have developed, new scholarly journals—such as *Environmental Impact Assessment Review, Environmental Management,* and the *Journal of Environmental Management*—have appeared, and learned societies—such as

3

the International Association for Impact Assessment and the Association of Environmental Professionals—have formed.

ECOLOGICAL IMPACT ASSESSMENT DEFINED

This book focuses on the prediction and evaluation of the effects of human activities on the structure and function of "natural" ecosystems—those close to their evolutionary state—though many of the concepts and methods also apply to heavily modified systems such as farms or urban areas (Figure 1.1). The realm of concern is not only the tangible features or "structure" of ecosystems (e.g., plants, animals, soil) but also the exchange of energy and materials between ecosystem components: the dynamics of interaction or "function" of ecosystems.

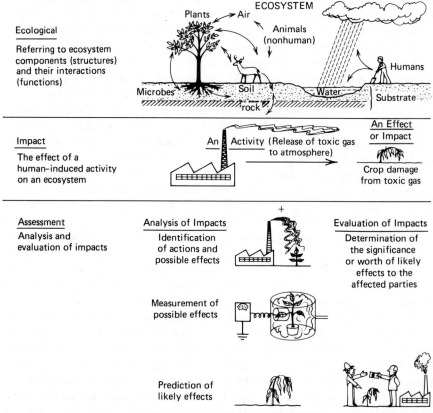

Figure 1.1. Definitions of words in the phrase "ecological impact assessment." Although the phrase is often associated with environmental impact statements, it can also refer to a wide range of predictive tasks within environmental planning.

A number of authors (Andrews et al. 1977, Dooley 1979, Lee 1982) have noted that the term "impact" has sometimes referred both to a human-induced *action,* or activity, and to its *effect* on ecosystems, or to the *effect* and its *significance* to human society. Action, effect, and significance are three distinct concepts in impact assessment; we will limit the term impact to the *effect* of a human-induced action on an ecosystem.

"Assessment" here refers to *analyzing* and *evaluating* impacts on ecosystems. *Analysis* is the *objective* task of identifying actions, taking measurements of baseline conditions, and predicting the changes to these baseline conditions that are likely to occur as a result of the actions. *Evaluation* is a *subjective* or normative task which depends on the application of human values. It involves determining the significance of the effects to the affected parties. In the case of effects on objects rather than people, the owners of those objects generally are taken to represent them.

For example, to evaluate the effect of air pollution damage to a crop, one would determine the significance of the damage to the farmer who owned the crop. Often the significance is measured in economic terms, such as the market value of the amount of the crop that was lost. Many objects in ecosystems are not marketed (e.g., mosquitoes), and damage to these objects must be evaluated in other ways. In the case of air pollution damage to a mosquito, the owner of the marshland on which it was killed may actually feel blessed by this event, in which case the significance of the damage may be evaluated as beneficial, though no monetary value may be placed on the benefit. A devout Jain or Zen Buddhist, who sanctifies all creatures including insects, would be less pleased (Watanabe 1974). Clearly the value placed on the significance of an effect depends on whose values are employed.

RATIONALES FOR ASSESSING ECOLOGICAL IMPACTS

The opportunity to identify costly and undesirable effects, and to modify projects in the design stage, is a chief rationale for conducting environmental impact assessments. In the United States as of 1970, and in at least 29 other countries since, an environmental impact statement (EIS) or report has been required for selected actions that are expected to have significant effects on the human environment. The cost of preparing such assessments at the federal and state level in the United States averages 1% or less (range: 0.1–5.4%) of the total cost of the proposed projects (Council on Environmental Quality 1976; Zigman 1978). Design modifications sometimes themselves lead to cost savings: design changes due to EIS's saved $35 million in construction costs for 49 sewage treatment plants in the United States.

The costs of ignoring potential ecological effects of human activities are well illustrated by the discharge of kepone pesticide into the James River of Virginia, a tributary of Chesapeake Bay. Kepone was discharged directly to the sewage treatment plant in Hopewell, Virginia, from the kepone manufacturing plant during 1966–1975, saving the manufacturer approximately $200,000 in pollution control costs (Miller 1982). Kepone killed sewage-digesting microorganisms, resulting in the discharge of inadequately treated sewage, and kepone, into the James River. A 150 km stretch of river was closed to fishing indefinitely (Council on Environmental Quality 1979), with sport and commercial fishery losses estimated at $20 million during 1975–1980 (Council on Environmental Quality 1980). The company has so far paid $13 million of the $160 million in health damage suits from workers and an additional $13 million in pollution fines. Two company executives were also convicted and fined (Council on Environmental Quality 1977).

Although economic indexes provide a means to quantify some social gains and losses, many impacts on social structures and natural resources cannot be adequately expressed by economic values. The realm of impact assessment embraces a larger set of changes in features of the human environment than those expressed by economic indexes. This book reviews a range of approaches for expressing and evaluating impacts quantitatively.

INTEGRATED IMPACT ASSESSMENT

A human action such as mining simultaneously affects both the natural and the social environment, not only displacing plants and polluting water but creating jobs and relocating people. Clearly a comprehensive assessment of mining impacts would have to consider both ecological and social effects, and the higher-order cumulative effects that result from their interaction.

In the early 1970s, when environmental impact statements were first being written in the United States as a result of the National Environmental Policy Act (1969), most EISs were limited to the webwork of effects on the natural environment. Through a series of legal challenges and administrative modifications, the scope of impact statements was gradually broadened to encompass a range of social and economic concerns (Figure 1.2, Table 1.1). Social impact assessment in the United States was also stimulated by the requirements of Sec. 122 of the River and Harbor and Flood Control Act of 1970 (U.S. Army Corps of Engineers 1973) and the requirement of the Department of Housing and Urban Development to prepare impact assessments for urban development plans.

Examination of the full social and ecological impacts of a proposed action requires a "holistic" approach, in the sense that examination of the effects on natural and social systems separately will not reveal the full scope of interactive effects. An example in which the secondary interactions between social and ecological effects played the crucial role in deciding the fate of a project occurred in the case of a proposed airport north of Everglades National Park, a major subtropical swamp forest area in Florida. The EIS indicated that the direct ecological effects of constructing and operating the

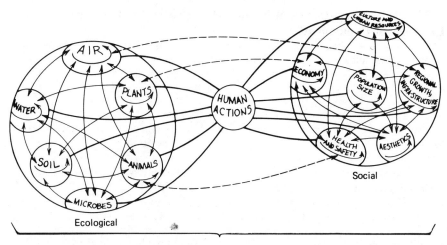

Integrated Impact Assessment

Figure 1.2. The realms of concern of ecological, social and integrated impact assessment. Direct effects of human actions on ecological and social systems are represented by heavy lines. The circular arrows within ecosystem or societal compartments represent internal dynamics (fluxes of energy, materials, resources). Arrows between compartments represent indirect, interactive, or higher-order effects.

airport on the nearby Everglades would be negligible. The secondary effects of the airport in encouraging urban development in the region, however, would in turn significantly alter the amount and quality of water flowing into the Everglades. It was on the basis of these long-term secondary effects that the U.S. Department of Transportation shelved the project (Convisser 1978).

The special skills required for an assessment of ecological impacts derive from a distinct, though overlapping, set of disciplines from those required for social impact assessment. In this book we limit focus to concepts and methods applicable to ecological analysis and evaluation. Readers interested in techniques of social impact assessment may consult books and symposia by Burkhardt and Itlleson (1978), Chalmers et al. (1982), Finsterbusch and Wolf (1977), Grigsby and Glickfield (1978), Leistritz and Murdock (1981), McEvoy and Dietz (1977), and bibliographies by Carley and Derow (1983), Clark et al. (1980), and Glickfield et al. (1977). Books discussing technology assessment include those of Arnstein and Christakis (1975), O'Brien and Marchand (1982), and Porter et al. (1980). A symposium that explores the commonalities among the various forms of impact assessment is that edited by Rossini and Porter (1983). The generic term "integrated impact assessment" has been proposed to refer to a study of the full range of ecological and social consequences of the introduction of a new technology, project, or program (Porter and Rossini 1983).

Table 1.1. Compartments of Ecological and Social Systems, with Examples of Components

Ecological Systems

Air
  Constituents (gases, particles)
  Energy and moisture content (described as climate, microclimate)
Water
  Chemical contents
  Volume of water
  Energy content

Soil and substrate
  Physical structure
  Chemical composition
  Microbes, plants, and animals
  Composition
  Abundance
  Distribution

Social Systems

Health and safety
  Crime levels
  Public risk of injury, health impairment, or death
  Psychological environment, including anxiety levels, personal comfort and enjoyment, privacy
Economy
  Employment
  Housing
  Commerce
  Cost of living
Cultural and urban resources
  Belief systems: religious, political; social values
  Identification: ethnic, racial, lifestyle, community
  Recreational and scientific resources
  Historical and archaeological resources
Aesthetic characteristics
  Odors
  Noise
  Atmospheric visibility

Visual qualities of landscape
Vibrations
Light quality and quantity
Regional growth and infrastructure
  Provision of social services, including police, fire protection, energy, sewage treatment, water supply, flood protection, solid waste disposal, health care, transportation, education
  Changes in land use
  Government plans and policies, including zoning policies, national and local plans, environment and land use laws and policies, antiquities and historic preservation acts, treaties and other governmental obligations
Population characteristics
  Birth and death rates
  Density and distribution
  In- and out-migration
  Age structure
  Sex ratio

IMPACT ASSESSMENT AND ENVIRONMENTAL PLANNING

An accurate prediction of impacts to a site cannot be made without a knowledge of the other proposed projects for the area and the intensity of direct and indirect stresses they will impose on the site. Thus the marine life in a bay may be able to resist severe impact from an oil tanker terminal or an oil refinery alone, but not the combined impacts of the two. For an accurate prediction of cumulative impacts to be made, impact analysts must be able to refer to some plan for the future development of the region. The plans in turn usually derive from a set of policies for regional development and national goals and values. To speak to the combined effects of independent proposals for a region, impact assessment is dependent on regional planning.

To be comprehensive and systematic, planning should proceed from the level of national goals to regional and local considerations. Thus an analysis of a proposed offshore oil drilling project would begin by considering alternative energy policies for the nation, then examine the cumulative effects of offshore oil drilling along the entire stretch of coast where lease sales are proposed, and finally consider the specific impacts of drilling at a particular site. This hierarchy of analytic steps is portrayed in Figure 1.3.

The preparation of impact statements for entire programs, with site-specific impact statements referring to the "programmatic EISs" for discussion of programwide issues, is encouraged by the Council on Environmental Quality (CEQ) guidelines (1978, Sec. 1508.26) in the United States. The process of nesting the site-specific EISs within broader programmatic ones is termed "tiering." Other examples of programmatic EISs for related projects include RARE II: Roadless Area Review and Evaluation (U.S. Forest Service), which classified 2900 roadless wilderness areas in National Forests in 38 states and Puerto Rico into those appropriate for wilderness preservation and those appropriate for development; and the Coal Loan

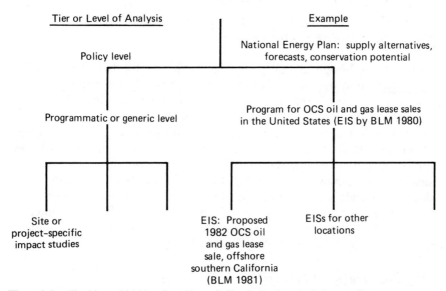

Figure 1.3. The hierarchical levels or tiers of planning and analysis from policy to program to site-specific project. The example at right is for outer continental shelf (OCS) oil drilling in the United States (BLM = Bureau of Land Management, U.S. Department of Interior.)

Guarantee Program (U.S. Department of Energy 1978), written to assess the impacts of implementing Federal Law P.L. 94-163. This law guarantees loans to small operators to mine primarily low-sulfur underground coal. The programmatic EIS considered the effects of opening 95 new mines and re-opening 160 existing mines. As individual mines opened, they would, in most cases, have to file further impact statements or reports.

THE PHASES OF ECOLOGICAL IMPACT ASSESSMENT

A possible sequence for conducting an impact assessment is illustrated in Figure 1.4. A set of questions that may guide the formulation of the pre-impact phases of an impact study is listed in Table 1.2. Comments on aspects of this table follow.

Phase I: Defining Study Goals

In countries where private consulting firms conduct many of the assess-ments (e.g., United States, Canada, Australia), study goals are usually de-fined before the contract is awarded to the firm. The public can play a useful role in defining study goals by communicating their long-range goals and desires, the alternatives they would be willing to consider, and the possible effects of proposed actions. "Scoping" meetings (see Chapter 2), which define the major issues to be investigated in the study, may be attended by representatives of public interest groups.

Phase II: Identifying Potential Impacts

Constraints on design of the study may be set by the boundaries of the project itself (how long it will last; how large an area it will encompass); by scientific limitations in the ability to measure or predict ecological changes; by administrative constraints; and by the boundaries of movements of spe-cies, materials and energy in the ecosystem in time and space (Beanlands and Duinker 1982). Programmatic EISs may be written to encompass larger boundaries and to assess the cumulative impact of a series of related projects. Other actions triggered by the proposed action must also be in-cluded in the study's scope (e.g., strip development of shops and houses by a new road, in turn requiring new social services). The U.S. Environmental Protection Agency has issued guidelines to assist in assessing the secondary impacts of sewage treatment plants and highways (Jameson 1976).

Defining Significant Impacts

Guidelines in the United States (CEQ 1978, Sec. 1508.25) note that "signifi-cance" of an effect will vary with the context within which the effect is

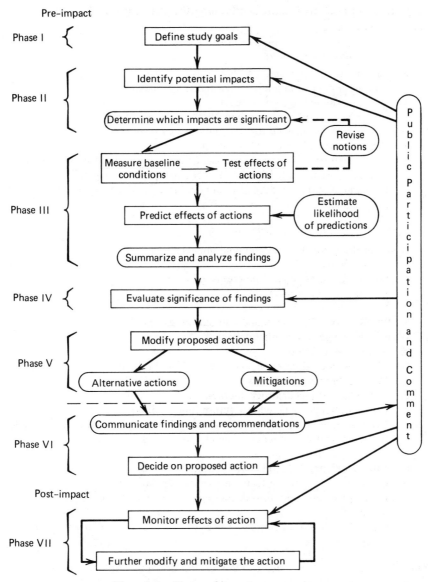

Figure 1.4. Phases of impact assessment.

evaluated (e.g., local vs. global scale) and that several contexts should be considered. Intensity of effect is an important element affecting significance. An impact is more likely to be significant the greater the magnitude of deviation from background levels, the larger the area affected, the longer the effect will last, or the greater the likelihood of occurrence (Andrews et al. 1977).

Table 1.2. Questions Useful in Planning the Pre-Impact Phases of an Impact Assessment

Phase I: Defining Study Goals

1. What information is needed, and how precise must it be for
   a. The proponent to minimize environmental impact?
   b. The government agency to reach a decision on approving the project?
   c. Concerned groups to know how they will be affected?
2. What resources are needed for the study? What resources are available?
   a. What expertise is needed? Available?
   b. How much time is needed for baseline and experimental studies? How much time remains before the project is supposed to begin?
   c. How much money is needed to conduct the proposed studies? How much is available?

Phase II: Identifying Potential Impacts

1. What are the boundaries of potential impacts?
   a. Area affected.
   b. Organisms or ecological functions affected.
   c. Duration of the project.
   d. Interval before effects occur.
   e. Duration of effects with and without mitigation.
2. What is the range of potential impacts?
   a. Major direct actions.
   b. Major ecological components (air, water, land, biota, structures) affected.
   c. Major ecological processes affected.
   d. Secondary or higher-order interactions.
   e. Indirect effects triggered at a future time or different place.
   f. Other actions (past, present, reasonably foreseeable future) that may add to the present action, causing cumulative effects.
3. Which potential impacts are most significant? Which effects will
   a. Violate existing laws, plans or policies?
   b. Cause major adverse effects on species population numbers?
   c. Cause major disruption to ecosystem processes, affecting species significantly?
   d. Cause health risks, economic losses or significant social disruption to people?

Phase III: Measuring Baseline Conditions and Predicting Significant Impacts

1. *Baseline Conditions:* What are the significant features of the ecosystem presently?
   a. What is the current pattern of fluctuation in population sizes for important species (measured over sufficient time to characterize the range of variation)?
   b. Which species are playing a dominant or critical role in maintaining ecosystem processes? What is their abundance, distribution, and functional behavior?

Table 1.2 (*Continued*)

c. What is the condition (quality, quantity, dynamics) of physical resources of the ecosystem?

d. What are the major pathways of interaction between ecological components?

e. What sources of stress from natural or human-induced sources already exist (fire, air pollution, grazing, etc.)? With what intensity and periodicity do these stresses occur?

2. *Predictions:* What will be the major effects of the proposed action? What is known from each of the following?

a. *Case Studies:* Extrapolation of effect from similar instances of disruption to the same or similar ecosystems elsewhere.

b. *Modeling:* Predictions from conceptual or quantitative models of ecosystem interaction.

c. *Bioassay and Microcosm Studies:* The effects of simulated disturbances on ecosystem components under controlled conditions.

d. *Field Perturbation Studies:* Response of a portion of the proposed project area to experimental disturbance.

e. *Theoretical Considerations:* Predictions of effect from current ecological theory.

3. *Estimation of Likelihood:*

a. What is the probability of occurrence of the predicted events?

b. How precisely can the magnitude and likelihood of impacts be estimated?

4. *Summarizing and Analyzing Findings:*

a. How can findings be summarized in tables, graphs, or indexes so that key findings emerge?

b. What is the ecological interpretation of the findings?

Phase IV: Evaluating Significance of Findings

1. How are the effects distributed among affected groups?

a. What is the nature and magnitude of impact on each affected group?

b. What weight shall be given to the concerns of each group?

c. What weight does each group give to the significance of predicted effects?

2. How well are goals achieved by the proposal?

a. Proponent's goals?

b. Governmental goals and policies?

c. Goals of affected groups?

3. What is the overall social significance of the predicted ecological effects?

a. How can effects be expressed in terms that allow meaningful comparison with other social goods, services, and values?

b. If monetary values are placed on normally unpriced goods and services, what features are inadequately evaluated by this procedure?

Phase V: Considering Alternatives to the Proposed Action

1. What alternatives to the proposed action exist?

a. What would be the effect of not proceeding with the project?

(continued on next page)

Table 1.2   (*Continued*)

---

   b.   What would be the effect of achieving ultimate project goals by an en-
        tirely different means (e.g., maintaining electrical service to a growing
        population by conserving energy rather than building a new power plant)?
   c.   What alternative designs could achieve project goals?
2. What steps could be taken to mitigate adverse environmental effects of the
   proposed project?
   a.   Could parts of the proposal be reduced or eliminated?
   b.   Could expected damage be repaired or rehabilitated?
   c.   Could ongoing management procedures be instituted to reduce damage?
   d.   Could affected components be replaced or owners compensated?
   e.   Could project design be modified to reduce effects?
   f.   Could effects be monitored, and provision made for future mitigation
        of project effects when the exact nature and extent of effects are better
        known?

The CEQ guidelines consider particularly important the effects on public
health or safety, on unique geographic areas (historic sites, park lands,
prime farmlands, wetlands, wild and scenic rivers, ecologically critical areas), and on locations or policies regulated by law. Furthermore actions or
effects that are likely to be highly controversial, involve unique or unknown
risks, or establish an important precedent are also considered to be significant.

Sometimes the levels of deviations from background have been defined as
"significant" by law (Andrews et al. 1977). Thus pollutant emissions that
will cause ambient concentrations to exceed set standards will violate pollution laws and result in clearly significant impacts. The majority of impacts,
however, do not clearly violate existing standards (Hanmer 1978); rather
they illuminate more subtle changes in which the judgment of significance
cannot be determined by law. For a further discussion of "significance" of
impacts as interpreted by U.S. courts, see Golden et al. (1979).

A major problem with arriving at an objective definition of "significance"
is that the term is itself normative and depends on the human values applied
in the judgment of significance. Beanlands and Duinker (1982), in a summary
of Canadian experience with ecological impact assessment, suggest that
what is considered important by decision makers, the public, and scientists
should be the guide to "significance." Although we might wish to divorce
the impact assessor completely from such judgments and leave them to the
decision maker, in practice this is usually neither wholly possible nor entirely helpful. In protecting society from abuse of this power of judgment,
the assessor should strive to make explicit the values being used in judging
significance, to err on the side of including matters whose significance is

uncertain, and to consult with government officials and representatives of the multiple publics regarding the values to be applied in judging "significance." Planning tools discussed in Chapters 4 and 5 are available to aid in this task.

Phase III: Predicting Significant Impacts

Measuring Baseline Conditions

Predictions should be presented in a form sufficiently precise to be useful to decision makers. The ecologist must first understand the pattern of natural fluctuations that occurs in ecosystem variables before determining whether a predicted (or measured) change in the ecosystem variable represents a significant departure from the normal range of variation.

Figure 1.5 illustrates fluctuations in numbers of salmon in a stream segment before and after a thermal discharge. Monitoring data taken after the project was completed indicate only that there are annual variations around a mean level and a small downward trend in population numbers.

The mean number of salmon present, however, is lower after the project than before. To determine the significance of this difference, we compare the range of variation within which 99% of the values fall, both before and after the project. In the first four years after the project the range of variation

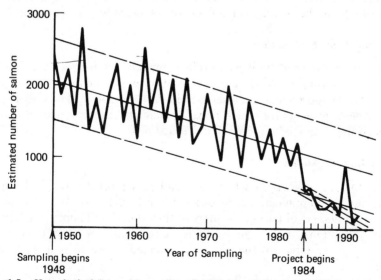

Figure 1.5. Hypothetical data on the number of salmon caught in a one-mile stretch of river in April of each year, before and after a power plant began discharging heated effluent into the river segment. The dashed line indicates the range of values within which 99% of the pre- or post-project observations fell. The thin solid line represents a linear regression, or trend line, through the scatter of points. The thick solid line connects individual observations. The dot represents the mean of pre- or post-project measurements.

overlaps with the pre-project range (even though three of the first four individual post-project values do not). In the fifth through eighth year, however, the envelopes of variation do not overlap, indicating a less than 1% chance that the range of variation in these post-project years can be considered part of the "normal" variation observed before the project began.

A downward trend in salmon numbers existed before the project began, but the slope of the trend is steeper after the project, indicating a 99% chance that the pre-project downward trend has been exacerbated by the project, at least after 1988. If we had sampled the preexisting environment for only one year before the project started, we could not detect any of these things. In fact in the first year after the project, salmon numbers went up compared to the previous year, and they reached their highest level in the previous eight in the sixth year after the project, a time when our statistical analysis indicated that the salmon levels were significantly lower than normal.

Another feature of this example is troubling. There are 37 years of pre-impact data, and only 8 years of post-impact data. This difference in sample size will influence the size of the envelopes of variation and hence of overlap. Given the existence of variation in ecological parameters, it would be desirable for pre- and post-impact data to be collected over extended periods of comparable length. If trends are linear and variation around the trend random, longer sample periods will narrow the confidence bands or "fiducial limits" around the data, permitting more precise statements of probability to be made. Such data also enable detection of other sources of ecosystem stress which can be distinguished from stresses due to project actions.

Testing Effects of Actions

Five avenues for predicting post-project impacts are use of case studies, models, laboratory studies, field perturbation experiments and ecological theory. These approaches are discussed in Parts IV and V. The post-impact monitoring of projects for which impact studies have been conducted is an aspect of the case study approach discussed next.

Post-Impact Audits

The number of projects in which pre- and post-project ecological impact studies have been published is not large (Table 1.3). The study of Tomlinson and Bisset (1982) is of particular interest: they found that approximately 5% of the predictions in the three studies examined were too imprecisely worded to be testable; of the predictions tested, close to half were inaccurate.

Other studies that examined some aspect of EISs following their completion include the reviews of ten EISs by ecologists in the United States, sponsored by the Institute of Ecology (Applegate and Baldwin 1973; Fletcher 1974; Fletcher and Baldwin 1973; Pearson et al. 1975, Smythe 1974,

**Table 1.3.   Post-impact Audits of Predictions Made in
Impact Assessments**

Type of Project	Comments	References
Motorcycle race in Mohave Desert, U.S.	Before and after photos and field measurements of damage to vegetation, soils, wildlife	U.S. Bureau of Land Management 1974, 1975; Bisset 1980
Nuclear power plants, U.S. (thermal effluents)	Effect of discharge on aquatic ecosystems compared to limits specified in operating licenses (12 plants)	Murarka et al. 1976
Nuclear power plants, U.S. (socioeconomic impacts)	Tested predicted effects on socioeconomic variables after several years of operation (12 plants)	Chalmers et al. 1982
Channelization, U.S.	2–3 year post-impact study of swamp channelization in Georgia	U.S. Soil Conservation Service, cited in Bisset 1980
Oil terminal, U.K.	Found 5% of predictions too imprecise to be tested; 8 of 24 audited predictions inaccurate	Tomlinson and Bisset 1982
Reservoir, U.K.	3 of 7 audited predictions inaccurate	Tomlinson and Bisset 1982
Steelworks, U.K.	7 of 11 audited predictions inaccurate	Tomlinson and Bisset 1982
Reservoir, U.S.	Studied social and ecological impacts of flood control reservoir in Illinois 10 years after construction	Burdge et al. 1980

Winder and Allen 1975). The reviews were largely based on reading of the EISs, but some site visits were also undertaken.

Sometimes similar projects in operation elsewhere can serve as case studies. In a Canadian example the Peace-Athabasca Delta Project Study Group (1973, cited in Beanlands and Duinker 1982) examined recovery of a drained wetlands in northern Saskatchewan to gain insight into the effect of similar actions to be instituted by the proposed delta project.

CEQ guidelines (Sec. 1505.3, 1978) require federal agencies to monitor the effects of proposed mitigation practices in "important cases," and to report on the progress of such mitigation measures to other agencies and the public on request. In practice, this provision has not yet been widely enforced, and the results are not normally published. Post-project follow-up studies represent a largely unexploited opportunity to enhance predictive skills.

Estimating Likelihood of Predictions

A distinction should be made between predictions arising from *projection* and those arising from *inference*. In making predictions, a scientist may extrapolate future events from a model of past and present trends. In such a *projection* or forecast from past trends (points 1 and 2 on Figure 1.6), one is assuming (1) that causal relations will continue to operate as they have in the past, and (2) that the model (or equation) being used to fit past data accurately reflects all sources of past and future variability in the magnitude of the event. In other words, we assume that the future will behave like the past and that we know accurately what the past has been like. In making predictions about complex systems such as ecosystems, however, we may not be able to arrive at a prediction through projection, either because (1) we may not have adequate past observations on which to build a model for extrapolative purposes or (2) we may have reason to believe that the future may not be like the past.

An example of a prediction from *inference* is the prediction of performance of a complex new technology, such as a nuclear power plant. Projections of operating conditions in the future are not possible because the past history of operation of the technology is not long enough to model a pattern of frequency for all types of accidents and malfunctions, including core meltdown (a most serious, but highly infrequent accident type). An attempt

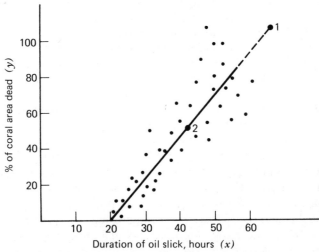

Figure 1.6. Predictions by projection from past observations. The solid line is a regression line through observed events of coral death, *y*, under particular duration of exposure of oil, *x* (hypothetical data). Point 1 is predicted by extrapolation, point 2 by interpolation; both are projections from past events.

at prediction from inference by Rasmussen et al. (U.S. Atomic Energy Commission 1974) involved, among other approaches, estimating the likelihood of malfunction of reactor components, based on past industrial performance records of previously used components. Clearly performance records for all components are not equally well known, and rate of failure due to interactions of the components in a new technology are known only from the limited trial experience with the new technology itself. The inference is made that extrapolations from analogous situations will provide accurate predictions. When we consider the performance of ecosystems and their components, our ignorance of past behaviors is often comparable. Because ecosystems are more complex than machines, predictions by inference about the interaction of a new component with the system are even less likely to be accurate.

Some questions of interest in making predictions are (1) How strongly is the event of interest related to the predictor variable? (2) What is the chance that the observed value will deviate from the predicted one by a given amount? (3) What is the chance that the event will occur in the first place? We may consider these questions in light of the example of death of coral reef area during an oil spill (Figure 1.6). A prediction such as "70% of the coral reef will die from exposure to an oil slick for 50 hours" might be derived from the straight line in Figure 1.6, if the latter is taken as a least-squares fit of a regression line through the scatter of points.

Such a simple linear regression equation can be taken as the best-fitting model for the observed data, if we assume that (1) the sum of squares of deviations of y-values from the regression line are at a minimum; (2) the error terms (e) are independent of each other and normally distributed with a mean value of zero and constant variance; (3) the values of X are fixed, and measured without error; (4) for given values of X, the corresponding values of Y are independently and normally distributed; the mean of Y values lies along a straight line. For detailed discussions of regression analysis see such statistical texts as Cochran (1963), Poole (1974), and Snedecor (1946).

The question of how strongly coral death (y) is related to duration of the oil slick (x) may be addressed by examining the correlation coefficient (r) of the regression. This coefficient indicates the strength of association between x and y. The probability (P) that the observed level of correlation would occur by chance alone can also be computed. These statistics tell us about the strength and significance of association between duration of oil slick and coral death.

The second question (chance of given deviation from prediction) can be addressed by examining so-called "fiducial" limits. Specifically the upper (l_u) and lower (l_l) limits within which 95% of the y-values fall can be calculated using the t-distribution with $n - 2$ degrees of freedom:

$$l_u = y + t_{0.05}S_y \tag{1.1}$$

$$l_l = y - t_{0.05}S_y$$

where

$$S_y = [(S_{y \cdot x})^2(1 + 1/n + x^2)/\Sigma x^2]^{1/2}$$

y = estimated value of y from the regression equation, given a specific x

$t_{0.05}$ = the value of t with $n - 2$ degrees of freedom, given a 5% probability of a larger value of t due to chance alone (tables of the value of t are available in many statistical texts, e.g., Snedecor 1946, p. 65)

$(S_{y \cdot x})^2$ = variance in the observed values of y from those predicted by regression on x

n = number of pairs of observations (x,y)

x = values of the independent variable

The fiducial limits then tell us the range within which the estimate of % dead coral will fall 95% of the time.

It should be borne in mind that simply because coral death can be predicted with a certain reliability by duration of oil slick does not establish that oil slicks are *causing* the coral death. We have merely a statistical association which can be used for predictive purposes; we have not demonstrated causation from these data alone.

The third question, the likelihood of occurrence of the predicted event (coral death), hinges on the question of causation. If oil slicks were the only cause of death, we could state from Figure 1.6 that below 20 hours of exposure to oil slicks, no coral would die. But we cannot say from information in Figure 1.6 alone whether other factors (e.g., climate, predators) might not be responsible for coral death instead or in addition. Causation is theoretically impossible to prove in complex systems. We may approach a theory of causation, however, by testing all reasonable hypotheses to explain the event, identifying those with the most evidence consistent with them and no evidence against them, demonstrating that the event has always occurred with the causal factor and never without it, and finally elucidating mechanisms of causation.

Phase IV: Evaluating Significance of Findings

Once data are collected, they must be summarized and key findings highlighted. Techniques for accomplishing this are discussed in Chapter 4. Many impact analyses stop at this point; having arrayed the findings, they leave to decision makers the task of evaluating the significance of findings. As noted earlier, it is appropriate to relegate the social judgment phase of impact assessment to that group of people empowered to represent social values. The task of evaluating a large data set is itself complex, however, and the planner can be of substantial assistance in pointing out to the decision maker the likely significance to be felt by each affected interest group from each impact. In other words, the planner can here play a role as information broker between affected interest groups and decision makers. The danger in this role bears repeating, however. Some evaluation techniques, including cost-benefit analysis, call on the evaluator (planner) to assign weights (in monetary units) to the value of goods and services. The decision maker must approach such an evaluation with a full understanding of the limitations imposed by having weights incorporated implicitly in the evaluation.

In the United States, CEQ guidelines (1978; Sec. 1502.15a, b) call for an evaluation of the significance of direct and indirect effects, and further imply that economic methods of evaluation can be used. Section 1502.23 of the guidelines specifically refers to the possible use of cost-benefit analysis in EISs and notes that when such analyses are conducted, the EIS should discuss "the relationship between that analysis and any analyses of unquantified environmental impacts, values, and amenities."

Phase V: Considering Alternatives to the Proposed Action

A major provision of impact assessment requirements in the United States and certain other countries is the requirement to compare the proposed project with "alternatives." The 1978 CEQ guidelines require agencies to include in their analysis the alternative of "no action." One meaning of no action is "no change," that is, the continuation of existing programs or policies in the absence of the proposed action. Thus a no-action alternative for proposed changes in a forest management plan would be the alternative of continuing the present direction and level of intensity of forest management (CEQ 1982, Appendix B). In the context of new, discrete projects, "no action" is the effect of not proceeding with the proposed project. In this case predictable consequences of no action should be addressed (CEQ 1982, Appendix B). For example, denial of a permit for construction of an urban fixed-rail system may lead to future demand for additional freeway construction and subsequent air pollution effects. Agencies must consider reasonable alternatives, which may include options outside their jurisdiction. For example, an alternative to a proposed power plant in meeting energy needs may be to increase efforts at promoting energy conservation, which may be handled by a different subagency; the alternative should still be considered.

The 1978 CEQ guidelines require the agency to identify the environmentally preferable alternative (in addition to the no-action alternative) as well as the agency's own preferred alternative. In each case the direct and indirect effects and their significance are to be compared among alternatives. In addition comparisons between alternatives are to examine possible conflicts with existing plans, policies, and controls and to examine the effects of proposed actions on the conservation of energy and of natural or depletable resources. In each case means to mitigate adverse environmental impacts, and the effects of these mitigations, are to be compared. Additional discussion on the generation of alternatives may be found in Chapter 3 and Andrews et al. (1977).

Encouraging the implementation of mitigating actions is one of the central goals of impact assessment legislation. The CEQ guidelines define mitigating actions to include reducing or eliminating parts of the action, repairing damages, instituting management practices over the life of the project, or substituting or replacing damaged resources or environments.

A critical research question for ecologists is the extent to which substitutions of humanly created ecosystems for those destroyed by development truly restore ecosystem structure and functions. Race and Christie (1982) examined this question for artificially created marshes which were constructed to mitigate the loss of other wetlands to urban development. Though comparative studies were scarce, Race and Christie found that often substantial differences, especially in animal density and composition, occurred between artificially created and natural wetlands. Quantitative information on these differences is important, since often decision makers approve creation of wetlands in substitution for those destroyed without precise knowledge of the ecological equivalence of the two systems. The California Coastal Zone Management Act requires an arbitrary substitution of four artificially created units of habitat for every unit of natural habitat destroyed, in the hope that this will establish an adequate margin of safety. Another practice that encourages substitution of artificially created wetlands for natural ones is the "mitigation banking" scheme developed in Oregon. Estuarine wetlands parcels are created or restored by the local authorities and sold to developers to mitigate development of other wetland parcels (Race and Christie 1982).

Phases VI and VII: Decision Making and Post-Impact Monitoring

The processes of decision making are discussed in Chapters 3, 4, and 5. If a decision has been made to proceed with a project, it is important to monitor the effects of the action and to allow for modification at the post-impact stage.

Sometimes the most appropriate mitigating actions are not known until the project is underway. Some recent developments in impact assessment attempt to deal with this problem. In the United Kingdom, for example, a major oil company regularly hires an ecologist to be on site during the construction phase of a project, to make on-the-spot recommendations regarding optimal mitigation measures (Elkington 1981). In the United States an impact report on an oil field in California provided, as part of the mitigation procedure, a phased drilling program. Monitoring of the impacts on vegetation and wildlife were to be conducted at early stages of drilling; as a result of findings on these impacts, subsequent siting and design of oil drilling structures would be modified (J. T. Gray, personal communication, 1982). Agency approval for subsequent portions of the project would be contingent on design modifications made as a result of initial post-project monitoring.

Beanlands and Duinker (1982) cite an additional benefit to industry of such post-impact monitoring: such data can help distinguish ongoing effects of the project from those due to other forces. For example, an oil spill on a beach near a nuclear power plant in Canada was cleaned with dispersal agents. The death of many marine organisms later observed was blamed on the oil and detergent; however, company monitoring data indicated that an alternative or additional factor of importance in the recent deaths was the unusually hot summer during the time of the spill.

While impact assessment has normally been conceptualized as stopping at the point of communicating findings and recommendations to decision makers, the full realization of environmental protection goals necessarily involves continually monitoring predicted effects and modifying actions as a result of findings. Only with the inclusion of this seventh phase will the process of impact assessment truly enter the realm of empirical science, in which predictions are tested and hypotheses revised as a result of observations.

REFERENCES

Andrews, R. N. L., Cromwell, P., Enk, G. A., Farnworth, E. G., Hibbs, J. R., and Sharp, V. L. (1977). Substantive guidance for environmental impact assessment. Inst. of Ecology, Washington, D.C.

Applegate, R., and Baldwin, M. F., eds. (1973). A scientific and policy review of the draft Environmental Impact Statement: Crow Ceded Area Coal Lease, Westmoreland Resources Mining Proposal. Inst. of Ecology, Washington, D.C.

Arnstein, S. R., and Christakis, A. (1975). *Perspectives on Technology Assessment.* Science and Technology, Jerusalem.

Beanlands, G. E., and Duinker, P. N. (1982). Environmental impact assessment in Canada: an ecological contribution. Inst. for Resource and Environ. Studies, Dalhousie Univ., Halifax.

Bisset, R. (1980) Problems and issues in the implementation of EIA audits. *Environ. Impact Assessment Rev.* **1**:379–396.

Burdge, R., et al. (1980). Ten years after: the social, economic, and biophysical impacts of Lake Shelbyville. Inst. for Environ. Studies, Univ. Illinois, Urbana.

Burkhardt, D. F., and Ittelson, W. H., eds. (1978). *Environmental Assessment of Socioeconomic Systems.* Plenum, New York.

Carley, M. J., and Derow, E. O. (1983). *Social Impact Assessment: A Cross-Disciplinary Guide to the Literature.* Westview, Boulder, Colo.

Chalmers, J., Pijawka, D., Branch, K., Bergmann, P., Flynn, J., and Flynn, C. (1982). Socioeconomic impacts of nuclear generating stations. Summary report on the NRC post-licensing studies. NUREG/CR-2750. U.S. Nuclear Regulatory Commission, Washington, D.C.

Clark, B. D., Bisset, R., Wathern, P. (1980). *Environmental Impact Assessment: A Bibliography with Abstracts*. Mansell, London.

Cochran, W. G. (1963). *Sampling Techniques*. 2nd ed. Wiley, New York.

Convisser, M. (1978). The environmental impact statement as a tool for transportation decision-making. In S. Bendix and H. R. Graham, eds. *Environmental Assessment: Approaching Maturity*. Ann Arbor Sci., Ann Arbor, Mich., pp. 257–263.

Council on Environmental Quality (1976). Environmental Impact Statements. An analysis of six years' experience by seventy federal agencies. GPO, Washington, D.C.

Council on Environmental Quality (1977). *Environmental Quality*. Eighth Ann. Rep. CEQ, Washington, D.C.

Council on Environmental Quality (1978). *Environmental Quality*. Ninth Ann. Rep. Appendix F. Regulations for implementing the procedural provisions of NEPA (40 CFR Parts 1500–1508). CEQ, Washington, D.C., pp. 760–798.

Council on Environmental Quality (1979). *Environmental Quality*. Tenth Ann. Rep. CEQ, Washington, D.C.

Council on Environmental Quality (1980). *Environmental Quality*. Eleventh Ann. Rep. CEQ, Washington, D.C.

Council on Environmental Quality (1982). *Environmental Quality*. Twelfth Ann. Rep. Appendix B. Forty most asked questions concerning CEQ's national environmental policy act regulations. CEQ, Washington, D.C., pp. 261–275.

Dooley, J. E. (1979). A framework for environmental impact identification. *J. Environ. Manage.* **9**:279–287.

Elkington, J. B. (1981). Converting industry to environmental impact assessment. *Environ. Conserv.* **8**:23–30.

Finsterbusch, K., and Wolf, C. P., eds. (1977). *Methodology of Social Impact Assessment*. Dowden, Hutchinson & Ross, Stroudsburg, Pa.

Fletcher, K., ed. (1974). A scientific and policy review of the draft environmental impact statement for the proposed Federal Coal Leasing Program. Inst. of Ecology, Washington, D.C.

Fletcher, K., and Baldwin, M. F., eds. (1973). A scientific and policy review of the Prototype Oil Shale Leasing Program Final Environmental Impact Statement. Inst. of Ecology, Washington, D.C.

Glickfield, M. J., Whitney, T., and Grigsby, J. E. (1977). A selective analytical bibliography for social impact assessment. Report IPM 1. Dept. of Civil Engineering, Stanford Univ., Stanford, Calif.

Golden, J., Ouellette, R. P., Saari, S., and Cheremisinoff, P. N. (1979). *Environmental Impact Data Book*. Ann Arbor Sci., Ann Arbor, Mich.

Grigsby, J. E., and Glickfield, M., eds. (1978). A symposium on social impact assessment and human services planning. Report IPM 2. Dept. of Civil Engineering, Stanford Univ., Stanford, Calif.

Hanmer, R. W. (1978). Concluding remarks regarding the review process. In S. Bendix and H. R. Graham, eds., *Environmental Assessment: Approaching Maturity*. Ann Arbor Sci. Publ., Ann Arbor Mich., pp. 225–228.

Jameson, D. L. (1976). Secondary impacts of urbanization on ecosystems. EPA, Washington, D.C.

Lee, N. (1982). The future development of environmental impact assessment. *J. Environ. Manage.* **14**:71–90.

Leistritz, F. L., and Murdock, S. H. (1981). *The Socioeconomic Impact of Resource Development: Methods for Assessment*. Westview, Boulder, Colo.

McEvoy, J., and Dietz, T. M. (1977). *Handbook for Environmental Planning: The Social Consequences of Environmental Change*. Wiley, New York.

Miller, G. T., Jr. (1982). *Living in the Environment*. 3rd ed. Wadsworth, Belmont, Calif.

Murarka, I. P., Ferrante, J. G., Daniels, E. W., and Pentecost, E. E. (1976). An Evaluation of Environmental Data relating to Seiected Nuclear Power Plant Sites. ANL/EIS-1 (Kewaunee), ANL/EIS-2 (Quad Cities), ANL/EIS-6 (Prairie Island). Argonne Natl. Lab., Argonne, Ill.

O'Brien, D. M., and Marchand, D. A., eds. (1982). *The Politics of Technology Assessment. Institutions, Processes, and Policy Disputes*. Lexington Books, Lexington, Mass.

Peace-Athabasca Delta Project Group (1973). The Peace-Athabasca Delta Project. Technical Rep. Information Canada, Ottowa.

Pearson, G. L., Pomeroy, W. L., Sherwood, G. A., and Winder, J. S. (1975). A scientific and policy review of the Final Environmental Impact Statement for the Initial Stage, Garrison Diversion Unit. Inst. of Ecology, Washington, D.C.

Poole, R. W. (1974). *An Introduction to Quantitative Ecology*. McGraw-Hill, New York.

Porter, A. L., Rossini, F. A., Carpenter, S. R., and Roper, A. T. (1980). *A Guidebook for Technology Assessment and Impact Analysis*. North Holland, New York.

Race, M. S., and Christie, D. R. (1982). Coastal zone development: mitigation, marsh creation, and decision making. *Environ. Manage.* **6**:317–328.

Rossini, F. A., and Porter, A. L., eds. (1983). *Integrated Impact Assessment*. Westview, Boulder, Colo.

Smythe, R. B., ed. (1974). A scientific and policy review of the draft Environmental Impact Statement, Wastewater Treatment and Conveyance System, North Lake Tahoe-Truckee River Basin. Inst. of Ecology, Washington, D.C.

Snedecor, G. W. (1946). *Statistical Methods Applied to Experiments in Agriculture and Biology*. 4th ed. Iowa State College Press, Ames.

Tomlinson, P., and Bisset, R. (1982). Impact audits and EIA. Project Appraisal for Development Control, Dept. of Geography, Univ. Aberdeen, Aberdeen, Scotland.

U.S. Army Corps of Engineers. (1973). Information supplement No. 1 to Sec. 122 Guidelines. CER 1105-2-105. Washington, D.C.

U.S. Atomic Energy Commission. (1974). Reactor safety study (WASH 1400). Washington, D.C.

U.S. Department of Interior, Bureau of Land Management. (1974). Final Environmental Impact Statement. Proposed Barstow–Las Vegas Motorcycle Race, October 1974. State Office, Sacramento, Calif.

U.S. Department of Interior, Bureau of Land Management. (1975). Evaluation Report. 1974 Barstow–Las Vegas Motorcycle Race. State Office, Sacramento, Calif.

Watanabe, M. (1974). The conception of nature in Japanese culture. *Science* **183:**279–282.

Winder, J. S., and Allen, R. H. (1975). The Environmental Impact Assessment Project: a critical appraisal. Inst. of Ecology, Washington, D.C.

Zigman, P. E. (1978). The California Environmental Quality Act and its implementation. In S. Bendix and H. R. Graham, eds. *Environmental Assessment: Approaching Maturity*. Ann Arbor Sci., Ann Arbor, Mich., pp. 53–70.

PART TWO

ENVIRONMENTAL LAW, PUBLIC POLICY, AND DECISION MAKING

2

ENVIRONMENTAL LAW:
PLANNING APPROACHES
AND ECOLOGICAL CONSTRAINTS

In this chapter we highlight the major legal strategies that have been used to achieve environmental goals in countries with impact assessment procedures. In addition we examine some legal and planning tactics used to implement the objectives set by law and make broad generalizations about their effectiveness in an ecological context. We first look at the basic elements of environmental impact analysis legislation and then environmental laws that deal with air and water quality, toxic substances, land use, species preservation, and noise.

We also discuss here methods of planning for the implementation of governmental policies. Such "policy planning" is to be distinguished from "physical planning," which concerns methods for designing the use of a particular portion of landscape to achieve specific objectives, such as where to site a road through a wilderness area to minimize damage to wildlife. Some principles of physical planning are discussed in Chapters 6 and 11.

A "goal" expresses a direction that the nation's laws and official actions are intended to follow, but has no official force of law. Lawmakers and administrative officials are not subject to legal suit if their actions fail to be consistent with goals expressed in law. In the United States *policies* are more specific statements of legislative directions that can be enforced. Failure to adhere to policies may result in civil suits against the government. In countries with a parliamentary system policies are less readily subject to court challenge. Policies may be further differentiated into a series of *objectives,* which in turn may be achieved by certain broad planning approaches (*strategies*). *Tactics* refer to the specific tools or procedures used to achieve an objective (Figure 2.1).

Environmental laws that are based on an erroneous conception of ecological processes are hampered in achieving their objectives as surely as laws that are based on erroneous assumptions about social responses to the law. Pollution laws vary in effectiveness, in part depending on the accuracy of the underlying conceptual model of how pollutants move through and are trans-

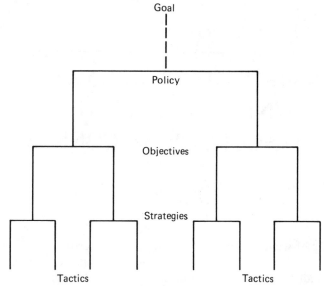

Figure 2.1.   A typical framework for establishing and implementing social intent through law.

formed by the biosphere. We therefore also examine in this chapter how the workings of ecosystems constrain the design of laws that seek to achieve environmental goals efficiently.

ENVIRONMENTAL IMPACT ASSESSMENT

Structure of Legislation

The ultimate aim of impact assessment legislation is typically to enhance the congruence of future actions with broad environmental goals. Such goals may already be adequately expressed in existing environmental legislation, or they may be included in the impact legislation itself. In the United States an aroused public called on Congress in the late 1960s to express the nation's environmental aspirations in more progressive and comprehensive terms than could be gleaned from the separate pollution and land management laws of the time. The expression of such broad goals as national policy was indeed the exclusive purpose of early drafts of the National Environmental Policy Act (NEPA) (Caldwell 1976). The final form of NEPA contains a broad, indeed lofty, statement. The broad structure of the act is shown in Figure 2.2.

EIA procedures are in various stages of development and application in many countries throughout the world (Table 2.1; see also Carpenter 1980; O'Riordan and Sewell 1981; Wandesforde-Smith 1980). Recent reviews of

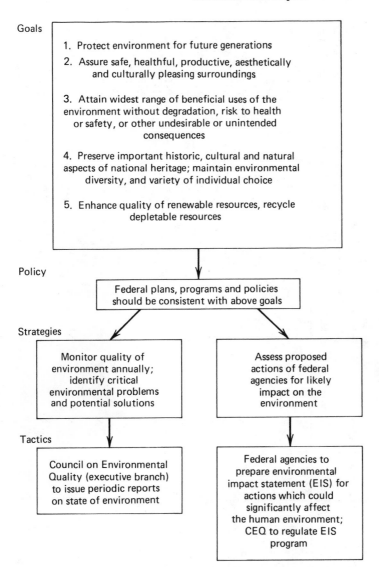

Goals

1. Protect environment for future generations

2. Assure safe, healthful, productive, aesthetically and culturally pleasing surroundings

3. Attain widest range of beneficial uses of the environment without degradation, risk to health or safety, or other undesirable or unintended consequences

4. Preserve important historic, cultural and natural aspects of national heritage; maintain environmental diversity, and variety of individual choice

5. Enhance quality of renewable resources, recycle depletable resources

Policy

Federal plans, programs and policies should be consistent with above goals

Strategies

Monitor quality of environment annually; identify critical environmental problems and potential solutions

Assess proposed actions of federal agencies for likely impact on the environment

Tactics

Council on Environmental Quality (executive branch) to issue periodic reports on state of environment

Federal agencies to prepare environmental impact statement (EIS) for actions which could significantly affect the human environment; CEQ to regulate EIS program

Figure 2.2.   Structure of U.S. environmental impact assessment legislation (National Environmental Policy Act 1969).

guidelines in various countries are those of Canter (1982) and Horberry (1983). Bimonthly updates of developments are reported in the *EIA Worldletter* (PADC Unit, Department of Geography, University of Aberdeen, and Environmental and Ground Water Institute, University of Oklahoma). Variations in approach to impact assessment legislation between countries are broadly summarized next.

Table 2.1. Status of Environmental Impact Assessment in Selected Countries.[a,b]

Country	Status of Program		Level of Jurisdiction		Scope and Focus[1]	References and Notes
	Proposed	Operational	National	Provincial or State		
Australia		x	x	x	E,Rs,Ps	2,3,4,5
Bangladesh		x	x		P,Rs	6,7
Belgium	x		x		Ps	6,21,23
Brazil		x		x	P,Rs	6
Canada		x	x	x	E,Rs	8,9,10
China, People's Republic of		x	(x)	x	H,P,E	11,12,16
Colombia		x	x		E	6
Denmark		x	(x)		Ps	6,7,13,23
Ecuador		x	x		P,E	6
Finland		x	x		P,Rc	6,7
France		x	x	x	E,Rc,Pr	14,23
German Democratic Republic		x	x		E,Rc	6,16
Germany, Federal Republic of		x	x	x	P,E	6,15,23
Ghana	x		x		E,Rc	6
Greece		x	x		E,Rs	6,7,14,16,23
Honduras	x		x		P,Rc	6,16
India		x	(x)	x	E,Rc	6,7,11,16
Indonesia	x			x	E,Rc	6,16
Ireland		x	x	x	P,E,Ps	6,14,23
Israel		x			E,Rs	6,16
Japan		x	(x)	x	P,E,Rs	17,18
Korea, Republic of	x		x		P	11
Luxembourg		x	x	x	E	14,23

Country				Type	References
Malaysia	x	(x)		E,Rc	6,11,16
Netherlands	x	x		H,P,E,Rc	6,19,23
New Zealand		x		H,P,E,Rc	20
Norway	x	x		Rc	6,7,14
Pakistan	x	x		P	6
Papua New Guinea	x	x		P,Rs	6,7,11
Phillipines	x	x		E,Rc	6,11
Poland		x		E,Rc	6
Spain		x			6,7
Sri Lanka	x	(x)			6,11
Sudan	x	x		H	6
Sweden	x	x		E,Rc	6,7,14
Switzerland	x	(x)		H,P	6,14
Thailand	x	x		E	6,11
USSR	x	x		H,P,E,Rc	6
United Kingdom	x	(x)		P,Rs	13,23
United States	x	x	x	P,E,Rc	22
Venezuela	x	x		E,Rs	6

a The nuances of difference between programs are necessarily obscured by summarization. In some cases impact "reports" are simply accounts accompanying pollution permit requests, or environmental plans; in other cases they are comprehensive statements of predicted impact. The symbol (x) = national guidelines available for optional use.

b References and notes to this table: 1. H = public health; P = specific pollutants, areas; E = Environmental protection, quality in general; Rs = resource management, selected resources; Rc = resource management, comprehensive; Pr = all major private projects included; Ps = selected categories of private projects included. 2. Fisher (1977). 3. Formby (1981). 4. Linacre (1976). 5. Hollick (1980). 6. Mayda (1981). 7. De facto part of an authorization procedure (permit, license). 8. Sewell (1981). 9. Plewes and Whitney (1977). 10. Cabinet-level guidelines at federal level, without full force of law. Ontario and Alberta have separate legislation. 11. Carpenter (1981). 12. EIA Worldletter (1983). 13. Lee and Wood (1978). 14. Prieur and Lambrechts (1980). 15. Kennedy (1981). 16. Incorporated as part of regional planning. 17. Hase (1981). 18. Shimazu and Harashima (1977). 19. Jones (1980). 20. Morgan (1983). 21. Tips and Deneef (1980). 22. Council on Environmental Quality (1978). 23. A draft directive exists to require uniform EIA procedures for major projects in the countries of the European Economic Community (Kennedy 1982).

When to Conduct Impact Assessments

In the United States, the lead agency decides whether a proposed action is to be assessed, but the decision is subject to legal challenge. In Canada the decision to perform an impact assessment is made by the federal agency and Environment Canada, or by a province-level environmental department (Sewell 1981). In Australia the federal minister in charge of environmental matters decides on the necessity of an EIS, though 12 fairly specific criteria guide this determination (Commonwealth of Australia 1975, Sec. 3.1.2.). In the United Kingdom assessments are discretionary; guidelines for conducting assessments of major industrial development projects have been prepared (Clark et al. 1980).

Where discretion is greater and opportunity for legal challenge to ministerial decision lacking the number of assessments commissioned has generally been less. In the first two and a half years following NEPA, 2962 draft EISs were filed in the United States, compared with 1 filed statement and 11 pending statements in Canada in the first two and a half years of its federal EIA program (CEQ 1976; Armour 1977). By the end of three and a half years the United States had accumulated 4227 draft EISs (federal level only), compared to 49 statements and 2 public inquiries conducted in Australia during the comparable period (CEQ 1976, Formby 1981).

Preliminary Environmental Studies

The preparation of a draft EIS or EIR in the United States typically takes 10 to 16 months (West 1982, CEQ 1976) and can cost tens of thousands or even millions of dollars (CEQ 1976) to complete. In 1975 the federal government estimated spending $163 million on EIS preparation (CEQ, 1976), or approximately $160,000 per EIS. Therefore, when uncertainty exists about whether the impacts of a proposed action will be significant, an agency typically conducts a preliminary study to determine whether an EIS is needed. In the United States at the federal level such a study is termed an *environmental assessment*. It describes the need for the proposal, highlighting major potential impacts and possible alternative actions and mitigations. Its main purpose is to provide a brief (several-page) analysis of potential impacts of the action, in order to determine whether a full EIS is needed.

If on the basis of the environmental assessment, the lead agency decides that a complete EIS is not required, a Finding of No Significant Impact will be issued. In cases where no study—neither an environmental assessment nor an EIS—is needed to determine that a proposed action will have no significant adverse environmental impact, the finding issued is termed a *categorical exclusion*. Similar provisions occur in most other impact assessment programs.

Avoiding Duplication of Effort

Sometimes an EIS would appear to require the same information that is called for in other planning and management documents. Several approaches to avoiding duplication have been used. In the United States the EIS can be integrated with any other relevant planning document at federal or state level (CEQ 1978 Reg. Sec. 1506.2). The U.S. Forest Service submits its forest management plan and draft EIS simultaneously (e.g., U.S. Forest Service 1980), with the EIS being used to comment on the impacts of the management plan. When federal and state impact statements are combined, they must satisfy requirements of both governmental levels. Certain states, such as California, have developed "certification programs" which exempt certain planning documents from state impact reporting requirements, upon review of the particular planning program (West 1982). In West Germany environmental planning activities and pollution laws and permits are categorically exempt from impact assessment requirements (Kennedy 1981). In Australia commonwealth/state agreements designate the types of actions subject to federal or state impact assessment requirements (Formby 1981).

Scoping

In the United States the first step a lead agency takes in preparing an EIS is usually to issue a Notice of Intent (NOI) to prepare an EIS. This notice is sent to other federal, state, and local agencies, the proponent, and relevant public groups and individuals. It sets forth the basic elements of the proposed action. Shortly after the NOI is circulated, a scoping meeting is called to identify the key environmental issues, possible mitigations, and additional parties who need to be informed of the process. In the implementation of the scoping provision, agencies have increasingly invited representatives of the public to participate in moderated discussions with agency officials and project proponents, in order to achieve early consensus on the key issues that need to be addressed in the environmental assessment. The American Arbitration Association (1980) has issued a report on this aspect of the scoping process (see also Chapter 3). In Canada the "screening" meeting is conducted by an assessment panel consisting of members of the federal department in charge of environmental matters (Environment Canada) and the proposing federal or provincial agency (FEARO 1978). Representatives from outside the public service can be included but in practice have not been (Sewell 1981).

Public Participation

The extent of public participation in the impact assessment process differs greatly, depending on the political traditions of each country. In the United States public hearings may be used during the scoping process or in solicitation of comments on draft EISs (or EIRs at state level). In Australia, Canada, and the United Kingdom, for major national environmental issues, a public inquiry may be held, witnesses called, and detailed reports written, whether or not an EIS has been prepared. The process may take several years (e.g., Mackenzie Valley Pipeline Inquiry, Canada, three years; Ranger Inquiry, Australia, two years). New South Wales, Australia, has used a structured hearing format in which an agenda is set, sometimes based on the

EIS, and public oral submissions on the agenda items received. Following formal submissions, less formal discussion may take place between the inquiry panel and the testifiers (Formby 1981). In Japan public hearings are a customary part of new proposal design and are most often conducted by the project proponent (e.g., industry) (Hase 1981); hearings are less common for government actions.

In the United States a draft EIS must be made accessible to the public and a minimum of 45 days allowed for public comment [CEQ 1978 Reg. 1506.10(c)]. Agencies are required to attach all substantive public comments to the final EIS and to respond to each comment received—by indicating how it has been used to modify the draft document, how the document already addresses the comment, or why the comment did not warrant further agency response. A notice of availability of all draft and final EISs prepared is published in the *Federal Register* and sent to interested organizations and may also be published in the *102 Monitor* and appropriate local news outlets.

In Australia at the federal level public comment on the draft EIS is solicited for a period of 28 days, but release of the document may be suspended by the environment minister upon request by the proponent. In Canada the government retains considerable discretionary power as to whether a draft or final statement is to be released for public comment (Sewell 1981). When written comments are solicited, they do not have to be responded to in the final document. In cases involving major controversy an Environmental Review Board, composed of members outside the federal public service, may be appointed to assess the EIS and receive public comment (Sewell 1981).

Role of Courts

In the United States the courts have played a critical role in defining and extending the scope and importance of impact statements. In 1980, 140 cases were filed in reaction to federal EISs alone (CEQ 1982a). Suits were brought by individuals or citizen groups (38%), environmental groups (24%), business groups (19%), property owners and residents (18%), local governments (12%), state governments (5%), Indian tribes and unions (3%), and federal agencies (1%), with some suits jointly filed. Most suits involve challenging either an agency's decision not to prepare an EIS (44%) or the adequacy of the EIS (43%). The most common federal actions that have been challenged in court have been highway construction projects (15%), projects affecting wetlands and water bodies (13%), urban renewal projects (13%), and energy projects (12%) (CEQ 1982a).

In parliamentary systems, such as Canada, Australia, the United Kingdom, and New Zealand, public access to the courts is currently somewhat less. Under Australian law, for example, a citizen must establish standing (i.e., right to bring suit) by establishing that the enjoyment of his or her property rights has been directly damaged. In the case of public nuisances (e.g., air pollution) a person must show that he or she will be more adversely affected by the pollution or other impact than anyone else (Westman 1973a; Higgins 1970, pp. 151–179, 355–362). Fairfax (1978) has pointed out that the expanded basis for public lawsuits on environmental matters in the United States developed initially from court decisions in 1968 and 1970. Litigation under NEPA, however, served to make the establishment of standing by citizens even easier (Liroff 1978). Before 1968–1970 the requirements in the United States regarding standing on environmental issues were not so different from that of commonwealth countries.

Decision Making

Two key questions are, To what extent must decisions be related to the findings of the environmental impact analysis? And to what extent should the basis for decision making be made public? In NEPA and the 1973 CEQ guidelines agencies were required only to ensure that their EISs adequately dealt with the topics they were mandated to cover. If an agency chose an action that was not environmentally preferable, this decision was not subject to challenge through NEPA; only the adequacy of the EIS in describing the action, impacts, and alternatives, as required by Sec. 102(c) of NEPA, could be challenged.

The 1978 regulations at first appear to require congruence of the agency decision with NEPA policy (Sec. 1505.1):

> Agencies shall adopt procedures . . . to ensure that decisions are made in accordance with the policies and purposes of the [National Environmental Policy] Act. Such procedures shall include but not be limited to:
> (a) Implementing procedures under section 102(2) to achieve the requirements of sections 101 and 102(1). . . .

Section 102(2) of NEPA refers to the preparation of EISs; Sec. 101 sets forth the national environmental goals and policy (Figure 2.2) and Sec. 102(1) requires other policies, regulations and laws to be administered in accordance with this policy. The significance of this regulation is that it would appear to provide under NEPA a legal basis for challenging not simply the adequacy of an EIS, but the actual decision of the agency if it does not conform with national environmental policy.

While this regulation appears clearcut, when viewed in the larger context of other laws and regulations to which an agency must adhere, it becomes arguable whether this requirement has precedence; indeed, the courts have generally ruled that NEPA does not supersede but is supplementary to other mandates. Within the 1978 CEQ regulations themselves, an agency must identify both the "environmentally preferable alternative" and its own preferred alternative, if different (Sec. 1502.14). The environmentally preferable alternative is the "alternative that will promote the national environmental policy as expressed in NEPA's Section 101" (CEQ 1982b, p. 263). Based on an analysis of the EIS, the agency must issue a concise record of decision (ROD) (Sec. 1505.2) setting forth its decision and the reasons for it. If the agency chooses an alternative other than the environmentally preferred one, it must state in the ROD "the reasons why other specific considerations of national policy overrode" choosing that alternative [Sec. 1505.2(b)]. By this last guideline, then, the agency has a basis to argue that other applicable mandates of public law supersede the NEPA Sec. 101 requirement.

How closely an agency's decision must adhere to the environmental policies of NEPA therefore becomes a point for debate regarding the relative importance of environmental vs. other, conflicting mandates of the agency and places the decision squarely back in the political arena. How this debate is ultimately resolved, either in the United States or in other countries, typically depends on how vigorously the public seeks to challenge agency decisions and what channels are open to them in doing so.

The NEPA requires that the record of decision be made public. In Australia at the federal level the officiating department must inform the proponent, but not the public at large, of the recommendations regarding the action (Hollick 1980). In Canada, whether the final decision is subject to appeal varies between provinces (Sewell 1981), as does the availability of the final EIA document.

Records of decision in the United States have a second important function beyond providing information to the public about the basis for the decision. Such records must identify the mitigation measures and the monitoring and enforcement programs that the agency has committed itself to adopt [CEQ Regs. Sec. 1505.2(c) 1978, 1982b, p. 273]. These become legal commitments, enforceable by suit from other agencies or the public (CEQ 1982b, p. 274).

International Impacts

Numerous questions arise concerning the applicability of domestic legislation to industrial or governmental activities abroad or in international waters, air or space ("the global commons"). Current policy in the United States requires federal agencies to prepare EISs or brief environmental assessments for agency actions that significantly affect the environment of a foreign nation or the global commons, including federal projects that involve the production of toxic or radioactive pollutants regulated in the United States (Executive Order 12114, January 4, 1979). Except in the case of global commons issues, such EISs are not subject to public comment and court action (Carpenter 1980). A court decision [*Natural Resources Defense Council, Inc.* v. *Nuclear Regulatory Commission* (Washington, D.C. Circuit)] ruled that export of nuclear power plants for which an environmental assessment had been prepared could be exempted from a requirement for full EIS on agency discretion. The overseas actions of companies based in the United States, or partially based in the United States (multinational corporations), are not subject to the NEPA regulation unless they involve federal government permits or other federal involvement.

A variety of international development agencies have adopted voluntary or mandatory EIA guidelines. In 1974 the U.S. Agency for International Development (USAID 1974, Printz 1978, USAID 1979) and the World Bank (World Bank 1974, 1975; see also Stein and Johnson 1979) adopted policies requiring consideration of environmental consequences of activities they fund in host countries and have detailed procedure for impact assessment. The British Overseas Development Administration similarly conducts a cost-benefit analysis on proposed projects, with more qualitative comment on environmental impacts (Overseas Development Ministry 1977; Abel and Stocking 1981). Guidelines have also been developed by various international organizations for internal use by member countries. The Organization for Economic Cooperation and Development has developed broad guidelines on the EIA by member nations of the heavily industrialized world (OECD 1979). The United Nations has issued impact assessment guidelines for agricultural development (U.N. Food and Agriculture Organization 1983) and for siting of industries in less industrialized countries (Horberry 1980).

Ecological Constraints and Public Policy

At least two principles arising from the nature of ecosystems are relevant to fashioning successful impact legislation, where success is defined as achievement of environmental protection goals. The first is that such legislation should require post-project monitoring of effects at least for major projects, because of the inherent difficulty in predicting accurately the behavior of ecosystems. A corollary of this is that provision should be made to modify or introduce mitigation actions not specified in the EIS when post-project monitoring reveals the need for such mitigations. The second is that the more comprehensive the potential scope of analysis, the more likely the EIS is to reflect accurately the true complexity of ecological and social interactions. To the extent that legislation restricts the issues that are to be considered in an impact analysis, it is likely to prejudge, often incorrectly, the impacts that are most significant in a particular case. A corollary of this is that in an effort to take a holistic approach to human-ecosystem interactions, it is important for EIS preparation to take place within a larger planning hierarchy (see Figure 1.3). Project-based EISs should be evaluated within the context of regional plans that have assessed the carrying capacity of the region's resources for the cumulative impacts of proposed actions and have identified areas of greater and less ecological suitability for particular purposes. In the absence of such larger plans, conducting an impact analysis is analysis without context, much like trying to predict the development of a child without knowing anything about the family and larger society within which it is to be reared. In the sections that follow we consider some of the ancillary legal and planning frameworks within which impact assessment takes place.

AIR QUALITY MANAGEMENT

Goals, Objectives, and Strategies

A frequent goal of air pollution control programs is to maintain a quality of air necessary to protect human health and welfare (Figure 2.3). Human welfare in this context refers to the protection of the health of domesticated and wild plants and animals, the prevention of soiling (corroding) of materials (paint, metal, etc.) and the maintenance of natural levels of visibility. This goal is usually translated into the policy objective of achieving and maintaining concentrations in the air ("ambient" concentration) of major air pollutants that are considered not to be harmful to people, other organisms, and materials. To determine such "safe" levels, scientific data of an epidemological and toxicological nature on the effects of various concentrations of air pollutants are compiled, and these used as criteria for setting ambient air quality *standards*. ("Standards" are the concentrations to be maintained to achieve the stated objectives.) Because the concentration of a pollutant at which damage to human health begins is not necessarily the same as the threshold for damage to other species and materials, separate standards are sometimes set for each of these broad target groups. In the United States the

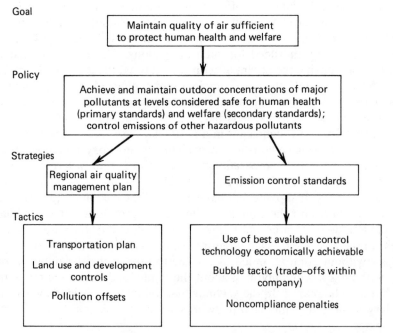

Figure 2.3.   Structure of U.S. air quality legislation (Clean Air Act 1970, and amendments).

standard necessary to maintain human health is termed the *primary standard;* to protect human welfare, the *secondary standard.*

The national standards for the seven major pollutants ("criteria" pollutants) thus controlled in the United States are indicated in Table 2.2, columns 3 and 4. The U.N. World Health Organization has also recommended levels necessary for protection of human health (Table 2.2, column 5). Furthermore the standards set for exposure to air pollutants in the work

Table 2.2. Ambient Air Quality Standards

Pollutant	Averaging Time	A. U.S. Primary Standard Levels	U.S. Secondary Standard Levels	B. WHO Primary Levels
Total suspended particulates (TSP)	Annual (geometric mean)	75 μg m^{-3}	60 μg m^{-3}	40 μg m^{-3} (98% of values
	24 hr[b]	260 μg m^{-3}	150 μg m^{-3}	below 120 μg m^{-3})
Sulfur oxides (SO$_x$)	Annual (arithmetic mean)	80 μg m^{-3} (0.03 ppm)	—	40 μg m^{-3} (0.015 ppm)
	24 hr[b]	365 μg m^{-3} (0.14 ppm)	—	(98% of values
	3 hr[b]	—	1300 μg m^{-3} (0.5 ppm)	below 120 μg m^{-3})
Carbon monoxide (CO)	8 hr[b]	10 mg m^{-3} (9 ppm)	10 mg m^{-3} (9 ppm)	10 mg m^{-3} (9 ppm)
	1 hr[b]	40 mg m^{-3}[c] (35 ppm)	40 mg m^{-3} (35 ppm)	40 mg m^{-3} (35 ppm)
Nitrogen dioxide (NO$_2$)	Annual (arithmetic mean)	100 μg m^{-3} (0.05 ppm)	100 μg m^{-3} (0.05 ppm)	
Ozone (O$_3$)	1 hr[b]	240 μg m^{-3} (0.12 ppm)	240 μg m^{-3} (0.12 ppm)	
Photochemical oxidants	1 hr (maximum)			120 μg m^{-3} (0.06 ppm)
	8 hr (arithmetic mean)			60 μg m^{-3} (0.03 ppm)
Hydrocarbons, nonmethane (RHC)[a]	3 hr (6–9 a.m.)	160 μg m^{-3} (0.24 ppm)	160 μg m^{-3} (0.24 ppm)	
Lead (Pb)	3 months	1.5 μg m^{-3}	1.5 μg m^{-3}	

Sources: U.S. EPA cited in CEQ 1982a; U.N. World Health Organization.

Note: μg m^{-3} = 10^{-6} grams per cubic meter; mg m^{-3} = 10^{-3} grams per cubic meter; ppm = parts of pollutant per million parts of air by weight. RCH = reactive (oxidant-forming) hydrocarbons.

[a] A non–health-related standard used as a guide for ozone control.
[b] Not to be exceeded more than once a year.
[c] EPA has proposed a reduction of the standard to 29 mg m^{-3} (25 ppm).

place are typically higher than those for other environments, sometimes by more than two orders of magnitude (Table 2.3, Commoner 1973). The fact that levels of the several standards differ reflects the complexity of the standard-setting process. Because a single number is designated on the basis of a large and continually changing data base and uncertainties in interpretation remain, expert judgment is used in deriving the standard (see Chapter 3 and Lowrance 1976).

The degree of harm caused by a pollutant actually depends on the "dose" to which an organism is exposed. The dose in turn is a function of both the duration of exposure at that concentration and the concentration of the pollutant (Chapter 9). Thus, in setting standards, both the average concentration of the pollutant and the period of exposure at that concentration are specified. The "arithmetic mean" of concentrations refers to the average of all values over the period; the "geometric mean" refers to the average of the logarithm of such values. The geometric mean reduces the weight given to very high values in influencing the average.

Two broad strategies are usually pursued, often jointly, to achieve and maintain the desired air quality standards. One is to implement a management plan for a region, within which a variety of tactics from discouraging vehicular travel to preventing the siting of new factories may be pursued; the other is to require emission control devices of particular stationary sources (e.g., factories) or mobile sources (e.g., autos) of pollution.

Requiring emission controls could be seen as simply a tactic within the larger management-plan framework, except that in a number of countries (e.g., Australia, West Germany; see Mangun 1979) emission controls are employed alone as the strategy for pollution control. Specific criteria for emission control are applied industry by industry. The objective in such countries is simply to reduce air pollution by technological means, rather than to achieve a particular ambient concentration nationally.

In the United States the two strategies are employed jointly. Emission controls are applied to stationary and mobile sources; if these controls are insufficient to achieve the national standards in the region (a so-called "non-

Table 2.3.   U.S. Occupational Standard

Pollutant	Averaging Time
Particulates	5000 μg m^{-3} (arithmetic mean)
SO$_2$	13,300 μg m^{-3} (5 ppm) (annual arithmetic mean)
CO	56 mg m^{-3} (50 ppm)—8 hr once per year
NO$_2$	10,000 μg m^{-3} (5 ppm) (annual arithmetic mean)

Source: U.S. Occupational Safety and Health Agency.

Note: μg m^{-3} = 10^{-6} grams per cubic meter; mg m^{-3} = 10^{-3} grams per cubic meter; ppm = parts of pollutant per million parts of air by weight. RHC = reactive (oxidant-forming) hydrocarbons.

attainment area''), additional management tactics are employed until the standards are achieved. Further, in those regions that have cleaner air than the primary standard, there is an additional policy under the Clean Air Act of 1970 (as amended, 1977) designed to prevent significant deterioration of the air quality. All such clean air areas are divided into one of three classes. Very little additional pollution is allowed in Class I areas, which include all international parks, wilderness areas, and national memorial parks in excess of 5000 acres and national parks in excess of 6000 acres; moderate deterioration of air quality is allowed in Class II areas, and more significant deterioration within Class III, as long as primary standards are still not exceeded. All lands not in Class I are designated Class II unless reclassified by the state governor. Levels of deterioration that are permitted are specified for each class and pollutant (Table 2.4.). For example, in a Class I region, new sources of pollution would be able cumulatively to raise the particulate matter above existing levels by only 5 μg m^{-3} annual geometric mean or 10 μg m^{-3} 24 hr maximum.

In addition to those pollutants for which ambient standards have been set by use of criteria documents, the Act authorizes the administrator of EPA to set emission standards for additional pollutants from stationary sources which are thought to be hazardous to human health.

Tactics to Achieve Ambient Standards

Tactics may be broadly characterized as those involving regulatory procedures (laws, permits, zoning, etc.), economic incentives (fines, emission

Table 2.4. Maximum Allowable Increases in Ambient Concentration by All New Sources in Class I and Class II areas

	Maximum Allowable Increase (μg m^{-3})	
	Class I	Class II
Total Suspended Particulates		
Annual geometric mean	5	19
24 hr maximum	10	37
Sulfur Dioxide		
Annual geometric mean	2	20
24 hr maximum	5	91
3 hr maximum	10	512

Note: Values for other criteria pollutants have not yet been promulgated.

charges, tax incentives, subsidies, etc.), information and volunteerism (labeling programs, public education programs), and government-induced technological changes (e.g., subsidized mass transit systems) (Mangun 1979).

Management Approaches

The experience of the United States in developing regional air quality management plans provides an instructive case study of the range of tactics used in the management-type strategy to achieve ambient standards. The U.S. Clean Air Act requires each state to prepare a state implementation plan (SIP) that outlines the process for achieving and maintaining primary and secondary standards for criteria pollutants throughout the state. Besides setting schedules for the application of emission control devices to existing and new stationary sources, such a plan can apply land use and transportation controls, among other tactics. Deadlines for achieving primary standards are set by Congress and those for secondary standards by the state, with EPA approval.

At the time of drafting the 1970 Clean Air Act, air pollutants in the United States were known to be transported on a regional scale. In order to encourage planning for air pollution control that conformed with natural boundaries of pollutant migration (an "airshed"), the Act required each state or interstate region to be divided into air quality control regions. For any air quality control region that had failed to attain an air quality standard, an additional air quality management plan (AQMP) was to be prepared. If such a region exceeded standards for motor vehicle-related pollutants (ozone or carbon monoxide) the AQMP was to include a transportation control plan.

An AQMP is to be coordinated also with regional water quality management (or "208") plans. One reason for this is that new housing developments cannot be built in the United States until adequate sewage treatment capacity has been provided. The sewage treatment capacity planned for in a water quality management plan will influence the level of housing and industry to be permitted; these in turn determine the level of future air pollution emissions. To guarantee maintenance of an air quality standard over a 20-year period (the normal planning horizon for an AQMP), such a plan must be coordinated with the water quality plan and with other community development plans that influence future growth in the region.

Numerous approaches to air quality maintenance planning exist. A few examples follow.

Land Use Controls and Emission Offsets

To maximize dispersal of the pollutants, a regional plan may zone the airshed to regulate where future stationary sources are to be located. This involves estimating maximum allowable emissions per zone as well as patterns of pollutant dispersion (see Chapter 7). The Clean Air Act has in effect required the zoning of the United States into air quality control regions. In areas exceeding national standards, the prevention of significant deterioration (PSD) rules discussed earlier apply, in which small increments in pollution are allowed depending on regional classification. In nonattainment areas no new emission sources may be located nor existing ones expanded unless there is a more than equivalent reduction in pollution by the cessation of some other emission in the region, so as to allow "reasonable further progress" toward attainment of the national standards.

Some growth in nonattainment areas has been achieved by the introduction of the *pollution offset* tactic in 1976. By this tactic a firm may expand its facility, or build a new plant, if it reduces the emission of pollutants at its existing facility by more than the amount of emission to be produced by the new facility. The amount of reduction to be achieved is governed by an offset trade-off ratio set by state or local authority. The pollution removed/ pollution added ratio is as high as 10:1 in some states (Lakhani 1982). Such an arrangement is known as an *internal offset*.

A firm that achieves a reduction in emissions greater than that allowed in the state implementation plan may bank this credit for its own future use in expansion or sell it as a marketable permit to some other firm that wishes to establish or expand a facility in the region. This arrangement is known as an *external offset*. In both internal and external offsets the new source is required to install control technology to attain the "lowest achievable emission rate" regardless of cost. It may then use the pollution credits to cover residual emission surpluses and meet the required trade-off ratio.

An interesting example of external offset was the proposal by an oil company to build an oil import terminal near Long Beach, California, in 1975. The company proposed to offset the hydrocarbon emissions from oil storage tanks by purchasing pollution control equipment for several small dry cleaning plants in the area that were emitting volatile organic cleaning compounds. To offset sulfur oxide emissions at the terminal, it proposed to purchase a sulfur dioxide scrubber for a utility some miles inland. Some problems with these arrangements included the fact that the hydrocarbon/ volatile organic compound offset was calculated on a weight basis, without considering the relative reactivity of the hydrocarbons and the volatile organics in combining with nitrogen oxides to form ozone. A second problem was that with the predominance of offshore winds in this region, the populated coastal city of Long Beach would receive the additional hydrocarbon (and subsequently ozone) pollution, while the less populated inland region would benefit from the pollution reduction at the inland utility.

A second example of an external offset also serves to illustrate potential strengths and weaknesses in the approach. A developer proposed a major housing development in Orange County, California, a nonattainment region. To offset the pollution to be created from increased auto traffic from residents of the development, the developer agreed to build a certain number of more moderately priced homes. These homes were thought to enable existing workers in the region to be able to afford to live closer to places of work, thus reducing commuting miles traveled and consequent air pollution. The problem with such an agreement is that there was no effective way to insure that the homes would in fact be occupied by local workers, either on initial sale or on resale.

Despite such problems studies of the application of the offset policy in the United States have indicated that it can achieve cost savings per unit of pollution reduction of some 20–80% (Lakhani 1982). During 1976–1979 more than 600 offset agreements were negotiated in the United States.

Technological Controls

In addition to the installation of emission control devices on motor vehicles and factory smokestacks, emission reduction can also be achieved by changing the process or manner of manufacture or extending controls to equipment not normally controlled (e.g., retrofitting gas or petrol pumps with vapor-lock rubber attachments to prevent release of hydrocarbon fumes during filling). One approach to choosing among such alternative tactics is to select those tactics that achieve the greatest pollution reduction per unit of cost, that is, are most cost effective. In Table 2.5 a list of tactics to achieve hydrocarbon reduction in the Los Angeles air basin is shown. To achieve a given total amount of emission reduction most cost effectively, one would apply the tactics listed in descending order until the total weight of emission reduction were achieved.

Frequently factors other than cost play a critical role in the acceptability of a control tactic. In Figure 2.4 alternative tactics for achieving particulate reduction associated with space heating and cooling are listed in a Baltimore, Maryland, Air Quality Management Plan (AQMP). Various factors influencing the ease of implementation of these tactics are arrayed in matrix form, facilitating review by decision makers.

Transportation Controls

Alternative designs and uses of transportation systems influence the kind and amounts of pollutants generated. For example, at higher speeds of continuous travel, motor vehicles emit a lower amount of hydrocarbons and carbon monoxide, and more nitrogen oxides; the reverse is true at lower speeds. With starts and stops motor vehicles also emit more of a variety of pollutants than with continuous travel. These phenomena have sometimes been used to argue for more high speed roads (freeways) as a pollution-reduction tactic. However, the replacement of surface streets by a freeway can encourage vehicular travel in the region; for example, vehicle miles traveled in the region increased 20% in a portion of Chicago after a freeway was built through the area. The replacement of an extensive trolley system by freeways and auto travel in southern California took place without thought to the impact on air quality (Crump 1962). By the 1950s fashion designers were proposing "smog veils" for the women of Los Angeles (Figure 2.5) to reduce intake of the heavy smog generated in part by automobile exhaust from freeway travel.

POTENTIAL CONTROL MEASURES FOR MAINTAINING AMBIENT AIR QUALITY STANDARDS FOR SUSPENDED PARTICULATES
DOMESTIC AND COMMERCIAL HEATING AND COOLING

SOURCE CATEGORY: DOMESTIC AND COMMERCIAL HEATING AND COOLING
18% TSP CONCENTRATION
13% PARTICULATE EMISSIONS

CONTROL MEASURES	POLICY INSTRUMENTS	PERCENTAGE RANGE OF EFFECTIVENESS: 0-2	2-5	5-10	10-25	25-50	50-100	% EMISSION REDUCTION	ug/m³ IMPROVED AIR QUALITY	ECONOMIC IMPLICATIONS (1. NOT EXPENSIVE 2. MODERATELY EXPENSIVE 3. VERY EXPENSIVE) — DIRECT COSTS: TO POLLUTER	TO CONSUMER	TO GOVERNMENT	OTHER (IDENTIFY)	ADMINISTRATIVE COSTS: PROGRAM MANAGEMENT	ENFORCEMENT	MONITORING	OTHER (IDENTIFY)	SOCIAL COSTS: AREA INCOME	AREA EMPLOYMENT	REGIONAL PRICES	OTHER (IDENTIFY)
IMPROVE DOMESTIC AND COMMERCIAL INSULATION	BUILDING CODES			●				1.0	1.1	1	2	1	-	1	1	1	-	1	1	1	-
CONTROL ROOM TEMP FOR A.C AND HEATING	ADVERTISING			●				1.0	1.1	1	2	1	-	1	1	1	-	1	1	1	-
CONCENTRATE NEW DEVELOPMENT AT DENSITIES WHICH ALLOW FOR MEASURE TO REDUCE EMISSIONS PER CAPITA OR PER UNIT OF PRODUCTION	INCREASING MULTI-FAMILY HOUSING AS OPPOSED TO DETACHED UNITS	●						1	1	2	2	1	-	1	1	1	-	1	1	1	-
REDUCE WINDOW AREA	BUILDING CODES		●					5	.5	1	1	1	-	1	1	1	-	1	1	1	-
INCREASE FUEL COSTS	TAXES; SURCHARGE	●						1	.1	1	1	2	-	1	1	1	-	1	1	1	-
DIURNAL ROOM TEMP.	SPECIFICATION		●					.5	.5	1	1	1	-	1	1	1	-	1	1	1	-
REDUCE ASH CONTENT OF FUEL	TAX POLICY			●				1.0	1.1	2	2	1	-	1	1	1	-	1	1	1	-
IMPROVE FURNACE DESIGN	IMPROVE DESIGN SPECS			●				1.0	1.1	2	2	1	-	1	1	1	-	1	1	1	-
IMPROVE MAINTENANCE OF HEATING COOLING SYSTEMS	REGULATORY CODE			●				1.0	1.1	1	2	1	-	1	1	1	-	1	1	1	-
MODIFY PILOT LIGHT	MODIFY DESIGN	●						1	.1	1	1	1	-	1	1	1	-	1	1	1	-
DESIGN HOME HEATING AND A/C SYSTEM AS UNIT	DESIGN SPECS CODE REQUIREMENT SUBSIDIES	●						1	1	1	1	1	-	1	1	1	-	1	1	1	-
ORIENTATION OF BUILDINGS AND WINDOWS	DESIGN SPECS CODE REQUIREMENTS SUBSIDIES		●					5	5	1	1	1	-	1	1	1	-	1	1	1	-
INSTALL CONTROL DEVICES ON SMALL COMBUSTION UNITS	REGULATIONS, IMPROVE DESIGN SPECS						●	9.8	10.8	1	1	1	-	1	1	1	-	1	1	1	-

CONTROL MEASURES	ADMINISTRATIVE CONSIDERATIONS — FLEXIBILITY: INFLEXIBLE	SOMEWHAT FLEXIBLE	MODERATELY FLEXIBLE	VERY FLEXIBLE	APPLICATION OF CONTROL: CONTINUOUS	NON-CONTINUOUS	SELECTIVE	UNIFORM	TIMING CONSIDERATIONS — YEARS BEFORE IMPLEMENTATION: 0-2	2-5	5-10	YEARS BEFORE EFFECTIVENESS REALIZED: 0-2	2-5	5-10	10+	POLITICAL CONSIDERATIONS — LEVEL OF IMPLEMENTATION: FEDERAL	STATE	REGIONAL	LOCAL	PRIVATE
IMPROVE DOMESTIC AND COMMERCIAL INSULATION	●				●		●		●					●					●	
CONTROL ROOM TEMP FOR A.C AND HEATING				●		●	●		●			●				●	●	●	●	●
CONCENTRATE NEW DEVELOPMENT AT DENSITIES WHICH ALLOW FOR MEASURE TO REDUCE EMISSIONS PER CAPITA OR PER UNIT OF PRODUCTION		●			●				●				●			●	●		●	
REDUCE WINDOW AREA	●				●		●		●					●		●	●		●	●
INCREASE FUEL COSTS		●			●		●		●					●					●	
DIURNAL ROOM TEMP.				●	●		●		●			●								●
REDUCE ASH CONTENT OF FUEL	●				●		●			●				●		●	●		●	●
IMPROVE FURNACE DESIGN	●				●		●			●				●		●				●
IMPROVE MAINTENANCE OF HEATING COOLING SYSTEMS			●			●	●		●			●							●	●
MODIFY PILOT LIGHT									●			●				●				●
DESIGN HOME HEATING AND A/C SYSTEM AS UNIT	●				●		●		●						●	●	●		●	
ORIENTATION OF BUILDINGS AND WINDOWS		●			●		●		●						●	●	●		●	
INSTALL CONTROL DEVICES ON SMALL COMBUSTION UNITS	●				●				●					●		●	●		●	

CONTROL MEASURES	LEGAL CONSIDERATIONS: LEGAL PRECEDENT EXISTING	NEW LEGISLATION REQUIRED	ENVIRONMENTAL EFFECTS (INTERMEDIA): OTHER GASES	WATER	SOLID WASTE	NOISE	ENERGY	PUBLIC RESPONSIVENESS: 1 LEAST ACCEPTABLE	2	3	4	5 MOST ACCEPTABLE	REFER TO REMARKS SECTION
IMPROVE DOMESTIC AND COMMERCIAL INSULATION	Y	Y	X			X	X	1	2	0	3	5	
CONTROL ROOM TEMP FOR A.C AND HEATING	Y	N	X				X	0	3	4	2	2	
CONCENTRATE NEW DEVELOPMENT AT DENSITIES WHICH ALLOW FOR MEASURE TO REDUCE EMISSIONS PER CAPITA OR PER UNIT OF PRODUCTION	Y	Y	X				X	9	0	1	0	0	
REDUCE WINDOW AREA	Y	Y	X				X	4	1	2	0	1	
INCREASE FUEL COSTS	Y	Y	X		X		X	1	2	4	1	0	
DIURNAL ROOM TEMP.	N	N	X				X	1	1	2	0	5	
REDUCE ASH CONTENT OF FUEL	Y	Y	X					1	5	1	1	2	
IMPROVE FURNACE DESIGN	N	Y	X				X	0	0	1	2	6	
IMPROVE MAINTENANCE OF HEATING COOLING SYSTEMS	Y	Y	X				X	0	1	1	1	6	
MODIFY PILOT LIGHT	N	Y					X	0	0	0	1	7	
DESIGN HOME HEATING AND A/C SYSTEM AS UNIT	Y	Y	X				X	1	1	2	3	3	
ORIENTATION OF BUILDINGS AND WINDOWS	Y	Y					X	1	1	1	3	5	
INSTALL CONTROL DEVICES ON SMALL COMBUSTION UNITS	Y	Y						7	0	1	1	2	

Figure 2.4. Extract on alternative tactics for controlling particulates, from the Air Quality Maintenance Plan for Baltimore, Maryland (U.S. Environmental Protection Agency 1974a).

Table 2.5.   Selected Cost-Effective Tactics for Achieving Hydrocarbon Emissions Control

Tactic	Cost ($ millions/yr)	HC Emission Reduction (tons/day)	Cost Efficiency ($/ton)
Modify new aircraft engines	2.5–3.0	5.0	45–54
Emission standards for new nonfarm heavy-duty off-road vehicles	1.18	4.8	60
Increase bicycle facilities	10	5.0	1250

Source: Southern California Association of Governments, 1978.

Less costly measures like timing of traffic lights on major thoroughfares may well be more cost effective than building freeways. Transportation plans can encourage voluntary measures such as carpooling, van pooling (rental of small bus or van for use by a group of commuters), or hitchiking. Often combinations of measures are more effective than isolated tactics. Thus, combining elevated bikeways with buses that have racks for bicycle storage to take bicyclists to and from bicycle ramp entrances are likely to be much more successful than fragmented bicycle path proposals.

Transportation planning is a major field of endeavor whose techniques are beyond the scope of this book. Interested readers may refer to such works as Altschuler (1981) and Meyer and Gomez-Ibañez (1981) for further information.

Tactics to Control Levels of Emission

Requiring the installation of uniform pollution control technology to reduce emissions is by far the most widely used tactic for air pollution control. The requirement is relatively clearcut and easy to enforce, and it requires no behavioral change on the part of the public. Disadvantages include the differential economic impact of the requirement on smaller firms and possible discouragement of innovation in meeting emission standards. Some of the arrangements for implementation of the tactic in the United States are discussed here.

Defining the Level of Technology

Under 1977 Clear Air Act amendments new sources of emission must achieve a level of control at least as good as that obtained by using the best

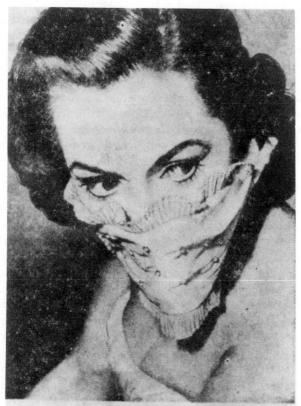

A LA MODE—Many beauties have turned the masks into stylish creations such as this fancy version trimmed in rhinestones, with frilly border providing an added touch of glamour.

Figure 2.5. Rhinestone-studded smog veils proposed for use in Los Angeles in the early 1950s. The veils were designed initially for use in London smog. Reprinted with permission from the *Los Angeles Herald Examiner,* December 6, 1953, p. 9.

technological system of continuous emission reduction. This requirement is often referred to as best available control technology (BACT). Such levels are defined by EPA for each industry, following solicitation of manufacturers' documents on proposed regulations. The system used as a basis for setting the performance standard must have been "adequately demonstrated" to be achievable in practice, although not necessarily routinely achieved within the industry prior to adoption (Mangun 1979). The administrator of EPA must evaluate the cost and inflationary impact of the proposed standard. For existing sources of pollution, each state must promulgate emission standards for such sources compatible with achieving and maintaining national ambient standards under their SIP. Emission of pollutants from existing and new sources for which ambient standards have not been set (so-called "hazardous pollutants") must comply with federal emission standards.

One problem with this approach to defining pollution control technology is that it alone provides little incentive to develop still more efficient performance at pollution reduction. The bankable or marketable pollution credit mentioned earlier is designed in part to provide such incentive.

A second problem is that a firm has little flexibility in achieving the overall emission reduction from a multistack factory, since the technology must be applied to each stack regardless of cost efficiency. To overcome this second problem, EPA introduced the single-plant "bubble" in 1979.

"Bubble" Tactic

A single large factory typically has many smokestacks arising from different machinery within the plant. The single-plant bubble tactic allows the firm to achieve an average emission reduction from all the stacks considered jointly, as though a bubble were placed over all the stacks of the plant and only the ambient air in the bubble measured for compliance with emission reduction requirements. This tactic allows considerable cost savings, as a firm may install superior technology, substitute fuel, or modify the process in a part of the plant where it is cheaper to do so, and permit another part of the plant to exceed normally allowable emission levels. Thus the bubble tactic is essentially an internal offset that applies to an existing factory, rather than to extensions or new sources in the region. For the 70 proposed bubbles as of February 1981, an average savings of $2 million per bubble was achieved (Lakhani 1982).

In 1980 the EPA extended the single-plant bubble concept to multiple plants owned by the same firm within a state (Federal Register 1980). Because these plants may be in substantially different areas or air quality control regions, approval of a multiplant bubble arrangement requires modification of the SIP, with the attendant public hearings.

Emission Charges and Noncompliance Penalties

Many economists have favored the concept of pricing each unit of pollution emitted and setting the price per unit of pollution so that the marginal cost of exceeding desirable emission levels exceeds the cost of installing pollution control technology to remove the excess pollution (e.g., Kneese and Schultz 1975). The cost of emissions can vary regionally depending on local air quality. Economists reason that by this means, each plant can operate with a combination of pollution and control that will be most economical for it while the overall ambient standard is achieved. The advantage of the approach is enhanced economic efficiency. Problems include the fact that clean air regions will be most economically advantageous for siting of plants, resulting in a significant deterioration of air quality in those regions. Second, determining the appropriate price to achieve desirable levels of ambient pollution is a difficult and complex task, in which errors (e.g., plant sitings) are not easily rectified. As a result it becomes more difficult to assure and control ambient standards in a region. Third, the system appears insensitive to the problem of long-range transport of pollutants from a region of clean air and cheap emission charges to an area of more polluted air and more expensive emission charges. In such a situation the distribution of costs and benefits is extremely inequitable.

Although such a system has not been implemented in the United States, a variation of it has been applied through the uniform national emission standards. The U.S. Clean Air Act Amendments of 1977 demand that stationary sources pay noncompliance penalties for failing to achieve emission standards by a stated deadline. The penalties are set equal to the cost savings from being out of compliance, so that firms have no economic incentive to continue polluting excessively (CEQ 1977).

Mobile Source Controls

The basic notion of requiring technology to maintain a specific level of emission per unit of fuel was applied by the U.S. Clean Air Act to motor vehicles. Vehicle engines were required to meet specific emission standards. In 1977 the Clean Air Act Amendments required transportation control plans in areas that had not attained ozone and carbon monoxide standards. Such plans must include a program for mandatory annual inspection and maintenance of devices for controlling emission levels from vehicles. Failure of states to comply with this requirement results in cut off of federal funds for highway and sewage treatment plant construction. Nevertheless, the history of adoption of requirements for emission control and inspection is replete with examples of postponement of deadlines, loosening of emission standards, and weakening of inspection requirements (see Miller 1982, Chapter 17). Such a history indicates that the selection of workable tactics for implementing national goals is an iterative process, involving a successive adjustment of approaches till an equilibrium is reached between broader social goals and the willingness of society to make the sacrifices to achieve them.

Ecological Constraints and Public Policy

As experience with the interactions of air pollutants with ecosystems accumulates, ways in which existing tactics for air pollution cleanup are incongruent with the workings of nature become increasingly apparent. One illustration concerns the long-range transport of pollutants. The 1970 U.S. Clean Air Act was designed on the assumption that air pollutant transport was, at most, regional. It was assumed that after a while, pollutants are diluted or assimilated to negligible form. Indeed, for some years the United States encouraged the building of tall smokestacks to "disperse" pollutants; this tactic only aggravated the problem of long-range transport.

Ecologists and atmospheric scientists have come to understand that many pollutants persist or are transformed to other toxic compounds and are carried over very long distances by trade winds and other upper atmospheric currents. Thus radionuclides from atmospheric nuclear testing in the 1950s in the South Pacific became concentrated shortly thereafter in the food chains of the Arctic (e.g., see Cook 1975). Sulfur and nitrogen oxides released from fuel burning in the midwestern and eastern United States wash out in the precipitation of the northeast and Canada, and similarly the British sulfur emissions precipitate out over Scandinavia (Likens et al. 1979). In the process the sulfur and nitrogen oxides are converted to sulfuric and nitric acids, acidifying the rain and snow and subsequently the ecosystems into which they percolate. Even the interstate air quality control regions provided for by the U.S. Clean Air Act are too small to provide for this distance of transport. As a result control measures that are adequate for one region end up penalizing another region downwind. New tactics to deal with long-range pollutant transport are needed.

One proposal in the United States has been to establish a large interstate superbubble in a major source region for sulfur oxides (east of the Mississippi River), within which a total emission reduction is to be achieved and within which states or regions may sell emission reduction credits; individual firms may also bank or sell emission reduction credits within five subdivisions of the interstate superbubble. The size of the region is a compromise between the problems of regions too small to be responsible for long-range transport problems and too large to be affected. However, acid deposition is now a national and international problem. Los Angeles, California, far from the proposed superbubble region, experienced acid fog of pH 2.5 in 1982 (Roberts 1982), an acidity greater than vinegar and greater than any recorded in the eastern United States.

A second ecological process poorly addressed by current air pollution control strategies is the problem of synergistic effects between pollutants. Concentrations of a pollutant that are harmless in isolation can become toxic in the presence of other pollutants. In theory such effects could be taken into account in setting ambient standards. This is done in the case of ozone, the product of the combination of HC and NO_x. But the many other synergisms that can occur in the body (e.g., see Table 7.6) are more inadequately dealt with in standard setting, partly because the relative proportions of such pollutants vary greatly on a regional basis and partly because their synergistic interaction occurs inside the body rather than in the atmosphere. By the standard-setting process, lower standards would have to be set to insure public health in the face of such synergisms.

In conducting environmental impact assessments, it is important for assessors to account for these ecological problems and not to assume that compliance with existing air pollution statutes is sufficient to protect human health and welfare.

WATER QUALITY MANAGEMENT

Because the control of water pollution has been proceeding for a considerably longer time than the control of air pollution or the writing of impact assessments, water law is considerably more profuse and complex. In the United States for example, at least 11 major federal laws govern water quality, quantity, and use (Table 2.6). A number of land use laws further affect water quality and water resources (e.g., Coastal Zone Management Act 1972, 1975; National Flood Insurance Act 1969, 1973). Beyond this, in many countries including the United States and British Commonwealth countries, common law (i.e., court precedents in the absence of regulation) governs rights to water withdrawal from surface and underground sources.

Goals, Objectives, and Strategies

Two alternative goals characterize water quality legislation in many industrialized nations. One goal is to maintain a quality of surface water sufficient for various specified human uses (drinking, fishing, industrial cooling, irrigation). This was the goal of the U.S. Water Quality Act of 1965 and continues to be the goal of water pollution legislation in Australia (Westman 1973b), the United Kingdom, and elsewhere. The second goal is the restoration and maintenance of the physical, chemical, and biological integrity of waterways. This is the goal of current U.S. law, the Federal Water Pollution Control Act Amendments (FWPCA 1972, 1977).

Because of these differences in goals very different policy objectives and strategies are pursued. Under the human-use goal the policy objective is to classify stream segments by their highest intended use and maintain stream water quality consistent with that use. An advantage is that segments are not maintained "cleaner than necessary" for intended human use, and pollution control costs are thus saved. The strategy pursued is to regulate discharges into the stream segment, such that the total effluent discharged does not cause deterioration of stream water quality beyond the designated level. Such a strategy requires a sophisticated ability to related physicochemical water quality parameters to their ecological effect and, further, from a knowledge of discharge quantities and models of river flow, to predict with precision the stream water concentrations at a point in the river. It requires a precise understanding of the relative "assimilative capacity" of water bodies for a broad mixture of pollutants. Such predictive abilities in some cases exceed current scientific and administrative capacity (Westman 1972a, 1977; Chapter 7).

The ecological "integrity" goal led to a policy objective of elimination of discharge of any wastes into waters and, in the interim, the restoration of waters capable of supporting a balanced population of fish, shellfish, and

Table 2.6. Major Federal Laws Affecting Water in the United States

Subject	Law	Year	Administering Agencies	Coverage Area
Surface water quality	Water Pollution Control Act Amendments	1972 1977	EPA, states, and Secretary of the Army	Controls discharge into surface waters; governs quality of such waters
Drinking water quality	Safe Drinking Water Act	1974, 1977	EPA and states	Regulates quality and protects supplies of drinking water (surfaces and underground)
Water quantity	Water Resources Planning Act	1965	Water Resources Council and Office of Management and Budget	Prepares regional or river basin plans to conserve, develop, and use water
Ocean dumping	Marine Protection, Research and Sanctuaries Act (1972)	1972	EPA and Secretary of the Army	Controls dumping of wastes and dredge soil into ocean from barges and other ships.
Navigable waters	River and Harbor Act	1899	Secretary of the Army, and Chief of Engineers	Governs dumping of water into, and the excavation of, navigable waterways
Scenic rivers	Wild and Scenic Rivers	1968 1974	States	Preserves rivers of special recreational and aesthetic value
Wetlands	Estuarine areas Wetlands (Executive Order 11190)	1968 1977	Department of Interior	Governs conservation and use of salty and freshwater wetlands
Marine oil pollution	Oil Pollution Act	1961 1973	Coast Guard	Governs discharge of oil at sea
	Water Pollution Control Act	1972 1977	EPA	Governs oil spills
Aquatic life	Migratory Bird Conservation	1929	Fish and Wildlife Service,	Governs conservation of migratory waterfowl and fisheries
	Fish and Wildlife Coordination	1934 1972	Department of Interior	
	Fish and Wildlife Act	1956 1974		

Sources: Golden et al. 1979; CEQ 1978; Federal Code

wildlife, and recreation in and on the water (FWPCA 1972). The strategy for achieving this objective was to apply technological effluent controls to point sources of discharge (outfall pipes from industries and sewage treatment plants) and best management practices to the control of nonpoint sources (dispersed sources of pollution such as vacant land and street surfaces which contribute pollutants to surface runoff during times of rain and snowmelt), as shown in Figure 2.6.

These differing goals and strategies rely on fundamentally different conceptions of the nature of aquatic ecosystems. The human-use strategy assumes that water bodies have a substantial ability to assimilate and degrade wastes, so that pollutants in one stream segment will have been assimilated by the time the next segment is reached (Figure 2.7). While water bodies have a limited capacity to assimilate pure organic wastes, respiring the carbohydrates to carbon dioxide and water, most other water pollutants (nutrients, toxic substances, pathogens) remain suspended in water, resuspendable from sediment, or incorporated in aquatic life for a time long enough to carry them well downstream from the source of discharge (Westman 1972a). Thus a human-use system can easily underestimate the extent of accumulation of slowly degraded pollutants as the river flows downstream. If such a system did not assume assimilative capacity, it would not be possible to have human uses requiring cleaner water downstream from segments that have

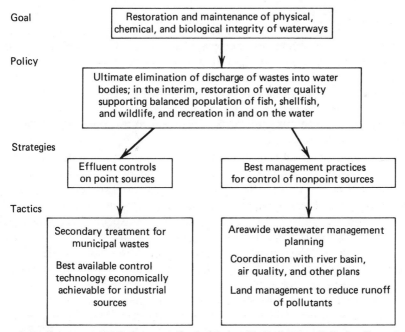

Figure 2.6. Structure of U.S. water quality legislation (Federal Water Pollution Control Act Amendments 1972, and subsequent amendments.)

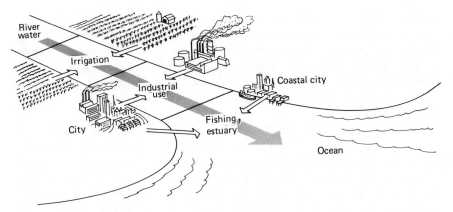

Figure 2.7. A typical stream segment classification by human use. Such a river use pattern typifies many coastal cities. Wastes must be assimilated as they flow downstream if the indicated human uses are to be safely maintained.

been allowed to degrade to a more polluted state. Dilution to acceptable concentrations occurs for some pollutants, but for many pesticides and heavy metals, the fate is rather to be concentrated in food chains to toxic levels (Chapter 7).

The ecological "integrity" strategy, by contrast, recognizes the extremely limited assimilative capacity of streams for many pollutant types, the tendency for aquatic ecosystems to bioaccumulate certain toxic substances, and the problems of accelerated eutrophication from nutrient discharge. It assumes that by the cessation of discharge of foreign matter, rivers will gradually cleanse, and by natural processes remain sufficiently clean to support the full range of human uses. The U.S. FWPCA (1972, 1977), however, did not attempt to impose a cessation of discharge into waters immediately. It set a goal of 1985 for the achievement of this aim, a goal that will clearly have to be postponed.

In the interim point discharges into water continue, using the best available technology economically achievable. Effluent control standards for point sources are set nationwide by EPA much as they are for air pollutants. Less technological management and planning strategies are also pursued, primarily for the control of nonpoint sources. An aspect of the human-use goal is retained for the interim and defined nationally: all receiving waters are to be capable of supporting balanced populations of aquatic life and human recreation. Thus the FWPCA has in fact elements of both major goal structures. Although the ecological "integrity" strategy is likely to result ultimately in cleaner streams, and may be most economic once conversion to land disposal and recycling of wastes is complete, it tends in the interim to be a more costly strategy since removal of wastes before water discharge is very expensive (Westman 1977).

Tactics

Management and Planning Approaches

A number of countries have used water quality management tactics success-fully to achieve cleanup goals. In the United States, under Sec. 208 of the FWPCA, an areawide wastewater management plan is drawn up for each state or interstate area with common watershed boundaries. The total ex-pected effluent discharges from point and nonpoint sources are examined, along with expected flow rates from river basins, and the effect of expected population growth and industrialization over the subsequent 20 years.

"Best management practices" for reducing nonpoint sources of pollution are to be used. Examples of such practices include revegetation of mined lands and liming of acid-forming coal mines; erosion control for farm and forest land; efficient street sweeping to capture the finer particles (<250 μ) which contain most of the toxic substances (Sartor and Boyd 1972); use of vegetated drains rather than curb drainage; the use of treatment ponds, lagoons, and groundwater aquifers to store and treat stormwater; the en-couragement of open space, lawns, and porous paving materials to enhance percolation of stormwater; the capture and holding or treatment of the first 1–3 cm of rain which contains the bulk of pollutants (Enviro-Control 1974); the replenishment of groundwaters to prevent salt water intrusion; and the encouragement of water conservation and reuse (Westman 1977). Similar river basin development plans are in operation in West Germany (Mangun 1979), Australia, and elsewhere.

One problem in the implementation of such plans has been that because point sources are required to install effluent control technology, there is no tendency to examine whether a portion of the funds thus expended could not better be spent on controlling nonpoint sources of pollution, which can contribute 30–80% of total urban oxygen-demanding pollutants and contain a level of toxic materials exceeding typical industrial discharges, and patho-gen levels exceeding that from the chlorinated effluent of treatment plants (Enviro-Control 1974, Colston 1974). In other words, the current structure of legislation, which administers point source controls uniformly, tends to inhibit the opportunity to set priorities among all pollutant sources (both point and nonpoint) by volume and toxicity and to spend pollution control funds most cost effectively when the two types of sources are jointly ranked (Westman 1977). In general, the management strategies for controlling nonpoint sources tend to be more cost efficient than the hard technology used for point source treatment (see e.g., Westman 1972b).

Technological Controls

"Best available control technology economically achievable" is required for point sources in the United States and is defined as that technology in use by the best performer in the industry. In West Germany effluent limitations are based on "generally recognized rules of technology" (Mangun 1979). In Australia the technology to be used is left to the discharger, so long as specified effluent limits are achieved.

The U.S. FWPCA attempted to encourage the use of land disposal following pretreatment as an alternative to conventional secondary treatment plants for municipal waste. Land disposal has been used successfully in parts of West Germany, France, Australia, the United States, and elsewhere (D'Itri 1982; Elazar et al. 1972; Josephson 1975; Westman 1972b, 1977). Both ecological and institutional problems, however, have hindered its widespread implementation to date (Westman 1977).

Residuals Charges

While not used in the United States except briefly in Vermont in the early 1970s, a system of residuals charges has been implemented elsewhere, particularly in West Germany. At present West German dischargers must pay a pollution charge for each noxious unit of effluent discharged. The unit is weighted for noxiousness by considering the amount of settleable solids, oxidizable substances, and toxic substances it contains. The system attempts to overcome the problem previously noted of determining the precise charge necessary to assure a given level of discharge, by specifying a maximum volume of effluent that can be discharged annually (Mangun 1979).

Ecological Constraints and Public Policy

To the extent that water pollution legislation continues to rely on erroneous assumptions about assimilative capacity of waters, achievement of cleanup goals will clearly be hindered. Stream water concentrations of pollutants are not the simple result of the dilution of effluent wastes; they are the complex result of fluxes of materials in the aquatic ecosystem. Most water laws to date have tended to deemphasize control of pollutants once they have entered ecosystems. The controls on nonpoint sources are generally less well defined and less well funded than those on point sources. As our awareness of the transformation and fate of pollutants in surface and groundwaters increases, heightened by unexpected instances of contamination of groundwater by toxic substances from landfill sites and nonpoint sources, water laws will need to be modified to take these ecological processes more directly into account.

TOXIC SUBSTANCES CONTROL

In the context of public policy, a toxic substance is one that causes illness or death to living organisms when present in the body in relatively minute concentrations. In the United States the control of toxic substances in the environment is governed in a variety of ways (Table 2.7). Seventeen agencies in the United States have responsibilities for the control of toxic substances (CEQ 1978). A similar complexity of governance of toxic substances exists in many countries (see e.g., Doern 1977 and Franson et al. 1977 for discussions of Canadian law). One reason for this is ecological: toxic substances move through ecosystem compartments using most of the same pathways as do nutrients, so that they become incorporated in the full array of media (air, water, soil) and organisms.

Most of the laws in Table 2.7 govern disposition of toxic substances: discharge, transport, and burial or incorporation into consumer products. The exception is the Toxic Substances Control Act (TSCA 1976), which provides for the pre-market testing of new chemical substances for potential toxicity and the additional testing of existing chemicals. Since there are over four million chemical compounds, and over 55,000 produced in the United States alone (CEQ 1978, 1982a; TSCA Chemical Inventory), the screening of these for toxicity must proceed on a selective basis in the first instances. The goals and tactics for implementing this Act are discussed next.

Goals, Objectives, and Strategies

The major goal of TSCA is the control of chemicals that pose "unreasonable risk" to human health or the environment, taking into account the chemical's toxicity and potential level of exposure, and the economic impact of limiting manufacture. The strategies for implementing this goal are threefold: (1) a review by EPA of current knowledge of the toxic effects of any new chemical proposed for manufacture and subsequent use, following notification by the manufacturer; (2) the testing of new or existing chemicals for which insufficient data exist to determine risk; (3) the control of existing chemicals that pose unreasonable risk to health or environment (Figure 2.8).

Tactics

Premanufacturing Screening

The manufacturer must provide information on the new chemical's identity, production volume, potential exposures and method of disposal, and any data on health and environmental effects. EPA must determine whether the new chemical poses unreasonable risk to human health or the environment. In the absence of new test data on the chemical, the EPA examines the chemical similarity of the substance to a class of compounds for which biological effects are better known. If the agency determines that insufficient information exists to determine reasonableness of risk but considers that an unreasonable risk may exist it can, within 90–180 days, prohibit or limit

Table 2.7. Major Federal Laws in the United States Controlling Toxic Substances.

Subject	Law	Year	Administering Agencies	Coverage Area
Manufacture and use				
Manufacture, use, and disposal of chemicals	Toxic Substances Control Act	1976	EPA	Requires testing of existing chemicals for toxicity (other than food additives, drugs, pesticides, alcohol, tobacco)
Air				
Discharge to atmosphere	Clean Air Act Amendments	1970, 1977	EPA	Sets emission standards for specified hazardous pollutants
Exposure in workplace	Occupational Safety and Health Act	1970	OSHA	Controls exposure to chemicals in workplace
Water				
Discharge to surface waters	Water Pollution Control Amendments	1972, 1977	EPA	Sets technological standards for discharge of 129 priority toxic pollutants in effluent
	Marine Protection, Research, and Sanctuaries Act	1972	EPA, Secretary of the Army	Governs content of toxic substances in wastes dumped at sea
Content in drinking water	Safe Drinking Water Act	1974, 1977	EPA	Regulates contaminants in drinking water
Land				
Disposition of wastes	Resource Conservation and Recovery Act (see also Toxic Substances Control Act)	1976	EPA	Governs handling of hazardous wastes during manufacture, transportation, treatment, storage, and disposal
	Comprehensive Response, Compensation, and Liability Act	1980	EPA	Governs liability and cleanup of leakage from disposal sites
Transportation				
General	Hazardous Materials Transportation Act	1970	DOT (Materials Transportation Bureau)	Governs transportation of toxic substances generally
By rail	Federal Railroad Safety Act	1970	DOT (Federal Railroad Administration)	Railroad safety
By ship	Ports and Waterways Safety Act	1972	DOT (Coast Guard)	Governs shipment of toxic materials

60

Table 2.7. *(Continued)*

Subject	Law	Year	Administering Agencies	Coverage Area
	Dangerous Cargo Act	1952	(Coast Guard)	
Pesticides				
	Federal Insecticide, Fungicide, and Rodenticide Act	1948, 1972, 1975	EPA	Pesticide sale and use
	Environmental Pesticide Control (see also food content)	1972, 1973	EPA	Pesticide registration and use
Consumer products				
Food, drugs, and cosmetics	Food, Drug, and Cosmetic Act and Amendments	1938	FDA	Governs content of toxic materials in food, drugs, and cosmetics
		1958		Further governs food additives; prohibits cancer-causing substances in foods
		1960	FDA	Governs color additives
		1962	FDA	Further governs drugs
		1968	FDA	Governs animal drugs and feed additives
		1976	FDA	Governs medical devices
	Sec. 346(a)	1954, 1972	EPA	Governs tolerances of pesticide residues in food
Meat	Wholesome Meat Act	1967	USDA	Controls food feed, and color additives and
	Wholesome Poultry Products Act	1968	USDA	pesticide residues in meat
Household products	Federal Hazardous Substances Act	1966	CPSC	"Toxic" household products
	Consumer Product Safety Act	1972	CPSC	Dangerous consumer products
	Poison Prevention Safety Act	1972	CPSC	Packaging of dangerous children's products
	Lead Based Paint Poison Prevention Act	1973 1976	CPSC	Use of lead paint in federally assisted housing

Sources: CEQ (1978, Table 14-3), CEQ (1977, 1982a), Golden et al. (1979).

Note: CPSC = Consumer Product Safety Commission; DOT = Department of Transportation; EPA = Environmental Protection Agency; FDA = Food and Drug Administration; OSHA = Occupational Safety and Health Administration; USDA = U.S. Department of Agriculture.

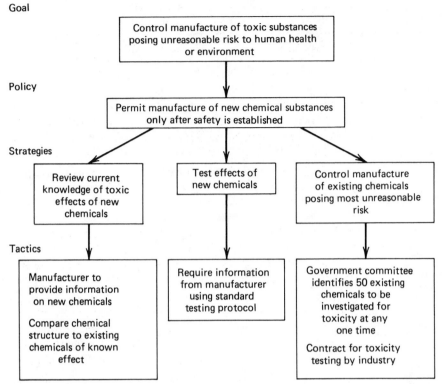

Figure 2.8. Structure of U.S. legislation for control of toxic substances (Toxic Substances Control Act 1976).

manufacture of the substance pending further testing by the manufacturer. About 700–800 new chemicals per year are thus screened (CEQ 1982a).

EPA expects to develop a procedure in the near future to allow further review of selected chemicals' effects once marketed (CEQ 1982a). Manufacturers must alert EPA of any information they obtain after marketing to indicate that the chemical may pose substantial risk of injury to health or environment.

Chemical Testing and Control

An Interagency Testing Committee recommends up to 50 existing chemicals at any one time that should be tested for toxicity. The EPA negotiates for the testing to be done by industry, following agreed-on protocol.

If, on examination of test data, the EPA determines that a chemical poses unreasonable risk, it may take a range of actions from prohibition to limitation or labeling. The identified risk may occur at any point from manufacture through distribution, use, and disposal.

Ecological Constraints and Public Policy

A number of ecological processes constrain the ability of existing toxic substance legislation to be maximally effective. In testing, one typically examines the effect of exposure to this chemical alone, thereby neglecting synergistic interactions it may have with other chemicals in the environment. Second, one typically tests the chemical on one or a few species of organisms, each in isolation, in a laboratory situation. The potential for observing the effect of the chemical on nontest species or in multispecies situations, where subtler effects of the chemical on competition, predation, and other ecological processes may be observed, is not utilized. Furthermore toxicological testing typically uses short exposure times at unusually high doses; long-term effects at lower doses in an ecosystem may well be different, as processes of natural selection operate on weakened species over longer periods. In addition the potential transformation of a substance into a more toxic one by microbial action or photochemical reaction in the environment (e.g., DDT transformation to the more toxic DDE by microbes; formation of ozone from hydrocarbons and nitrogen oxides) is not tested by normal toxicological procedures. Biospheric processes of interaction with the chemical are not normally tested. The possible destruction of the stratospheric ozone layer by fluorocarbons (freons) from aerosol cans and refrigeration equipment and the production of nitrous oxide from bacterial decomposition of nitrogen fertilizers are effects that were discovered long after these chemicals were introduced and probably would not have been discovered by toxicological testing at the time. Predicting the widespread leakage and incorporation of an industrial organic compound (e.g., PCBs) into ecosystems and its subsequent bioaccumulation would require both microcosm and ecosystem testing.

To point out these ecological constraints is not to detract from the very useful information to be obtained from toxicological data nor to criticize TSCA, which filled a major need in toxic substances control. Rather, it is to recall that ecological and biospheric processes can transform and interact with chemicals in vast and complex ways that often confound human efforts to predict them by quick laboratory methods. As such these processes constrain the ability of existing legislation to achieve its goals. In addition to toxicological testing, longer-term, *in situ* testing and monitoring of the effects of toxic substances in microcosms, ecosystems, and larger segments of the biosphere continue to be needed (see Chapter 9).

LAND USE PLANNING AND CONTROL

Perhaps more than for any other type of environmental legislation the extent of government control on uses of land varies with the political values and traditions of each country. The basic rationale of land use legislation is that activities on a piece of privately owned land may generate damage or nuisance (erosion, air pollution, etc.) to adjacent property owners or others; some social control is necessary to prevent such damage to those who are not reaping the benefits of use of the land. In the case of publicly owned lands the major rationale for land use control is to insure that activities on public grounds do not prevent the use of the land for the purposes designated.

In the United Kingdom land use planning and control has been based on the Town and Country Planning Act of 1947. Under the 1947 law planning authorities (on County and County Borough Councils) had to grant permission before any significant change in land use occurred. The planning authorities drew up a development plan for each region and a detailed town map, indicating present and proposed future uses (over roughly the next 20 years). The plan was subject to public comment, ministerial approval, and periodic update (5+ years) (Rhind and Hudson 1980). In amendments to the act in 1968 and 1971, the development plans were renamed structure plans, and town maps were renamed local plans. These new plans were intended as broader community plans, incorporating the spatial distribution of jobs, people, communication routes, recreation, conservation, green belts, and development (Figure 2.9). In developing structure plans, a series of "strategy plans" are made to map the desired locations of such elements as population, employment, or parkland (Figure 2.10).

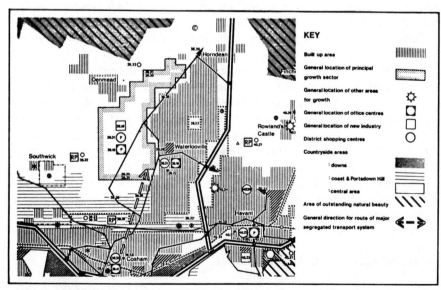

Figure 2.9. A section of a Structure Plan diagram for the area around Waterlooville, Hampshire County, England. Reproduced, with permission, from R. Brown in *The Planner,* Vol. 60, 1974, official journal of The Royal Town Planning Institute, London, U.K.

RWAB Sub-Region: Proposed Strategy

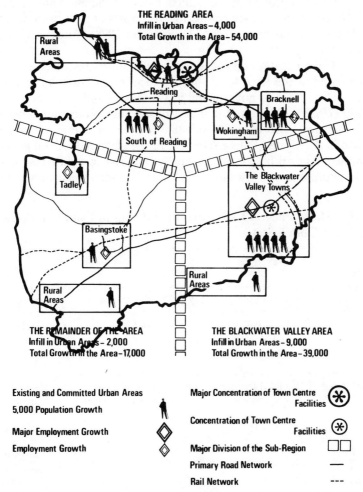

THE READING AREA
Infill in Urban Areas – 4,000
Total Growth in the Area – 54,000

Rural Areas

Reading

Bracknell

Wokingham

South of Reading

Tadley

The Blackwater Valley Towns

Basingstoke

Rural Areas

Rural Areas

THE REMAINDER OF THE AREA
Infill in Urban Areas – 2,000
Total Growth in the Area – 17,000

THE BLACKWATER VALLEY AREA
Infill in Urban Areas – 9,000
Total Growth in the Area – 39,000

Existing and Committed Urban Areas

5,000 Population Growth

Major Employment Growth

Employment Growth

Major Concentration of Town Centre Facilities

Concentration of Town Centre Facilities

Major Division of the Sub-Region

Primary Road Network —

Rail Network - - -

Figure 2.10. A strategic plan diagram for the potential location of future population, employment, and urban facilities in the Reading-Wokingham-Aldershot-Basingstoke area of southeastern England. Reproduced, with permission, from J. Glasson and M. Elson in *The Planner,* Vol. 62, 1976, official journal of The Royal Town Planning Institute, London, U.K.

Permission for changes in existing land use must still be obtained from the planning authorities. Although such plans map out a desired distribution of land uses, little can be done at a local level to ensure that desired proposals for land use change are forthcoming. Thus planning authorities take on an essentially one-sided role of denying proposals incompatible with desired uses (Rhind and Hudson 1980). The U.K. government has, however, taken several initiatives to influence the location of development. Most notably, by the New Towns Acts of 1946 and 1965 development corporations have been able to acquire land and build on it according to land use plans reviewed by the public and approved by the minister (see Rhind and Hudson 1980, Hall 1973, Cherry 1974). Somewhat analogous efforts at new town development have been undertaken in France (Rubinstein 1978), Australia, Brazil, and elsewhere.

In the United States controls on private land use that do exist are primarily generated at the local level through local plans and zoning ordinances. About 36% of the local government jurisdictions employed land use controls by 1980 (USDA 1981, cited in The Conservation Foundation 1982). As in most countries there is also a substantial body of common law governing land use in the United States. The role of the state and federal governments in land use control has been principally to assist in policy planning (Table 2.8). In addition to the laws listed in Table 2.8, such legislation as the Clean

Table 2.8. Major Federal Laws Directly Relating to Land Use Planning and Control in the United States

Subject	Law	Year	Administering Agencies	Coverage
Management policy for public lands	BLM Organic Act or Federal Land Policy and Management Act	1976	DOI, Bureau of Land Management	Federal public lands to be managed under a multiple-use, sustained yield principle
Cultural Sites				
Historic sites and buildings	National Historic Preservation Act; Executive Order 11593	1966 1976	DOI, National Park Service	Maintains inventory of approved and eligible historic sites and structures and requires impact assessment before developing them
Interesting land preservation	Antiquities Act	1906		Allows preservation by executive order of lands of extraordinary scientific, historic, or cultural value
Water Margins				
Coastal zone	Coastal Zone Management Act	1972 1975	Coastal states, DOC	Federal grants provided to states for coastal zone planning; federal actions in coastal zone must be approved by States; estuarine sanctuaries authorized

Table 2.8. (*Continued*)

Subject	Law	Year	Administering Agencies	Coverage
Flood plains	National Flood Insurance Act Executive Order 11988	1969 1973 1977	HUD; DOT; DOD (Army Corps); DOI (USGS); SCS; NOAA; TVA	Requires flood plans and subscriptions to federal flood insurance program to be eligible for federal flood disaster financial assistance
Riparian margins	Wild & Scenic Rivers	1968 1974	States	Authorizes control of land use adjoining designated wild and scenic rivers
Estuaries	Estuarine areas Executive Order 11190	1968 1977	DOI	Requires federal agencies to protect estuarine areas and balance need for conservation and development
Forests				
National forests	Multiple-Use, Sustained Yield Act Forest and Rangeland Renewable Resources Planning Act National Forest Management Act Public Rangelands Improvement Act	1960 1974 1976 1978	USDA, Forest Service	National forests to be managed for sustained yield of timber, recreation, range, watershed, fish, and wildlife
Nonfederal forests	Cooperative Forestry Assistance Act	1978	USDA, Forest Service	Provides federal funds for planning on nonfederal forests
Wilderness and Parks				
Wilderness areas	Wilderness Act	1964 1972 1975	USDA, Forest Service	Prohibits roads and vehicles from areas in National Wilderness Preservation System
National Parks	General Revision Act; National Parks and Recreation Act	1891 1978	DOI, NPS	Sets aside national forests, parks, and wilderness areas;[a] 1978 act extends National Trails Systems (established 1968); establishes National Historic Trails

(continued on next page)

Table 2.8.   (*Continued*)

Subject	Law	Year	Administering Agencies	Coverage
Wildlife refuges	Executive Order 1014	1903 1966	DOI, Fish and Wildlife Service	Establishes national wildlife refuges for conservation, wildlife, and recreation.
Land Disturbance				
Off-road vehicles	Executive Order 11644	1972		Controls use of off-road recreational vehicles on public lands
Mining on public lands	General Mining Law	1872	DOI	Regulates mining on public lands (hard rock mining requires no prospecting permit, easy purchase, and no royalty payments; surface mine reclamation is required only for coal)
	Mineral Leasing Act	1920		
	Surface Mining and Reclamation Act	1977		

Sources: Golden et al. (1979), CEQ (1978, 1979, 1980).

Note: A number of additional U.S. laws influence community development and land use indirectly. See text for examples. DOI = Department of Interior; USDA = U.S. Department of Agriculture; DOC = Department of Commerce; HUD = Department of Housing and Urban Development; DOD = Department of Defense; USGS = U.S. Geological Survey; SCS = Soil Conservation Service; NOAA = National Oceanic and Atmospheric Administration; TVA = Tennessee Valley Authority; DOT = Department of Transportation.

[a] National Park System was established in 1872.

Air Act of 1970 and 1977 and Federal Water Pollution Control Act Amendments of 1972 and 1977 have important indirect effects on community growth, since they restrict development in regions that have not attained primary air quality standards or do not have secondary wastewater treatment capacity for new developments. The 1978 CEQ guidelines [Sec. 1500, 8a(2)] on EIA under NEPA require a description of the relationship between the proposed action and land use plans, policies, and control in the affected area. The EIS must explain, but not necessarily resolve, areas of conflict and inconsistency between the action and existing land use plans.

About a third of the land in the United States is owned by the federal government. The bulk of public land (78%) is within the control of the Bureau of Land Management and the USDA Forest Service (National Forests). Land under the jurisdiction of these agencies is managed for multiple use and sustained yield. "Sustained yield" refers primarily to timber harvesting and range grazing in National Forests, and "multiple uses" to recreation, watershed maintenance, fish and wildlife maintenance and sport harvest, and protection of wilderness. The remainder of the federal land is managed for rather specific purposes: national parks, monuments, and preservation of wilderness (see Lemons and Stout 1982); wildlife refuges for conservation, recreation and wildlife; wild and scenic rivers and national trails for recreation; historic sites for cultural appreciation. For a concise review of federal legislation regulating public land use in the United States, see CEQ (1980, Chapter 7). For a broader treatment of land use in the United States see, for example, Jackson 1981.

Goals, Objectives, and Strategies

At the policy level, land use planning goals are normally designed (1) to specify and maintain particular land uses, consistent with public goals and values, and/or (2) to limit uses that are incompatible with the ecological processes of the land. In the United States, for example, most federal public lands are governed by the first policy goal (Figure 2.11).

A number of federal land use laws are designed simply to assist states in limiting ecologically inappropriate land uses. For example, legislation dealing with flood plains, wetlands, and the coastal zone encourages states to identify these areas and develop plans to restrict uses in them to actions that are compatible with the ecological processes of the area. The Department of Housing and Urban Development administers grants for urban development planning. Such comprehensive plans must incorporate environmental considerations and be accompanied by an environmental impact assessment if any Federally supported projects would result (U.S. Department of Housing and Urban Development 1977; see also guidebook to areawide environmental assessment, U.S. Department of Housing and Urban Development 1984). Thus the typical strategy employed to enforce state planning is either to offer federal financial assistance to conduct the planning (e.g., coastal zone planning, urban development planning) or to withhold related federal funds on failure to conduct the planning (e.g., floodplain management).

A broad range of tactics have been employed in different countries to achieve land use control. A few of the major tactics are described below.

Tactics

Comprehensive Plans

Comprehensive plans set forth broad guiding principles for future development of a region. At their best they are accompanied by maps indicating natural hazards and other ecological constraints and attempt to limit devel-

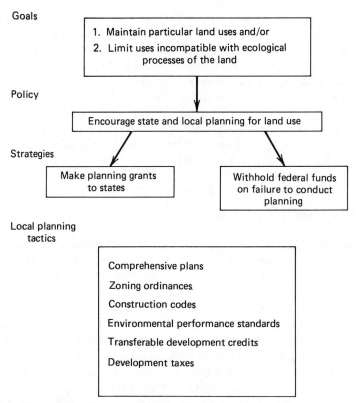

Figure 2.11.   Structure of land use legislation in the United States.

opment to the estimated carrying capacity of the region's natural resources (Chapter 6). Typically, the decisions on development are made by a planning board, which refers to the comprehensive plan for policy guidance. Environmental impact assessments are increasingly prepared in association with such plans in the United States (U.S. Department of Housing and Urban Development 1977; Hall 1977), Canada (Walker 1977), and elsewhere.

Zoning Ordinances

Zoning ordinances specify the exact locations in a region where particular land uses will be acceptable (urban, agricultural, park) and such matters as density of building or inhabitants and nature of activity (commercial, residential, industrial). They can also be used to restrict the extent and nature of development in hazardous areas such as flood plains, unstable slopes, and fire-prone regions (Figure 2.12).

Subdivision Controls and Construction Codes

Subdivision ordinances and construction codes regulate design, building, and grading practices within a zone. Thus they may specify such items as

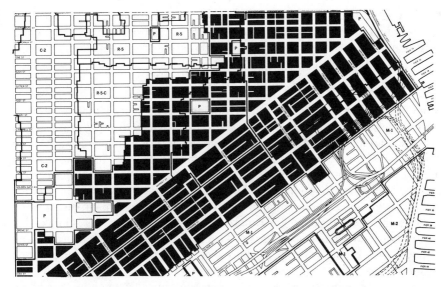

Figure 2.12. A section of a zoning map for downtown San Francisco and a section of the key to zoning districts (San Francisco Department of City Planning 1966). Excerpts from San Francisco City zoning code:

	New Districts Proposed		Existing Districts to Remain			
	C–3–G Downtown General Commercial	R–5–C Residential- Commercial	R–4 High Density Residential	R–5 Highest Density Residential	C–2 Community Business	P Public Use
Permitted uses	Apartments permitted without conditional use ap- proval. Certain business services permitted. Drive-in uses prohib- ited, except for service stations.	R–5 uses permitted on any floor C–2 uses permitted on or below the ground floor	Apartments, institutions, nonprofit clubs Hotels and professional offices as conditional uses.	Apartments, institutions, nonprofit clubs Hotels and professional offices as conditional uses.	Retail and consumer services, offices, entertain- ment, other community business uses, clubs, institutions, apartments	Standards, where applicable, are gener- ally deter- mined in individual cases by the City Planning Commission

height and setback of buildings, amount and angle of grading on slopes, drainage and open space requirements. Such regulations can be designed to help minimize the effects of natural hazards on urban development (see, e.g., Griggs and Gilchrist 1983).

Environmental Performance Standards

More flexible than zoning and subdivision ordinances, environmental performance standards specify the ultimate level of some ecological process that must be retained after development is completed. For example, a standard in a county in Georgia required that storm water runoff following development not exceed the natural amount of runoff in the area. In measuring performance, a storm of an intensity that occurs once every two years on average is to be used, and an area immediately downstream from the project is to be used as the zone from which runoff is measured. Failure to achieve this performance level can trigger the need for further mitigation measures, using a bond posted by the developer (Thurow et al. 1975). The advantages of such an approach are that the requirement is more closely tied to the ultimate land use goal, and developers have greater flexibility in designing how to achieve the performance standard.

Transferable Development Credits

By a tactic of transferable development credits, each land parcel is assigned a development right according to its zoned level of use at the time of onset of controls. In Los Angeles, California, for example, transferable development credits have been used since 1970 to regulate land use within five miles of the coast under the California Coastal Zone Management Act. For example, in order to build houses on a parcel zoned for wilderness, a developer must purchase the development right from an existing parcel that is zoned residential. If the latter parcel already has houses on it, the houses must be torn down; if the parcel is zoned for residences but no houses are yet built, by selling the development right, the parcel owner forfeits the right to build on it and retires the parcel to a less-intensive zoning classification. Subdivision rights sold for about $35,000 per acre in the 1970s (M. J. Glickfield, personal communication, 1982). The California Coastal Commission serves as a clearing house for purchasing, banking, selling, and trading transferable development credits. Such a system is useful for preserving a given mix of land uses while allowing new development to occur in the region.

Development Taxes

Development or land assessment taxes levy a sliding fee on land according to the intensity of its use. Great Britain has periodically taxed profits arising from major changes in land use (e.g., from agricultural to urban; see Rhind and Hudson 1980). Property or valuation taxes, which are widely used in the western world to tax property according to the value of developments on it, are a variant of the same concept applying to very minor changes in land use (e.g., adding a room to a house). Such taxes can act to rein the pace of development while helping fund the increased public costs of providing social services (sewage treatment, roads, schools, etc.) to inhabitants of more intensively used land.

Ecological Constraints and Public Policy

To the extent that land use plans and controls proceed with attention to the ecological processes of the landscape, they are clearly more likely to achieve the ultimate goals of maintaining land in a desired condition. All of the tactics described here could be used to achieve environmentally sensitive land use, or could abuse this purpose by disregarding ecological constraints on land development. With land perhaps more clearly than with any other ecosystem compartment, the natural processes of erosion, sedimentation, flooding, wildfire, avalanche, slope failure, volcanic eruption, earthquake, and subsidence impose constraints on its development. In Chapter 6 we discuss ways in which these constraints of the landscape can be recognized and planned for.

BIOLOGICAL CONSERVATION

Goals, Objectives, and Strategies

Three goals that have been pursued in different kinds of legislation designed for biological conservation are (1) the management of game species for sustained yield, (2) the conservation of examples of endangered or threatened species, and (3) the preservation of balanced populations of species in their native habitats.

Depending on which of these goals is pursued, quite different strategies are employed (Figure 2.13). To achieve the first goal of managing populations for sustained harvest, wildlife refuges may be established with a management program to control levels of hunting and culling. A number of international conventions (Table 2.9) on the harvesting of whales, fur seals, polar bears, and fishes are of this type; in the United States the management of wildlife refuges for game similarly involves harvest quotas. Such a strategy may permit relatively sustained harvest for some years, but because the size of the managed population may be different than the natural average, populations maintained too close to their threshold may be more vulnerable to major fluctuations in climate or other resource factors (see Glantz 1980). Such strategies require both excellent wildlife management skills and cooperation among the harvesters. Historically these requisites have not been easy to achieve (e.g., whale management, Figure 2.14).

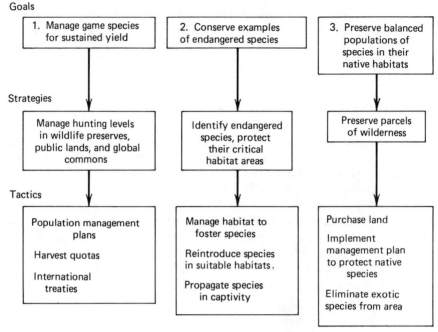

Figure 2.13. Structure of laws for biological conservation in the United States.

The second goal of conserving examples of endangered species is usually pursued by identifying species that are rare, threatened, or endangered; these are categories of increasing scarcity, respectively. The major remaining locations of these species, the "critical habitats," are identified and preserved from further direct disturbance. Extremely endangered species are sometimes introduced or reintroduced to habitats where they are believed capable of surviving, or attempts are made to propagate and maintain the species in zoos, gardens, wildlife parks, or seed or sperm banks. The U.S. Endangered Species Act (Table 2.10) is of this type.

Table 2.9. Major International Conventions and Treaties Pertaining to Biological Conservation

Subject	Law	Year First Completed	Coverage
Endangered species	Convention on International Trade in Endangered Species of Wild Fauna and Flora	1973	Controls trade of a list of species threatened by overharvest; encourages habitat preservation

Table 2.9. (*Continued*)

Subject	Law	Year First Completed	Coverage
	Migratory Species Treaty for Wild Animals	1979	Requires protection of certain migratory vertebrates
	Convention on the Protection of the World Cultural and Natural Heritage	1972	Lists specific sites for international protection
	Antartic Treaty	1959	Agreement concerning use and study of Antartica
	Convention on Nature Protection and Wildlife Preservation in the Western Hemisphere	1941	Encourages protection of habitats of native species in the Americas
Marine and arctic mammals	Interim Convention for Conservation of North Pacific Fur Seals	1957	Prevents harvest of fur seals in the open sea
	International Convention for the Regulation of Whaling	1946	Sets annual quotas on harvest of whale species
	Agreement on the Conservation of Polar Bears	1973	Studies and controls harvest of polar bears
Birds and game mammals	Convention for the Preservation of Migratory Birds	1916	Agreement between United States and Canada
	Convention for the Protection of Migratory Birds and Game Mammals	1936	Agreement between United States and Mexico
	Convention for the Protection of Migratory Birds and Birds in Danger of Extinction and their Environment	1972	Agreement between United States and Japan

Source: CEQ 1980; 1982a.

Note: Many other bilateral and multilateral agreements exist, especially regarding fishing.

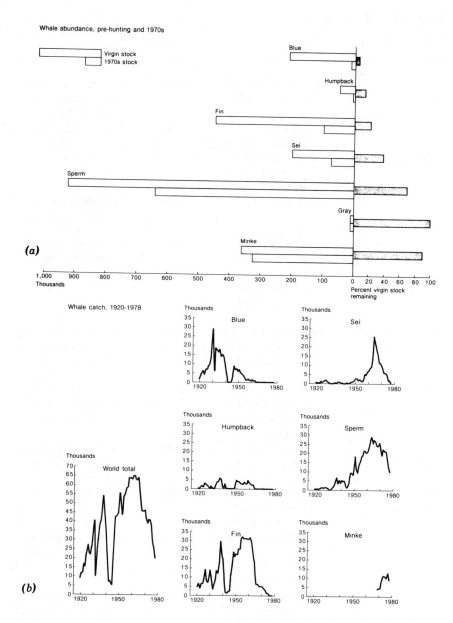

Figure 2.14 (a) Whale population numbers, by species, pre-hunting and 1970s, and (b) whale catch, 1920–1978. The International Whaling Commission has set quotas on whale harvest since 1947 and banned hunting of blue, humpback, bowhead, and gray whales (except for small, subsistence catches of the latter three). Except for the California gray whale, there is as yet little sign of population recovery. World total harvest figures reflect both declining stocks and IWC quotas (46,600 whales in 1971; 13,900 in 1981). The blue, humpback, fin, and sei whales, with lowest remaining stocks, were the earliest to be hunted. Sperm and Minke whales are more recent targets. Reproduced from Figure 13-18 of U.S. Council on Environmental Quality (1981).

Table 2.10. Major Federal Laws Pertaining to Biological Conservation in the United States

Subject	Law	Year	Administering Agencies	Coverage
Birds	Migratory Bird Conservation	1929	DOI, Fish and Wildlife Service	Protects and conserves migratory game, insectivorous birds, and habitats of endangered species
Fish and wildlife management	Fish and Wildlife Coordination	1934	DOI, Fish and Wildlife Service	Coordinates federal, state, and local plans for hunting, fishing, and conservation
	Fish and Wildlife Act	1956 1974	DOI, Fish and Wildlife Service	Authorizes management of fish and wildlife
	Fish and Wildlife Conservation Act	1980	DOI, Fish and Wildlife Service	Coordinates federal and state plans for conserving nongame vertebrates.
	Protection of Bald and Golden Eagles	1972	DOI, Fish and Wildlife Service	Protects bald and golden eagles from disturbance
Endangered species	Endangered Species Act	1973, 1978, 1982	DOI, Fish and Wildlife Service, DOC, USDA	Requires protection of critical habitat for endangered or threatened species
	Executive Order 11911			
	Lacey Act Amendments	1981		Prohibits interstate and international commerce in wild animals and plants illegally obtained
	Marine Mammal Protection	1972		

Source: Golden et al. 1979; CEQ 1982a.

Note: See also Table 2.8 for park and wilderness preservation laws, and Table 2.9 for international treaties and conventions. The United States has signed all agreements in Table 2.9, except the Migratory Species Treaty for Wild Animals. DOI = Department of Interior; DOC = Department of Commerce; USDA = U.S. Department of Agriculture.

The third goal seeks to preserve not just species but the ecosystem of which they are a part. Such a goal preserves not only populations but the processes, including behaviors, that characterize the interaction between native species and between species and their habitat. Legislation preserving wilderness areas (see, e.g., Table 2.8) pursues this third goal.

Tactics

Harvest Quotas

An important issue in establishing numerical values for annual harvest of species managed for "sustainable yield" is whether such a quota should be set at the apparent maximum or threshold value for sustained yield or whether some lower value, allowing a margin for variation in species numbers, should be permitted. The real problem is determining the value needed to maintain population numbers in the long term in the face of natural fluctuations in population numbers. Evidence from fisheries management indicates that the optimum value for long-term management is at a level of harvest somewhat less than that apparently sustainable in a shorter term (Krebs 1972; see Chapter 8). Indeed, in 1974 the International Whaling Commission agreed to allow a margin of safety in setting its harvest quotas (CEQ 1978, p. 340).

Preserving Critical Habitats of Endangered Species

The ecological problems in implementing this tactic are manifold. First, there is the question of what is the minimum area necessary to maintain a healthy breeding population of a species. For top carnivores such as condors and lions, the feeding territory may encompass a hundred square miles or more. For territorial species, density within a habitat is likely to be low. Thus the minimum area of critical habitat needed is often quite large, based on feeding habits and minimum densities needed for mating (Shaffer 1981). Beyond this the size of a minimum healthy breeding population may be determined by behavioral constraints. For example, herring gulls do not begin courtship behavior until a minimum flock of at least 20 gulls has gathered (so-called "social facilitation").

An additional constraint is the size of a gene pool necessary to prevent the buildup of expressed recessive characteristics (often lethal or maladaptive) through excessive inbreeding. These problems are discussed further in Chapter 8.

A great deal of research and paperwork is required to declare a species endangered. For example, as of 1983, 63 plant species in the United States had been placed on the federal endangered species list, yet it was estimated that at least 1200–2500 species of plants were in jeopardy (Elias 1983, CEQ 1980). Less than 15 percent of the listed endangered species had been assigned recovery plans (CEQ 1980). There is an additional tendency to concentrate on large, furry or familiar creatures to the exclusion of others. No microorganisms, for example, are listed (Table 2.11).

Despite these problems the "endangered species" approach to biological conservation is widespread. For a list of U.S. federal endangered and threatened species, consult the Federal Register and Golden et al. (1979); see Ayensu and De Fillips (1978) for plants. Additional species are listed by certain states. Endangered species in Europe are listed by the Council of Europe (1977), and plant species discussed by Walters (1976) for Europe and by Elias (1983) for the USSR. A series of "Red Books" published by the International Union for the Conservation of Nature and Natural Resources (IUCN) list endangered and threatened species worldwide.

Captive Propagation

When the appropriate habitat for a species no longer exists or is threatened, the only alternative remaining for preservation of the species is propagation in the captivity of zoos, botanical gardens, and seed or sperm banks. The problems of preserving genetic diversity are acute (Senner 1980). To help mitigate this problem, many zoos and wild animal parks exchange offspring bred in captivity. This is of course only a short-term solution as the population in the world's zoos represents a small and finite gene pool that must be enriched by additions from the wild to avoid ultimate inbreeding depression. Further the techniques for inducing breeding in many captive species are not

Table 2.11. Number of Animals and Plants Designated or Proposed as Endangered or Threatened Species by the U.S. Federal Government

Category	Endangered			Threatened		
	United States	Foreign	Total	United States	Foreign	Total
Mammals	32	242	274	3	20	23
Birds	66	158	224	3	0	3
Reptiles	13	61	74	10	4	14
Amphibians	5	8	13	3	0	3
Fishes	33	15	48	12	0	12
Snails	2	1	3	5	0	5
Clams	23	2	25	0	0	0
Crustaceans	1	0	1	0	0	0
Insects	7	0	7	4	1	5
Plants	63	2	65	7	3	10
Total	245	489	734	47	28	75

Source: U.S. Fish and Wildlife Service, Endangered Species Technical Bulletin 5(8):16, 1980; endangered plant numbers in United States from Elias (1983).

Note: Number of critical habitats listed = 35.

yet known. Many of the behavioral characteristics of the species in the wild are lost, as any visitor to a zoo quickly observes. Such loss further inhibits the possibility of reintroduction of such species to the wild (Campbell 1980). Reintroductions to the wild also carry the risk that species may migrate to unintended areas or spread disease to native populations (Campbell 1980).

The problems of sperm and seed banks are even greater. First, it should be realized that preservation of sperm only saves the costs of shipping or maintaining a male in captivity for a temporary period; it does not eliminate the need to maintain live populations to carry out the actual breeding, rearing, and production of new generations. As for seed banks, seeds of most tropical forest tree species and many temperate plant species have no dormancy, and lose viability within a few weeks. For those seeds that can be stored, germination and propagation requirements must be known. Further the seeds must be grown every few years to produce new seed, as few species retain viability for longer than a decade or two (Went 1969). The number of individuals that can be grown is small and genetic maintenance of variability is therefore a major problem. Many tree species do not set seed for 20 years or more, so that time, space, and cultivation all become costly procedures. Seed banks are not therefore a satisfactory solution to species preservation in the long term.

Ecosystem Preservation

The maintenance of intact ecosystems in preserve status avoids most of the problems of the other tactics but has problems of its own in relation to disturbances from preserve edges, and maintenance of species in small, isolated territories. These problems will be discussed further in Chapter 11. For detailed discussion of tactics of species preservation, see Ehrlich and Ehrlich (1981), Frankel and Soulé (1981), and Soulé and Wilcox (1980).

Ecological Constraints and Public Policy

The greater the isolation of a species from its ecosystem context, the more extensive are the problems in preserving the species indefinitely. For these reasons programs that emphasize preservation of species rather than of their habitat are expensive and often least effective in accomplishing species preservation goals. Furthermore, to save a species once its population has been reduced to an endangered level is much more difficult, due to "genetic bottlenecking" (i.e., the loss of gene pool as a population shrinks; Franklin 1980; Chapter 8). Furthermore species in vastly contracted habitats are likely to require costly intervention for their management, whereas species able to survive in protected habitats of adequate size normally need less management intervention. Thus from the point of view of cost efficiency, species preservation strategies could do well to focus on preserving species not yet endangered, by the mechanism of wilderness or habitat preservation.

Efforts at preserving endangered and threatened species have occasionally shown encouraging results, but many species continue to decline despite significant governmental efforts (Figure 2.15; CEQ 1981).

NOISE CONTROL

Goals, Objectives, Strategies

Noise has been implicated as a causal factor in a variety of human health effects beyond hearing loss, from hypertension and other symptoms of stress in adults to weight loss in fetuses (CEQ 1979). In addition loud noise can interfere with a range of human activities such as sleep, communication, and

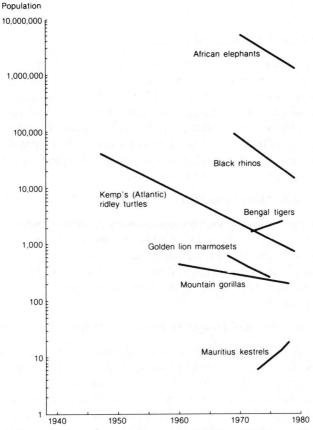

Figure 2.15. Population of selected endangered and threatened species, 1947–1979. Reproduced from Figure 13-19 of U.S. Council on Environmental Quality (1981).

mental concentration. There is also evidence for interference with communication and other behavior patterns in wildlife (Fletcher and Busnel 1978). Thus the major goal of noise reduction legislation typically embraces the broad concerns of human health and welfare. A widely used strategy for noise control in Europe, Japan, the United States, and elsewhere is the establishment of noise emission limits on manufactured products (OECD 1978). The OECD (1978) has sought to standardize noise emission standards for new products between member countries so that such standards will not interfere with trade. A second strategy involves mitigation measures to insulate the human ear from the ambient noise; most such measures are technological in nature (e.g., highway walls for noise deflection). The third major strategy involves a variety of planning and design measures to separate major noise sources from the locations where people spend their time, or to control the time of day or night when noise is generated (Figure 2.16).

Tactics

Product Emission Standards and Labels

Designing products to reduce their noise output (cushioning noisy parts, enclosing product in box, etc.) sometimes reduces the effectiveness of the product for its original purpose and adds to cost of manufacture. For such a tactic to be adopted by manufacturers, noise emission standards need to be set uniformly by law over a wide proportion of the market, or some incentive to enhance sale of the product must be introduced. Thus a companion tactic to noise emission standards is labeling the products with information for the consumer on the amount of noise generated.

Technological Mitigation Measures for Ambient Noise

Included in this tactic are such measures as earphones or plugs, acoustic tile or carpeting, and noise barriers along roadsides or factories, concrete and brick walls, or wide plantings of vegetation. This tactic is often the least cost-effective of the three discussed here (see, e.g., CEQ 1979) because the potential receptors that must be insulated from a single noise source are typically so numerous.

Spatial and Temporal Design Measures to Mitigate Ambient Noise

One major approach is to zone adjacent portions of the landscape for compatible uses. Thus a factory or golf course would be more compatible next to an airport than would be a hospital, school, or group of houses. Sometimes airport authorities buy adjacent land to insure future compatible uses; or they buy air "rights" or easements. The problem with the latter approach is that although the current resident of an adjacent home is financially compensated for the disturbance, future buyers of the property will typically not be

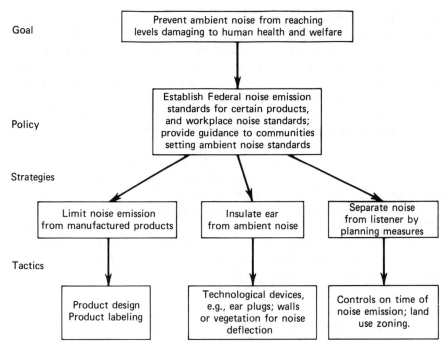

Goal

Prevent ambient noise from reaching
levels damaging to human health and welfare

Policy

Establish Federal noise emission
standards for certain products,
and workplace noise standards;
provide guidance to communities
setting ambient noise standards

Strategies

Limit noise emission
from manufactured products

Insulate ear
from ambient noise

Separate noise
from listener by
planning measures

Tactics

Product design
Product labeling

Technological devices,
e.g., ear plugs; walls
or vegetation for noise
deflection

Controls on time of
noise emission; land
use zoning.

Figure 2.16. The broad structure of noise control efforts in the United States (Noise Control Act of 1972, other Federal regulations, and local ordinances). The goal of protecting wildlife from adverse noise effects is implied by NEPA and State impact assessment legislation.

compensated. A temporal design approach widely used around airports is to limit the number of flights, or the runways and associated flight paths, used at night, following the principle that noise of a given level is more annoying during periods of rest and sleep than during active work.

A common ordinance to achieve ambient noise standards is to establish a "property line standard" designating the maximum noise level that will be allowed at property boundaries. Such an approach allows the property owner to mix design, siting, and technological measures in meeting the property line standard. It is analogous to "environmental performance standards" discussed under land use control tactics.

Ecological Constraints and Public Policy

The biggest problem in meeting the goal of protecting public welfare by noise legislation is that noise effects on wildlife are poorly known, often assumed to be insignificant, and therefore not accorded a high priority in most noise control programs. Noise levels can cause many of the same stress symptoms in laboratory animals that are caused in humans (U.S. Environmental Protection Agency 1974b, Ames 1978, CEQ 1979). Furthermore interference with communication signals between animals (bird calls, whale songs, insect sounds, etc.) by human-generated noise can have major disruptive effects on population-level behavior such as schooling and mating (Ellis et al. 1978, Myrberg 1978). Even more insidious is the fact that the sensitivity to various frequencies of sound on the part of animal species differs in most cases from that for humans, so that an action whose sound-generating properties are mitigated only in human hearing frequencies may prove to be quite damaging to species sensitive to ultralow or high frequency sound. The same consumer product, (e.g., fan, blender) which has been controlled for noise emissions in the range of human hearing may prove extremely disturbing to a pet dog or bird. Ultrahigh or low frequency sounds emitted by certain transmission lines may affect wildlife in ranges both audible and inaudible to people (Ellis et al. 1978, Lee and Griffith 1978). As research on the effects of noise on nonhuman species increases, we may reasonably expect to see revisions in approaches to noise control to account for the sensitivities of other species and thus to enhance achievement of public welfare goals.

CONCLUDING REMARKS

Ecological impact assessment involves in part evaluating predicted impacts in light of the variety of relevant laws and plans. At its best an EIA may also encourage environmental standards that exceed minimum existing legislative requirements for achieving social goals. To limit one's biological impact analysis to a listing of endangered species present, for example, is to ignore the value of preserving the habitat of nonendangered species in achieving the larger social goals of biological conservation. Similarly, to ignore the effects of noise on wildlife simply because this is not directly regulated by noise pollution legislation is to ignore the broader objectives of environmental policies that seek to preserve ecosystems in a healthy state. Thus in impact analysis, narrow and specific legislative requirements should be seen as minimum requirements to be met by a proposed action; the broader social goals that the laws seek to achieve must also be considered.

REFERENCES

Abel, N., and Stocking, M. (1981). The experience of underdeveloped countries. In T. O'Riordan and W. R. D. Sewell, eds. *Project Appraisal and Policy Review.* Wiley, Chichester, pp. 253–295.

Altschuler, A. (1981). *The Urban Transportation System: Politics and Policy Innovation.* The MIT Press, Cambridge, Mass.

American Arbitration Association (1980). *Improving EIS Scoping.* Washington, D.C.

Ames, D. R. (1978). Physiological responses to auditory stimuli. In J. L. Fletcher and R. G. Busnel, eds. *Effects of Noise on Wildlife*. Academic Press, New York, pp. 23–45.

Armour, A. (1977). Understanding environmental assessment. *Plan Canada* **7**(1): 9–18.

Ayensu, E. S., and DeFillips, R. A. (1978). Endangered and threatened plants of the United States. Smithsonian Institution and World Wildlife Fund, Washington, D.C.

Brown, R. (1974). Linking local planning with structure planning. *The Planner* **60**:505–508.

Caldwell, L. K. (1976). The National Environmental Policy Act: retrospect and prospect. *Environ. Law Reporter* **6**:50030–50038.

Campbell, S. (1980). Is reintroduction a realistic goal? In M. E. Soulé and B. A. Wilcox, eds. *Conservation Biology: An Evolutionary-Ecological Perspective*. Sinauer, Sunderland, Mass., pp. 263–270.

Canter, L. (1982). The status of EIA in developing countries. Environ. and Ground Water Inst., Univ. Oklahoma, Norman.

Carpenter, R. (1980). Using ecological knowledge for development planning. *Environ. Manage*. **4**:13–20.

Carpenter, R. (1981). Balancing economic and environmental objectives: the question is still, how? *Environ. Impact Assessment Rev.* **2**:175–188.

Cherry, G. E. (1974). *The Evolution of British Town Planning*. Heath and Reach, London.

Clark, B. D., Bisset, R., and Wathern, P. (1980). *Environmental Impact Assessment: A Bibliography with Abstracts*. Mansell, London.

Colston, N. V., Jr. (1974). Characterization and treatment of urban land runoff. Rep. 670/2-74-096, EPA, Washington, D.C.

Commoner, B. (1973). Workplace burden. *Environment* **15**(6):15–20.

Commonwealth of Australia (1975). Administrative procedures under the Environmental Protection (Impact of Proposals) Act, 1974–1975.

Cook, E. (1975). Ionizing radiation. In W. W. Murdoch, ed. *Environment: Resources, Pollution and Society*. Sinauer, Sunderland, Mass., pp 297–323.

Council of Europe (1977). List of rare, threatened and endemic plants in Europe. Nature & Environ. Ser. 14. Strasbourg, France.

Council on Environmental Quality (1976). Environmental impact statements. An Analysis of Six Years' Experience by Seventy Federal Agencies. GPO, Washington, D.C.

Council on Environmental Quality (1977). *Environmental Quality*. Eighth Ann. Rep. Washington, D.C.

Council on Environmental Quality (1978). *Environmental Quality*. Ninth Ann. Rep. Appendix F. Regulations for implementing the procedural provisions of NEPA (40 CFR Parts 1500–1508). Washington, D.C.

Council on Environmental Quality (1980). *Environmental Quality*. Eleventh Ann. Rep. Washington, D.C.

Council on Environmental Quality (1981). *Environmental Trends*. Washington, D.C.

Council on Environmental Quality (1982a). *Environmental Quality 1981*. Twelfth Ann. Rep. Washington, D.C.

Council on Environmental Quality (1982b). *Environmental Quality 1981*. Twelfth Ann. Rep. Appendix B. Forty most asked questions concerning CEQ's national environmental policy act regulations. Washington, D.C.

Crump, S. (1962). *Ride the Big Red Cars: How Trolleys Helped Build Southern California*. Crest, Los Angeles.

D'Itri. (1982). *Land Treatment of Municipal Wastewater*. Ann Arbor Sci., Ann Arbor, Mich.

Doern, G. B. (1977). Regulatory processes and jurisdictional issues in the regulation of hazardous products in Canada. Sci. Council Canada Background Study 41. Supply & Services Canada, Ottowa.

Ehrlich, P., and Ehrlich, A. (1981). *Extinction: The Causes and Consequences of the Disappearance of Species*. Random House, New York.

EIA Worldletter (1983). People's Republic of China. 1(1):4. PADC Unit, Dept. Geogr., Univ. Aberdeen, Aberdeen, Scotland.

Elazar, D. J., Schlesinger, J., Locard, J., Stevens, R. M., and Stevens, B. A. (1972). *Green Land–Clean Streams: The Beneficial Use of Waste Water through Land Treatment*. Center for Study of Federalism, Temple Univ., Philadelphia, Pa.

Elias, T. S. (1983). Rare and endangered species of plants—the Soviet side. *Science* **219**:19–23.

Ellis, D. H., Goodwin, J. G., Jr., and Hunt, J. R. (1978). Wildlife and electric power transmission. In J. L. Fletcher and R. G. Busnel, eds. *Effects of Noise on Wildlife*. Academic Press, New York, pp. 81–104.

Enviro-Control, Inc. (1974). Total urban water pollution loads: the impact of stormwater. CEQ, EPA and HUD, Washington, D.C.

Fairfax, S. K. (1978). A disaster in the environmental movement. *Science* **199**: 743–748.

Federal Environmental Assessment Review Office (1978). Guide for environmental screening. Ottowa.

Federal Register (1980). Proposed rulemaking: revision to the Rhode Island state implementation plan. **45**(208):70513.

Fisher, D. E. (1980). *Environmental Law in Australia*. Univ. Queensland Press, St Lucia.

Fletcher, J. L., and Busnel, R. G., eds. (1978). *Effects of Noise on Wildlife*. Academic Press, New York.

Formby, J. (1981). The Australian experience. In T. O'Riordan and W. R. D. Sewell, eds. *Project Appraisal and Policy Review*. Wiley, Chichester, pp. 187–225.

Frankel, O. H., and Soulé, M. E. (1981). *Conservation and Evolution*. Cambridge Univ. Press, Cambridge.

Franklin, I. R. (1980). Evolutionary change in small populations. In M. E. Soulé and B. A. Wilcox, eds. *Conservation Biology: An Evolutionary-Ecological Perspective*. Sinauer, Sunderland, Mass., pp. 135–149.

Franson, R. T., Lucas, A. R., Giroux, L., and Kenniff, P. (1977). Canadian Law and the Control of Exposure to Hazards. Sci. Council Canada Background Study 39. Supply & Services Canada, Ottowa.

Glasson, J., and Elson, M. (1976). Practice review. Reading–Wokingham–Aldershot–Basingstoke sub-regional study: report of the study team. *The Planner* **62**:88–89.

Golden, J., Ouellette, R. P., Saari, S., and Cheremisinoff, P. (1979). *Environmental Impact Data Book*. Ann Arbor Sci., Ann Arbor, Mich.

Griggs, G. B., and Gilchrist, J. A. (1983). *Geological Hazards, Resources, and Environmental Planning*. 2nd ed. Wadsworth, Belmont, Calif.

Hall, P., ed. (1973). *The Containment of Urban England*. Vols. 1, 2. Allen & Unwin, London.

Hall, R. C. (1977). MEIRS—a method for evaluating the environmental impacts of general plans. *Water, Air, Soil Poll.* **7**:251–260.

Hase, T. (1981). The Japanese experience. In T. O'Riordan and W. R. D. Sewell, eds. *Project Appraisal and Policy Review*. Wiley, Chichester, pp. 227–251.

Higgins, P. F. P. (1970). *Elements of Torts in Australia*. Butterworths, Sydney.

Hollick, M. (1980). EIA in Australia: the federal experience. *Environ. Impact Assessment Rev.* **1**:330–336.

Horberry, J. (1980). United Nations Environmental Programme: guidelines for industrial environmental impact assessment and environmental criteria for the siting of industry. *Environ. Impact Assessment Rev.* **1**:208–210.

Horberry, J. (1983). Environmental guidelines survey: an analysis of environmental procedures and guidelines. Joint Environ. Serv., Intl. Inst. Environ. & Development, London, and Intl. Union Conserv. Nature, Gland, Switzerland.

Jackson, R. H. (1981). *Land Use in America*. Wiley, New York.

Jones, M. G. (1980). Developing an EIA process for the Netherlands. *Environ. Impact Assessment Rev.* **1**:167–180.

Josephson, J. (1975). Green systems for wastewater treatment. *Environ. Sci. Tech.* **9**:408–409.

Kennedy, W. V. (1981). The West German experience. In T. O'Riordan and W. R. D. Sewell, eds. *Project Appraisal and Policy Review*. Wiley, Chichester, pp. 155–185.

Kennedy, W. V. (1982). The directive on environmental impact assessment. *Environ. Policy Law* **8**(3):84–95.

Kneese, A. V., and Schultze, C. L. (1975). *Pollution, Prices and Public Policy*. Brookings Inst., Washington, D.C.

Krebs, C. J. (1972). *Ecology. The Experimental Analysis of Distribution and Abundance*. Harper & Row, New York.

Lakhani, H. (1982). Air pollution control by economic incentives in the U.S.: policy, problems and progress. *Environ Manage* **6**:9–20.

Lee, J. M., Jr, and Griffith, D. B. (1978). Transmission line audible noise and wildlife. In J. L. Fletcher and R. G. Busnel, eds. *Effects of Noise on Wildlife*. Academic Press, New York, pp. 105–168.

Lee, N., and Wood, C. (1978). Environmental impact assessment of projects in EEC countries. *J. Environ. Manage.* **6**:57–71.

Lemons, J., and Stout, D. (1982). National parks legislative mandate in the U.S.A. *Environ. Manage.* **6**:199–207.

Likens, G. E., Wright, R. F., Galloway, J. N., and Butler, T. J. (1979). Acid rain. *Sci. Amer.* **241**(4): 43–51.

Linacre, E. (1976). Views of the environment. *Search* **7**:227–230.

Liroff, R. A. (1978). The effectiveness of NEPA. *Science* **202**:1036, 1038.

Lowrance, W. W. (1976). *Of Acceptable Risk*. Kaufmann, Los Altos.

Mangun, W. R. (1979). A comparative evaluation of the major pollution control programs in the United States and West Germany. *Environ. Manage.* **3**:387–401.

Mayda, J. (1981). Environmental assessment as an instrument for the development and implementation of environmental law. United Nations Environmental Program, Nairobi, Kenya.

Meyer, J. R., and Gomez-Ibañez, J. A. (1981). *Autos, Transit and Cities*. Harvard Univ. Press, Cambridge, Mass.

Miller, G. T., Jr. (1982). *Living in the Environment*. 3rd ed. Wadsworth, Belmont, Calif.

Morgan, R. K. (1983). The evolution of environmental impact assessment in New Zealand. *J. Environ. Manage.* **16**:139–152.

Myrberg, A. A., Jr. (1978). Ocean noise and the behavior of marine animals: relationships and implications. In J. L. Fletcher and R. G. Busnel, eds. *Effects of Noise on Wildlife*. Academic Press, New York, pp. 169–208.

OECD (1978). Reducing noise in OECD countries. Org. Econ. Coop. Dev., Paris.

OECD (1979). Environmental impact assessment: analysis of the environmental consequences of significant public and private projects. Paris.

O'Riordan, T., and Sewell, W. R. D., eds. (1981). *Project Appraisal and Policy Review*. Wiley, Chichester.

Overseas Development Ministry (1977). A guide to the economic appraisal of projects in developing countries. Her Majesty's Stationery Office, London.

Plewes, M., and Whitney, J. B. R., eds. (1977). *Environmental Impact Assessment in Canada: Processes and Approaches*. Inst. Environ. Studies, Univ. Toronto, Toronto.

Prieur, M., and Lambrechts, C. (1980). Model outline environmental impact statement from the standpoint of integrated management or planning of the natural environment. Council of Europe, Nature & Environ. Ser. 17. Strasbourg, France.

Printz, A. C., Jr. (1978). Environmental considerations for U.S. assistance programs in developing countries. In S. Bendix and H. R. Graham, eds. *Environmental Assessment: Approaching Maturity*. Ann Arbor Sci., Ann Arbor, Mich., pp. 43–49.

Rhind, D., and Hudson, R. (1980). *Land Use*. Methuen, London.

Roberts, L. (1982). California's fog is far more polluted than acid rain. *BioScience* **32**:778–779.

Rubinstein, J. M. (1978). *The French New Towns*. Johns Hopkins Univ. Press, Baltimore, Md.

Sartor, J. D., and Boyd, G. B. (1972). Water pollution aspects of street surface contaminants. Rep. R2-72-081. EPA, Washington, D.C.

Senner, J. W. (1980). Inbreeding depression and the survival of zoo populations. In M. E. Soulé and B. A. Wilcox, eds. *Conservation Biology. An Evolutionary-Ecological Perspective.* Sinauer, Sunderland, Mass., pp. 209–224.

Sewell, W. R. D. (1981). How Canada responded: the Berger inquiry. In T. O'Riordan and W. R. D. Sewell, eds. *Project Appraisal and Policy Review.* Wiley, Chichester, pp. 77–94.

Shaffer, M. L. (1981). Minimum population sizes for species conservation. *BioScience* **31**:131–134.

Shimazu, Y., and Harashima, R. (1977). Environmental assessment–management system for local development project: cases in Japan. *J. Environ. Manage.* **5**:243–258.

Soulé, M. E., and Wilcox, B. A., eds. (1980). *Conservation Biology. An Evolutionary-Ecological Perspective.* Sinauer, Sunderland, Mass.

Southern California Association of Governments (1978). Draft Air Quality Management Plan. Los Angeles, Calif.

Stein, R. E., and Johnson, B. (1979). *Banking on the Biosphere? Environmental Procedures and Practices of Nine Multilateral Development Agencies.* Heath, Lexington, Mass.

The Conservation Foundation (1982). *State of the Environment, 1982.* Washington, D.C.

Thurow, C., Toner, W., and Erley, D. (1975). Performance controls for sensitive lands: a practical guide for local administrators. Rep. 600/5-75-005. EPA, Washington, D.C.

Tips, W. E. J., and Deneef, R. (1980). Environmental impact assessment in a closed government system: the case of Belgium. *Environ. Impact Assessment Rev.* **1**:432–435.

U.N. Food and Agriculture Organization (1983). Environmental impact analysis and agricultural development. FAO Environ. Paper No 2. Unipub, New York.

U.S. Agency for International Development (1974). Environmental assessment guidelines manual. GPO, Washington, D.C.

U.S. Agency for International Development (1979). Environmental and natural resource management in developing countries. A report to Congress. Vols. 1, 2. GPO, Washington, D.C.

U.S. Department of Agriculture (1981). Soil and Water Resources Conservation Act 1980, Appraisal, Part I. Soil, water and related resources in the United States: status, condition and trends. Washington, D.C.

U.S. Department of Housing and Urban Development (1977). Integration of Environmental Considerations in the Comprehensive Planning and Management Process. Washington, D.C.

U.S. Department of Housing and Urban Development (1984). Areawide environmental assessment: a guidebook. Worldletter Publ., Environ. & Groundwater Inst., Univ. Oklahoma, Norman.

U.S. Environmental Protection Agency (1974a). Development of a trial air quality maintenance plan using the Baltimore Air Quality Region. Rep. 450/3-74-050. EPA, Washington, D.C.

U.S. Environmental Protection Agency (1974b). Information on levels of environmental noise requisite to protect public health and welfare with an adequate margin of safety. GPO, Washington, D.C.

U.S. Forest Service (1980). Land management plan and draft environmental impact statement. Alpine lakes area, Wenatchee and Mt. Baker–Snoqualmie National Forests. GPO, Washington, D.C.

Walker, J. (1977). Town of Oakville prepares environmental plan. *Plan Canada* **17**(1):35–37.

Walters, S. M. (1976). The conservation of threatened vascular plants in Europe. *Biol. Conserv.* **10**:31–41.

Wandesforde-Smith, G. (1980). International perspectives on EIA. *Environ. Impact Assessment Rev.* **1**:53–64.

Went, F. W. (1969). A long-term test of seed longevity. II. *Aliso* **7**:1–12.

West, H. D. (1982). An analysis of California's system of environmental regulation. Environ. Sci. & Engr. Doctoral degree report, Univ. Calif., Los Angeles.

Westman, W. E. (1972a). Some basic issues in water pollution control legislation. *Amer. Sci.* **60**:767–773.

Westman, W. E. (1972b). The future for land disposal of water wastes. *Operculum* **2**(3):73–80.

Westman, W. E. (1973a). Environmental impact statements—boon or burden? *Search* **4**:465–470.

Westman, W. E. (1973b). Queensland's clean water regulations. *Operculum* **3**(3): 3–5.

World Bank (1974). Environmental, health and human ecologic considerations in economic development projects. Washington, D.C.

World Bank (1975). Environment and development. Washington, D.C.

3

ENVIRONMENTAL
DECISION MAKING

One widely held ideal in modern democracies is that the decision-making process should be an optimizing one. Thus a desired solution is that which maximizes social welfare, while being as equitable as possible in distributing costs and benefits among the various segments of society or "multiple publics." Both "means" and "ends" of the decision-making process are important foci of concern. Thus another common ideal is that decisions should be reached by desirable (e.g., democratic or participatory) processes, even if the solution is not optimal for social welfare (White and Hamilton 1983). In practice, many decisions appear to result more from compromise among powerful interests and a desire not to depart from past precedent than from an optimizing, holistic process (Friesema and Culhane 1976). A decision may thus differ from the rational, optimizing model for both technical and philosophical reasons: limited knowledge, time, money, or motivation, on the one hand, and a desire to reach a popular decision that preserves political traditions, on the other.

Impact assessors typically facilitate decision making by developing information and catalyzing discussion. In this chapter we discuss five broad topics of relevance to rational decision making: fashioning alternatives, estimating risk, finding optimum solutions, involving the public, and resolving conflicts.

FASHIONING ALTERNATIVES

As conceptualization of a proposal proceeds from policy level through programs to site-specific projects (Figure 1.3), the range of possible alternatives narrows. It is therefore important to begin fashioning alternatives at the broadest level of conceptualization not yet constrained by prior decisions. For example, a site-specific project (e.g., nuclear power plant) should consider higher level policy options (alternatives for altering energy supply or demand) if (and only if) these have not already been decided for the particular case. At the project-specific level modifications to the scale, design, or scope of the project constitute reasonable alternatives, along with the alternative of no action. The opinions of clients, the public, other experts, and the literature are useful, indeed usually essential, to generating an acceptable range of alternatives.

Once a range of possible project alternatives has been established, one must determine what possible impacts may arise from each. The range of possible impacts is initially boundless, and some limits to the analysis are needed. In one sense the fall of a leaf in a forest can be said to be interconnected with the welfare of the globe; a hierarchy of decreasing *significance* of effects, however, exists and can normally be identified. Typically the effects of a site-specific project are likely to be greatest on the local community; hence most effort should be devoted to detailing these impacts. The effects on the larger scales of society should also be identified, with detail proportional to their significance.

Panel-of-Experts Approach

There are several steps to be followed in generating impact scenarios, especially for an analysis of larger scales of impact. Suppose we wish to examine the ecological impacts of the operation of a nuclear power plant on a scale larger than the immediate air environment and watershed. One option is to enlist a group of experts to help devise alternative hypotheses about possible ecological interactions at the larger scales. The experts might be invited to conduct a "brainstorming session" (see e.g., Earwicker 1974), perhaps consulting a previously prepared checklist (see Chapter 4) to trigger the memory on possible compartments of impact and interaction. In situations where an ecological model is to be built, attention is often best focused by asking what must be known about ecosystem compartment x and its dynamics in order to predict how ecosystem compartment y will be affected (Hilborn et al. 1980). For example, litter decomposition rate may be the most relevant substrate-compartment property to describe rate of plant growth in a nutrient-poor forest. For additional discussion of methods of creative thinking and problem solving, see, for example, Harrison and Bramson (1982).

Suppose by these means three possible impacts are identified, such as the transport of radionuclides by stratospheric winds and subsequent fallout, the impact of atmospheric heat releases on micro- and macroclimate (fog incidence, changes to air temperature, precipitation and global heat balance), and the effect of chlorine from thermal effluent on aquatic food chains in receiving waters (see, e.g., Dickson et al. 1977). In the next step the consequences of these events are postulated: the capture and bioconcentration of particular radionuclides of particular species to x concentration, and subsequent toxic effects; the shift in local, regional, and global climatic zones of precipitation and thermal regime; the reduction in primary and secondary productivity in the stream. Although some information from literature, expert judgment, and modeling is available on these issues, much uncertainty will surround any precise predictions of occurrence and magnitude of events.

A third step is therefore to estimate the likelihood of event occurrence and the uncertainty surrounding the predictions of magnitude. Table 3.1 lists along the diagonal the estimated (really "guesstimated") probabilities of occurrence of each of the three postulated initial events. In the panel-of-experts approach these probabilities may be proferred by experts jointly, or independently, and then averaged. In the latter approach, known as the Delphi technique, each expert's independently solicited estimate is averaged with that of the others, and the median and range of estimates are conveyed to each expert. For estimates that do not fall in the median quantile, for example, the expert may be asked to explain the basis for his or her estimate, and these reasons are also conveyed anonymously to each other expert. The experts have an opportunity to change their estimate based on this new information, and a new median is computed. This iterative process can continue until little or no further change in estimate occurs. The value of the Delphi technique lies in its minimization of the role of personalities in influencing the judgment of other experts. Some subtle pressure, albeit anonymous, nevertheless remains for estimates to converge to the median (Earwicker 1974; see also Bakus et al. 1982, McAllister 1980).

Once the likelihood of occurrence of each separate event has been "guesstimated" in this way, one may also determine the likelihood of occurrence of pairs of events (off-diagonals in Table 3.1). In estimating the probabilities of joint occurrence of pairs of events in a matrix such as Table 3.1, the problem arises that once certain probabilities are specified, the remainder must fall within certain bounds in order for the probability values to be internally consistent mathematically. These constraints may be computed (Sarin 1978) and communicated to the experts at a certain point in the Delphi process. The bounds of higher-order probabilities (in our example, the occurrence of all three events simultaneously) can then be computed by similar means; the allowable range within which such probabilities may lie will now be so narrow that the best-guess value can usually be chosen by a nonexpert

Table 3.1.   Hypothetical Example of Estimated Probabilities of Occurrence of Postulated Regional and Global-Scale Impacts from a Nuclear Power Plant

Possible Impacts	Stratospheric Transport	Climatic Change	Chlorine Toxicity
Stratospheric transport of radionuclides	0.6		
Climatic changes from heat discharge	0.3	0.5	
Chlorine toxicity to stream organisms	0.1	0.2	0.2

(Nair and Sarin 1979). One now has an estimate of the likelihood of seven scenarios, involving all combinations of the occurrence of one, two, or three of the postulated events. These scenarios may be ranked from most to least likely.

Because the probabilities were determined by expert guesswork, with only the additional constraint of internal mathematical consistency, they may of course be inaccurate reflections of "reality." Although there is no way to ensure the accuracy of such estimates, it is possible to ask what effect a given change in the estimate of one of the probabilities would have on the relative likelihood of scenarios. Such a process is known as *sensitivity analysis*.

In our example, suppose the likelihood of climatic change were changed from 0.5 to 0.6. This scenario would now be equally likely as that of stratospheric transport, and the scenarios involving joint occurrence of climatic change with stratospheric transport and chlorine toxicity would increase in likelihood but not change rank. Were the likelihood of chlorine toxicity to be increased by 0.1, from $p = 0.2$ to $p = 0.3$, this would also raise the likelihood of pairwise events involving chlorine toxicity. But since the latter events are all of low likelihood, they would not have changed sufficiently to influence the ranking of the most likely three or four scenarios. Thus an "error" or change in the estimation of the likelihood of climatic change is more important than an equivalent change in the likelihood of chlorine toxicity in influencing the most likely scenarios. As a result one would want to concentrate further effort on improving the accuracy of the estimate of probability of climatic change.

When variables are interdependent in nonlinear ways and scenarios involve quite different mixes of variables, sensitivity analysis can be a major aid in identifying those estimates that are critical to the observed outcome and therefore deserve further study. An analogous procedure (estimation of likelihood, sensitivity analysis) could, in theory, be used to determine standard errors of estimate associated with the magnitude of events, and these values converted into fiducial limits or confidence intervals, provided possible magnitudes are normally distributed (see Chapter 1).

As we have noted, this whole procedure is based on guesswork by experts. While some will argue that an educated guess is superior to no information, others will argue that such a procedure conveys a false sense of precision to statements that have no direct empirical origin. Both points of view have merit. This method can be worse than useless if experts are asked to make guesses for which they do not have a substantial empirical basis. Certainly it is highly unlikely that an expert would have sufficient information on which to base a guesstimate of the standard error of estimate of the magnitude of, say the iodine-131 concentration in the thyroid of a small mammal 5 km down the fallout pathway, unless a substantial number of strictly comparable measurements had been made. Thus generating information on the likelihood of alternative scenarios can only be responsibly done in situations where substantial past experience on comparable actions has accumulated; even then, appropriate caveats about the guesswork nature of estimates should be emphasized to decision makers.

The problems with use of panel-of-experts information will be further discussed in Chapter 4 (see also McAllister 1980). For an application of the foregoing procedure to forecasting the growth of the solar-electric energy industry, see Sarin (1979); for an application to predicting global climate change to the year 2000, see National Defense University (1978). For further discussions on converting expert judgment to quantitative form and other aspects of decision theory, see Hammond et al. (1978), Kates (1978), Keeney and Raiffa (1976), and Raiffa (1968).

Interactive Computer Modeling Approach

Another approach to scenario generation, which has been increasingly applied in ecological impact assessment, has been the use of interactive computer models of the ecosystem under study. Once ecosystem components and the patterns of possible interaction have been postulated by experts, these relationships can be converted into qualitative expressions or *conceptual models* describing the relationship between compartments. These expressions clarify what kind of data are needed to convert the expressions into quantitative equations. Such data are then obtained from the literature, from short-term observations, from monitoring of effects of comparable actions, or from the project itself in its early stages. A computer model of ecosystem operation can then be constructed and run, and the effects of various impacts, mitigation measures, and management strategies tested with the model to generate scenarios in quantitative or graphical form. This combination of procedures is known as adaptive environmental assessment and management and has been developed primarily by C. S. Holling, R. Hilborn, C. J. Walters, and colleagues at the Institute of Animal Resource Ecology of the University of British Columbia (Holling 1978). It has been applied to more than 25 environmental problems in Canada and the United States (Environment Canada 1983).

An example of this approach is the examination of salmon fishery management along the Canadian Pacific coast (Argue et al. 1983). A variety of salmon types are fished commercially and by sport fishers in the Strait of Georgia, British Columbia, and the streams flowing into this estuary. At the time of model development (1976–1978) commercial fishing was regulated, but sport fishing had not been. Upriver stocks of chinook salmon had declined to less than half their 1950 levels, and similar declines in coho salmon in the Fraser River had occurred. Biologists and modelers from the University of British Columbia and the Canadian Department of Fisheries and Oceans collaborated during a workshop in sketching an initial model of salmon population dynamics, outlining the growth in size of wild and hatchery-raised individuals of chinook and coho in each year class, as well as their natural mortality, sport and commercial catch rates, rate of emigration from the strait (for feeding or spawning), and loss as sublegal fish (shakers) caught and subsequently dying after return to the water. Data on each of these processes were then obtained by the Canadian Department of Fisheries and Oceans from their records (Figure 3.1). Changes in each population size were calculated by the computer over 15-day intervals for a 20-year period. Model predictions were then made and examined, the model revised, and new predictions made about 15 times (iterative steps). Some relationships were revised, new indicators (e.g., economic value of sport fishery) added, and a variety of management techniques (minimize size limits on fish caught, seasonal fishing restrictions, bag limits) explored.

The model was discussed with interested citizen and managerial groups and further reviewed. The model was then used to develop and explain new management policy for the salmon populations. The model required one and a half years of calendar time, two and a half person-years of effort, and $8000 in computer time to develop (Buckingham 1980). The commercial and sport fishing industry it studied is a multimillion dollar enterprise.

Although this approach has great value in developing an ongoing tool for environmental management, and enhancing ecological understanding among biologists, managers, and the public, it also has its limitations. Primary among them is that models can only be built readily concerning ecosystems for which extensive data are available. The "adaptive management" aspect

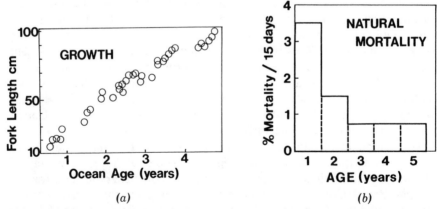

(a)   (b)

Figure 3.1. Examples of data used as input in the construction of an ecological model of salmon fishery in the Georgia Strait of British Columbia. (a) Salmon growth is determined in the model by a graph of fish length vs. fish age, derived from purse seine sampling data. (b) Natural mortality varies over the life of the salmon (chinook illustrated). Because the model calculates population changes over 15-day intervals, data are incorporated in the model to indicate the percent of each age class dying during each 15-day period. Reprinted from S. Buckingham, 1980, The Strait of Georgia Chinook and Coho Fishery User Guide, with permission of the University of British Columbia.

of the approach advocates that, when data are not available, data should be collected from a project once it has begun to operate, with plans to modify the project when management information from the model becomes available (Hilborn et al. 1980; Hilborn and Walters 1981). In the case of managing ecosystems the "adaptive management" philosophy would encourage managers to allow deviations from an existing management policy (over- and underfishing, or lower or higher water flow over a dam) in given years, in order to accumulate knowledge useful in developing a long-term management plan (Hilborn and Walters 1981). The risk involved is that either modifications will not be undertaken once the project is built, or that the "test" management schemes will do irreparable harm to the ecosystem.

A second limitation of this approach is that models are always simplifications of complex systems and, as such, may fail to predict with consistent accuracy the true behavior of the system they model. Techniques for predicting the future of variable systems will always suffer from uncertainty about the outcome, however. Panel-of-experts attempts to predict the future are usually assumed to be based on implicit extrapolations from past experience which are akin to regression models without explicit quantitative expression. (For a discussion of intuition vs. extrapolation in risk estimation, see Kates 1978, Ch. 2.) Although panel-of-expert scenarios are quicker and cheaper to generate, and require less information, the kinds of questions that can be answered with comparable precision are substantially fewer. A useful test of the value of the modeling approach is to determine how well predictions are met in reality ("validation"). Simulation models of ecosystems have proved useful in impact prediction in certain cases, provided considerable validation-testing and revision has taken place (Hall and Day 1977, Holcomb Research Institute 1976, Holling 1978, Mitsch et al. 1982, U.S. Environmental Protection Agency 1979).

ESTIMATING RISK

Risk Characterization and Evaluation

The term "risk" is increasingly used in the assessment literature to refer to the probability of occurrence of a certain level of impact (e.g., Lowrance 1976, Mishan 1971, Whyte and Burton 1980). In common parlance, "risk" is often used synonymously with "hazard" to mean simply exposure to a danger ("walking that tightrope is a risk"); however, most risk analysts have defined "risk" to include the concept of likelihood of damage. ("The risk in walking that tightrope is that you have a 2% chance of falling and breaking your neck.") It is in this latter sense that we will use the term. The distinctions made by Mishan (1971) in the economic context are useful here:

Risk can be determined when all possible future outcomes and their respective probabilities of occurrence are known

Uncertainty exists when all possible outcomes are known, but their probabilities of occurrence are unknown.

Incomplete knowledge exists when not all outcomes are known.

Table 3.2 presents a hypothetical example in which conditions of risk may be described (columns A and C). An economic evaluation of the risk is also presented. In this analysis, the expected economic cost of each outcome is estimated (column B), and the expected value of annual costs is computed in monetary terms as the average of costs of each outcome, weighted by their relative probabilities of occurrence (column D, bottom). This expected value ($55,000) represents an average cost to be expected a priori on the basis of the risks involved. When the annual expected monetary benefit of introducing the action (pesticide application) is compared with expected annual cost, the procedure is termed risk-benefit analysis. One may also express the "costs" in nonmonetary units, such as number of reported illnesses or other indexes.

In risk analysis the evaluation of the significance of the risk results in a determination of "safety." Indeed, the distinction between empirical observation and judgment of significance appears several times in the language of decision theory (Table 3.3). In this dichotomy the concepts of "impact" and "risk" are analogous, except that by convention "impacts" usually refer to effects on the natural environment and human welfare, whereas "risk" usually refers to effects on human health.

A number of useful books on environmental risk analysis may be consulted for thorough discussions of the definition, estimation and evaluation

Table 3.2. Hypothetical Example of a Condition Where Risk May Be Calculated

Column A Human Health Effects:	Column B Annual Cost $\times$ 10^3	Column C Probability of Occurrence	Column D Annual Expected Cost $\times$ 10^3
Mild toxicity to 20,000 farm workers	$ 60,000	50%	$30,000
Severe toxicity to 20,000 farm workers	100,000	25	25,000
No toxic effects	0	25	0
Average annual expected cost			$55,000

Note: Column A presents all possible human health outcomes of a proposal to introduce a new pesticide on the market.

Table 3.3. Observational (Empirical) versus Judgmental (Normative) Terms in Decision Theory

Observational Terms (Measurement)		Judgmental Terms (Evaluation)	
Effect or impact	Degree of disturbance to preexisting environmental condition, sometimes including probability of occurence	Significant effect or impact	A level of disturbance judged important to human welfare
Risk	Probability and severity of adverse effects on human health	Safety	Degree of acceptability of risk
Efficacy	Probability and degree of beneficial effect (e.g., effectiveness of a drug)	Benefit	Degree to which efficacy is considered desirable
Distribution of costs	The degree to which different parties must surrender resources or accept damages	Equity	How fair the distribution of costs is judged to be

Source: Based in part on W. W. Lowrance (1976), *Of Acceptable Risk: Science and the Determination of Safety* (William Kaufmann, Inc.), pp. 94–95, and W. J. Petak (1980), *Environ. Manage.* **4**:287–295 (Springer-Verlag, New York). Adapted with permission.

of environmental risk, including Fischhoff et al. (1981), Kates (1978), Lowrance (1976), Schwing and Albers (1980), and Whyte and Burton (1980). A discussion of the judicial interpretation of risk-minimizing requirements in U.S. environmental law may be found in Ricci and Molton (1981).

Risk Perception

The field of risk perception examines social attitudes toward kinds and levels of risk. In assessing the social acceptability of a new risk (e.g., nuclear power), one may compare it to risks whose social acceptability is better appreciated. For example, in Figure 3.2 the radiation health risk to a nuclear power plant worker is compared to natural background levels and to other familiar risks (e.g., medical uses, consumer goods). In addition the risks to workers in the nuclear power industry may be compared with risk levels to workers in alternative energy-producing industries (Table 3.4). While risk-

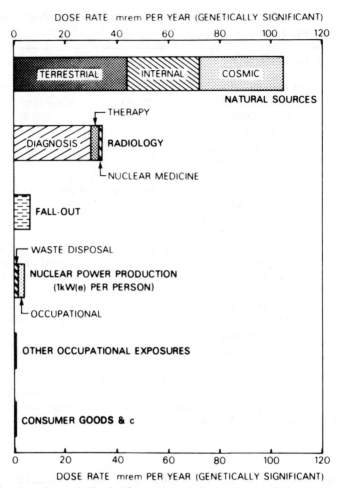

DOSE RATE mrem PER YEAR (GENETICALLY SIGNIFICANT)

Figure 3.2. Annual genetically significant dose rate for low level radiation, averaged through the whole population (Canada). Reprinted from Aiken et al. (1977), with permission of the Ministry of Energy, Mines and Resources, Ottawa, Canada.

risk comparisons are useful, they can also be deceptive in isolation. In the example just given it should be made clear to the decision maker that radiation doses tend to be additive in effect, so that the dose received by a nuclear power plant worker must be added to his or her dose from background radiation and other sources; furthermore, in considering the acceptability of radiation from nuclear power, risks not just to workers but to the public and to ecosystems must be considered.

A distinction can be made between the social acceptability of voluntary and involuntary risks. In the nuclear power example, though a worker may be willing to accept voluntarily the additional radiation dose in exchange for

Table 3.4. Deaths from Accidents during the Production of Energy from Coal, Oil/Gas, and Nuclear Power in the United Kingdom.

Energy Source	Operation	Deaths/Gigawatt/Year From Accidents
Coal	Extraction	1.4
	Transport	0.2
	Generation	0.2
	Total	1.8
Oil/gas	Extraction	0.3
	Transport	Insignificant
	Generation	None reported
	Total	0.3
Nuclear	Extraction	0.1
	Transport	Insignificant
	Generation	0.15
	Total	0.25

Source: Data from U.K. Health and Safety Commission, 1978. *The Hazards of Conventional Sources of Energy.* Her Majesty's Stationery Office, London.

the benefits of the job, a person living in the vicinity of the plant and deriving electrical energy from a windmill on his or her property may be much less willing to accept an ambient radiation dose from the plant, both because the dose is being received involuntarily and because no compensatory benefit is being received. Starr (1972) has attempted to quantify the relationship between risk level, benefit level, and freedom of choice about acceptance of risk (Figure 3.3). He finds that people are generally more willing to accept risks voluntarily than involuntarily for a given level of benefit. Otway and Cohen (1975) have reanalyzed Starr's data with regression analysis, after omitting natural disasters from the category of involuntary risks, and suggested that at higher levels of benefit, the distinction between willingness to accept voluntary vs. involuntary risks diminishes. Starr has summarized his findings on social acceptability of risk this way (Figure 3.4) (1972, pp. 38, 41, cited in Kates 1978, pp. 43–44):

1. Rate of death from disease is an upper guide in determining the acceptability of risk—somewhat less than 1 in 100 years.
2. Natural disasters ("acts of God") tend to set a base guide for risk—somewhat more than 1 in a million years—similar to the intrinsic "noise" level of physical systems. Man-made risks at this level can be considered almost negligible, and can certainly be neglected if they are several magnitudes less.

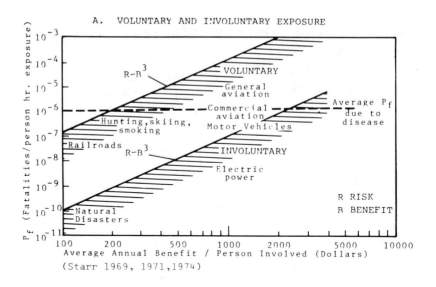

A. VOLUNTARY AND INVOLUNTARY EXPOSURE

(Starr 1969, 1971, 1974)

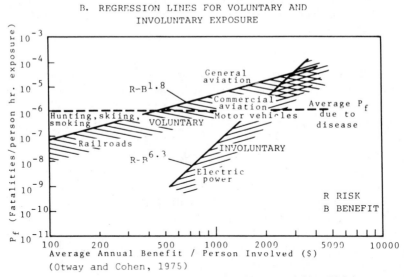

B. REGRESSION LINES FOR VOLUNTARY AND INVOLUNTARY EXPOSURE

(Otway and Cohen, 1975)

Figure 3.3. Risk versus benefit for voluntary and involuntary activities. Risk is measured as probability of fatality (P_f), derived from number of fatalities per person-hours of exposure. Benefit is measured as the average annual expenditure on the activity by each person. (*a*) Best-fitting lines drawn by eye with error bands (shading) to show their approximate nature. (*b*) Linear regression lines through the same data set after deleting natural disasters from the category of involuntary risks. Part (*a*) from Starr (1972), reprinted with permission of the National Academy of Sciences Press. Part (*b*) from Otway and Cohen (1975), reprinted with permission of the International Institute for Applied Systems Analysis, Laxenburg, Austria.

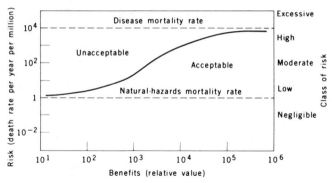

Figure 3.4. A postulated model of the benefit-risk relationship for involuntary exposure. Reprinted with permission from C. Starr and C. Whipple (1980). Risks of risk decisions, *Science* **208**:1114–1119. Copyright 1980 by the American Association for the Advancement of Science.

3. As would be expected, societal acceptance of risk increases with the benefits to be derived from an activity. The relationship appears to be non-linear, with [Starr's] study suggesting that the acceptable level of risk is an exponential function of the benefits (real and imaginary).

4. The public appears willing to accept voluntary risks roughly 1,000 times greater than involuntary exposure risks.

Social acceptability of risk is likely to vary from culture to culture and between groups within a society. Factors such as gender, age, income, and education will also influence how a person perceives risk. Therefore the precise numbers introduced into the literature on this subject are of limited applicability. Hohenemser et al. (1983) have recently shown that people tend to respond to hazards differently, depending particularly on their perception of the likely spatial extent of potential damage, the potential for sudden death of large numbers of people or wildlife, the persistence of the hazard over time, and the delay before adverse effects appear. In the small sample of people (34) questioned by Hohenemser et al., the actual annual human mortality of a hazard was not strongly correlated with how the risk was perceived, but the study needs to be replicated with a much larger sample size.

Still another way to illustrate the variation in perceived risk on the part of different sectors of the public is shown in Figure 3.5. Here it is seen that different people exhibit wide differences in the distance they would be willing to accept between their homes and an industrial installation (Popper 1983). The items illustrated in this figure exemplify locally unwanted land uses (LULUs). The figure reminds us that we are often more willing to accept actions when we receive the benefits and others are exposed to the risks.

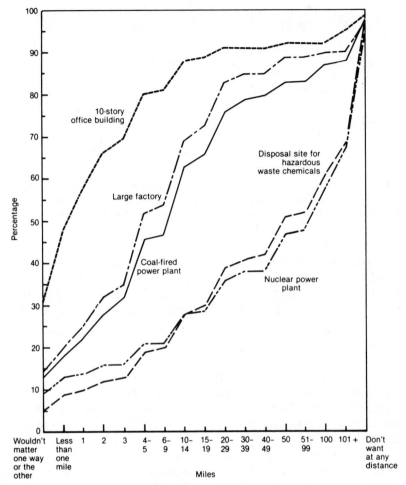

Figure 3.5.   Cumulative percentage of people willing to accept new industrial installations at various distances from their homes. Reprinted from Council on Environmental Quality, Department of Agriculture and Environmental Protection Agency (1980). Public Opinion on Environmental Issues: Results of a National Public Opinion Survey. Government Printing Office, Washington, D.C.

FINDING OPTIMUM SOLUTIONS

Single-Objective Optimization

The steps we may take to achieve environmental goals are almost always subject to constraints. Thus the developer of a new power plant who has the goal of minimizing the cost of pollution control in order to maximize profit is subject to constraints from pollution laws. For example, air pollution law may specify that the percent of sulfur dioxide removed from stack gases in a coal-fired power plant exceed 85% and that the primary standard for sulfur dioxide not be exceeded beyond the margin of the factory site. To achieve the goal of cost minimization subject to these constraints, the utility owner can choose one or more options: for example, installing different kinds of pollution control technology, using fuel of different sulfur contents, varying rate of production in the factory and hence volume of stack gases generated, and considering site locations of varying background levels and pollution dispersion conditions. The process of maximizing some function (in this case, cost) by manipulating options subject to constraints is known as *optimization*. The function we seek to maximize or minimize (cost) is called the *objective function*. The options we can manipulate are called *control variables* (e.g., types of control technology); the *constraints* (e.g., 85% removal) in our example are set by law.

In our example the developer will presumably decide to remove the minimum amount of SO_2 required (85%) subject to the further constraint of achieving the required ambient concentration at the factory margin. For the latter the owner would normally estimate dispersion of the pollutant using a model that takes into account variations in atmospheric conditions at various sites (see Chapter 7). This may result in the need to take extra precautions (e.g., use of lower sulfur fuel) during times of adverse atmospheric conditions.

In order to find solutions that meet the requirements of the objective function and fall within the specified constraints, we can build a mathematical model describing the relationship between control variables. The following hypothetical example (adapted from Haith 1982, Ch. 2) will illustrate the model-building process and the search for optimal model solutions. Let us define

x = level of coal burned (in units of 10^4 kilograms per week)
y = quantity of stack gas treated for SO_2 (in units of 10^4 kg/wk)

Suppose we determine that for every kg of coal burned, 75% of the stack gases are put through a pollution control device, and from this emission stream 97% of the SO_2 is removed. Since the efficiency e of the pollution control device is 97%, $e = 1 - 0.03y$. The fraction of SO_2 not removed is $1 - e = 0.03y$; when multiplied by the initial amount of gas entering the controlled stack, the amount of SO_2 released from the stack becomes $y(1 - e) = 0.03y^2$. The total emissions are $(3x - y) + (0.03y^2)$. Figure 3.6 illustrates these relationships.

We seek to express the objectives of maximizing profit and meeting pollution control standards in terms of x and y.

Let us define the profit Z as equal to utility revenue minus cost of production and of pollution control. Suppose for every $1.30 of revenue per kg of coal burned per week, $0.90 must be spent in production and $0.10 in pollution control.

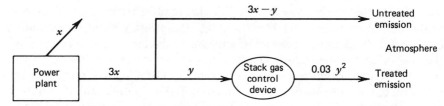

Figure 3.6. Mass fluxes of sulfur dioxide from the power plant. $4x$ = total coal burned; y = proportion of coal gas treated for SO_2 removal. Modified from Fig. 2.1 of Haith (1982), with permission of John Wiley and Sons, New York.

Then, since x and y have units of 10^4 kg/wk,

$$Z = (1.3)10^4x - (0.9)10^4x - (0.10)10^4y \qquad (3.1)$$

or

$$Z = 4000x - 1000y \qquad (3.2)$$

Suppose that no more than 10,000 kg/wk of the total stack gas mixture (treated and untreated) $(3x - y + 0.03y^2)$ can be emitted per week to meet the ambient primary standard at the power plant boundary. This is an oversimplification, since meeting the ambient standard will depend on the mix of treated and untreated stack gases, as well as atmospheric dispersion conditions.

Suppose further that the plant cannot burn more than 5.5×10^4 kg/wk of coal or treat more than 14×10^4 kg/wk of stack gases. Then we can write the optimization problem as follows:
Maximize

$$Z = 4000x - 1000y \qquad (3.3)$$

such that

$$3x - y + 0.03y^2 \leq 10 \qquad (3.4)$$

$$x \leq 5.5 \qquad (3.5)$$

$$y \leq 14.0 \qquad (3.6)$$

Furthermore it is obvious that none of the mass flows should be zero or less, so

$$x \geq 0 \qquad (3.7)$$

$$y \geq 0 \qquad (3.8)$$

$$3x - y \geq 0 \qquad (3.9)$$

Figure 3.7 illustrates these constraints on x and y graphically.

The "feasible solutions" are all pairs of x and y values that satisfy all the constraints (hatched area of Figure 3.7). The "optimal solutions" are those pairs of x, y values that maximize the objective function (profit, in this case). A variety of approaches exist for finding such "optimal solutions," all using an iterative approach, that is, trying possible solutions and refining the guess by successive elimination of less successful solutions. The approach may be informal and graphical, or it may involve such mathematical and computerized techniques as Lagrange multipliers, various search algorithms, and linear and dynamic programming. A good recent text on optimization techniques is that of Haith (1982), but many other books on the subject exist (e.g., Brebbia 1976, Converse 1970, Gass 1975).

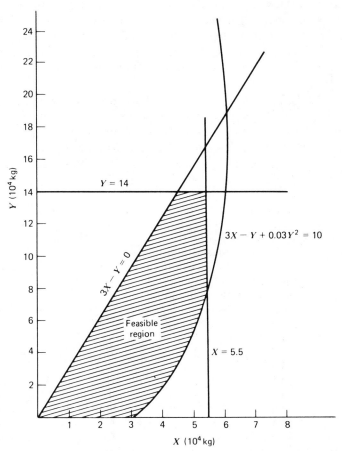

Figure 3.7. Constraints on solutions to the power plant problem discussed in the text. Reprinted, with permission, from Haith (1982). Copyright © 1982 by John Wiley and Sons, New York.

In the example of Figure 3.7 we may use a simple trial-and-error approach to find one or more optimal solutions. Examination of the objective function

$$Z = 4000x - 1000y \qquad (3.3)$$

suggests that a value of x as large as possible and a value of y as small as possible within the feasible region would maximize Z. This immediately limits optimal solutions to values along the right-hand curved portion of Figure 3.7, since other values would have either larger y or smaller x values. To evaluate pairs of values along this curve, we may solve Eq. 3.4 for x,

$$x = \frac{10 + y - 0.03y^2}{3} \qquad (3.10)$$

and substitute values of y between 0 and 8 (from Figure 3.7) in the equation. The x values increase monotonically from $x = 3.33$ at $y = 0$ to $x = 5.36$ at $y = 8$. However, we note that when $y = 5$, $x = 4.8$, resulting in a total emission $(3x - y + 0.03y^2)$ of 10.15, which exceeds the constraint (10). Therefore we need to examine x and y values in the context of the total emission constraint (Table 3.5).

Table 3.5 indicates that a value of coal burned $x = 4.5 \times 10^4$ kg/wk would yield the largest profit while meeting the constraint on total gaseous emissions. The values $x = 4.5 \times 10^4$ kg/wk (coal burned) and $y = 4 \times 10^4$ kg/wk (stack gases subjected to SO$_2$ control) then constitute the optimal solution.

To find optimal solutions, we needed to be able to express all the relation-

Table 3.5.   Outcomes of Alternatives Selected to Satisfy the Power Plant Problem

Alternative	Coal Burned x (10^4kg/wk)	Stack Gases Subjected to SO$_2$ Control y(10^4 kg/wk)	Total Emissions $3x - y + 0.03y^2$ (10^4 kg/wk)	Profits $Z = 400x - 1000y$ ($/wk)
1	3.3	0	9.9	13,200
2	3.6	1	9.8	13,400
3	3.9	2	9.8	13,600
4	4.2	3	9.9	13,800
5	4.5	4	10.0	14,000
6	4.7	5	9.9	13,800
7	4.9	6	9.8	13,600
8	5.1	7	9.8	13,400
9	5.3	8	9.8	13,200

Source: Adapted with permission from Table 2.2 of Haith (1982). Copyright © 1982, Wiley, New York.

ships between control variables quantitatively and to quantify all constraints. Further the constraints had to be sufficiently compatible that a feasible range of solutions occurred. If the control of total emissions had been set so stringently that the curve $3x - y + 0.03y^2$ had shifted above $3x - y = 0$ in Figure 3.7, there would have been no feasible region, and no solution that satisfied all constraints. In the latter case some further negotiation and bargaining would have to occur to change constraints in order to create a feasible region. In the case where not all relationships can be quantified, some implicit method of decision making, such as the panel-of-experts approach, must be used.

Thus a number of conditions must exist before optimization models can be built, and this approach to decision making cannot be universally applied. Nevertheless, there are many examples of successful application of optimization techniques to environmental problems. For example, optimization models have been built to find an oil tanker size that best makes the trade-off between small but frequent oil spills from many small tankers and the less frequent but larger spills from large tankers (Sarin and Scherr 1975); to design wastewater treatplant facilities that minimize cost of construction and operation while meeting certain effluent treatment standards (Haith 1982, Ch. 3); to find rates of pesticide accumulation that do not harm wildlife while maximizing crop protection for farmers (Haith 1982, Ch. 2); and to site power plants sufficient to satisfy regional power demand while minimizing exposure of people to pollution (Emanuel et al. 1978).

Multiple Objective Decision Making

In the previous example we started with two objectives: to minimize pollution and to maximize profit. In order to convert this to a single-objective optimization problem, we converted pollution minimization into a constraint: pollution must not exceed a certain emission level, leaving profit to be maximized. An alternative approach when two objectives exist is to solve for each separately and examine the trade-off function as each objective is made a constraint for the other (see Emanuel et al. 1978).

How do decision makers choose an "optimal" solution when objectives conflict? Earlier we noted that discussion and alteration of constraints must occur to create a feasible region in a single-objective optimization problem. In the case of multiple objectives, however, another alternative is available apart from attempting to convert second and subsequent objectives into constraints. The alternative is to elicit from decision makers some relative preferences or weights attached to achieving each objective and to seek solutions in which the weighted sum of objectives is maximized. This process is known as *multiple-objective decision making* (see, e.g., Bell et al. 1977, Cohon 1978, Keeney and Raiffa 1976, Nijkamp 1980, Wilhelm 1975). The term applies when the values that fit the weighted decision index (and associated objective functions) vary continuously. For example, scenarios that vary the amount of sulfur fuel to use, or the amount of coal to burn, along a continuum characterize multiple-objective problems. When scenarios that satisfy the objective functions vary by discrete intervals (e.g., install one type of pollution control equipment or another, each with fixed costs and performance levels), the process is termed *multiple-criteria* decision making (see, e.g., Bakus et al. 1982, Delft and Nijkamp 1977, Guigou 1974, Nijkamp 1980, Starr and Zeleny 1977, Thiriez and Zionts 1976). Many of the techniques of multidimensional decision making are drawn from the same battery of multivariate statistical techniques used to weight and array biological data (e.g., principle components analysis, multidimensional scaling; see Chapter 10 and Nijkamp 1980).

One of the great values of multidimensional decision-making procedures is that the different objective functions do not have to be in comparable units. This is important in environmental decision making. Thus one may be able to compare trade-offs between pollution control and cost without finding a mathematical relationship relating pollution control (in units of pollution concentration) to cost (in monetary units). In single-objective optimization problems one has to be able to identify such quantitative interrelationships. In multidimensional decision making the general approach is to find the optimal value for each objective function separately, in its own units. One then standardizes values relative to their own optimum or normalizes to the mean value to achieve unitless values, which are then weighted according to decision makers' preferences (often called "revealed preferences"). The deviation of these weighted solution scenarios from some ideal solution (in which all objectives are given maximum score and weight) can then be computed using one of the resemblance functions discussed in Chapter 10 (see Zeleny 1976) or by a number of other approaches reviewed in Nijkamp (1980).

Multicriteria Decision Making: A Case Study

An example of a multicriteria decision-making approach is concordance analysis (Nijkamp 1980, Ch. 11; Nijkamp and Vos 1977). Nijkamp (1980) illustrated the use of a version of this technique in evaluating alternative land reclamation (wetlands-filling) and development projects in the Netherlands. An area called Markerwaard was proposed to be reclaimed from the inland sea by closing of a dike, and various developments, including housing, roads, farms, and an airport, were proposed for the newly drained or reclaimed land. Five plan scenarios were developed, ranging from a no action alternative to intensive development. Twelve decision criteria were identified following public debate.

The following steps are involved in concordance analysis:

1. Identify the various decision criteria, such as changes in employment,

changes in land use, changes in bird species richness, as a result of each plan scenario.

2. Determine the most desirable outcome for each decision criterion considered separately, averaged from a panel of experts. For example, based on a Delphi process, experts may decide that the most desirable plan would result in a 20% increase in employment, a 10% shift from agricultural to urban land use, and a 0% change in bird species richness. These values constitute the "norm vector."

3. Determine from interest groups the positive or negative weights (on a 1–10 ordinal scale) to be attached to decision criteria when the level of outcome exceeds or is less than the norm value for a particular scenario. For example, each percentage increase in employment above 20% (the norm) will be weighted positively by multiplying by 8; each percentage decrease in employment below 20% will be weighted negatively by multiplying by -2.

4. Compare pairs of plans for their respective deviations from the norm vector. A concordance index is calculated as the sum of the weighted preferences for all decision criteria in which one plan is preferred over the other, divided by the sum of weighted preferences for all decision criteria. A "preference" occurs when one plan exceeds the norm value for a decision criterion in a positive direction more than the other plan. A deviation in the negative direction is termed a "penalty." An analogous discordance index is calculated for all pairwise comparisons among decision criteria in which a negative deviation or penalty occurs. Thus a concordance index is calculated for all those criteria in which plan 1 exceeds the norm more than plan 2, and another concordance index is calculated for all those criteria in which plan 2 exceeds the norm more than plan 1. Similarly, two discordance index values are calculated for the set of penalties that is greater in plan 1 than 2 and in plan 2 than plan 1.

5. Compare plans to see whether concordance or discordance indexes between plans differ by an arbitrary threshold amount or more. Plans in which index values indicate superiority beyond a threshold value are indicated by a 1 in the concordance or discordance dominance matrix. These matrixes show whether a given plan is preferable to one or more of the others, based either on high weighted preferences (concordance index) or low weighted penalties (discordance index).

6. A final comparison between plans is made by entering into a total dominance matrix those plans that appear superior (indicated by 1) in both the concordance and disconcordance dominance matrixes (see Table 3.6).

Thus concordance analysis can be used to reduce a large number of alternative scenarios to a more manageable number of alternatives preferred be-

**Table 3.6.   The Total
Dominance Matrix**

	a_1	a_2	a_3	a_4	a_5
a_1		1	1	1	0
a_2	0		0	0	0
a_3	1	1		1	0
a_4	0	1	0		0
a_5	0	0	0	0	

Source: Reprinted with permission from Table 11.6 of Nijkamp (1980). Copyright © 1980, Wiley, New York.

Note: Letters a_1–a_5 represent each of five alternative plans.

cause their deviations from the preferred norm are least. In Nijkamp's (1980) example two of the five plans could be eliminated by the total dominance matrix (plans a_2 and a_5 in Table 3.6).

The values of such multicriteria decision-making procedures are that they make explicit the criteria for decision making, allow the opinions of different interest groups to be used separately in identifying preferred choices, and permit comparison between decision criteria without requiring the building of a mathematical model of interrelationships. Also the amount of quantitative information about impacts that is required is not as large as would be required for a single-objective optimization model, since information is only required on which to make judgments of an ordinal (relative rank) nature.

The problems with such procedures include the fact that the decision outcome is sensitive to many decisions that were taken in constructing the calculation and that are not obvious from an examination of the final outcome. For example, the number and kinds of criteria chosen influence the outcome. If ten rather than one economic criteria were used, it is likely that their combined weight in the decision would increase. Bird conservation concerns in Nijkamp's (1980) example were expressed as "number of species added or lost" rather than "number of individuals of each species added or lost," thereby neglecting effects on bird population numbers short of extinction. No other species were considered. The choice of the norm vector and of weights is highly dependent on the individuals consulted and is sensitive to how the sample of those polled is constructed. The selection of weights is sensitive to how different individuals use the ordinal scale (some people will use only a portion of the scale as their complete range, others use only the extremes, etc.). A partial mitigation to these problems is to conduct a sensitivity analysis to see how sensitive decision outcomes are to small changes in the norm vector or weights. Finally, such quantitative procedures share with other such techniques the problem that they give an air of objectivity to outcomes that incorporate much subjective and "best-guess" information. These problems will be further discussed in Chapters 4 and 5.

The concordance technique closely resembles evaluation methods if decision criteria are considered simply evaluation criteria. A distinction between the two, however, is that through the dominance matrix the concordance method proceeds to select preferable outcomes, that is, make a decision, whereas evaluation techniques normally simply indicate the extent to which alternatives meet certain evaluation criteria (i.e., stopping with the construction of concordance and discordance matrices). As will be seen in Chapters 4 and 5, this distinction is not always a sharp one.

INVOLVING THE PUBLIC

Multiple Publics

We have made reference to public input in impact assessment in various contexts thus far. We have noted that the public may play a role in identifying the range of impacts of a proposal, in evaluating the relative significance of impacts, in commenting on the fairness and completeness of draft impact statements, and in monitoring the effects of projects and highlighting the need for further mitigation measures. If "the public" is defined to include everyone who is not directly responsible for planning and decision making on the proposed action, then clearly the public may be disaggregated into a series of multiple publics, ranging from organized to unorganized interest groups, including people who will be affected by the action, and independent experts.

Not all members of the public will be interested in or significantly affected by the action. The process of "involving the public" begins by identifying those multiple publics that are, or should have, a substantial interest in the proposed action, based on the extent to which they, or interests they represent, will be affected ("stakeholders"). Such a list will vary from project to project, but will typically include other governmental agencies, commercial interests, organized public-interest groups (environmental, neighborhood, labor, tribal, religious, public service, etc.), and individuals who are not organized in such groups (residents, technical experts, interested individuals, other affected parties).

Public Participation Approaches

Useful input from the public requires that the public be informed on the proposed project, alternatives, and major potential impacts. Just how this information is conveyed can itself influence the nature of the feedback from the public. Highly technical information will be useful to only one subsection of the public (technical experts); information in particular media (magazines, radio, TV) will reach different subsections of the public. Furthermore how public input is received (written, oral; public demonstration, hearing, workshop, etc.) will greatly influence who participates, what kind of information is received, and what pressure will be exerted on decision makers.

The public generally participates in commenting and decision making in proportion to the extent to which it is affected and to the extent to which solicited comments will influence decisions. Thus one way to classify public participation methods is by the extent to which they provide power to the public over the ultimate decision. Public participation approaches at four levels of increasing public power can be recognized (Collins 1978, Lang and Armour 1980): information-feedback approaches, consultation, joint planning, and delegated authority. Characteristics of these approaches are summarized in Table 3.7. Bishop (1973) has categorized a range of public participation techniques by the degree and level of communication that occurs and the kinds of information transfer (from informing to evaluating) achieved by each (Table 3.8). The extent to which these techniques are successful in their communication objectives is much dependent on the level of commitment by both organizers and participants to the stated objectives of the communication exercise.

Still another important way to view public participation techniques is by the sector of multiple publics that will be encouraged to participate by each technique. A classification on this basis is presented in Table 3.9. Tables 3.7–3.9 are based on subjective experience rather than quantitative, empirical study. One empirical study (Wood 1978) examined the extent to which different public participation techniques are used in highway planning in the United States. Of the 30 techniques employed in 901 surveyed cases of public participation efforts, Wood (1978) found that the most often used were public hearings, information meetings, and legal notices (43% of the cases). These are techniques involving low levels of delegation of power and relatively low levels of representativeness.

Many factors influence the degree of representativeness of response to different public participation techniques. A person who works away from home during the day is not likely to participate in daytime hearings or advisory meetings or to be home to respond to pollsters by phone or at the door. Busy people will also be less likely to answer questionnaires, biasing responses to the unemployed, underemployed, and retired. Media releases in major newspapers miss an entire segment of the public who do not regularly read newspapers or read newspapers other than the big circulation dailies. Comments received in field offices and by ombudspeople are usually from highly involved members of the public. A further difficulty with such comments is that they are not readily summarized objectively, since they are being filtered through the ombudsperson. The time, place, and manner of

Table 3.7.   Public Participation Approaches Classified by Degree of Power Given to Public in Decision Making

Approaches	Extent of Public Power in Decision Making	Advantages	Disadvantages
Information Feedback			
Slide or film presentation, information kit, newspaper account, notices, etc.	Nil	Informative; quick	No feedback; presentation subject to bias
Consultation			
Public hearing, ombudsperson or representative, etc.	Low	Allows two-way information transfer; allows limited discussion	Does not permit ongoing communication; somewhat time-consuming
Joint Planning			
Advisory committee, structured workshop, etc.	Moderate	Permits ongoing input and feedback; increases education and involvement of citizens	Very time-consuming; dependent on what information is provided by planners
Delegated Authority			
Citizens' review board, citizens' planning commission, etc.	High	Permits better access to relevant information; permits greater control over options and timing of decision	Long-term time commitment; difficult to include wide representation on small board

Sources: Terms used for approaches are from Collins (1978) and Lang and Armour (1980).

conduct of polls will of course greatly influence the sector of public polled. Polls in shopping centers will reach a different subset of the public than phone polls, mail polls, ballot initiatives, etc. A citizens' advisory committee that fails to compensate participants for time and transportation costs favors representatives who are paid by the group they represent (e.g., commercial interests).

Table 3.8. Public Participation Approaches Classified by Degree, Level, and Type of Communication

Communication Characteristics				Planning Objectives					
Degree of Two-way Communication	Level of Public Contact Achieved	Ability to Handle Specific Interest	Public Participation Technique	Inform/ Educate	Identify Problems/ Values	Get Ideas/ Solve Problems	Feedback	Evaluate	Resolve Conflict/ Consensus
2	1	1	Public hearings	X	X		X		
2	1	2	Public meetings	X	X		X		
1	2	3	Informal small group meetings	X	X	X	X	X	X
2	1	2	General public information meetings	X					
1	2	2	Presentations to community organization	X	X		X		
1	3	3	Information coordination seminars	X			X		
1	2	1	Operating field office		X	X	X	X	
1	3	3	Local planning visits		X		X	X	
1	3	1	Class action litigation	X		X	X		X
2	2	1	Information brochures and pamphlets	X					

		Technique						
1	3	Field trips and site visits	X		X			
3	2	Public displays	X	X		X	X	X
2	2	Model demonstration projects	X			X	X	X
3	1	Material for mass media	X					
1	2	Response to public inquiries	X					
3	1	Press releases inviting comments	X		X	X		
1	1	Letter requests for comments		X	X	X	X	
1	3	Workshops	X	X	X	X	X	X
1	3	Advisory committees	X	X	X	X	X	
1	3	Task forces	X	X	X		X	
1	3	Employment of community residents	X	X				X
1	3	Community interest advocates		X	X	X		X
3	3	Ombudsperson or representative	X	X	X	X	X	
2	1	Environmental impact statement review by public	X	X	X	X	X	

Source: Reprinted from Bishop (1973).

Note: 1 = low; 2 = medium; 3 = high.

Table 3.9. Public Participation Approaches Classified by Types of Most Likely Participants and Degree of Representativeness of Multiple Publics Typically Achieved

Type of Participant Favored	Public Hearings	Structured Workshop	Legal Notices	Information Meetings	Citizen Advisory Committees	Polls	Ombudsperson or Field Office	Written Comment
Highly literate	X		X					X
Highly verbal		X			X		X	
Employed during the day								
Blue-collar worker				X		X	X	
White-collar worker	X	X	X	X	X	X	X	X
Management			X			X		X
Houseperson or unemployed				X	X	X	X	
Highly informed, involved	X	X	X	X	X		X	X
Degree of representativeness of multiple publics	1	3	1	2	2	3	1	2

Note: 1 = low; 2 = medium; 3 = high.

Structured workshops, in which citizens representing a variety of multiple publics can be chosen by planners and invited to a series of information-feedback and joint-planning sessions, can achieve good representation from a variety of groups and can inform the different participants to a comparable level. Structured workshops are one of the better techniques judged on these criteria, but they will tend to bias against busy people or those who are not highly motivated. Some groups are almost always underrepresented by most public participation techniques: children, the incapacitated, those in remote areas, those who cannot communicate in the language being used. Special efforts must be undertaken if these groups are to be adequately represented. Further, as noted in Table 3.9, many public participation techniques differentially attract white-collar workers or the highly involved. Very often the major ecological impacts of projects are on nonhuman species and inanimate objects (earth, air, water). For these to be represented, environmental groups and citizens must speak on their behalf (see, e.g., Stone 1974), not only from the point of view of benefits to people but of the welfare of nature itself (see Ehrenfeld 1978; Chapter 5).

No single public participation technique fulfills all the criteria discussed. Before embarking on a public participation program, therefore, the planner should consider what groups she or he wishes most to reach, what level of power regarding decision making is to be delegated, and what kind of feedback is desired. As such, a public participation program may be planned in a highly manipulative way to elicit particular kinds of responses. The only real safeguard against unethical abuse of public participation programs is for those concerned with the quality of public input to evaluate the program in terms of its likely biases and to attempt to supplement these inputs with comments from other sources designed to countermand the bias.

RESOLVING CONFLICT

Major sources of conflict in environmental disputes include the existence of competing resource demands, differences in human values regarding the relative worth of resources, and inadequate knowledge or understanding of the costs, benefits, and risks involved in proposed actions. Processes such as public hearings and written comments on draft impact statements typically provide a forum for voicing differences but are not designed to find mutually acceptable solutions. The impact assessment process in the United States has sometimes led to the ensnarling of proposed actions in lengthy litigation in the adversary environment of the courtroom or to a process of review of impact statements which consists more of attack and defense than of search for acceptable alternatives. In the public inquiry process of British Commonwealth countries, testimony is received with little opportunity for witnesses to assist in fashioning compromises or acceptable alternatives. For these reasons other techniques for environmental conflict resolution have been pioneered in recent years, in part adapted from labor-management negotiating techniques and in part newly developed.

Types of Environmental Conflict Resolution

Five categories of environmental conflict resolution are defined in Table 3.10.

Conflict anticipation is most useful at the earliest stages of project planning for a controversial proposal or problem. The scoping process in impact assessment is an example of conflict anticipation that serves to identify potential problems so that they may be studied and mitigated and allows

Table 3.10.   Types of Environmental Conflict Resolution

Type	Definition	Examples
Conflict anticipation	A third party identifies potential disputes before opposing positions are fully identified	Scoping or screening process in impact assessment identifies likely problems and affected groups
Joint problem solving	Ongoing group meetings discuss and clarify issues and resolve differences; agreements reached are informal	Structured workshops; adaptive environmental assessment; environmental planning citizens' advisory committees
Mediation	Formal negotiations between empowered representatives of constituencies; mediator facilitates but does not impose settlement	Technical meetings to seek settlements; facilitator uses a variety of negotiating and mediating techniques
Policy dialogues	Meetings to discuss and resolve differences between conflicing policy-making agencies; results become advisory to official policymaking bodies	Interagency advisory committees; ad hoc meetings between members of different governmental agencies
Binding arbitration	Formal arguments presented by opposing parties; arbitrator imposes settlement that parties have previously agreed to abide by	Labor-management contract arbitration; court arbitration hearings

Source: Based in part on information in Bellman et al. (1981).

affected parties to become involved in the assessment process. ROMCOE, the Center for Environmental Problem Solving in Boulder, Colorado, conducted a conflict anticipation project concerning energy in three Colorado communities. The center identified a community member to run the project and helped organize energy fairs and projects to increase community awareness of energy problems and solution alternatives (Bellman et al. 1981).

Joint problem solving involves reaching informal agreements among concerned parties for consideration and possible adoption by decision makers. Typically this process starts at the early stage of problem solving, when issues are still being defined, and continues through the full process of decision making. A citizens' advisory committee associated with development of an air quality maintenance plan would be an example; such a committee would normally consist of citizens, public-interest and commercial-interest group representatives, and public officials. Such committees operated throughout the United States during preparation of air quality maintenance plans and areawide wastewater management plans. The city of San Francisco resolved disputes over expansion of its airport by forming a temporary Joint Authorities Board among affected jurisdictions and developing an airport development plan with the use of citizens' advisory committees and various public hearings and presentations (Bellman et al. 1981). In Queensland, Australia, the State Land Use Committee is such an example in which citizens and government department representatives jointly explored land use problems and solutions. The adaptive environment assessment and management technique (Holling 1978) is also a form of joint problem solving involving the use of computer models as part of the scenario-generating process.

Environmental mediation is a briefer and more formal process of negotiation among officially recognized representatives of affected constituencies. It is usually undertaken after a conflict is fully developed, parties can be identified, and there is a shared willingness among parties to attempt negotiation. In a conflict situation parties are only likely to be willing to negotiate if they feel that they cannot achieve their own objectives without unacceptably high costs (Bellman et al. 1981). The mediator may be asked to initiate and facilitate negotiations, clarify areas of agreement and disagreement, and suggest possible solutions and ways to implement them. Nevertheless, the final agreement typically must be ratified by a separate decision-making body to become binding. One example of a mediation agreement is a settlement after 18 years of dispute among several utility companies, environmental groups, and government agencies over pumped-storage facilities on the Hudson River, New York. Under the settlement the utilities agreed to halt construction of an existing plant, donate land for park use, provide $12 million to study effects of the power plants on aquatic life, provide a fish hatchery, and use cooling towers on future plants. In exchange, utilities did not have to build cooling towers on three existing plants, at considerable cost savings (Bellman et al. 1981). Another example involved arriving at agreement on a proposed solid waste disposal site after an initially proposed site in Riverside County, California, had been opposed by citizens. A mediation panel of county officials, site opponents, business interests, public service organizations, waste haulers, and two community waste disposal district representatives was endorsed by the decision-making board. The final agreement reached by mediation was later accepted unanimously by the decision makers (Bingham 1982).

Parties enter mediation voluntarily, and the mediator serves at their pleasure. The major role of the mediator is to facilitate negotiation. One technique that a mediator often encourages among participants is *active listening*. In this technique a respondent paraphrases the contents of the previous speaker's remarks before proceeding with his or her own statement. This technique ensures that communication has been accurate, reassures the previous speaker that his or her remarks have been understood, and may avoid repetition by the initial speaker. For example, if speaker 1 says "I oppose this solid waste site because it will cause the neighborhood to smell bad." Speaker 2, using active listening, might respond "I understand that you don't want this site because you are worried about the smells in the vicinity, but I would point out that the refuse is continually covered by soil so that no odors emanate beyond the site boundary."

A second mediation technique sometimes used by the facilitator is *dialectical scanning*. In this technique each party is asked to state his or her position in turn. The facilitator then attempts to reach consensus on the areas of agreement or close agreement and areas of disagreement that have minor significance. These issues are laid aside, and discussion focuses on remaining major areas of disagreement. This technique helps to determine the priority of issues for discussion.

Beyond such techniques a thorough knowledge of group dynamics is extremely helpful to a mediator. People often display behavior in groups that differs from their behavior in one-to-one situations. Social psychologists have studied group dynamics extensively, and much of this information has been used to manage group meetings more effectively. Virtually all environmental decisions pass through a stage of group discussion at one or more points, and some decisions are made entirely by groups. Thus a knowledge of the psychological dynamics of group behavior is quite relevant to environmental decision making. Some additional reading on this subject may be found in Berne (1963, 1964), Cartwright and Zander (1968), Edney and Harper (1978), and Guzzo (1982).

On the more general subject of environmental mediation and conflict management, useful works include books by Lake (1979), Mernitz (1980), and Rivkin (1977); symposia edited by Baldwin (1978) and Marcus and Emrich (1981) and in the Environmental Professional (1980); an annotated bibliography by Bingham et al. (1981); and additional bulletins and a quarterly newsletter, *Resolve,* published by The Conservation Foundation, Washington, D.C.

Policy dialogues resolve differences regarding governmental policies by providing informal forums for discussion and subsequent advice to agencies. The discussants may be representatives from different agencies on an inter-agency panel, or they may be outside experts who submit a report to policy-makers. In the United States the Congressional Office of Technology As-sessment and the National Academy of Sciences submit independent policy analyses to Congress and the Executive branch. A private conservation group, for example, formed a "toxic substances dialogue group" with repre-sentatives from industry and environmental groups and submitted its find-ings to EPA, which responded with considerable interest (Bellman et al. 1981). In Australia the Federal Advisory Committee on the Environment, composed of technical experts, industry, labor, and environmental groups, conducted such policy dialogues in the mid 1970s, reporting on policy issues to the Minister for Environment.

Binding arbitration, in which a settlement imposed by the arbitrator has the force of law on participating parties, is not currently in widespread use for resolving environmental disputes. Nevertheless, to the extent that arbi-tration hearings are part of the legal system, they may be used during aspects of litigation over environmental issues. The "public inquiry" in Canada, Australia, and the United Kingdom is similar in having a hearing officer who makes final recommendations, but these are advisory only; the Australian "Mines Courts," which adjudicate disputes over conditions of mining leases play something of an arbitration role, but the court findings are subject to the policy of the Department of Mines and are thus not independent of political influence. Proposals for an "environmental court," which would have judges with scientific expertise to rule on environmental disputes, and for a "science court" (Task Force of the Presidential Advisory Group on Antici-pated Advances in Science and Technology 1976), which would report and reach judgments on disputed statements of fact, have been made but not implemented. Part of the problem with the role of judges in resolving envi-ronmental disputes is that conflict resolution is not a matter of determining "the truth" of a matter or determining its legality. As noted at the outset, the conflicts typically concern competition for resources or differences in values regarding the relative worth of resources. To resolve these conflicts in a way that satisfies disputants requires each party to feel that he or she has achieved the most that is possible under the circumstances. This is a subjec-tive judgment on the part of disputants, most likely to be determined follow-ing an extended process of negotiation. A court settlement has not involved litigants in the negotiating and accepting of compromise that are involved in the other forms of environmental conflict resolution; as a result a court order may well result in appeal or protest rather than true conflict resolution.

In this chapter we have considered a variety of processes in environmen-tal decision making. Clearly the degree to which mathematical models can be used in decision making depends on the amount of explicit empirical data available. Techniques for decision making range from those that require little explicit information (mediation, panel of experts) to those that require a great deal (model building, optimization), with some techniques occupying a middle ground (multiobjective decision making). A decision-making technique should make maximum use of available empirical information. At the same time, because many of the conflicts concern values, decisions should explicitly examine the implications for different multiple publics. The choice of decision-making tools can have a major effect on whether the decision achieves its objectives and proves acceptable to the multiple publics.

REFERENCES

Aiken, A. M., Harrison, J. M., and Hare, F. K. (1977). The management of Canada's nuclear wastes. Rep. EP77-6. Energy Policy Sector, Ministry of Energy, Mines and Resources. Ottawa.

Argue, A. W., Hilborn, R., Peterman, R. M., Staley, M. J., and Walters, C. J. (1983). Strait of Georgia Chinook and Coho Fishery. *Canadian Bull. Fisheries Aquatic Sci.* **211**:1–91.

Bakus, G. J., Stillwell, W. G., Latter, S. M., and Wallerstein, M. C. (1982). Decision-making: with applications for environmental management. *Environ. Manage.* **6**:493–504.

Baldwin, P. (1978). Environmental mediation: an effective alternative? RESOLVE, Center for Environmental Conflict Resolution, Palo Alto, Calif.

Bell, D., Keeney, R. L., and Raiffa, H. (1977). *Conflicting Objectives in Decisions.* Wiley, New York.

Bellman, H., Bingham, G., Brooks, R., Carpenter, S., Clark, P., and Craig, R. (1981). Environmental conflict resolution: practitioners' perspective on an emerging field. In *Environmental Consensus,* Winter issue, pp. 1, 3–7. RESOLVE, Center for Environmental Conflict Resolution, Palo Alto, Calif.

Berne, E. (1963). *The Structure and Dynamics of Organizations and Groups.* Ballantine Books, New York.

Berne, E. (1964). *Games People Play.* Ballantine Books, New York.

Bingham, G., ed. (1982). Update. In *Resolve,* Spring issue, p. 7. The Conservation Foundation, Washington, D.C.

Bingham, G., Vaughn, B., and Gleason, W. (1981). Environmental conflict resolution: an annotated bibliography. RESOLVE, Center for Environmental Conflict Resolution, Palo Alto, Calif.

Bishop, A. B. (1973). Public participation in environmental impact assessment. Paper presented at Engineering Foundation Conference on Preparation of Environmental Impact Statements, New England College, Henniker, N.H., July 29–August 3.

Brebbia, C. A., ed. (1976). *Mathematical Models for Environmental Problems.* Wiley, New York.

Buckingham, S. (1980). The Strait of Georgia Chinook and Coho Fishery: user's guide. Inst. of Resource Ecology, Univ. British Columbia, Vancouver.

Cartwright, D., and Zander, A. (1968). *Group Dynamics: Research and Theory*. 3rd ed. Harper & Row, New York.

Cohon, J. L. (1978). *Multiobjective Programming and Planning*. Academic Press, New York.

Collins, D. (1978). A view from the other side: citizens participate in planning urban housing. *Urban Forum* 3:14–23.

Converse, A. O. (1970). *Optimization*. Krieger, New York.

Delft, A. van, and Nijkamp, P. (1977). *Multicriteria Analysis and Regional Decision-Making*. Nijhoff, The Hague.

Dickson, K. L., Cairns, J., Jr., Gregg, B. C., Messenger, D. I., Plafkin, J. L., and van der Schalie, W. H. (1977). Effects of intermittent chlorination on aquatic organisms and communities. *J. Water Poll. Control Fed.* **49**:35–44.

Earwicker, J. (1974). The future of planning: an exploration using futures techniques. *The Planner* **60**:650–652.

Edney, J. T., and Harper, C. S. (1978). The commons dilemma: a review of contributions from psychology. *Environ. Manage.* **2**:491–507.

Ehrenfeld, D. (1978). *The Arrogance of Humanism*. Oxford Univ. Press, New York.

Emanuel, W. R., Murphy, B. D., and Huff, D. D. (1978). Optimal siting of energy facilities for minimum air pollution exposure on a regional scale. *J. Environ. Manage.* **7**:147–155.

Environment Canada (1983). Review and evaluation of adaptive environmental assessment and management. Ministry of Supply and Services Canada, Vancouver.

Environmental Professional (1980). Environmental mediation and conflict management. Vol. 2(1). Pergamon, New York.

Fischhoff, B., Lichtenstein, S., Slovic, P., Derby, S. L., and Keeney, R. L. (1981). *Acceptable Risk*. Cambridge Univ. Press, Cambridge.

Friesema, H. P., and Culhane, P. J. (1976). Social impacts, politics and the environmental impact statement process. *Natural Resources J.* **16**:339–356.

Gass, S. I. (1975). *Linear Programming: Methods and Applications*. 4th ed. McGraw-Hill, New York.

Guigou, J. L. (1974). *Analyse des données et choix à critères multiples*. Dunod, Paris.

Guzzo, R. A., ed. (1982). *Improving Group Decision Making in Organizations. Approaches from Theory and Research*. Academic Press, New York.

Haith, D. A. (1982). *Environmental Systems Optimization*. Wiley, New York.

Hall, C. A. S., and Day, J. W., Jr., eds. (1977). *Ecosystems in Theory and Practice: An Introduction with Case Histories*. Wiley-Interscience, New York.

Hammond, K. R., Klitz, J. K., and Cook, R. L. (1978). How systems analysts can provide more effective assistance to the policymaker. *J. Appl. Syst. Anal.* **5**:111–136.

Harrison, A. F., and Bramson, R. M. (1982). *Styles of Thinking: Strategies for Asking Questions, Making Decisions, and Solving Problems*. Anchor Press/Doubleday, Garden City, N.Y.

Hilborn, R., Holling, C. S., and Walters, C. J. (1980). Managing the unknown:

approaches to ecological policy design. In *Biological Evaluation of Environmental Impacts*. Rep. FWS/OBS-80/26. Council on Environmental Quality and U.S. Dept. of Interior, Fish & Wildlife Service, Washington, D.C., pp. 103–113.

Hilborn, R., and Walters, C. J. (1981). Pitfalls of environmental baseline and process studies. *Environ. Impact Assessment Rev.* **2**:265–278.

Hohenemser, C., Kates, R. W., and Slovic, P. (1983). The nature of technological hazard. *Science* **220**:378–384.

Holcomb Research Institute (1976). *Environmental Modeling and Decision-Making: The US Experience*. Praeger, New York.

Holling, C. S., ed. (1978). *Adaptive Environmental Assessment and Management*. Wiley, New York.

Kates, R. W. (1978). *Risk Assessment of Environmental Hazard*. SCOPE Rep. 8. Wiley, New York.

Keeney, R. L., and Raiffa, H. (1976). *Decision Analysis with Multiple Objectives: Preferences and Value Tradeoffs*. Wiley, New York.

Lake, L. (1979). *Environmental Mediation: The Search for Consensus*. Westview, Boulder, Colo.

Lang, R., and Armour, A. (1980). *Environmental Planning Resourcebook*. Lands Directorate, Environment Canada, Ottowa.

Lowrance, W. W. (1976). *Of Acceptable Risk*. Kaufmann, Los Altos, Calif.

Marcus, P. A., and Emrich, W. M., eds. (1981). Environmental conflict management: working papers. Council on Environmental Quality and U.S. Dept. of Interior, Washington, D.C.

McAllister, D. M. (1980). *Evaluation in Environmental Planning: Assessing Environmental, Social, Economic, and Political Trade-offs*. The MIT Press, Cambridge, Mass.

Mernitz, S. (1980). *Mediation of Environmental Disputes*. RESOLVE. Center for Environmental Conflict Resolution. Palo Alto, Calif.

Mishan, E. J. (1971). *Cost-Benefit Analysis*. Allen & Unwin, London.

Mitsch, W. J., Bosserman, R. W., and Klopatek, J. M. (1982). *Energy and Ecological Modeling*. Elsevier, Amsterdam.

Nair, K., and Sarin, R. K. (1979). Generating future scenarios—their use in strategic planning. *Long Range Planning* **12**:57–61.

National Defense University (1978). *Climate Change to the Year 2000*. Dept. of Defense, Washington, D.C.

Nijkamp, P. (1980). *Environmental Policy Analysis: Operational Methods and Models*. Wiley, New York.

Nijkamp, P., and Vos, J. B. (1977). A multicriteria analysis for water resources and land use development. *Water Resources Research* **13**:513–518.

Otway, H. J., and Cohen, J. J. (1975). Revealed preferences: comments on the Starr benefit-risk relationship. Rep. RM-75-54. Intl. Inst. for Applied Systems Analysis, Schloss Laxenburg, Austria.

Petak, W. J. (1980). Environmental planning and management: the need for an integrative perspective. *Environ. Manage.* **4**:287–295.

Popper, F. J. (1983). LULUs. *Resources* **73**:2–4.

Raiffa, H. (1968). *Decision Analysis*. Addison-Wesley, Reading, Mass.

Ricci, P. F., and Molton, L. S. (1981). Risk and benefit in environmental law. *Science* **214**:1096–1100.

Rivkin, M. D. (1977). Negotiated development: a breakthrough in environmental controversies. The Conservation Foundation, Washington, D.C.

Sarin, R. K. (1978). A sequential approach to cross-impact analysis. *Futures* **10:** 53–62.

Sarin, R. K., and Scherer, C. R. (1976). Optimal oil tanker size with regard to environmental impact of oil spills. *J. Environ. Economics Manage.* 3:226–235.

Schwing, R. C., and Albers, W. A., Jr., eds. (1980). *Societal Risk Assessment: How Safe is Safe Enough?* Plenum, New York.

Starr, C. (1972). Benefit-cost studies in sociotechnical systems. In *Perspectives on Benefit-Risk Decision Making*. Natl. Acad. Engr., Washington, D.C., pp. 17–42.

Starr, C., and Whipple, C. (1980). Risks of risk decisions. *Science* **208**:1114–1119.

Starr, M. K., and Zeleny, M., eds. (1977). *Multiple Criteria Decision Making*. North Holland, Amsterdam.

Stone, C. D. (1974). *Should Trees Have Standing? Toward Legal Rights for Natural Objects*. Kaufmann, Los Altos, Calif.

Task Force of the Presidential Advisory Group on Anticipated Advances in Science and Technology (1976). The Science Court experiment: an interim report. *Science* **193**:653–656.

Thiriez, H., and Zionts, S., eds. (1976). *Multiple Criteria Decision Making*. Springer, Berlin.

U.S. Environmental Protection Agency (1979). Environmental modeling catalogue: abstracts of environmental models. Manage. Info. & Data Systems Div., EPA, Washington, D.C.

White, I. L., and Hamilton, M. R. (1983). Policy analysis in integrated impact assessment. In F. A. Rossini, and A. L. Porter, eds. *Integrated Impact Assessment*. Westview, Boulder Colo., pp. 39–55.

Whyte, A. V., and Burton, I., eds. (1980). *Environmental Risk Assessment*. SCOPE Rep. 15. Wiley, New York.

Wilhelm, J. (1975). *Objectives and Multi-Objective Decision Making under Uncertainty*. Springer, Berlin.

Wood, W. M. (1978). Public involvement techniques utilized in highway transportation planning. In S. Bendix and H. R. Graham, eds. *Environmental Assessment: Approaching Maturity*. Ann Arbor Sci., Ann Arbor, Mich., pp. 205–213.

Zeleny, M. (1976). The theory of displaced ideal. In M. Zeleny, ed. *Multiple Criteria Decision Making*. Springer, Berlin, pp. 153–206.

PART THREE

SUMMARIZING AND EVALUATING IMPACTS

4

QUANTITATIVE APPROACHES

At the outset of an impact study the investigator faces the often bewildering task of identifying and organizing potential impacts in a systematic way. *Impact identification and summarization* techniques discussed in the first part of this chapter are useful in helping to think comprehensively about the full range of possible impacts and in summarizing and highlighting potentially significant impacts. Five main classes of quantitative impact identification techniques exist: (1) checklists, (2) matrices, (3) networks, (4) map overlays, and (5) ad hoc methods. Map overlays, a cartographic approach to identifying landscape segments with a particular set of environmental characteristics, are discussed with other land use techniques in Chapter 6. Ad hoc methods typically identify impacts by brainstorming, and characterize them qualitatively or summarize them in tables (Rau and Wooten 1980). Because checklists are a more organized version of this process, ad hoc methods will not be considered further.

Later in the chapter we discuss techniques for summarizing the judgments made about the significance of ecological impacts to human society. These *evaluation* techniques are classified into two types: (1) approaches that aggregate the values of the multiple publics into a single index or judgment, and (2) approaches in which the values of the multiple publics remain distinct. We also discuss identification techniques that have incorporated some minimal evaluation.

IMPACT IDENTIFICATION TECHNIQUES

Checklists

Checklists are standard lists of the range of impacts associated with a particular type of project. Table 4.1 excerpts ecological impacts from a checklist used for federal transportation projects in the United States (Little 1971). A large number of such checklists have been prepared for particular kinds of projects (e.g., for water resources projects, see Multiagency Task Force 1972; for highway routes, see Adkins and Burke 1974; for nuclear power

Table 4.1.   Checklist of Potential Ecological Impacts during Construction Phase of a Transportation Project

Noise Impacts	Construction Phase
I. Noise impacts	x
A. Public health	
B. Land use	
II. Air quality impacts	x
A. Public health	
B. Land use	
III. Water quality impacts	x
A. Groundwater	
1. Flow and water-table alteration	
2. Interaction with surface drainage	
B. Surface water	
1. Shoreline and bottom alteration	
2. Effects of filling and dredging	
3. Drainage and flood characteristics	
C. Quality aspects	
1. Effect of effluent loadings	
2. Implication of other actions, such as	
a. Disturbance of benthic layers	
b. Alteration of currents	
c. Changes in flow regime	
d. Saline intrusion in groundwater	
3. Land use	
4. Public health	
IV. Soil erosion impacts	x
A. Economic and land use	
B. Pollution and siltation	
V. Ecological impacts	x
A. Flora	
B. Fauna (other than humans)	

Source: Excerpted from a checklist of environmental impacts used by the U.S. Department of Transportation (Little 1971).

Note: The x indicates impact may be beneficial or adverse, depending on circumstances.

plants, see U.S. Atomic Energy Commission 1973). Some checklists have been computerized so that from input on the nature of the project, a list of likely impacts will be generated. Balbach and Novak (1975) describe such a system for military housing construction projects.

The computer has also been used to automate other aspects of impact identification and data summarization. For example, MERES, which is a program of the U.S. Department of Energy, computes pollutant outputs, given specifications on the size and nature of a power plant. Some recent computer-aided impact identification and data summarization systems are discussed by Strand et al. (1983) and reviewed extensively by Riggins (1980–81).

The main advantage of a checklist is that it promotes thinking about the array of impacts in a systematic way and allows concise summarization of effects. Problems include the fact that checklists may be too general or incomplete; they do not illustrate interactions between effects; the same effect may be registered in several places under headings that overlap in content ("double counting"); and the number of categories to be reviewed can be immense, thus distracting attention from the most significant impacts. The identification of effects is qualitative and subjective; for example, "water quality will be adversely affected." As such, these predictions cannot be tested empirically with precision. Furthermore, no statements of likelihood of occurrence are being made. Because of the subjective nature of estimates, checklists will often not be filled out identically by different investigators.

Matrix Methods

Checklists are one-dimensional lists of potential impacts of an action. They are readily expanded to two-dimensional matrices by listing a range of actions associated with the project along the second axis. Thus in Table 4.1 the checklist of impacts could be repeated for the design and operational phases of the transportation project as well as the construction phase. In Figure 4.1 a matrix for an urban renewal project in Hawaii is illustrated. "Positive" impacts are defined subjectively as those that reduce urban blight or other preexisting urban problems; "negative" impacts disturb existing environmental amenities.

Leopold Matrix and Component Interaction Matrix

One of the better known matrices is that prepared by Leopold et al. (1971) for construction projects. One hundred possible project actions are listed on one axis, 88 human and natural environmental elements on the other (Table 4.2). A 10-point scale is suggested to score levels of impact; positive and negative impacts are identified with a "+" or no sign, respectively. Leopold et al. (1971) suggest that the scoring of impacts be done separately for "magnitude" and "importance" of each impact (see Figure 4.3). Magnitude is defined as the degree, extensiveness, or scale of impact (how large an area, how severely affected); importance as the human significance of the impact. The scoring of "importance" is a normative or evaluative process, whereas the scoring of "magnitude" can be relatively objective or empirical. As an evaluation process a single importance score fails to reflect the differing values of the multiple publics, and is consequently an example of an evaluation technique that aggregates public values.

ELEMENTS		Residential Relocation	Business Relocation	Demolition, Grading, Construction	Interim Period (Temporary Uses)	New Utilities In Place	New Residential Buildings	New Commercial Buildings	Parking Structures	Parks and Open Space	Historical Preservation	Modifications to Street System
		ACTION PERIOD				**EFFECTS OF COMPLETED ACTIONS**						
PHYSICAL	Soil & Geology	✷	✷	✷	✷	✷	✷	✷	✷	●	✷	✷
	Sanitary Sewer Systems	✷	✷	O	O	●	●	●	✷	✷	✷	●
	Water Systems	✷	✷	O	O	●	●	●	✷	✷	✷	●
	Vegetation	✷	✷	O	O	✷	●	●	✷	●	✷	✷
	Animal Life	✷	✷	✷	✷	✷	✷	✷	✷	O	✷	✷
	Air Quality	✷	✷	O	✷	✷	◯	◯	◯	●	●	✷
	Adjacent Land Use	✷	✷	◯	◯	✷	●	✷	✷	●	●	X
	Storm Drainage	✷	✷	O	O	●	●	●	✷	●	✷	●
	Transportation System — Streets	✷	O	◯	O	●	●	●	●	✷	✷	●
	Transportation System — Public Transportation	✷	✷	O	O	✷	X	X	X	✷	X	X
	Transportation System — Pedestrian	O	O	◯	O	✷	●	●	●	●	X	X
	Open Space	✷	✷	✷	✷	✷	●	O	O	●	X	X
SOCIOECONOMIC	Demand for Ancillary Services	●	●	●	O	✷	●	●	✷	✷	●	●
	Tax Base	✷	✷	✷	O	●	●	●	●	✷	X	✷
	Health & Safety	✷	✷	O	O	●	●	●	✷	●	●	●
	Neighborhood Viability	O	◯	◯	◯	✷	●	●	●	●	●	X
	Residents	O	◯	◯	◯	●	●	●	●	●	●	X
	Public Schools	✷	✷	O	O	✷	●	✷	✷	●	●	X
	Police Services	O	O	◯	O	●	●	●	●	X	✷	X
	Fire Services	O	O	◯	O	●	●	●	●	X	●	X
AESTHETIC	View	✷	✷	O	◯	✷	●	●	O	●	O	✷
	Historic Structures	✷	✷	O	O	●	✷	✷	X	●	●	✷
	Amenity	O	O	O	O	●	●	●	●	●	●	X
	Neighborhood Character	O	O	O	◯	●	●	●	O	●	●	X

IMPACTING ACTIONS

LEGEND

O	indicates a minor negative impact.
◯	indicates a major negative impact.
•	indicates a minor positive impact.
●	indicates a major positive impact.
X	indicates an undetermined impact.
✷	indicates no appreciable impact.

Figure 4.1. Matrix of impacts in a proposed urban renewal project in Pauahi, Hawaii (U.S. Department of Housing and Urban Development 1974).

Table 4.2. List of Matrix Elements for Construction Projects

Part 1: Project Actions

A. Modification of regime
   a. Exotic flora or fauna introduction
   b. Biological controls
   c. Modification of habitat
   d. Alteration of ground cover
   e. Alteration of groundwater hydrology
   f. Alteration of drainage
   g. River control and flow modification
   h. Canalization
   i. Irrigation
   j. Weather modification
   k. Burning
   l. Surface or paving
   m. Noise and vibration

B. Land transformation and construction
   a. Urbanization
   b. Industrial sites and buildings
   c. Airports
   d. Highways and bridges
   e. Roads and trails
   f. Railroads
   g. Cables and lifts
   h. Transmission lines, pipelines, and corridors
   i. Barriers, including fencing
   j. Channel dredging and straightening
   k. Channel revetments
   l. Canals
   m. Dams and impounds
   n. Piers, seawalls, marinas, and sea terminals
   o. Offshore structures
   p. Recreational structures
   q. Blasting and drilling
   r. Cut and fill
   s. Tunnels and underground structures

C. Resource extraction
   a. Blasting and drilling
   b. Surface excavation
   c. Subsurface excavation and retorting
   d. Well drilling and fluid removal
   e. Dredging
   f. Clear cutting and other lumbering
   g. Commercial fishing and hunting

D. Processing
   a. Farming
   b. Ranching and grazing
   c. Feed lots
   d. Dairying
   e. Energy generation
   f. Mineral processing
   g. Metallurgical industry
   h. Chemical industry
   i. Textile industry
   j. Automobile and aircraft
   k. Oil refining
   l. Food
   m. Lumbering
   n. Pulp and paper
   o. Product storage

E. Land alteration
   a. Erosion control and terracing
   b. Mine sealing and waste control
   c. Strip-mining rehabilitation
   d. Landscaping
   e. Harbor dredging
   f. Marsh fill and drainage

F. Resource Renewal
   e. Reforestation
   b. Wildlife stocking and management
   c. Groundwater recharge
   d. Fertilization application
   e. Waste recycling

(continued on next page)

Table 4.2. (Continued)

G. Changes in traffic
    a. Railway
    b. Automobile
    c. Trucking
    d. Shipping
    e. Aircraft
    f. River and canal traffic
    g. Pleasure boating
    h. Trails
    i. Cables and lifts
    j. Communication
    k. Pipeline

H. Waste emplacement and treatment
    a. Ocean dumping
    b. Landfill
    c. Emplacement of tailings, spoil, and overburden
    d. Underground storage
    e. Junk disposal
    f. Oil well flooding
    g. Deep well emplacement
    h. Cooling water discharge

    i. Municipal waste discharge including spray irrigation
    j. Liquid effluent discharge
    k. Stabilization and oxidation ponds
    l. Septic tanks, commercial and domestic
    m. Stack and exhaust emission
    n. Spent lubricants

I. Chemical Treatment
    a. Fertilization
    b. Chemical deicing of highways, etc.
    c. Chemical stabilization of soil
    d. Weed control
    e. Insect control (pesticides)

J. Accidents
    a. Explosions
    b. Spills and leaks
    c. Operational failure

Others
    a.
    b.

Part 2: Natural and Human Environmental Elements

A. Physical and chemical characteristics
    1. Earth
        a. Mineral resources
        b. Construction material
        c. Soils
        d. Landform
        e. Force fields and background radiation
        f. Unique physical features
    2. Water
        a. Surface
        b. Ocean
        c. Underground
        d. Quality
        e. Temperature
        f. Recharge
        g. Snow, ice, and permafrost

    3. Atmosphere
        a. Quality (gases, particulates)
        b. Climate (micro, macro)
        c. Temperature
    4. Processes
        a. Floods
        b. Erosion
        c. Deposition (sedimentation, precipitation)
        d. Solution
        e. Sorption (ion exchange, complexing)
        f. Compaction and settling
        g. Stability (slides, slumps)
      • h. Stress–strain (earthquake)
        i. Air movements

B. Biological conditions
    1. Flora
        a. Trees
        b. Shrubs
        c. Grass
        d. Crops

        e. Microflora
        f. Aquatic plants
        g. Endangered species
        h. Barriers
        i. Corridors

Table 4.2. *(Continued)*

2. Fauna
   a. Birds
   b. Land animals including reptiles
   c. Fish and shellfish
   d. Benthic organisms
   e. Insects
   f. Microfauna
   g. Endangered species
   h. Barriers
   i. Corridors

C. Cultural factors
1. Land use
   a. Wilderness and open spaces
   b. Wetlands
   c. Forestry
   d. Grazing
   e. Agriculture
   f. Residential
   g. Commercial
   h. Industrial
   i. Mining and quarrying
   e. Unique physical features
   f. Parks and reserves
   g. Monuments
   h. Rare and unique species or ecosystems
   i. Historical or archaeological sites and objects
   j. Presence of misfits
2. Recreation
   a. Hunting
   b. Fishing
   c. Boating
   d. Swimming
   e. Camping and hiking
   f. Picnicking
   g. Resorts
4. Cultural status
   a. Cultural patterns (life style)
   b. Health and safety
   c. Employment
   d. Population density
3. Aesthetics and human interest
   a. Scenic views and vistas
   b. Wilderness qualities
   c. Open space qualities
   d. Landscape design
5. Man-made facilities and activities
   a. Structures
   b. Transportation network (movement access)
   c. Utility networks
   d. Waste disposal
   e. Barriers
   f. Corridors

D. Ecological relationships, such as:
   a. Salinization of water resources
   b. Eutrophication
   c. Disease—insect vectors
   d. Food chains
   e. Salinization of surficial material
   f. Brush encroachment
   g. Other

Others
   a.
   b.

Source: From Leopold et al. (1971).

Note: Part 1 lists project actions (arranged horizontally); Part 2 lists natural and human environmental elements (arranged vertically in matrix).

The Leopold matrix suffers from the same problems earlier identified for checklists, since it is really only a series of checklists for different actions. Indeed, due to the added number of cells (a possible maximum of 8800, though only relevant cells need to be completed, and a "reduced matrix" presented), the method is time-consuming, and its key results difficult to conceptualize quickly. As with checklists the reasoning or empirical data on which scoring was based are entirely implicit, except to the extent that they are revealed in accompanying explanatory prose. A careful examination of Table 4.2 reveals numerous examples of potential double counting (e.g., in Part 2, water quality A2d and fishing C2b or swimming C2d), a heavy emphasis on biophysical effects (67 of 88 entries in Part 2), and limited opportunity to exhibit secondary effects (section D in Part 2).

On the last point, an Environment Canada (1974) team developed a method to trace secondary effects or interdependencies, termed the Component Interaction Matrix. Each environmental element (e.g., Part 2 of Table 4.2) is listed on both sides of a new matrix and cases of dependency noted (e.g., impacts on birds B2a are dependent in part on impacts on trees, shrubs and grasses B1a, b, c,). Environment Canada actually used a different list of environmental elements than that used by Leopold et al. (1971). By squaring this matrix using linear algebra, second-order dependencies could be noted (e.g., birds of prey are affected by effects on vegetation because they are dependent on small mammals, which in turn are dependent on vegetation). Higher-order dependencies can be identified by further matrix multiplication. Ross (1976) further modified the method by having experts weight each dependency for its importance in influencing biological productivity. The use of biological productivity as the sole criterion of significance and the expense of using experts are both problems (see also Bisset 1980).

A number of variations to the basic matrix approach described here can be introduced. Depending on how detailed the list of actions or effects is, what criteria are used to score impacts (magnitude, importance, duration, probability of occurrence, feasibility of mitigation), and what type of scale is used for scoring, quite different final impressions regarding severity of impact can be conveyed. Before describing the summarization of matrix results in a grand index, we must note a few properties of scales.

Types of Scales

Indexes are intended to represent the difference between a set of objects by reference to a *scale*. Four of the most common types of scales are nominal, ordinal, interval, and ratio (Table 4.3; see Coxon 1982 for four additional scale types). Although numbers can be used to represent objects on any of these scales, the scales differ by the way in which the numbers correspond to the underlying objects represented. Mathematical operations which make sense for some scales do not necessarily make sense for others. For example, suppose we observe that three ball players are wearing shirts numbered 80, 110, and 162. If these are numbers on a nominal scale, in which the

Table 4.3. Four Types of Scales, Permissible Mathematical and Statistical Operations, and Examples

Scale	Nature of Scale	Permissible Mathematical Transformations	Measures of Location	Permissible Statistical Procedures	Examples
Nominal	Classifies objects	One-to-one substitution	Mode	Information statistics	Classifying species, numbering soil types
Ordinal	Ranks objects	Equivalence to another monotonically increasing or decreasing function	Median	Nonparametric statistics	Ranking land use plans in order of preference
Interval	Rates objects in units of equal difference	Linear transformation	Arithmetic mean	Parametric statistics	Time (in hours and minutes); temperature (°F or °C)
Ratio	Rates objects in units of equal difference and equal ratio	Multiplication or division by a constant or other ratio-scale value	Geometric, harmonic, or arithmetic mean	Parametric statistics	Height, weight

Note: Each scale can be transformed mathematically by means indicated in previous rows as well.

numbers are simply being used to distinguish shirts in a bin, the average of the three numbers (112) would be meaningless (except for a trivial fact about the particular shirt numbers). Suppose, however, that these numbers actually represented the weights of the three players. The weight scale is a ratio scale, and the average of the numbers would represent the mean weight of the three players.

Nominal scales classify objects—for example, (1) hot, (2) warm, and (3) cold. *Ordinal* scales rank objects in order—(1) hottest, (2) next hottest, (3) next hottest after previous item, and so on. Ordinal scales do not convey how *much* hotter one object is than another but simply indicate relative order on a temperature scale. *Interval* scales rate the quantitative degree of difference between objects—for example, 50°F, 40°F, 35°F—in relation to each other. Note that the Fahrenheit (°F) scale of temperature has no absolute meaning, since the point at which water freezes could be set at 250°F rather than 32°F initially. What is important is that equal changes in temperature are represented by equal numerical intervals throughout the scale. *Ratio* scales indicate the quantitative degree of difference between objects in relation to some absolute starting point—for example, 50°K, 40°K, 35°K, where 0°K (Kelvin) is the point at which molecular motion ceases.

The mathematical manipulations that are meaningful with these scales differ. Nominal and ordinal values cannot meaningfully be added, subtracted, multiplied, divided, or averaged. It is meaningless, for instance, to average the first, third, and fourth choice to get the $2\frac{2}{3}$ choice. Furthermore the values cannot be tested for fit to a normal distribution, since increments between values have not been measured. Consequently ordinal values must be analyzed by nonparametric techniques (e.g., Mosteller and Rourke 1973, Torgerson 1958). Interval and ratio values can be averaged; they can also be linearly transformed ($y' = ay + b$; $a > 0$), and analyzed by parametric statistics (e.g., standard deviation, *t*-test, *F*-test). The *differences* between interval scores can be added, subtracted, multiplied, or divided, although the interval scores themselves should not be. Numbers on a ratio scale can be subjected to all arithmetic manipulations. Thus one can compute a mean grand index score by averaging 9, 12, and 21 (= 14) if the numbers represent rates on an interval or ratio scale, but not if they represent ranks on an ordinal scale or classes on a nominal one.

How can one tell if a scale is nominal, ordinal, interval, or ratio? One must examine what relation the numbers on the scale bear to each other and to the objects they purport to describe. For numbers a, b, and c to have interval properties, the sum of intervals between a and b, and between b and c, must equal the interval between a and c. For the numbers to have ratio properties, $a/b \times b/c$ must equal a/c as well. In the absence of these consistency checks, scales should not be *assumed* to have interval or ratio properties. It is usually simple to distinguish nominal from ordinal scales merely by whether the scales are being used to classify or rank. For further discussion of consistency checks and scaling, see, for example, Dawes (1972), Maranell (1974) and Stevens (1974).

Sometimes scales are constructed in which an equal interval on the scale corresponds to a logarithmic change in the underlying objects rated. The pH scale of acidity and the decibel scale, dB(A), in acoustics have this property and are called logarithmic interval scales (Stevens 1974). Thus a change from pH 3 to 4 represents a 10-fold decrease in acidity, from pH 3 to 5 a 100-fold decrease. To average values on a logarithmic scale, one should compute the geometric mean by taking the $antilog_{10}$ of the pH (or other scale) value, averaging these arithmetically and retransforming the result to log_{10} scale. One must also be careful not to compare ratios computed using different bases. Thus, for example, one cannot use percentage values in a chi-square test unless all percentage values were computed using the same base. Percentages with different bases must be converted to absolute (original) values, or some other means must be used to weight for differences in absolute size of numbers (Snedecor 1946).

Grand Indexes

When matrices contain many completed cells the overall result is often complex and difficult to grasp. It is also difficult to compare two large matrices prepared for project alternatives to ascertain net differences. Assessors have sought to summarize the results of the various cells of a matrix by combining them into a single number or *grand index*. This is usually done by summing positive and negative cell contents, and sometimes weighting cells, rows, or columns, to achieve a net result.

In the case of computing grand indexes from matrix scores, the discussion of scaling has highlighted that ordinal values cannot be summed or averaged. This is also true for values supposedly on an interval or ratio scale that do not have true interval or ratio properties. If the underlying intervals in an ordinal scale are actually consistent in size, they can be transformed to have equal-interval properties, provided the underlying transformation curve can be discerned from the scale user (Figure 4.2; see Dawes 1972, Maranell 1974 for techniques).

To illustrate a grand index calculation, consider Figure 4.3 in which two Leopold-type matrices are shown for alternative airport construction, operation, and maintenance plans. To calculate the grand index for each matrix, we weight each magnitude by its corresponding importance and sum, subtracting the unsigned products from the positive products. We could also disaggregate to the extent of comparing air quality effects across actions for each plan, for example, or effects on all environmental elements of the construction phase. In Figure 4.3 the grand index would indicate that Plan 2 is preferable, as its weighted cumulative adverse impacts are less; indeed, it would have a net beneficial impact in this example. Note that such computations are only mathematically permissible if the matrix scores lie on an

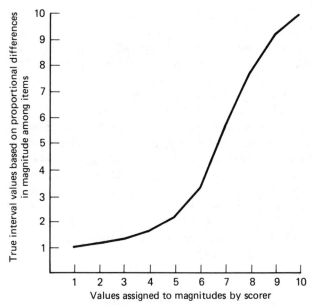

Figure 4.2. A hypothetical transformation curve relating a scorer's use of an interval scale to a true (proportional) interval scale.

interval or ratio scale, which has not been demonstrated, and indeed would be difficult to prove.

The advantage of the grand index is that it summarizes large quantities of data into a single index for decision making. The main disadvantage is that the relative contributions of different elements and actions to the outcome are obscured. In Figure 4.3 it can be seen that the beneficial effects of airport maintenance on the vegetation element, combined with the great importance given to this cell, contributed most to the outcome. In larger matrices the contributing factors will not always be obvious. Note also that, while in our example effects were given different importance weights in the two plans, this will not normally be the case when two comparable projects are compared. If weights for Plan 1 were used in Plan 2, the grand index for Plan 2 would be 31; Plan 2 would still be preferable. A sensitivity analysis could be performed on assignment of magnitudes and importance weights to determine which values, when changed by a constant amount, have the greatest effect on the outcome. The study would then profit most from further collection of data to determine these "sensitive" values more precisely.

Networks

Sorensen Networks

An alternative to the Component Interaction Matrix for illustrating the secondary and subsequent effects of actions on environmental elements is to

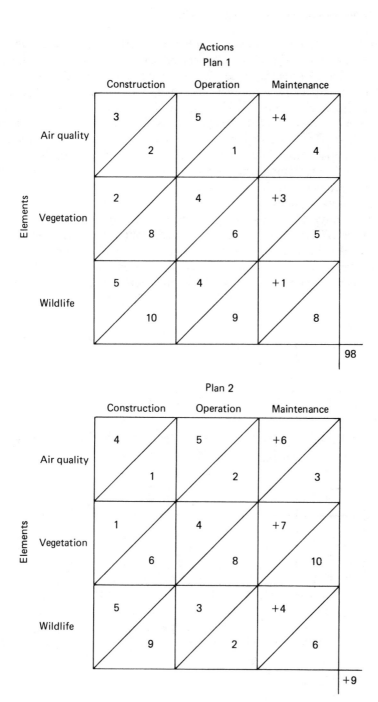

Figure 4.3. Comparison of matrices showing effects of alternative airport construction, operation, and maintenance plans on environmental elements. Magnitude score, upper left cell; importance score, lower right cell. Grand index, matrix lower right. Beneficial effects are shown with a ''+'', adverse effects with no sign.

construct a network tracing such effects (Sorensen 1971, 1972). Figure 4.4*a* illustrates a portion of such a network, developed for one of several land uses in the coastal zone (Sorensen 1971). Figure 4.4*b* shows the cause-effect links diagrammetrically. Rau (1980) has suggested that a grand index can be computed for such a network if a magnitude and importance score is assigned to each impact and the probability of occurrence of each impact is known. These scores are shown in Table 4.4 along with the expected environmental impact index for each branch, computed by multiplying each magnitude score by its importance and by its probability of occurrence, and summing these per branch. Indexes for each branch can be further summed to achieve a grand index.

The advantage of a network approach is that it permits clear tracing of higher-order effects of initial actions; indeed, mitigation and control measures can also be illustrated (Figure 4.4*a*). One problem encountered in applying the network approach is that many higher-order effects can be postulated that are actually unlikely to occur. Obtaining reliable information

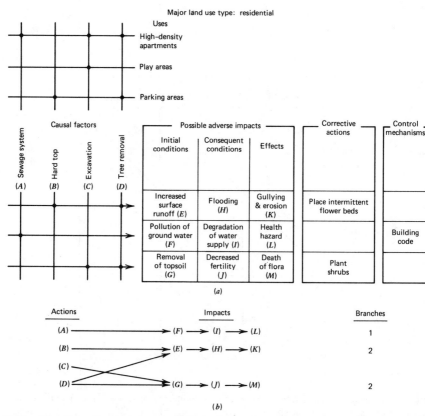

Figure 4.4. Example of a network of effects. Letters in parentheses in part (*b*) correspond to items labeled in part (*a*). Adapted from Sorensen (1971).

Table 4.4.   Calculation of Branch Indexes and Grand Index for Network Illustrated in Figure 4.4

Impacts	Impact Scores (1–10 Interval Scale)		Probability of Occurrence
	Magnitude	Importance	
E	5	3	$B \rightarrow E$ (0.8); $D \rightarrow E$ (0.7)
F	2	5	$A \rightarrow F$ (0.5)
G	3	4	$C \rightarrow G$ (0.3); $D \rightarrow G$ (0.4)
H	4	5	$E \rightarrow H$ (0.7)
I	2	9	$F \rightarrow I$ (0.6)
J	2	5	$G \rightarrow J$ (0.8)
K	3	7	$H \rightarrow K$ (0.7)
L	2	10	$I \rightarrow L$ (0.9)
M	1	6	$J \rightarrow M$ (0.8)

Environmental Impact Indexes

Branch 1 (2) (5) (0.5) + (2) (9) (0.6) + (2) (10) (0.9) = 33.8
Branch 2 (5) (3) (0.8) + (5) (3) (0.7) + (4) (5) (0.7) + (3) (7) (0.7) = 51.2
Branch 3 (3) (4) (0.3) + (3) (4) (0.4) + (2) (5) (0.8) + (1) (6) (0.8) = 21.2
Grand network index   33.8 + 51.2 + 21.2 = 106.2

on probabilities of occurrence (by expert judgment, see Chapter 3) is difficult and often not possible due to lack of information on which to base a judgment. Furthermore it is difficult to ensure use of true interval or ratio scales for scoring magnitude and importance. As with all grand index approaches, the final index value may obscure important uncertainties in the component data.

Systems Diagrams

The network or "stepped matrix" approach of Sorensen as modified by Rau (1980) permits the rating of each impact in common units which are scaled relative to each other by the importance rating. The scaling of impacts that are initially in quite different units (e.g., comparing water quality in units of BOD to parkland loss in hectares) on a 10-point importance scale is certainly a highly subjective process, making the construction of a true interval scale virtually impossible. H. T. Odum (1971) suggested that for analyzing impacts on ecosystems, the effects of actions on energy fixation and flux among ecosystem compartments is a vital index to ecosystem functioning. Therefore energy (in kilocalories) may be a useful common unit for measuring effects, avoiding the need for rescaling from other units. Odum (1971) and Odum and Odum (1976) diagramed the flow of energy in natural and human ecosystems using electronic circuitry symbolism (Figure 4.5), and Odum

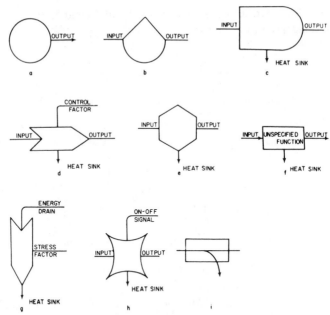

Figure 4.5. Symbols of the energy circuit language. (*a*) Causal force—source of energy or materials to the system, originating from outside the system; (*b*) passive storage—a storage of energy or materials within the system; (*c*) green plant—normally used to illustrate photosynthesis, but used in regional diagrams to represent entire ecosystems; (*d*) work gate—intersection of two pathways where one flow makes possible a second; (*e*) self-maintaining consumer—combination of storage and work gate symbols whose response is autocatalytic, such as an animal, city, industry; (*f*) unspecified function—function unknown or insignificant; (*g*) stress—represents a loss of potential energy; an inverted work gate with energy from the system being drained into a heat sink by an environmental stress; (*h*) switch—represents flows that have only on and off states; the pathway is open or closed until some threshold is exceeded; (*i*) sensor—monitors a flow rate or storage level, controlling the input of a quantity in proportion to the amount sensed. Reprinted from Gilliland and Risser (1977) with permission of Elsevier Scientific Publishing, Amsterdam; cf. Odum 1972a.

(1972b) suggested its potential for application to impact assessment. Davis (1975), Zucchetto (1975), and Gilliland and Risser (1977) applied the approach to impact studies. For example, in Figure 4.6, a systems diagram illustrating ecological effects of a missile firing range in White Sands, New Mexico, has expressed most effects on organisms in energy units (kilocalories) (numbers 1–22 in Figure 4.6) reflecting their energy production and respiration. The comparability of units is not complete, however, since other effects, such as noise (number 25, decibels), radiation (number 23, curies), air pollutants (number 61, g/yr) and water and sewage (number 23, liters/yr) are in noncomparable units.

Despite the attractiveness of tracking ecological impacts in a common natural unit (energy), the method has numerous limitations. First, many

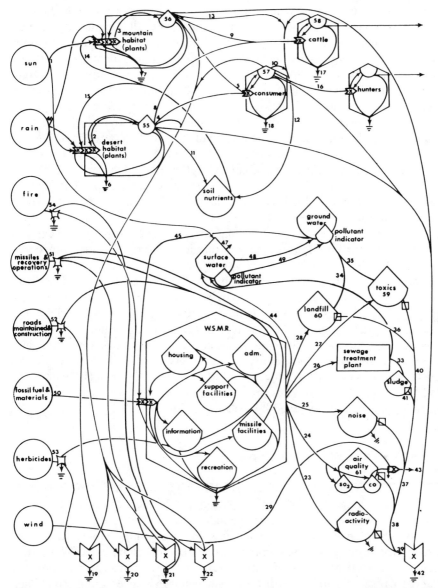

Figure 4.6. Systems diagram indicating the interactions of New Mexican missile range activities with nature, and the stresses on nature caused by those activities. Numbers correspond to those in Table 1 of Gilliland and Risser (1977), where a description of each pathway and its numerical value are given in kcal yr^{-1} or other units. Reprinted from Gilliland and Risser (1977) with permission of Elsevier Scientific Publishing, Amsterdam.

impacts are still expressed in noncomparable units, and some effects (e.g., impact on landscape aesthetics) are not quantified at all. Second, energy flux is only a very indirect indicator of the importance of the other functions carried out by organisms such as nitrogen fixation, pollination, and soil binding. Thus when Gilliland and Risser (1977) express stress to natural ecosystems from herbicides in kcal/yr, they consider only the loss of vegetative biomass and neglect such secondary impacts as increased soil erosion, death of soil organisms, and repercussions for other animals in the food chain. Similarly, they express fire effects in terms of kcal/yr of heat released by fire, which tells a decision maker nothing about which species are killed, what long-term compositional changes in the ecosystem will occur, and other features of post-fire recovery. Further they express nutrient flows in terms of the energy of decomposition, neglecting the significance of particular nutrient ratios to growth (Chapter 7).

Gilliland and Risser (1977) argue that energy is an appropriate indicator of other ecosystem functions, since it is the driving force for such functions. This is like describing human activity by our daily caloric flux: too much important detail about what we actually *did* during the day is not described. The method has its greatest value in describing changes in productivity (energy flux) of ecosystems. Gilliland and Risser (1977) also describe a hydrologic model using circuitry symbolism to describe water flows on the missile range. So long as the model uses units of water to describe only water flow, it is quite reasonable and useful. Additional criticism of energy systems analysis may be found in Gilliland and Risser (1977), Hyman (1979–80), and McAllister (1980). More conventional approaches to use of energy units in impact analysis are reviewed by Folsom (1980).

IMPACT EVALUATION TECHNIQUES

Many of the impact identification techniques just discussed contain aspects of subjective judgment about the importance of impacts to society. Thus the Leopold matrix and its variants (including Rau's modification to network analysis), which involve rating impacts by their "importance" or significance, are hybrids of identification and evaluation techniques. When both empirical and normative information is combined in a single grand index, the decision maker is unable to distinguish empirical observation from judgment, much less to modify the outcome by using alternative evaluation values. Techniques that clearly identify multiple publics and their values in the final outcome, and separate data on evaluation from those on impact identification, avoid this obscuring of information. Weighting-scaling techniques, which make more explicit the basis for the rating scales, are a step closer to fully explicit evaluations. Even closer are techniques which fully disaggregate the views of multiple publics.

Approaches that Aggregate Public Values

Weighting-Scaling Techniques

Environmental Evaluation System

One of the early and better-known weighting-scaling techniques is the Environmental Evaluation System (EES; Whitman et al. 1971, Dee et al. 1973) developed at Batelle Laboratories for use with water resources projects. The basic approach is to measure the environmental impact of actions on 78 environmental components (Figure 4.7), convert these values to common units using scalars, weight the scaled impacts by importance values, and sum the products to calculate a grand index. This index is compared to the grand index calculated for environmental baseline conditions.

The novel step in this method is the use of explicit *scalars* to convert environmental measurements into common units. In this case the common

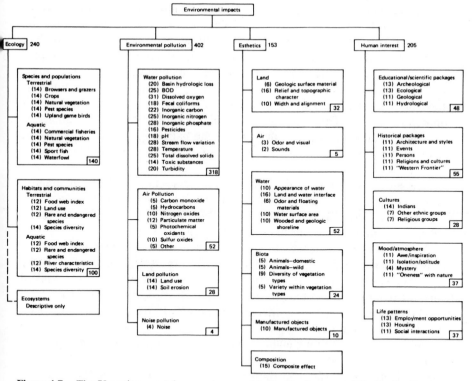

Figure 4.7. The 78 environmental parameters used in the Environmental Evaluation System. Importance value weights are in parentheses. Total of weight units for all environmental categories (boxes, bottom right) sum to 1000. Reprinted from Dee et al. (1973), *Water Resources Research* **9**:523–535, copyrighted by the American Geophysical Union.

units are an interval scale from 0 to 1, termed an "environmental quality scale." The curves used to transform measurements from their original units to the environmental quality score are called "scalars" or "value functions." Some examples of scalars used in EES are shown in Figure 4.8. To be comparable between environmental parameters, the scalars should have certain properties: the end points of the parameter axis (abscissa) should range from "worst" to "best" possible values from the point of view of environmental quality; in moving along the environmental quality scale from, say, 0.0 to 0.6, a comparable change in environmental quality should occur in units of one environmental parameter as another (see O'Banion 1980). For example, using the scalars of Figure 4.8, a change from 0 to 6 species/1000 individuals becomes comparable to a change from no value to low value significance of architectural style on a historic building, from 0 to 5 mg/liter of dissolved oxygen in stream water, and from a static, turbid stream to a very slow-moving clear stream. To compare "apples" and "oranges" in this way is clearly a questionable endeavor.

A further examination of any one of these scalars shows that they have been designed to apply to the "general case," whereas the form of the scalar is likely to differ drastically from one case to the next. For example, the "species diversity" curve (which is actually a species richness curve; see Chapter 11) reflects a particular range of species richness that will be appropriate for one ecosystem and not another, since species richness varies drastically between community types (see, e.g., Whittaker 1975). Furthermore the scalar assumes that an increase in species richness improves environmental quality. This assumption is only likely to have some validity in cases where an ecosystem from which some native species have been lost experience a recolonization by these natives. It will certainly not be the case when an intact ecosystem experiences invasion of an exotic pest species. Clearly such a scalar is worse than meaningless when applied universally. Even if new scalars are developed for each project, the information on which to make such comparisons is usually scant, and considerable subjective judgment is incorporated in their construction.

The EES system proceeds to weight each scaled environmental parameter by an importance value, indicated in parentheses in Figure 4.7. These importance values were determined by expert judgment using the ranked pairwise comparison technique and the Delphi procedure to divide 1000 initial weighting units among the 78 parameters. In the ranked comparison technique (see Maranell 1974), pairs of parameters are compared for relative importance, and the total weights are divided among available units proportionately. To illustrate, if 48 of the 1000 units are to be devoted to educational/scientific features, and these 48 points are to be divided among four weighting factors (indicated in the educational/scientific package of Figure 4.7), experts are asked by how much archaeological variables exceed in importance ecological ones, or vice versa, and their opinions are modified by the Delphi procedure (Chapter 3). In the present case the two variables were

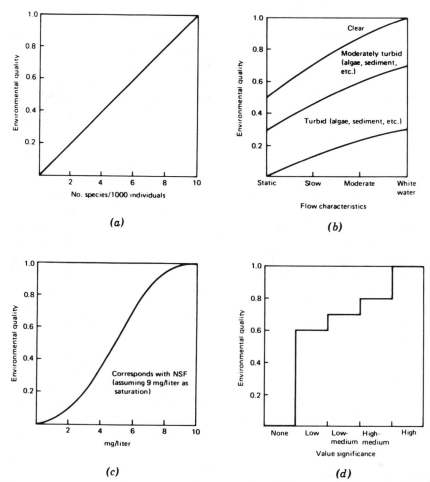

Figure 4.8. Examples of scalars used in the conversion of raw scores to the environmental quality index for four of the 78 environmental parameters in the Environmental Evaluation System. (*a*) Species diversity; (*b*) Stream flow variation; (*c*) Dissolved oxygen; (*d*) Architecture and styles. Reprinted from Dee et al. (1973), *Water Resources Research* **9**:523–535, copyrighted by the American Geophysical Union.

considered equally important, and each considered $\frac{1}{5}$ more important than either geological or hydrological variables, which in turn were ranked of equal importance. The final rating, then, gave 13 points each to archaeological and ecological values, 11 points each to geological and hydrological features.

Although these numbers have the patina of scientific respectability, they are based on the opinions of the judges, who cannot represent all of the multiple publics. Thus an EES is flawed both by its failure to reflect the disaggregated opinions of multiple publics and by its overly general application of scalars based on dubious scientific criteria in many cases. Further problems of most grand index or matrix methods which reappear here include double counting of categories (e.g., vegetation impacts appear under ecology, aesthetics, and human interest), arbitrary and incomplete nature of categories (e.g. mood/atmosphere is represented by four categories, but effects on the consumer price index is altogether absent), the burial of subjective judgment within numerical scores, and the failure to illustrate secondary and higher-order interactions. For further discussion and criticism of EES, consult Clark et al. (1978, 1980), Lee (1982), and McAllister (1980; Ch. 12), among others.

Ecological Rating Systems

Another variant of the weighting-scaling approach consists of developing criteria for rating some ecosystem attribute (e.g., individual trees or entire forests; Helliwell 1967), multiplying scaled scores to form a grand index, and weighting the index by some arbitrary monetary equivalent. Thus in Table 4.5 seven criteria for rating woodlands (a–g) are proposed; a four-point scalar is developed for each; and the final grand index units are equally weighted at £10 per unit. Analogous schemes for rating individual trees (Helliwell 1967), wildlife habitats (Helliwell 1969), and natural areas (Gehlbach 1975, Helliwell 1973, Tans 1974, Wright 1977) have been proposed.

Several problems with such an approach are evident: (1) the criteria used in evaluation form an arbitrary and incomplete list; (2) each factor is weighted equally, whereas factors may hold different weights for different publics; (3) as with other scaling approaches there is an untested equivalence assumed between scalars for different factors; (4) the monetary equivalent used as a constant weighting factor has no objective basis in economic theory or market price (Chapter 5). Note that Helliwell's index (Table 4.5) is similar to an EES, except for the assignment of a monetary equivalent, rather than variable weights, to rating scores.

One ecological rating index that has attracted particular attention is Helliwell's scarcity index (1973, 1974a; see Park 1980, Sinden and Windsor 1981). Helliwell proposes to rate the value of a potential nature reserve by the number of individuals of each species in it, weighted in turn by the rarity of each species in relation to the region or country as a whole. Helliwell suggests that as a species becomes rarer, its value should increase exponentially; similarly, smaller parks should be valued less than larger ones.

The exact mathematical relationships assigned by Helliwell are completely arbitrary. Helliwell (1973, 1974a) suggests that an 85% reduction in the number of individuals of a species in a park area will halve the value of the park for that species. He assumes that there is a linear equivalence between number of individuals and park area (not a particularly good assumption, see Chapters 10, 11), and suggests an equation $R = SI \times A^{0.36}$ to express this relationship (upper curve, Figure 4.9), where R = relative value

Table 4.5. Ecological Index for the Valuation of Forests or Woodlands

Factor	Units			
	1	2	3	4
a. Area visible (acres); (length of perimeter (chains) × 0.25; or visible acreage if greater)	0.5–1.5	1.5–5	5–20	20–100
b. Position in landscape	Secluded	Generally visible	Prominent	Very prominent
c. Average daylight viewing population (taken as 1% of population of urban areas; or 1 person per vehicle on roads)	(0–1) Remote	(1–20)	(20–100)	(100+) Near to one or more well-trafficked routes
d. Presence of other trees and woodlands (or other features of similar interest)	Densely wooded	Some other woods (usually with hedgerow trees or features)	Hedgerow trees only: or few other woods or features only	No other woods, trees, or features of interest
e. Accessibility	No access to wood or land adjacent	Access to land adjacent	Access to wood, but difficult	Readily accessible to large numbers of people
f. Species and state of crop	Young plantation of 1 species; and derelict woodland	Mixed plantation	Semimature or natural woodland	Mature or irregular woodland
g. Any special value	No special value (i.e., most woods)	Local beauty spot	Well-known beauty spot	Feature of widespread fame or screening eyesore

Source: Reprinted from D. R. Helliwell (1967), The amenity value of trees and woodlands, *Arboric. J.* **1**:128–131, with permission of AB Academic Publishers, Berkhampsted.

Note: Units for each factor to be multiplied together. Scale goes from 0 to 16,384 units. Equate 1 unit to £10 value.

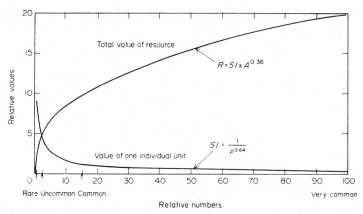

Figure 4.9. Relationship between the number of individual units and the total value of a resource that is becoming scarce. See footnote (*) for corrections to SI formula. Reprinted, with permission, from D. R. Helliwell (1973), Priorities and values in nature conservation. *Journal of Environmental Management* 1:85–127. Copyright: Academic Press, Inc. (London) Ltd.

of the species, *SI* is the scarcity index for the species, and *A* is the park area or number of individuals.*

To derive a scarcity index, Helliwell (1973) sought to express the weight or value of one individual as a function of the population of a species remaining (p_i) in the region in which the reserve occurs. If the value of a reserve increases proportional to the number of individuals in it (A) raised to the power 0.36, then the worth of the number of individuals remaining in the region might also be said to increase in value as $p_i^{0.36}$. Helliwell then reasons that the value of one individual can be taken as the reciprocal, $p_i^{-0.36}$. This reciprocal (see footnote to previous paragraph) is the scarcity index (*SI*) by which the number of individuals in an area is weighted to assign a relative value to the species (Figure 4.9). Helliwell (1973) suggested assigning some monetary equivalent per unit score (£10,000 per unit) to convert relative values to monetary units. Sinden and Windsor (1981) attempted to make the assignment of this weight slightly less arbitrary by using a panel of experts to determine it.

Although weighting a species by rarity in evaluating a potential preserve is a reasonable criterion, Helliwell's approach unfortunately incorporates too many arbitrary judgments to justify application. In addition the amount of data required is large and not readily available in most countries outside of Great Britain. Further the comparison of study sites of different shape presents enormous problems (Helliwell 1974b). As we will demonstrate in Chapter 10, more straightforward and theoretically sound ecological meth-

* Two errors flaw Helliwell's mathematics. To achieve an 85% reduction in number of individuals, the exponent of A should be 0.23 rather than 0.36. Also $p_i^{-0.36}$ is mistakenly converted to $1/p_i^{0.64}$ rather than $1/p_i^{0.36}$.

ods are available for comparing sites of different composition and weighting for rare species.

Optimum Pathway Matrix, and Others

Other weighting-scaling techniques exist. One is the optimum pathway matrix of E. P. Odum et al. (Institute of Ecology 1971; Odum et al. 1976), which incorporates a sensitivity analysis of weighting factors. An example of its application, in conjunction with the EES, to assess alternative navigation routes is provided by Canter et al. (1974). The method is also described in Canter (1977).

Another weighting-scaling technique is the Water Resources Assessment Methodology (WRAM) (Solomon et al. 1977; see also Canter 1979) developed for the U.S. Army Corps of Engineers. The WRAM has the virtue of not attempting to aggregate separate impacts into a grand index; a field test on alternative flood control projects in Louisiana is described by Richardson et al. (1978). Still another method, proposed by Sondheim (1978), uses a panel of experts to determine magnitude scores and a separate panel of public representatives to determine importance scores; the method is illustrated with a Canadian case study of replacement of a dam in a provincial park. A variety of sources review in greater detail quantitative summarization techniques and evaluation techniques that aggregate public values. Among these are Bisset (1980), Canter (1977), Clark et al. (1978, 1980), Golden et al. (1979), Nichols and Hyman (1980), Rau (1980), and Whitlatch (1976).

Approaches That Disaggregate Public Values

We consider here evaluation methods that explicitly document the distribution of impacts among affected groups. Two such techniques that have found wide use involve expressing impacts, where possible, in monetary units. The techniques are the planning balance sheet (Lichfield 1966; see also Lichfield et al. 1975) and the goals-achievement matrix (Hill 1968; see also McAllister 1980). The assumptions underlying expressing impacts in monetary terms are discussed in Chapter 5 and will not be considered here. The two methods exemplify ways to disaggregate impacts but are by no means the only possible approaches. Other approaches, termed trade-off matrices, are also described as alternative methods that overcome certain disadvantages of the planning balance sheet and goals-achievement matrix.

Planning Balance Sheet

In the planning balance sheet approach, parties to the transactions (i.e., actions which generate effects) are classified as producers or consumers. The producers may be a company, an individual, an activity, or a place (e.g., airport, factory); the consumers are the parties affected by the transaction (e.g., nearby residents, other species). In a table the annual benefits and costs to producer and consumer sectors from each plan alternative are quantified, either in monetary terms or in physical units (kilograms of pollutant discharged; number of illnesses resulting per year, etc.). Where quantification is not possible, the item is footnoted to a verbal description of impact. Net advantages may be compared between plan alternatives. As can be seen from the simplified example in Table 4.6, many effects cannot be quantified in monetary or physical units. Even the items that can be valued in monetary units are often inadequately estimated. For example, in Table 4.6, noise effects on airport residents are estimated from projected declines in property values to residents' homes from airport noise. Even if the changes in property values are correctly estimated, such an estimate neglects other costs of noise such as health effects and disturbance to conversation and communication. Effects of noise and plane collisions on other species, such as birds, are difficult to predict and to express in quantitative units for the table. Items incorporated by footnote (e.g., effects of noise and plane collisions on nearby wildlife) tend to be given less emphasis by decisionmakers, as they are not obvious on inspecting the table. A considerable subjectivity is exercised by the evaluator in defining producer and consumer categories (e.g., airport workers have been omitted from consumer groups, and all nonhuman species aggregated into a single category).

Goals-Achievement Matrix

The goals-achievement matrix disaggregates parties to the transaction but does not distinguish producers from consumers. Rather it simply examines

Table 4.6. Airport Operation Using the Planning Balance Sheet

	Plan 1		Plan 2		Net Benefit (+) or Cost (−)	Net Advantage
	Benefit	Cost	Benefit	Cost		
Producer						
Airplane traffic	$4.0	$3.0	$4.5	$3.0	+$0.5	Plan 2
Consumers						
Air travelers	—[a]	$4.0	—[a]	$4.5	—	—
Nearby residents	0	$1.2[b]	0	$1.3	−$0.1	Plan 1
Plants and wildlife	0	—[c]	0	—[c]	—	—

Note: See Lichfield (1966) and McAllister (1980) for discussion. Annual costs and benefits in millions.
[a] Benefits of air travel to consumers not quantifiable.
[b] Noise effects converted to monetary units by estimating loss of property value to houses and other buildings resulting from increased noise.
[c] Effects of noise and plane collisions on nearby wildlife not quantified but expected to be significant.

how each group benefits or loses from each impact. Impacts are further categorized or grouped as they contribute toward or detract from some "community goal" such as clean air or quiet surroundings. Each community goal, and each affected group, is assigned a weight by a panel of experts. The costs and benefits can be expressed in monetary and/or physical units, but with the use of noncomparable units, no grand index can be calculated. Thus

Table 4.7. Evaluating Airport Operation Plans Using the Goals-Achievement Matrix

		Plan 1: Large Airport					
Community goal:		Airplane travel		Quiet neighborhood			
Goal weight:		3		2			
	Group Weight	Impact: Large Number of Flights		Group Weight	Impact: Increased Noise (Large)		Grand Index
Affected Groups		Benefits	Costs		Benefits	Costs	
Air travelers	3	+2	0	1	0	0	
Nearby residents	2	0	0	3	0	−2	
Plants and wild-life (represented by environmental groups)	1	0	−2	2	0	−2	
Weighted totals		+18	−6		0	−20	−8
		Plan 2: Smaller Airport					
Community goal:		Airplane travel		Quiet neighborhood			
Goal weight:		3		2			
	Group Weight	Impact: Small Number of Flights		Group Weight	Impact: Increased Noise (Small)		Grand Index
Affected Groups		Benefits	Costs		Benefits	Costs	
Air travelers	3	1	0	1	0	0	
Nearby residents	2	0	0	3	0	−1	
Plants and wild-life	1	0	−1	2	0	−1	
Weighted totals		9	−3		0	−10	−4

Note: See Hill (1968) and McAllister (1980) for discussion. Benefits and costs are ranked on a five-point ordinal scale, +2 = maximum progression toward goal; (0) = no change; −2 = maximum regression from goal.

often an ordinal scale is used. In Table 4.7 a five-point scale of progression toward (+) or regression from (−) community goals is used.

The airport example is illustrated in goals-achievement format in Table 4.7. Notice that the importance weight assigned to affected parties differs depending on the community goals being evaluated. Thus effects on nearby residents are given greater weight in assessing impacts on the goal of a quiet neighborhood, whereas effects on air travelers are given greater weight in assessing effects of increased numbers of flights. Although in the table the relative costs and benefits to affected parties does not change between plans, and the importance weights for groups and goals remain constant between plans, Plan 2 emerges as preferable (lower negative grand index) because the combined number of costs, and their weights, to impacted parties exceed the benefits to benefiting parties, and Plan 2 scales down benefits and costs proportionately. A sensitivity analysis on the assignment of importance values would be helpful in clarifying the key components that influence the grand index. In Great Britain, where the goals-achievement matrix has been used extensively to evaluate structure plans (McAllister 1980, p. 164), planners have solicited importance weights from different groups (e.g., other planners or various community groups) in different trials of the matrix, as an alternative to sensitivity analysis.

The goals-achievement matrix contains many of the drawbacks of previous matrix approaches to the extent that it relies on interval scales without true interval properties, which are then weighted and summed, contrary to mathematically permissible procedures. Furthermore the classification of impacts within "community goals" may be advantageous in some situations to highlight the larger social issues, but it will be a cumbersome constraint in other situations where impacts can most simply be described as such. Its main advantage is in disaggregating the list of affected parties and noting the separate effects of each impact on each group.

Priority–Trade-Off-Scanning Matrix

Davos (1977) has noted that each affected group in a goals-achievement matrix prefers a different weight for each community goal and hence the choice of community goal weights by each affected group could differ between plans. Also group choices about the compromises they are willing to accept differ. Davos suggests that three matrixes of preference be generated. The first is the goals-achievement matrix. The second, which Davos (1977) calls the "goal-priority–trade-off matrix," expresses the community goal weights each affected group would assign for each different plan. In addition each group is asked how its evaluation of goal weights between plans would change if it were willing to change its priority among goals so that, say, goal *b* were now required to be more important than goal *a*. The consequent weight changes are recorded in the goal-priority–trade-off matrix, and the process repeated for a third and fourth choice among goal priorities. A third matrix, called the interest-priority–trade-off matrix, compiles the preferences of each group for plans other than its first choice ("interest trade-offs"), and the plans each group would be willing to accept based on its willingness to compromise with the values of other groups, in view of their strong bargaining positions ("imposed trade-offs"). Armed with all this information, the analyst can choose options that optimize achievement of all goals, generate least opposition, or satisfy some other decision criterion (Davos 1977). The so-called priority–trade-off-scanning (PTS) approach has been applied to an energy planning problem (Davos et al. 1979).

Obviously the application of the PTS approach requires enormous amounts of information on value weights, some of it difficult to generate from groups with uncompromising positions. Furthermore all of this information becomes highly aggregated toward the end of the analysis, making it very difficult for a decision maker to trace the particular judgments that influenced the outcome. The PTS is an example of a highly complex assessment method that will be difficult for decision makers and the public to understand, and whose results may therefore not be trusted by them. On the positive side the method does attempt to make explicit the value trade-offs that are available for seeking acceptable compromises. A technique like environmental mediation (Chapter 3), which involves face-to-face contact, is an alternative method that can elicit such compromises.

Simple Trade-Off Matrix

Herson (1977–78) has pointed out that when impacts occur in noncomparable units, one can simply compare plans by noting how they differ in relation to each impact in the respective units. Thus plan A may differ from plan B in generating 200 tons/day less of an air pollutant, while costing $200,000 more to construct. He terms the compilation of such lists "trade-off analysis."

Developing this theme, and combining it with a matrix of "affected groups" versus "impacts," we may construct what might be called a *simple trade-off matrix*. In such a matrix we express each impact in monetary and physical units and in verbal, qualitative terms to permit maximum information content for use by decision makers. The way in which each impact has been calculated can be explained in footnotes to the matrix cells and the sources of information indicated. A hypothetical example is shown in Table 4.8.

The simple trade-off matrix does not involve assigning weights to impacts or affected groups; this is left to decision makers. Its advantages lie in clear exposition of the effects of each impact on each multiple public in whatever terms are appropriate, both monetary and otherwise. The simple trade-off matrix also permits maximum flexibility in structuring the matrix, incorporating available information, and avoiding the use of planners' judgment in determining the importance of impacts. The simple trade-off matrix in Table 4.8 has incorporated all the information from both the planning balance sheet and goals-achievement matrix, plus additional information.

The disadvantages of the simple trade-off matrix lie in the size of the

Table 4.8. Evaluating Airport Operation Plans Using the Simple Trade-Off Matrix

Affected Groups	Impacts of Increased Air Travel				Noise			
	Benefits		Costs		Benefits		Costs	
	Plan 1	Plan 2	Plan 1	Plan 2	Plan 1	Plan 2	Plan 1	Plan 2
Air Travelers								
Monetary								
Physical	40,000 additional flights[c]	50,000 additional flights[c]	$4.0 M[a]	$4.5 M[a]				
Qualitative								
Nearby Residents								
Monetary							$1.2 M[d]	$1.3 M[d]
Physical								
Qualitative	Some added convenience of air travel by proximity to airport						Interruption to conversation; possible health effects; less adverse than Plan 2	Health and communication effects worse than Plan 1

160

Plants and Wildlife

Monetary				
Physical				
Qualitative	Fewer birds killed by collision with planes	More birds killed by collision with planes	Noise effects on wildlife significant but less than Plan 2[e]	Noise effects worse than Plan 1[e]

Airplane Companies

Monetary	$4.0 M[a]	$4.5 M[a]	$3.0 M[b]	$3.0 M[b]
Physical	40,000 flights[c]	50,000 flights[c]		
Qualitative				

[a] Retail sales of plane tickets. Source: Document A.
[b] Cost of operating planes. Source: Document B.
[c] Source: Report C.
[d] Predicted loss in property values. Source: Report D.
[e] No quantification available. Source: Report E.

resulting matrix, and the difficulty of summarizing "net" benefits or costs, since a grand index cannot be calculated. As we have shown, the calculation of a grand index will always be a mixed blessing: by aggregating impacts, it diminishes the amount of information presented to the decision maker. The advantage of summarization occurs at the expense of permitting decision makers to make independent judgments on the importance (value weights) to be assigned to impacts.

CONCLUDING REMARKS

Existing impact identification and evaluation techniques present a variety of choices for organizing and summarizing information. Techniques involving a grand index maximize summarization, presenting a clear "answer" to the decision maker. However, in calculating grand indexes, the judgments used become hidden in the calculation, denying the decision makers access to the judgment process. Techniques that use ordinal scales to rank impacts and subsequently calculate a grand index also carry out impermissible mathematical operations; ordinal data should not be added and should be analyzed using only appropriate methods of nonparametric statistical analysis.

Techniques that aggregate public values fail to analyze the equity or distributional effects of an action; for this reason evaluation techniques using disaggregated public values are more informative. Indeed, they require explicit consideration of effects that might otherwise be ignored.

The use of scalars, though attractive in reducing impacts to common units, is fraught with difficulties. It is rare that the shape of a scalar can be truly documented on the basis of empirical studies. Furthermore the scalar incorporates judgment in its design and as a result is an aggregated, highly value-implicit, evaluation technique. It is easier for a decision maker to apply his or her own weights to apples and oranges when they are presented as such, than when they have both been scaled to some generic fruit using panel-of-experts scalars. In defense of scalars, they do permit summarization and comparison of large amounts of otherwise noncomparable data.

Quantitative impact assessment techniques exist on a continuum from those that simply summarize to those that also evaluate impacts. In using these techniques, the assessor should clearly distinguish the two processes and make explicit the source of judgments used in evaluation. Even so, all quantitative assessment techniques can be criticized from one or another perspective: either they summarize too much or not enough; attempt to quantify based on inadequate (subjective) data, or remain too qualitative; are arbitrary and incomplete in their selection of impacts to include, or are too exhaustive; take away too much judgment from decision makers, or leave too much to be decided.

In view of this multitude of trade-offs some analysts have suggested that quantitative assessment techniques not be used at all. Their concern is that elaborate quantification and manipulation of data prevent the public and decision makers from following the steps in reasoning and challenging judgments (Bisset 1980, Hollick 1981, Lee 1982). This criticism is certainly a valid one. On the other hand, leaving impacts in qualitative, unsummarized form often makes precise comparison of the impacts of alternatives impossible. The middle ground seems to lie in using techniques that do not aggregate information unduly and that clearly identify the sources of information and lines of reasoning. The simple trade-off matrix represents one such middle-ground technique; others will be appropriate depending on particular purposes.

To date, evaluation techniques have been used primarily in large planning efforts; they are not routinely used in environmental impact statements because they are costly and time-consuming to perform well. Nevertheless, evaluation is clearly an essential step in impact assessment. If an explicit evaluation is not performed by assessors, a totally implicit evaluation will be made by decision makers. Clearly explicit procedures, which identify the values of the multiple publics and the significance of effects of actions on each, provide a better basis for rational analysis of the basis of a decision.

REFERENCES

Adkins, W. G., and Burke, D. (1974). Interim report: social, economic and environmental factors in highway decision-making. Res. Rep. 148-4. Texas Transp. Inst., Texas A & M Univ., Austin, Tex.

Balbach, H. E., and Novak, E. W. (1975). Field investigation of an environmental computer system. In B. Hutchings, A. Forrester, R. K. Jain, and H. Balbach, eds. *Environmental Impact Analysis: Current Methodologies and Future Directions*. Dept. Architecture, Univ. Illinois, Urbana, Ill., pp. 131–136.

Bisset, R. (1980). Methods for environmental impact analysis: recent trends and future prospects. *J. Environ. Manage.* **11**:27–43.

Canter, L. W. (1977). *Environmental Impact Assessment*. McGraw-Hill, New York.

Canter, L. W. (1979). *Water Resources Assessment Methodology and Technology Source Book*. Ann Arbor Sci., Ann Arbor, Mich.

Canter, L. W., Risser, P. G., and Hill, L. G. (1974). Effects assessment of alternate navigation routes from Tulsa, Oklahoma, to vicinity of Wichita, Kansas. Univ. Oklahoma, Norman, Okla.

Clark, B. D., Bisset, R., and Wathern, P. (1980). *Environmental Impact Assessment: A Bibliography with Abstracts*. Mansell, London.

Clark, B. D., Chapman, K., Bisset, R., and Wathern, P. (1978). Methods of environmental impact analysis. *Built Environ.* **4**:111–121.

Coxon, A. P. M. (1982). *The User's Guide to Multidimensional Scaling*. Heinemann, Exeter, N.H.

Davis, M. (1975). Future directions in environmental impact analysis. In B. Hutchings, A. Forrester, R. K. Jain, and H. Balbach, eds. *Environmental Impact*

Analysis: Current Methodologies and Future Directions. Dept. Architecture, Univ. Illinois, Urbana, Ill., pp. 153–158.

Davos, C. (1977). A priority-tradeoff-scanning approach to evaluation in environmental management. *J. Environ. Manage.* **5**:259–273.

Davos, C., Smith, C. J., and Nienberg, M. W. (1979). An application of the priority-tradeoff-scanning approach: electric power plant siting and technology evaluation. *J. Environ. Manage.* **8**:105–125.

Dawes, R. M. (1972). *Fundamentals of Attitude Measurement.* Wiley, New York.

Dee, N., Baker, J. K., Drobny, N. L., Duke, K. M., Whitman, I., and Fahringer, D. C. (1973). Environmental evaluation system for water resource planning. *Water Resources Research* **9**:523–535.

Environment Canada (1974). An environmental assessment of Nanaimo Port alternatives. Ottawa.

Folsom, B. (1980). Energy impact analysis. In J. G. Rau and D. C. Wooten, eds. *Environmental Impact Analysis Handbook.* McGraw-Hill, New York, Ch. 5.

Gehlbach, F. R. (1975). Investigation, evaluation, and priority ranking of natural areas. *Biol. Conserv.* **8**:79–88.

Gilliland, M. W., and Risser, P. G. (1977). The use of systems diagrams for environmental impact assessment. *Ecol. Modell.* **3**:188–209.

Golden, J., Ouellette, R. P., Saari, S., and Cheremisinoff, P. N. (1979). Techniques for aiding in the assessment process. *Environmental Impact Data Book.* Ann Arbor Sci., Ann Arbor, Mich., Ch. 2.

Helliwell, D. R. (1967). The amenity value of trees and woodlands. *Arboric. J.* **1**:128–131.

Helliwell, D. R. (1969). Valuation of wildlife resources. *Regional Studies* **3**:41–47.

Helliwell, D. R. (1973). Priorities and values in nature conservation. *J. Environ. Manage.* **1**:85–127.

Helliwell, D. R. (1974a). The value of vegetation for conservation I: Four land areas in Britain. *J. Environ. Manage.* **2**:51–74.

Helliwell, D. R. (1974b). The value of vegetation for conservation II: M1 Motorway area. *J. Environ. Manage.* **2**:75–78.

Herson, A. (1977–78). Trade-off analysis in environmental decision-making: an alternative to weighted decision models. *J. Environ. Systs.* **7**:35–44.

Hill, M. (1968). A goals-achievement matrix for evaluating alternative plans. *J. Amer. Inst. Planners.* **34**:19–28.

Hollick, M. (1981). The role of quantitative decision-making methods in environmental impact assessment. *J. Environ. Manage.* **12**:65–78.

Hyman, E. L. (1979–80). Net energy analysis and the theory of value: is it a new paradigm for a planned economic system? *J. Environ. Systs.* **9**:313–324.

Institute of Ecology (1971). Optimum pathway matrix analysis approach to the environmental decision-making process. Univ. Georgia, Athens, Ga.

Lee, N. (1982). The future development of environmental impact assessment. *J. Environ. Manage.* **14**:71–90.

Leopold, L. B., Clarke, F. E., Hanshaw, B. B., and Balsley, J. R. (1971). A procedure for evaluating environmental impact. Geological Survey Circular 645. U.S. Dept. Interior. Washington, D.C.

Lichfield, N. (1966). Cost-benefit analysis in town planning: a case study of Cambridge. Cambridgeshire and Isle of Ely County Council, U.K.

Lichfield, N., Kettle, P., and Whitbread, M. (1975). *Evaluation in the Planning Process,* Pergamon, Oxford.

Little, Arthur D., Inc. (1971). Transportation and environment: synthesis for action. Impact of National Environmental Policy Act of 1969 on the Department of Transportation. Prepared for Dept. Transportation. 3 vols. Cambridge, Mass.

Maranell, G. M., ed. (1974). *Scaling: A Sourcebook for Behavioral Scientists.* Aldine, Chicago.

McAllister, D. M. (1980). *Evaluation in Environmental Planning: Assessing Environmental, Social, Economic, and Political Trade-offs.* The MIT Press, Cambridge, Mass.

Mosteller, F., and Rourke, R. E. K. (1973). *Sturdy Statistics: Nonparametric and Order Statistics.* Addison-Wesley, Reading, Mass.

Multiagency Task Force (1972). Guidelines for implementing principles and standards for multiobjective planning of water resources. Review draft. U.S. Dept. Interior, Bureau of Reclamation. Washington, D.C.

Nichols, R., and Hyman, E. (1980). A review and analysis of fifteen methodologies for environmental assessment. Water Research & Technology Report. U.S. Dept. Interior. Washington, D.C.

O'Banion, K. (1980). Use of value functions in environmental decisions. *Environ. Manage.* 4:3–6.

Odum, E. P., Bramlett, G. A., Ike, A., Champlin, J. R., Zieman, J. C., and Shugart, H. H. (1976). Totality indexes for evaluating environmental impacts of highway alternatives. In Transportation Research Record No. 561, Trans. Res. Board, Nat'l. Academy of Sciences, Washington, D.C., pp. 57–67.

Odum, H. T. (1971). *Environment, Power and Society.* Wiley-Interscience, New York.

Odum, H. T. (1972a). An energy circuit language for ecological and social systems: its physical basis. In B. C. Patten, ed. *Systems Analysis and Simulation in Ecology II.* Academic Press, New York, pp. 139–211.

Odum, H. T. (1972b). Use of energy diagrams for environmental impact statements. In *Tools for Coastal Management.* Proc. Conf. Marine Technology Soc., Washington, D.C., pp. 197–213.

Odum, H. T., and Odum, E. C. (1976). *Energy Basis for Man and Nature.* McGraw-Hill, New York.

Park, C. C. (1980). *Ecology and Environmental Management: A Geographical Perspective.* Westview, Boulder, Colo.

Rau, J. G. (1980). Summarization of environmental impact. In J. G. Rau and D. C. Wooten, eds. *Environmental Impact Analysis Handbook.* McGraw-Hill, New York, Ch. 8.

Rau, J. G., and Wooten, D. C., eds. (1980). *Environmental Impact Analysis Handbook.* McGraw-Hill, New York.

Richardson, S. E., Hansen, W. J., Solomon, R. C., and Jones, J. C. (1978). Preliminary field test of the Water Resources Assessment Methodology (WRAM). Tensas River, Louisiana. Misc. Paper Y-78-1. U.S. Army Corps of Engineers, Vicksburg, Miss.

Riggins, R. E. (1980–81). Comprehensive computer-aided environmental impact analysis. *J. Environ. Systs.* **10**:81–91.

Ross, J. M. (1976). The numeric weighting of environmental interactions. Occ. Paper 10. Lands Directorate, Environment Canada, Ottowa.

Sinden, J. A., and Windsor, G. K. (1981). Estimating the value of wildlife for preservation: a comparison of approaches. *J. Environ. Manage.* **12**:111–125.

Snedecor, G. W. (1946). *Statistical Methods Applied to Experiments in Agriculture and Biology.* 4th ed. Iowa State College Press, Ames, Iowa.

Solomon, R. C., Colbert, B. K., Hansen, W. J., Richardson, S. E., Canter, L., and Vlachos, E. C. (1977). Water Resources Assessment Methodology (WRAM)—Impact Assessment and Alternative Evaluation. Tech. Rep. Y-77-1. U.S. Army Corps of Engineers, Vicksburg, Miss.

Sondheim, M. W. (1978). A comprehensive methodology for assessing environmental impact. *J. Environ. Manage.* **6**:27–42.

Sorensen, J. C. (1971). A framework for identification and control of resource degradation and conflict in the multiple use of the coastal zone. Masters thesis. Dept. Landscape Architecture, Univ. California, Berkeley, Calif.

Sorensen, J. C. (1972). Some procedures and programs for environmental impact assessment. In R. B. Ditton and T. L. Goodale, eds. *Environmental Impact Analysis: Philosophy and Methods.* Univ. Wisconsin Sea Grant Program, Madison, Wisc.

Stevens, S. S. (1974). Measurement. In G. M. Maranell, ed. *Scaling: A Sourcebook for Behavioral Scientists.* Aldine, Chicago, pp. 22–41.

Strand, R. H., Farrell, M. P., Goyert, J. C., and Daniels, K. L. (1983). Environmental assessments through research data management. *J. Environ. Manage.* **16**:269–280.

Tans, W. (1974). Priority ranking of biotic natural areas. *Michigan Botanist* **13**:31–39.

Torgerson, W. S. (1958). *Theory and Methods of Scaling.* Wiley, New York.

U.S. Atomic Energy Commission (1973). General environmental siting guides for nuclear power plants. Draft. AEC, Washington, D.C.

U.S. Department of Housing and Urban Development (1974). Final Environmental Impact Statement for Pauahi urban renewal project, Hawaii. Rep. R-15, EIS-HI-73-0851-F. Washington, D.C.

Whitlatch, E. E. (1976). Systematic approaches to environmental impact assessment. *Water Res. Bull.* **12**:123–138.

Whitman, I. L., Dee, N., McGinnis, J. T., Fahringer, D. C., and Baker, J. K. (1971). Design of an environmental evaluation system. Rep. PB-201 743. Battelle Columbus, Columbus, Ohio.

Whittaker, R. H. (1975). *Communities and Ecosystems.* 2nd ed. Macmillan, New York.

Wright, D. F. (1977). A site evaluation scheme for use in the assessment of potential nature reserves. *Biol. Conserv.* **11**:293–305.

Zuccheto, J. (1975). Energy-economic theory and mathematical models for combining the systems of man and nature, case study: the urban region of Miami, Florida. *Ecol. Modell.* **1**:241–268.

5

ECONOMIC APPROACHES

In addition to the abstract rating scales discussed in Chapter 4, monetary units may be used to evaluate resource transactions. Monetary units have a familiar meaning to decision makers and the public in relation to many other items of value and can therefore be compared with the large subset of valued items that have market prices. A problem with monetary units is that they are difficult or impossible to apply in a universally acceptable way to resources that are not typically marketed (e.g., human life, bacteria, wind). Because using monetary units for evaluation is attractive, it is particularly important to understand the assumptions and limitations of the methods employed.

SOCIAL THEORY UNDERLYING ECONOMIC EVALUATION

Utilitarianism is the social philosophy in which the enjoyment derived by an individual from participation in the larger society is seen to be quantifiable as the sum of the resources the person uses and enjoys. "Resources" are broadly defined as goods, services, and amenities used, whether marketed or not. The nineteenth-century philosopher, Jeremy Bentham, proposed quantifying the resource benefits enjoyed by an individual in units which he called "utils." Thus the utility of resources enjoyed by an individual could in theory be measured as the quantity and quality of goods, services, and other amenities enjoyed or consumed over a period of time, each measured in utils. This notion proved impractical because of the difficulty of measuring "utility" in absolute units. The *relative* utility of objects, however, can be more readily established by ranking objects in order of preference. Monetary units are used as measures of relative utility.

The basis for using money to measure relative utility lies in the notion that the market price of a good is the amount that consumers are collectively willing to pay for the last increment of the good. At a higher price, consumers would not buy as many of the good; at a lower price, they would buy more of the good (Figure 5.1). It is assumed that the consumer's relative preference for alternative goods (as measured by price) is a measure of the relative utility of the goods to consumers. Thus the ratio of the prices of two

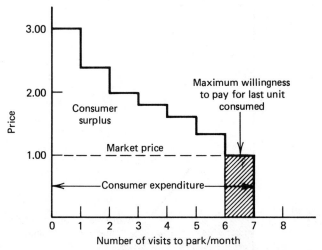

Figure 5.1. A demand schedule for park entrance, showing willingness to pay for the last visit per month and the consequent market price. The lower area in the histogram represents the total amount spent on entry fees by all consumers during the month. The upper area reflects the surplus utility gained which was not reflected in the market price ("consumer surplus").

goods should equal the ratio of their marginal utilities, where "marginal utility" is the gain in utility to the consumer from the last item purchased (McAllister 1980). Since prices are determined by their marginal utilities, the utility of items purchased before the last increment is greater than its price, resulting in a surplus of utility to the consumer ("consumer surplus," Figure 5.1). Notice that the fact that individuals value items differently has been sidestepped by considering only the aggregate behavior of all consumers toward the marketed item. When a resource is given a price that underestimates the true marginal utility to the consumer, additional consumer surplus results (Figure 5.2). This becomes an important concern when we attempt to estimate prices for items that are not marketed. Various methods have been proposed to derive surrogate or "shadow" prices for goods not directly marketed so that the total relative utility derived by the individual could be expressed in monetary terms.

Monetary units are not good measures of utility for several reasons: (1) The market price of a resource does not necessarily reflect the value of the resource to the individual; it reflects only the balance between supply of, and collective demand for, the resource in a free market with perfect competition for resources. (2) Some items of value are not marketed and therefore have no direct market price, yet they are valuable (clean air, attractive scenery, etc.). Also methods for deriving "shadow" prices are imperfect and sometimes impossible to apply. (3) The value placed on a resource varies between individuals; thus one person may be willing to pay handsomely for a package of cigarettes, while another would not be willing to pay for it at all. Underly-

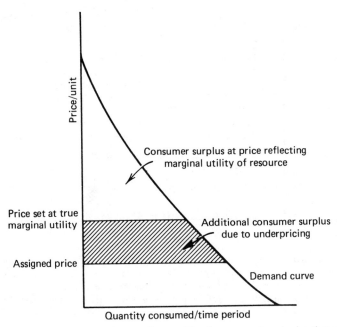

Figure 5.2. The additional consumer surplus resulting from an assigned price that undervalues the marginal utility of a natural resource (hatching). The discrete values expressed as a "demand schedule" by histogram in Figure 5.1 are here expressed as continuous values on a demand curve.

ing attempts to equate resource values to their market prices is the assumption that goods and services of equal price are substitutable in value to society (but not necessarily to any one individual).

The "util," then, cannot be simply equated with a monetary unit. The "util" is conceptually useful in its own right as a common unit for evaluating the enjoyment derived from resources, much as the "environmental quality unit" was used in Chapter 4 to scale a variety of environmental amenities. However, because the value of an amenity varies between individuals, it is not possible to assign a universal value in "utils" to a good, service, or amenity.

Despite the fact that the individual utility of a good, service, or amenity cannot be directly measured, the concept of utility has been used widely in political economics and philosophy. Partly, this has been accomplished by assuming that a person's relative preference for resources permits estimation of relative utility (Hicks 1939, McAllister 1980), and partly by speaking only of aggregate social utility. The social welfare of an action or good is often defined as the sum of the utility derived by individuals from it, with each person's utility assumed equally important to that of the next person. Within the political traditions of western society (see, e.g., McAllister 1980, Ch. 3), resource managers often assume that the best management options are those that maximize social welfare. This assumption, however, ignores the distribution of benefits or utility among individuals in society.

If a resource action makes one or more people better off without making anyone worse off, neoclassical economists call the action a "Pareto improvement." In cases where the action makes some people better off and others worse off, it is impossible to measure the extent of change in utility because of the lack of a measure of individual utility. If society were able to agree on some measure of utility by universal assent, one could then modify the Pareto criterion to say that social welfare is improved if the total gains in utility among those who benefit exceed the total losses in utility by those adversely affected. The latter reformulation is known as the "potential Pareto criterion" (from Kaldor 1939; Hicks 1939; see Mishan 1976, Ch. 59; McAllister, Ch. 6). Notice that the distribution of benefits is being ignored, and we are assuming that utility can be measured.

At this point many neoclassical economists seeking to find a social welfare criterion for use by resource decision makers take a bold leap. They say, let us assume that we can measure relative utility in monetary units, and that the distribution of benefits among members of society can be ignored for individual decisions. Then the potential Pareto criterion can be expressed in monetary terms (the "social welfare test" of Pigou 1932) as follows: social welfare is increased if those who gain monetarily by the resource action can fully compensate those who lose by it and still have some money left.

This social welfare test has the great advantage that it can actually be quantified and implemented. In order to maximize social welfare, resource managers can compare the costs (including potential compensation to losers) and benefits from alternative actions and choose the action in which the net stream of benefits over the lifetime of the project will be greatest. This procedure is called cost-benefit analysis.

In using the social welfare test, however, the following assumptions have been made: (1) Losers are willing to accept financial compensation for their losses, and the full value of losses can be expressed in monetary terms. (2) Losers know the value of what they are losing at the time of the transaction. (3) There does not have to be equity in the distribution of gains and losses; the benefits may accrue to one party, the losses to another.

Furthermore the following problems are not directly addressed: (1) Future generations of losers may not value the loss in the same way. (2) Species and objects other than people cannot be compensated directly for their losses, nor can they be consulted on whether they are willing to partake of the transaction. (3) Even in the case of people, compensation for losses may not actually occur. (4) Different members of society will value the losses (or compensation) differently; for example, the poor and the rich may place a different value on obtaining an additional $100.

In what follows we discuss the application of cost-benefit analysis (CBA) to decision making and steps that can be taken to mitigate some of the problems we have enumerated. Cost-benefit analysis is most readily applied when the entire set of resources being developed or lost is marketed and therefore has established market prices. Problems arise when the value of nonmarketed resources must be expressed in monetary terms for the sake of analysis. Farnworth et al. (1981) call the value of items with market price "Value I." The value generated by some "shadow pricing" technique for a nonmarket item is termed "Value II." In some cases values occur that are not marketed (e.g., pain from illness) and for which existing shadow-pricing techniques seem inappropriate or inapplicable. Such a "nonmonetizable" value is termed "Value III." The problems with CBA center on finding appropriate techniques for ascertaining Value II items; the incompleteness of economic analyses arise in part from the exclusion of Value III items.

EVALUATION USING MARKET PRICES

Free market prices will bear their closest relation to marginal utility if they are indeed reflective of the willingness to pay for the last increment of a good. Such prices are most likely to occur under conditions of vigorous competition and full compensation for resource use. The lack of vigorous competition (e.g., a monopoly in production) will result in a higher price for the product than would occur in a free market. In cases where "free goods," such as the atmosphere, are polluted without compensation to those who breathe it, an uncompensated cost external to the cost of production has been generated, and the market price of the product will be artificially low since it does not reflect the willingness to pay for clean air.

Perfect market conditions are an ideal rarely if ever achieved in practice. Thus the prices we see in the marketplace cannot be assumed to be completely reflective of social utility, and their use in CBA is problematic for this reason. For example, if we assume that the current price of a liter of gasoline (petrol) is a good measure of its social utility, we are assuming that as the price goes up, perfectly substitutable goods are available, that the supply of oil and its wholesale price are totally unregulated, and that there is no oil cartel interfering with free market competition. Since these are not valid assumptions in this case, the use of this price in CBA cannot be assumed to result in maximizing social welfare.

The Changing Value of Resources over Time

The use of current market price to reflect a cost or benefit assumes that the worth of a resource in the future will be the same as at present. Not only can the price change in the future, but the value placed on the resource in the future by present or future generations may change. Benefits received in future years are typically not as valuable to people as cash received today. People can invest money they have today and be earning interest on it so that the value of $1 today is actually equal to $1 plus compounded interest several years from now. The value of invested money at time t can be calculated by the compound-interest formula:

$$V_t = V_0(1 + r)^t \qquad (5.1)$$

where

V_t = value at time t
V_0 = present value
r = present rate of interest
t = time in years

Thus a dollar benefit paid several years from now can be argued to be worth correspondingly less today. Rearranging Eq. 5.1 to express the present value of a future benefit,

$$V_0 = \frac{V_t}{(1 + r)^t} \qquad (5.2)$$

When used in the context of Eq. 5.2, r is called the *discount rate*. Economists have suggested that the problem of forgoing present interest on benefits paid in the future can be overcome by discounting future benefits using Eq. 5.2.

In Table 5.1 we can see that an office building which will accrue benefits over a 50-year period will have a benefit/cost ratio > 1 at a discount rate of 6% but not at one of 8%. Higher discount rates devalue future benefits more.

Table 5.1. Example of Cost and Benefits of Constructing an Office Building Which Will Accrue Annual Net Benefits for 50 Years, Using Two Rates of Discount of Future Benefits

Initial cost of construction: 28×10^6
Net benefits: 12×10^6/yr for 50 years
Discounted future benefits at 6%:

$$V_0 = \frac{V_t}{(1 + r)^t} = \frac{\$12 \times 10^6/\text{yr} \times 50 \text{ yr}}{(1 + 0.06)^{50}} = \frac{\$600 \times 10^6}{18.42} = \$32.6 \times 10^6$$

Discounted future benefits at 8%:

$$V_0 = \frac{\$12 \times 10^6/\text{yr} \times 50 \text{ yr}}{(1 + 0.08)^{50}} = \frac{\$600 \times 10^6}{46.90} = \$12.8 \times 10^6$$

The procedure of discounting accounts well for the problem of forgone interest. However, it does not directly address the problem of the value placed on a resource by future generations. Indeed, one effect of discounting is to assume that benefits of a resource are worth less to future generations than they are to us today. Discounting makes the tacit assumption that we will be the beneficiaries of all future benefits, rather than subsequent generations, or chooses to ignore the problem of intergenerational equity altogether. This problem is particularly severe for exploitation of nonrenewable natural resources. To assume that oil, or an endangered species like the humpback whale, can be exploited to exhaustion or extinction by us because the value placed on these resources by future generations will be much less (i.e., discounted) is to make a questionable assumption on behalf of the unborn (see also Ehrlich and Ehrlich 1981).

Another effect of discounting is that the present value of a future benefit or cost becomes extremely small for time horizons greater than 25–50 years. Thus discounting has the effect of estimating only the value of the lost resource to people living in the next 25 or 50 years and ignores losses to succeeding generations. This assumption is also clearly questionable for exhaustible, nonrenewable resources. Most people may be willing to forgo concern about generations living 1000 years hence, but 50 years seems unduly short as a horizon of concern for such matters as the preservation of species or valuable minerals. In a survey of 1795 American undergraduates during 1976–1982, 52% felt that civilization will "last" less than 200 years before it is vastly diminished or destroyed (Figure 5.3) (cf. Westman 1976). While the expectation of doomsday is a different matter from the horizon of concern for future generations, the response to this question may be some indication of a deeply held, intuitive planning horizon of concern many people maintain at a subconscious level.

The value to be chosen for a discount rate for CBA is unclear. Some would choose prevailing prime interest rates, a problematic procedure if interest rates are expected to fluctuate over the life of the project. Others feel that because the discount rate involves a social judgment regarding the values of future generations, it should be set by a political process in the legislature. Indeed, the U.S. Congress sets discount rates for use in CBA for public works projects by the U.S. Army Corps of Engineers. Myers (1977, 1979) has suggested that discount rates for projects with conservation value (e.g., tropical forest preserves) be set lower than others so that benefits to future generations would be more highly valued.

Still another approach toward selecting a discount rate is to ask: What is the most productive alternative use of the resource? The discount rate could be set equal to the rate of return on investment for this alternative use. For example, suppose grazing land is slated for housing development with an expected rate of return on the housing construction costs of 20% per year. The interest on the investment if placed in a savings account might be 10%. But the next most productive use of land may be to plant it to vegetable

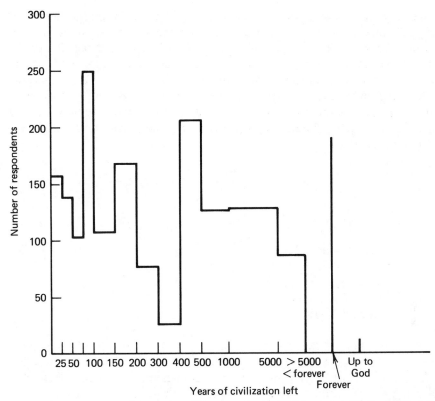

Figure 5.3. Responses to the question, "How long do you think human civilization will last till it is vastly diminished or destroyed?" Respondents comprised 1795 undergraduates during 1976–82. Question was asked at the beginning of an introductory environmental studies course.

crops at an expected rate of return of 15%. The discount rate for the housing development might then be set at 15% to represent the interest lost in failing to pursue the next-most-productive investment. The return forgone in the next-most-productive use of capital is called the *opportunity cost* of an investment. Whereas basing discount rates on opportunity costs is simple in principle, in practice it is not always easy to determine what the most productive alternative use of a resource will be, nor to determine with confidence the rate of return on such an investment. Both the circumstances of application of the discount rate and the manner in which the rate is set remain controversial. For further discussion, see such texts as McAllister (1980), Mishan (1976), and Pearce (1976).

Grand Indexes in Cost-Benefit Analysis

The cost and benefit data from a proposed action may be summarized by a variety of grand indexes, suitable for use with different decision criteria. The discussion that follows draws on McAllister (1980, Ch. 7) and Mishan (1976, Ch. 2). Some hypothetical data on the costs and benefits of three sizes of proposed power plant are provided in Table 5.2. The total benefits of the project will be the sum of annual benefits over the productive life of the project (in this case, 10 years), discounted annually ($r = 5\%$). The total costs will be the initial construction cost plus the annual operating cost over the 10-year period. In this case only the operating costs are discounted. The *net benefits* will be the difference between total benefits and costs, both discounted; that is,

$$\text{Net benefits} = \sum_{t=1}^{n} \frac{B_t}{(1 + r)^t} - \left[C_0 + \sum_{t=1}^{n} \frac{C_t}{(1 + r)^t}\right] \qquad (5.3)$$

where

r = discount rate
n = total number of years action will yield benefits
C_0 = initial cost (first year)
C_t = continuing annual costs
B_t = annual benefits

The net benefits will be a useful decision criterion when the only concern is the absolute amount of benefit. However, in view of opportunity costs (the loss of interest on capital not invested elsewhere), it is more likely that decision makers will be interested in the benefit per unit of investment (the benefit/cost ratio):

$$\text{Benefit/cost ratio} = \frac{\sum_{t=1}^{n} (B_t/(1 + r)^t)}{C_0 + \sum_{t=1}^{n} (C_t/(1 + r)^t)} \qquad (5.4)$$

Notice that we are assuming construction takes one year. If construction takes several years, the second and subsequent years' construction costs should also be discounted.

By the net benefit criterion, the 1000-MW plant would be best; by the benefit/cost ratio, the 1000-MW plant would still be chosen (Table 5.2), although the degree of difference between plants appears not as great. One might prefer information in the form of the net benefit/cost ratio (Eq. 5.3 divided by denominator of Eq. 5.4):

$$\text{Net benefit/cost ratio} = \frac{\sum_{t=1}^{n} (B_t/(1 + r)^t) - \left[C_0 + \sum_{t=1}^{n} (C_t/(1 + r)^t)\right]}{C_0 + \sum_{t=1}^{n} (C_t/(1 + r)^t)} \qquad (5.5)$$

Table 5.2. **Ways to Express Relationship of Costs to Benefits, Illustrated for Hypothetical Data on a Power Plant of Three Possible Sizes**

Plant Size (Generating Capacity, MW)	Initial Data				Net Benefits	Cost-Benefit Indexes		
	Construction Cost, C_0	Operating Cost/yr, C_t	Total Benefits/yr, B_t	Discounted Total Benefits Over 10 Years		Total Benefit/Total Cost Ratio	Net Benefit/Net Cost Ratio	Capital/Output Ratio
500	1.0	0.3	0.4	2.5	−0.4	0.9	−0.1	1.6
750	2.0	0.4	0.5	3.1	−1.4	0.7	−0.3	3.3
1000	2.5	0.5	1.0	6.1	0.6	1.1	0.1	0.8

Note: Assumes 1 year of construction, 10 years of operation, and a discount rate of 5%. Amounts are expressed as 10^6 monetary units.

which will be 0 or negative when costs exceed benefits, whereas the benefit/ cost ratio will always be positive. If b represents total benefits and c total costs (appropriately discounted), the indexes of Eqs. 5.3–5.5 are of the form $b - c$, b/c, and $b - c/c$, respectively.

Sometimes the main concern is the amount of initial capital that must be invested relative to net benefits in subsequent years. In this case information may be usefully presented as the capital/output ratio, in which initial costs are divided by discounted benefits minus discounted operating costs:

$$\text{Capital/output ratio} = \frac{C_0}{\sum_{t=1}^{n} ((B_t - C_t)/(1 + r)^t)} \tag{5.6}$$

As an alternative index, economists calculate the discount rate (i) which will equate the present value of benefits with the present value of costs, so that

$$\sum_{t=1}^{n} \frac{B_t}{(1 + i)^t} = \sum_{t=0}^{n} \frac{C_t}{(1 + i)^t} \tag{5.7}$$

Such a discount rate is called the *internal rate of return*. The internal rate of return does not separate initial costs from continuing costs, as costs are summed starting the first year $(t = 0)$. If we wish to separate initial from continuing costs, we may calculate the rate of return at which initial costs equal the present value of net benefits during operation, that is,

$$C_0 = \sum_{t=1}^{n} \frac{B_t - C_t}{(1 + i)^t} \tag{5.8}$$

In this case i is termed the *investment rate of return*.

These discount rates may then be compared with rates of return from other investment opportunities, in order to determine whether the proposed action is the most lucrative use of the money relative to other investments. If the investment rate of return from building a dam, say, is less than the rate of return from investment in municipal bonds, it would suggest that the dam is not as profitable as the alternative investment opportunity.

In the next sections we examine the difficulties of determining costs and benefits for resources without market prices (Values II and III). These resources must be priced in order to be included in CBA.

EVALUATION USING SHADOW PRICES FOR NONMARKETED RESOURCES

A variety of techniques exist for attempting to determine a price that reflects the utility of nonmarketed goods, services, and amenities. All such techniques rest on questionable assumptions. Nevertheless, they provide a means to compare apples and oranges, that is, marketed versus nonmarketed goods. Thus they have the same ultimate goal as do scaling approaches discussed in Chapter 4.

The "costs" of environmental disruption may be viewed as the loss of the free benefits of nature in the undisrupted state. Figure 5.4 illustrates a range of such benefits. Some of nature's benefits have market prices, namely, such free "goods" of the ecosystem's structure as timber, certain species of wildlife, soil and minerals (left side of Figure 5.4). Ecosystem structure is defined as the tangible items (plants, animals, soil, air, water) of which it is composed. Even the recreation of walking in a forest is enjoyment of a benefit derived from ecosystem structure (the size and spatial arrangement of trees). By contrast, the functions of ecosystems, the dynamics of exchange of mass and energy, are nature's free services (right side of Figure 5.4). These are typically not sold in the marketplace and have traditionally been totally neglected in cost-benefit analyses.

The benefits of nature to people can be categorized as direct and indirect. The direct benefits usually arise from the enjoyment or harvest of ecosystem structural features (e.g., food, medicine, fiber, shade, recreation from plants and animals), the indirect benefits more typically from ecosystem functions (e.g., gas exchange, radiation balance, pest regulation; see Ehrlich and Ehrlich 1981). More typically in economic evaluations we are concerned with the benefits of preserving nature in the unimpacted state as opposed to the benefits of resource development (with the accompanying disturbance to parts of nature). From this perspective the benefits of pollution control arise

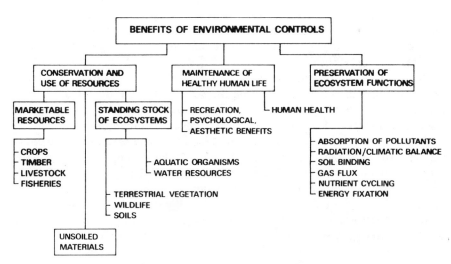

Figure 5.4. Three major categories of benefit from reducing environmental damage. Effects on ecosystem structure at left, on function at right. From Westman and Conn (1976).

from the lessened damages to the free goods and services of nature (*direct benefits*) and from the reduced costs of repairing damages (*indirect benefits*) (Westman and Conn 1976). Thus benefits of reduced oxidant pollution include both the reduced oxidation of house paint, and consequent enjoyment of fresher house appearance, and the reduced frequency with which the house must be repainted.

Damage, Repair, and Replacement Costs

One approach to quantifying the economic "benefits" of maintaining nature in the unimpacted state is simply to estimate the costs of damage, repair, or replacement of nature's goods and services upon disturbance. Thus, for example, when a highway is built through pastureland, one of the free services lost is the gaseous pollution-absorption capacity of the pasture plants. Using experimental data (Inman et al. 1971), Westman (1977) estimated a net loss of pollution absorption of approximately 442 kg ha^{-1} yr^{-1} of carbon monoxide for each hectare of San Bernardino freeway built through pasturelands. This figure is the difference between the carbon monoxide (CO) absorbed by the pasture (1046 kg ha^{-1} yr^{-1}) and by the freeway surface (604 kg ha^{-1} yr^{-1}). The costs of this loss in a free service could be estimated by the corresponding damage to crops, materials, and health from the CO pollution (damage-costs approach) or by the cost of equipment to remove an equivalent amount of CO emissions to the region (replacement-cost approach). A repair-cost approach is not feasible in this example, since it is not possible to repair the damage to plants, wildlife, and people caused by CO pollution. The costs calculated represent only part of the costs of the pasture loss, as other free services will also be lost, such as the absorption of other pollutants, binding of soil, and maintenance of a certain radiation balance (Westman 1977). Marketable goods lost will include the value of the products from the grazing stock that are sold.

The repair-cost approach to evaluating nature's free services is illustrated by considering the costs of repairing the damage to pine trees from ozone ("smog") pollution in the San Bernardino Mountains east of Los Angeles (Westman 1977). In 1972 the U.S. Forest Service estimated that 57% of the trees in a 4000 ha area of this forest were in a declining phase due to ozone-related mortality (see Figure 5.5). In calculating repair costs, Westman (1977) assumed that 50% of this area would be replaced by herbaceous successional vegetation, that erosion would be comparable to that on a nearby hillside where native chaparral had been replaced with grasses (Rice et al. 1969, Rice and Foggin 1971), and that the resulting eroded sediment would be trapped equally in debris basins, sewers, and street edges. Using 1976 figures for sediment removal costs from such structures (Ateshian 1976), he estimated that the annual repair cost from loss of the soil-binding function would be $27 million per year. This figure is substantially larger

Figure 5.5. Photochemical oxidants reaching the conifer forests in the San Gabriel Mountains northeast of Los Angeles. Arrows show top of mountain range scarcely visible through the smog.

than that actually spent, implying that dams, sewers, creek beds, and estuaries are filling with sediment without the complete level of dredging or sweeping necessary to remove all sediment. The year after the calculation was published, the San Bernardino Mountains experienced a particularly wet year. The clogged creek beds overflowed, causing $5.2 million in flood damages to houses and other structures at the base of the mountains in San Bernardino County (U.S. Army Corps of Engineers 1978). This figure may represent a belated damage-cost estimate attributable at least in part to smog damage to the pines and resulting erosion.

Estimates derived by each of the damage-, repair-, and replacement-cost approaches are likely to differ. For example, the damage that could be done by a moving, devegetated dune which had been strip mined can be much greater than the cost of replacing the vegetation in a rehabilitation program.

In the San Bernardino forest example the repair-cost estimate ($27 million/ yr) was greater than the damage-cost estimate ($5.2 million in one year). Not all damages were evaluated, however: the damages to estuarine fisheries downstream, the lost productivity of marketable timber and of nonmarket species, and so on. Nor was the damaged forest ecosystem itself repaired. The only things repaired were the human structures clogged with sediment. Thus, although damage-, repair-, or replacement-cost approaches may be useful in providing some range of estimates of lost goods and services, the estimates will vary with the approach used and will, in any event, be an incomplete tally of the full range of effects from the lost goods and services (see Westman and Conn 1976, Westman 1977, Freeman 1979, and Hyman 1981 for further discussion). Thus the economist's temptation to use the lowest estimate as the most "efficient" estimate of cost must be tempered by recognition of the incomplete nature of all of the estimates.

Damage-, repair-, or replacement-cost approaches usually fail to recognize the importance of the structure and interconnectedness of ecosystems. Thus the damages to sublittoral marine communities from an oil spill on the Santa Barbara, California, coast in 1969 (Sorenson 1974, Westman and Conn 1976) were estimated from the cost of replacing the dead organisms by reseeding the algae, scattering tank-reared shellfish, and buying sea lions from zoos. Such an approach ignores the fact that not all the organisms can be successfully reared in captivity, not all are replaced (e.g., bacteria), not all will survive in the natural environment, and the population structure, behavioral attributes, and genetic composition of animals reared in captivity will differ from those killed in the wild. Also not all ecosystem functions can be replaced by technology. For example, what inventor can yet lay claim to a machine that regulates global climate (Westman 1977)?

Economic Surrogates

The money people spend on ancillary goods and services needed in order to enjoy some of nature's free goods and services is sometimes taken as a "surrogate" or artificial substitute for the true value of the nonmarket items themselves (Hyman 1981).

Travel-Cost Approach

The amount people spend on travel to and from a recreational site and any park permit fee they pay are sometimes used to estimate the economic worth of the park itself (e.g., Clawson 1959, Pearse 1968). Some analysts have included the value of a person's time in travel to the recreational site (Knetsch and Davis 1966). The "travel cost" method has no way of incorporating costs for people who are already paying more to live close to the resource amenity (see Thibodeau and Ostro 1981).

In an application of the travel-cost approach to evaluating the recreational benefits of wildlife, Everett (1979) surveyed 1425 day visitors to the Dalby Forest (North Yorkshire Moors National Park, England). He determined the visitation rate, V, for each concentric zone around the forest as

$$V = \frac{N}{P} \qquad (5.9)$$

where

N = number of cars visiting the Dalby forest area from that zone over the survey period (day visits only)

P = population of that zone

The cost of travel, C, from each zone was based on mileage, car engine size, and fuel costs. Results (visitation rate vs. travel cost) are plotted in Figure 5.6. The demand curve for visits was fitted by seeking constants (a, b, and k) for an equation of the form

$$V = a + \frac{b}{(C + k)} \qquad (5.10)$$

The constants a and b were determined by regression analysis; k was chosen to maximize the correlation coefficient, r. The values $a = -0.233$, $b = 109.7$, and $k = 179.1$ gave a correlation coefficient of $r = 0.98$ to the data in Figure 5.6. This equation was then used to produce a simulated demand schedule (Figure 5.7) of visitations for varying entrance fees. The curve predicts that with no entrance fee, 125,000 people yr^{-1} will visit the forest; at £2.6 (approximately $3.90), no one will visit. The area under the curve (£69,000, or approximately $104,000) represents the total amount people would be willing to pay above the amount they are already paying. Everett (1979) adjusted this figure for travel costs of visitors who stay overnight or just pass through, revenue from visitors, and costs of park operation. The resulting adjusted annual value of the Dalby Forest was computed as £98,000 (approximately $147,000).

In a questionnaire visitors to the forest were asked the extent to which their trip was motivated by an interest in the area's wildlife; responses were scored on a 1–10 scale (Everett 1977, 1979). The mean proportion of the total recreational experience attributable to interest in wildlife was 25% which, multiplied by the total value of the forest, gave a recreational value of wildlife for the forest of £24,000 (or $37,000) for the 1975–76 year.

Everett (1979) notes that visitors of different income and occupation have different likelihoods of visiting the forest, yet the travel-cost approach assumes that willingness to pay is proportional only to distance from the amenity. The travel-cost approach also assumes that people will react to an increase in entrance fee in the same way as to an increase in travel cost.

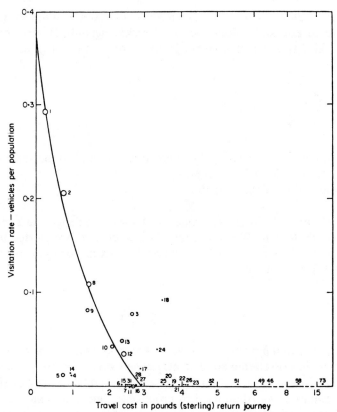

Figure 5.6. Visits to park versus travel cost. The visitation rate from each zone of distance is plotted against mean travel cost for day visitors to Dalby Forest. ○ represents >500 visits; o represents 100–500 visits; ● represents <100 visits. Numbers relate to county districts or counties. Reprinted, with permission, from R. D. Everett (1979), The monetary value of the recreational benefits of wildlife, *Journal of Environmental Management* **8**:203–213. Copyright: Academic Press, Inc. (London) Ltd.

Despite such limitations the travel-cost approach has been relatively widely used (see, e.g., Smith and Kavanagh 1969, Usher 1973, 1977).

Property Values, Wage Differentials, and Related Expenditures

Changes in the value of property adjacent to a site, such as a lake, due to some environmental impact, such as water pollution, may be taken as a surrogate for the actual damage costs to the lake. For example, Lind (1973) found that the value of land, and the houses on it, increased after an adjacent polluted lake was cleaned up. He estimated the benefits of the clean lake accordingly. Problems with the property-value approach include the fact that it is hard to allocate the portion of a property value due to the environmental amenity with confidence. Double counting may occur if house prop-

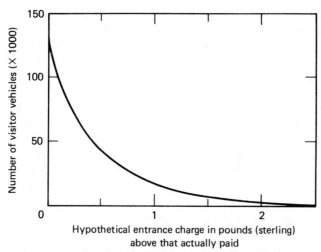

Figure 5.7. Simulated demand schedule showing the number of visiting vehicles to the Dalby Forest area for hypothetical entrance charges, over that already charged. Reprinted, with permission, from R. D. Everett (1979), The monetary value of the recreational benefits of wildlife, *Journal of Environmental Management* 8:203–213. Copyright: Academic Press, Inc. (London) Ltd.

erty values are used with such other benefit measures as value of recreational amenities in the neighborhood, which are already supposedly reflected in housing property values (Westman and Conn 1976).

Another surrogate approach has used the differences in wages of laborers in areas of low and high environmental quality, on the assumption that people will only work in more polluted or degraded areas if they are compensated by higher wages. In practice, wage differentials are only poorly correlated with environmental quality, as so many other factors influence prevailing wages in an area.

Another example of the "economic surrogate" is the use of related expenditures as an index of the value of natural resources. Thus the value of nongame birds may be evaluated by the expenditures on travel to birdwatching sites, binoculars, bird guidebooks, and membership in environmental conservation organizations, or on birdseed and birdfeeders in the case of backyard birdwatchers (see, e.g., Payne and De Graaf 1975). Related expenditures are used as an economic surrogate on the theory that they reveal preferences for nonmarket items through economic approaches.

Numerous problems exist with economic surrogate approaches, however. The poor may value natural resources as much as the rich, but refrain from spending as much to travel or buy related items (fishing rods, tents, etc). because they do not have as much income. Natural resources which people will not directly spend money to enjoy (e.g., centipedes, bacteria) are undervalued by this approach, as are the services they perform (e.g., decomposition). In the travel-cost approach the amount people actually spend, or are prepared to spend to visit a natural area, does not include the value of other natural functions performed by the area's ecosystems when the visitor is not there (soil binding, pollution absorption, etc.), nor can it deal with the free global services provided by places most people do not visit (e.g., Amazon, Antarctica).

Some economists (e.g., Sinden and Worrell 1979, Hyman 1981) argue that people might be willing to pay more to enjoy a natural resource than they currently have to spend to do so, and therefore related expenditures are not a true index of the actual value ("utility") of natural resources to people; in other words, the expenditures reflect consumer surplus. To overcome the problem of prices below true value, economists have attempted to determine what people would be willing to pay above what they are currently paying (if anything) on a hypothetical basis. In the Dalby Forest travel-cost example, a willingness-to-pay curve was derived from existing travel-cost information. The procedures discussed next are applicable where such empirical data are unavailable.

Hypothetical Valuation

Consumers may be asked, in an interview setting, what they would be willing to pay to obtain each of several levels of a resource (e.g., visit to a park with one vs. two or three different scenic elements: meadows only vs. meadows plus mountains and waterfalls). The result of such a hypothetical valuation method is a demand curve for park scenic elements. In the bidding game approach the interviewer gives the consumer a starting bid and asks whether the consumer would be willing to pay that amount for the amenity (e.g., 350 days per year of clean air). If the answer is "yes," the bid is raised, and the question repeated. When a price is reached that the consumer is not willing to pay, the bid is lowered slightly to fine-tune the estimate. In the use-estimation game the interviewer holds the price constant and asks about the amount of good required at that price, or the possibility of a substitution. For example at $20 per visit, how often would you visit the park with meadows, mountains, *and* waterfalls? Would you be willing to substitute several trips to the park that has only meadows? How many trips for $20?

Such interview situations are difficult to accomplish without introducing inadvertent biases in the answers obtained. The level of starting bid influences the nature of responses obtained (see Hyman 1981), as does information on how the money is to be obtained (e.g., entry permit fees vs. non-park-related purposes).

It is also difficult for consumers to provide accurate answers of their true behavior in the hypothetical survey situation (Hyman 1981). If people expect their answers to influence the actual price eventually charged, for example, they will tend to keep their estimates of willingness to pay purposely lower than what they will actually be willing to pay. Typically people also give very different estimates when asked "How much would you be *willing to pay* to have 350 rather than 300 days of clean air?" and "How much would you be *willing to accept in compensation* to have 300 rather than 350 days of clean air?" This is because in the first case people must have the disposable income to purchase a "free good," whereas in the second they are relinquishing a free good at no economic expense. Also the answer is likely to differ depending on whether the person already enjoys a resource (50 extra days of clean air) which is being taken away or is being offered a resource not previously enjoyed.

Meyer (1976) asked residents near the Fraser River in Canada about their hypothetical economic preferences in relation to preservation of environmental amenities in the region. Each respondent was asked the following four questions for particular amenities such as fishing, boating, swimming, and preservation of scenic vistas:

1. What would you pay (to enjoy swimming, fishing, etc.)?
2. What would I have to pay you to give it up?
3. If you were making a community decision, how would you reallocate the budget for recreation on the Fraser River?
4. If you were a judge and someone had been arbitrarily excluded from the activity listed below for one year, what dollar damages would you award?

In addition to these alternative ways of asking about the value of using the resources, people were also asked what they would be willing to accept in compensation for permanent loss of the recreational resources ("additional option demand value": an "option demand" is the amount people are willing to pay to retain potential access to a resource). Table 5.3 indicates significant differences in responses to all questions except 3 and 4, and generally higher values assigned to permanent loss of the resource. Fraser River residents thus expect a greater amount in compensation for loss of the resource than they would be willing to pay to gain it. Which of these values to use as a shadow price is unclear. Whether any estimate will reflect ultimate consumer behavior is unknown and will depend in part on the extent to which the people surveyed represent the relevant consumers.

A somewhat different approach to hypothetical valuation is called "threshold analysis" (Hyman 1981). In a method described by Krutilla and Fisher (1975), decision makers (or multiple publics) are asked how much the benefits of development would have to be in order to leave them indifferent about preserving the resource intact. They are further asked whether they think the actual amenity benefits exceed that value. In the example given, decision makers felt benefits of preserving Hells Canyon, Idaho, from flooding for a hydroelectric project would have to be at least $40,000–$150,000 in the first year. They further decided that the present value of the canyon in its

Table 5.3. A Comparison of Four Direct Questioning Methods

	Use Value		Additional Option Demand Value	
Question Type	Mean	95% Confidence Interval	Mean	95% Confidence Interval
1. Annual amount you would pay	$ 1,099	± 993	$ 2,894	±1,698
2. Annual amount you would accept	20,961	±6,028	27,079	±4,744
3. Community decision making, annual budget	11,833	±4,793	14,833	±3,858
4. Judicial award per annum	11,683	±5,559	10,519	±3,013

Source: From Tables I and IV of P. A. Meyer (1976). Used with permission of Parks Canada.

intact state exceeded these amounts. A major problem with this approach is that it is based purely on conjecture regarding the true value of the intact resource. People generally are not aware of the full scope of nature's goods and services (e.g., decomposition processes, radiation balance, gas exchange), nor are they able to quantify the worth of these items in dollar terms in a meaningful way in the abstract.

LIMITATIONS OF ECONOMIC EVALUATION METHODS

The discussion of shadow pricing has illustrated some of the difficulties in evaluating nonmarket goods in economic terms. Some of the general problems encountered include the following: (1) Different methods (e.g., damage costs vs. repair costs) result in different economic estimates for the same resource; the degree of incompleteness of each estimate is usually unknown, making it difficult to choose between them. (2) Because nature's goods and services are free to begin with, and their ecological value not fully appreciated, many shadow-pricing methods underestimate the true value of the resource to people (resulting in higher consumer surplus, see Figure 5.2). (3) Because people value money differently, a price fails to reflect the differing weights attached to the evaluation unit (money). (4) Because people value natural resources differently but prices reflect aggregated or average social utilities, the differing values attached to a resource by different publics are not separately indicated. (5) Some of nature's goods and services are not readily evaluated in economic terms by existing methods either because they are too complex and incompletely known (e.g., global climate) or because they are not considered exchangeable for money (e.g., human life); these items are often excluded from economic analyses, making such analyses incomplete.

Point 5 deserves further comment. It addresses the issue of the Value III items of Farnworth et al. (1981). To be sure, economists determined to perform "complete" cost-benefit analyses have attempted to evaluate items that others would argue cannot or should not be evaluated. For example, a variety of approaches to evaluating human life have been developed. Some argue that this is appropriate since society implicitly places values on human life by the risks imposed on individuals for the sake of social welfare (e.g., operators of and residents near a coal-fired or nuclear power plant). Others argue that because most individuals are not willing to exchange their life for money, attempts at economic evaluation are based on a false assumption, namely, that this "good" is exchangeable for money.

Valuation of Human Life

Approaches to evaluating the worth of human life are reviewed by Freeman (1979), Hyman (1981), Westman and Conn (1976), and Zeckhauser (1975). One of the more controversial techniques involves equating the value of human life to the discounted value of expected future wages lost when the person dies (e.g., Ridker 1967). Such an approach of course fails to value retired people or housepersons and obviously does not account for the value of personal feelings. Hypothetical valuation approaches ask people what they would be willing to pay, or accept in compensation, for a change in the risk of death (e.g., how much they would pay to disinfect drinking water that may be contaminated with lethal pathogens, or accept in compensation for living next to a nuclear power plant). This approach has the advantage that questions are similar to those people face every day in the context of taking risks as part of living. However, the approach has all the problems of willingness-to-pay methods. The price may not be an accurate reflection of relative utility, especially since many people tend to assume "it can't happen to me" and are willing to take risks that undervalue their true sense of self-worth. Their attitude may change abruptly when the risk becomes a certainty, and they contract, for example, a fatal waterborne disease. Prices for willingness-to-accept compensation will typically be higher than those from willingness-to-pay questions. Bids will vary with a person's age, income, perceived benefits from the risky activity, understanding of the true level of risk, and other factors (Hyman 1981; see also the discussion on risk analysis in Chapter 3).

Other techniques include the "related expenditures" approach, in which individual and social expenditures on safety precautions are used to evaluate human life. Society spends quite different amounts to lower risk of death from different activities. For example, Table 5.4 illustrates the implicit valuation of human life in three industries in the United Kingdom, based on rate of occupationally induced death and amounts spent on safety precautions in each industry. The difference in the shadow prices of human life in Table 5.4 implies the presence of additional consumer surplus over that which would exist at prices equal to marginal utility, at least in the case of agriculture and steel handling; indeed, possibly all three estimates are "underpriced." The whole idea of estimating the "marginal utility" of a human life seems intractable, even preposterous, when confronted directly. The shadow prices shown in Table 5.4 have resulted not from any explicit calculation of marginal utility but from a series of decisions probably based on considerations other than a comparative risk of death or any other explicit evaluation approach to human life.

The problem of evaluating an "intangible" or Value III item like human life is not different in kind from developing shadow prices for other nonmarket goods. Because of the highly variable range in values (from zero to infinity) assigned to human life by individuals, however, evaluating such an item in monetary units emphasizes in an extreme way the shortcomings of economic evaluation methods and their assumptions (willingness to accept monetary compensation, aggregation of individual relative utilities to determine a single price, use of marginal utility to set price, with consequent generation of unvalued "consumer surplus," etc.). Probably the most critical assumption underlying all economic evaluation methods is that people are willing to exchange the utility or worth of the goods and services of nature for money.

Natural Resources That Are Not For Sale

A variety of nonmarket values can be enumerated for natural objects and services, including their value for recreation, aesthetic enjoyment, scientific study, education, stabilization of ecosystems, and storage of genetic stock (see, e.g., Ehrenfeld 1976, Ehrlich and Ehrlich 1981, Myers 1979). One can attempt an economic equivalent for any or all of these values using the shadow-pricing techniques discussed earlier. The only declaration that would firmly characterize these as nonmonetizable values (Value III) is that we are not willing to exchange these assets for money, because we are not willing to sell them at any cost (value is infinite), they have no known value (present value is zero), or because they are not ours to sell.

The belief or determination that there are natural resources not appropriately exchangeable in the marketplace is a particular human value, held by some, but not by most members of current western society. The ecologist C. S. Elton (1958) characterizes this belief as fundamentally religious, although it can be viewed simply as an ethical value. Ehrenfeld (1976) quotes Elton as saying that one reason why natural diversity should be conserved is

Table 5.4.   Comparative Risk of Death, Expenditures on Safety, and Implicit Monetary Values Placed on Human Life in Three Industries in the United Kingdom.

Sector	Annual Risk of Death per 1000 Workers	Average Outlay (£/Worker)	Implicit Valuation of Human Life, £
Agriculture	0.197[a]	3	15,000
Steel handling	0.216[b]	50	230,000
Pharmaceuticals	0.020[b]	210	10,500,000

Source: Modified from Sinclair (1972, p. 50).

[a] Sources: Ministry of Agriculture, Fisheries and Food, 1966–68 average.

[b] Annual Report of the Chief Inspector of Factories, 1969.

"because it is a right relation between man and living things" (Elton 1958, p. 143). Ehrenfeld (1978) also notes the ethical argument that species and communities should be conserved "because they exist and because this existence is itself but the present expression of a continuing historical process of immense antiquity and majesty. Long-standing existence in nature is deemed to carry with it an unimpeachable right to continued existence." Aesthetic arguments in favor of species preservation have also been made (see, e.g., Ehrenfeld 1976). White (1967) has traced some of the changing views toward nature in the Hebrew and Christian traditions, though even today there is not unanimity of view within a culture (see Watanabe 1974).

It is common for hunter-gatherer societies to have viewed nature as something that cannot be owned by humans and that could only be harvested by trade with the guardian spirits through sacrifices or other forms of appeasement. Thus Chief Seattle of the Dwamish tribe, on declining to sign a treaty selling tribal land to the United States in 1855, is reported to have said:

> Our land is more valuable than your money. It will last forever. It will not even perish by the flames of fire. As long as the sun shines and the waters flow, this land will be here to give life to man and animals. Therefore we cannot sell this land. It was put here for us by the Great Spirit and we cannot sell it because it does not belong to us.

By contrast, when we undertake to market emission rights (Chapter 2), we are selling clean air—a formerly public resource available free of charge—to industry for a price. We thus assume that the public is willing to trade clean air for money (and indirectly, other goods and services). This is also an anthropocentric act, since other species using the air are neither consulted nor compensated. Such an exchange could not occur if predominant public values declared clean air a nonmarketable resource.

Incomplete Knowledge of Ecosystems

Even if we are willing to express natural values in monetary terms, another major problem with placing a monetary value on natural resources is that we may not know enough about the value of the resource to us to estimate its true worth. For example, about 80,000 species of edible plants are known, of which only about 150 have ever been cultivated on a large scale. Of these, only about 20 species (and their varieties) produce about 90% of the world's food (Webb 1980). Much of the information on potential economic uses of plants (including uses for fiber, pharmaceuticals, and other purposes) is not widely known, has been forgotten, or was never known. Given this situation it is inevitable that people will assign a shadow price to noncultivated plants that undervalues their potential worth to present or future generations.

In addition some ecological processes are so complex that their valuation in economic terms inevitably involves oversimplification, sometimes to the point that any use of such figures in decision making would be dangerous. An important case in point is evaluation of the economic worth of global climatic regulation. Much interest in this topic has arisen in the past decade as a result of evidence that net increases in carbon dioxide and particulate release, and decreases in stratospheric ozone, may induce changes in global precipitation, temperature, and radiation quantity and quality. D'Arge (1974) and others have estimated economic costs of such climatic changes on human health, agriculture and forestry, wages, energy demand, and other items. Estimates are derived by assuming hypothetical changes in climate and using some surrogate of this change (e.g., changes in food yield) as input to a global computer simulation of population, agriculture, and other economic sectors (e.g., Liverman 1983; see also Barney 1980, Meadows et al. 1982).

Even given that climatic changes are not directly modeled, the estimates of impact on economic sectors resulting from these models can be seriously in error. In a comparison of actual data with predictions on regional food production from one such global model, Liverman (1983) found model estimates in error by 70–100% and more. Such model estimates have value in learning about interactions among component variables. Estimates will inevitably be subject to large variations, however, induced both by highly variable elements in the system modeled and by variables not included in the model. These factors inevitably limit the reliability of economic estimates and their usefulness in economic evaluation.

Even at the more modest scale of a forested ecosystem, the task of evaluating the structure and function of the ecosystem implies that we know fully what the ecosystem is doing for us and what that is worth to us. As noted earlier, the worth of ecosystem structure is generally better appreciated than that of ecosystem functions. To evaluate the worth of nitrogen fixation, nutrient and soil retention, gas exchange, radiation balance, hydrological balance, pollution absorption, biomass growth, decomposition, and other ecological processes for any given segment of landscape taxes present ecological knowledge beyond its bounds. To take only one example, the rate of gaseous exchange of pollutants (absorption and release) is only presently known for a few dozen species and soil types (see, e.g., Inman et al. 1971, Smith 1981, Westman and Conn 1976) under a few environmental conditions. The ecological data to evaluate this function fully are simply not available. Even ecosystem structure is incompletely known. To evaluate the worth of the insect fauna, or soil fungi, when many of these species have never even been described taxonomically, taxes human knowledge beyond current limits.

The reader may be tempted to conclude from this chapter that economic evaluation methods are so fraught with difficulties and questionable assumptions that the entire set of methods discussed should be avoided. It is well to recall that a similar list of problems associated with noneconomic quantitative evaluation techniques was elucidated in Chapter 4. The need for evaluation itself, however, is inescapable. In the absence of explicit evaluations, implicit evaluations will be performed by decision makers. Given the imperfections of explicit evaluation techniques, the analyst is left to choose methods whose assumptions and data requirements seem most appropriate to the task at hand. Beyond this it is important to remind decision makers of the assumptions and limitations of the evaluation methods being used, as part of the process of presenting the evaluation. The use of sensitivity analysis, and of several different evaluation methods simultaneously, can help to reveal the assumptions and limitations more graphically.

An elaborate cost-benefit analysis was prepared for the study of where to site a third London airport (Roskill Commission 1971, Adams 1971). Decision makers, however, chose a different option than that suggested by CBA. This decision reflected not necessarily a rejection of the rational mode of decision making (Chapter 3) but perhaps an awareness that the economic evaluation method did not reflect the full spectrum of human concerns and that the relative weights placed on different issues were not precisely reflected by their monetary values. Cost-benefit analysis did play a useful role as one informational input, since its assumptions and limitations were at least partially understood. The ever-present danger with any evaluation is that decision makers will accept numbers as an objective rationale for a decision, when such numbers merely reflect a quantification of particular human values.

There is much to the field of environmental or welfare economics that has not been discussed in this chapter. For more detailed treatments of environmental and welfare economics generally, see Freeman (1979), Hjalte et al. (1977), Hufschmidt and Hyman (1982), Hyman (1981), Krutilla and Fisher (1975), Möler and Wyzga (1976), Pearce (1976), Seneca and Taussig (1979), Sinden and Worrell (1979), and Westman and Conn (1976). For discussions of environmental cost-benefit analysis in particular, see Ahmad (1981), American Institute of Certified Public Accountants (1977), Bohm and Henry

(1979), Dasgupta and Pearce (1972), Karam and Morgan (1976), and Mishan (1976). For critiques of environmental cost-benefit analysis see, for example, Anderson (1974), Ghiselin (1978), McAllister (1980), Muller (1974), and Price (1977). For examples of applications of economic evaluation techniques to air pollution, see Graves and Fishelson (1979–80) and Ridker (1967); to recreation, see Knetsch and Davis (1966) and Meyer (1976); to wetlands, see Gosselink et al. (1973), Greeson et al. (1979), Larson (1974, 1976), Lonard et al. (1981), U.S. Army Corps of Engineers (1979), and Wharton (1970).

REFERENCES

Adams, J. G. U. (1971). London's third airport. *Geographical J.* **137**:468–504.

Ahmad, Y. J. (1981). Cost-benefit analysis and environmental decision-making: a note on methodology. *Industry Environ.* **2**:2.

American Institute of Certified Public Accountants (1977). Environmental cost-benefit studies. New York.

Anderson, S. O. (1974). Concepts and methods of cost-benefit analysis. In T. G. Dickert and K. R. Domeny, eds. *Environmental Impact Assessment: Guidelines and Commentary.* Univ. Ext., Univ. California, Berkeley, Calif., pp. 89–105.

Ateshian, K. H. (1976). Comparative costs of erosion and sedimentation control measures. In Proc. 3rd Fed. Inter-Agency Sedimentation Conf., Sec. 2, Water Resources Council, Denver, Colo., pp. 13–23.

Barney, G. O., ed. (1980). The Global 2000 Report. U.S. Council on Environmental Quality, Washington, D.C.

Bohm, P., and Henry, C. (1979). Cost-benefit analysis and environmental effects. *Ambio* **8**:18–24.

Clawson, M. (1959). Methods of measuring the demand for and value of outdoor recreation. Reprint No. 10. Resources for the Future, Washington, D.C.

D'Arge, R. C. (1974). Economic impact of climatic change: introduction and overview. 3rd Conf. Climatic Impact Assessment Program. Rep. DOT-TSC-OST-74-15. Dept Transportation, Washington, D.C.

Dasgupta, A. K., and Pearce, D. W. (1972). *Cost-Benefit Analysis: Theory and Practice.* Harper & Row, New York.

Ehrenfeld, D. W. (1976). The conservation of non-resources. *Amer. Sci.* **64**:648–656.

Ehrenfeld, D. W. (1978). *The Arrogance of Humanism.* Oxford Univ. Press, New York.

Ehrlich, P., and Ehrlich, A. (1981). *Extinction: The Causes and Consequences of the Disappearance of Species.* Random House, New York.

Everett, R. D. (1977). A method of investigating the importance of wildlife to countryside visitors. *Environ. Conserv.* **4**:227–231.

Everett, R. D. (1979). The monetary value of the recreational benefits of wildlife. *J. Environ. Manage.* **8**:203–213.

Farnworth, E. G., Tidrick, T. H., Jordan, C. F., and Smathers, W. M., Jr. (1981).

The value of natural ecosystems: an economic and ecological framework. *Environ. Conserv.* **8**:275–282.

Freeman, A., III. (1979). *The Benefits of Environmental Improvement.* Johns Hopkins Univ. Press, Baltimore, Md.

Ghiselin, J. (1978). Perils of the orderly mind: cost-benefit analysis and other logical pitfalls. *Environ. Manage.* **2**:295–300.

Gosselink, J. G., Odum, E. P., and Pope, R. M. (1973). The value of the tidal marsh. Center for Wetlands Research, Louisiana State Univ., Baton Rouge, La.

Graves, P. E., and Fishelson, G. (1979–80). Benefits of pollution control: the SO_2 case. *J. Environ. Systs.* **9**:231–258.

Greeson, P. E., Clark, J. R., and Clark, J. E., eds. (1979). *Wetland Functions and Values: The State of Our Understanding.* Tech. Publ. Series, Amer. Water Resources Assn., Minneapolis, Minn.

Hicks, J. R. (1939). *Value and Capital.* Clarendon, Oxford.

Hjalte, K., Lidgren, K., and Stahl, I. (1977). *Environmental Policy and Welfare Economics.* Transl. C. Wells. Cambridge Univ. Press, Cambridge.

Hufschmidt, M., and Hyman, E., eds. (1982). *Economic Approaches to Natural Resource and Environmental Quality Analysis.* Tycooly, Wicklow, Ireland.

Hyman, E. L. (1981). The valuation of extramarket benefits and costs in environmental impact assessment. *Environ. Impact Assessment Rev.* **2**:227–258.

Inman, R. E., Ingersoll, R. B., and Levy, E. A. (1971). Soil: a natural sink for carbon monoxide. *Science* **172**:1229–1231.

Kaldor, N. (1939). Welfare propositions of economics and interpersonal comparisons of utility. *Econ. J.* **49**:549–552.

Karam, R. A., and Morgan, K. Z., eds. (1976). *Energy and the Environment: Cost-Benefit Analysis.* Energy: An Intl. J., Suppl. 1. Pergamon, New York.

Knetsch, J., and Davis, R. (1966). Comparison of methods for recreation evaluation. In A. V. Kneese, and S. Smith, eds. *Water Research.* Johns Hopkins Univ. Press, Baltimore, Md.

Krutilla, J. V., and Fischer, A. C. (1975). *The Economics of Natural Environments: Studies in the Valuation of Community and Amenity Resources.* Johns Hopkins Univ. Press, Baltimore, Md.

Larson, J. S., ed. (1974). *A Guide to Important Characteristics and Values of Freshwater Wetlands in the Northeast.* Pub. 31, Water Resources Research Center, Univ. Massachusetts, Amherst, Mass.

Larson J. S., ed. (1976). *Models for Assessment of Freshwater Wetlands.* Pub. 32, Water Resources Research Center, Univ. Massachusetts, Amherst, Mass.

Lind, R. (1973). Spatial equilibrium, the theory of rents and the management of benefits from public programs. *Quart. J. Econ.* **87**:188–207.

Liverman, D. (1983). The use of a simulation model in assessing the impacts of climate on the world food system. Ph.D. dissertation, Univ. California, Los Angeles, Calif.

Lonard, R. I., Clairain, E. J., Jr., Huffman, R. T., Hardy, J. W., Brown, L. D., Ballard, P. E., and Watts, J. W. (1981). Analysis of methodologies used for the assessment of wetlands values. Tech. Rep., Water Resources Council, Washington, D.C.

McAllister, D. M. (1980). *Evaluation in Environmental Planning. Assessing Environmental, Social, Economic, and Political Trade-offs*. The MIT Press, Cambridge, Mass.

Meadows, D. H., Richardson, W., and Bruckman, G. (1982). *Groping in the Dark: A History of the First Decade of Global Modeling*. Wiley, New York.

Meyer, P. A. (1976). Obtaining dollar values for public recreation and preservation. Tech. Rep. Ser. PAC/T-75-6. Environment Canada, Vancouver.

Mishan, E. J. (1976). *Cost-Benefit Analysis*. 2nd ed. Praeger, New York.

Möler, K. G., and Wyzga, R. E. (1976). Economic measurement of environmental damage. OECD, Paris.

Muller, F. G. (1974). Benefit-cost analysis: a questionable part of environmental decisioning. *J. Environ. Systs.* **4**:299–307.

Myers, N. (1977). Discounting and depletion: the case of tropical moist forests. *Futures* **9**:502–509.

Myers, N. (1979). *The Sinking Ark*. Pergamon, New York.

Payne, B. R., and De Graaf, R. M. (1975). Economic values and recreational trends associated with human enjoyment of nongame birds. In Proc. Symp. Management of Forest and Range Habitats for Nongame Birds. Gen. Tech. Rep. WO-1. U.S. Forest Service. Washington, D.C., pp. 6–10.

Pearce, D. (1976). *Environmental Economics*. Longman, New York.

Pearse, P. H. (1968). A new approach to the evaluation of non-priced recreational resources. *Land Econ.* **44**:87–99.

Pigou, A. C. (1932). *The Economics of Welfare*. Macmillan, London.

Price, C. (1977). Cost-benefit analysis, national parks and the pursuit of geographically segregated objectives. *J. Environ. Manage.* **5**:87–97.

Rice, R. M., Corbett, E. S., and Bailey, R. G. (1969). Soil slips related to vegetation, topography, and soil in southern California. *Water Resources Research* **5**:647–659.

Rice, R. M., and Foggin, G. T., III. (1971). Effects of high intensity storms on soil slippage on mountainous watersheds in southern California. *Water Resources Research* **7**:1485–1496.

Ridker, R. (1967). *Economic Costs of Air Pollution: Studies in Measurement*. Praeger, New York.

Roskill Commission (1971). Report of the Commission on the Third London Airport. Her Majesty's Stationery Office, London.

Seneca, J. J., and Taussig, M. K. (1979). *Environmental Economics*. 2nd ed. Prentice-Hall, Englewood Cliffs, N.J.

Sinclair, T. C. (1972). A cost-effectiveness approach to industrial safety. Commission on Safety and Health at Work. Res. Pap. Her Majesty's Stationery Office, London.

Sinden, J. A., and Worrell, A. (1979). *Unpriced Values—Decisions without Market Price*. Wiley-Interscience, New York.

Smith, R. J., and Kavanagh, N. J. (1969). The measurement of benefits of trout fishing: preliminary results of a study at Grafham Water, Great Ouse Water Authority, Huntingdonshire. *J. Leisure Res.* **1**:316–332.

Smith, W. H. (1981). *Air Pollution and Forests. Interactions between Air Contaminants and Forest Ecosystems.* Springer-Verlag, New York.

Sorenson, P. E. (1974). Economic evaluation of environmental damage resulting from the Santa Barbara oil spill. Unpublished rep. Univ. California, Santa Barbara, Calif.

Thibodeau, F. R., and Ostro, B. D. (1981). An economic analysis of wetland protection. *J. Environ. Manage.* **12**:19–30.

U.S. Army Corps of Engineers (1978). Report on floods of February and March 1978 in southern California. Los Angeles district, Los Angeles, Calif.

U.S. Army Corps of Engineers (1979). *Wetland Values. Concepts and Methods for Wetland Evaluation.* Res. Rep. 79-R1. Inst. for Water Resources, Washington, D.C.

Usher, M. B. (1973). *Biological Management and Conservation: Ecological Theory, Application and Planning.* Chapman and Hall, London.

Usher, M. B. (1977). Coastline management: some general comments on management plans and visitor surveys. In R. S. K. Barnes, ed. *The Coastline.* Wiley, New York, pp. 291–311.

Watanabe, M. (1974). The conception of nature in Japanese culture. *Science* **183**:279–282.

Webb, L. J. (1980). Natural systems—their scientific value to man. *Qld. Conserv. Council Bull.* (October):3–5.

Westman, W. E. (1976). Doomsday expectations. *Science* **193**:720.

Westman, W. E. (1977). How much are nature's services worth? *Science* **197**:960–964.

Westman, W. E., and Conn, W. D. (1976). *Quantifying Benefits of Pollution Control: Benefits of Controlling Air and Water Pollution from Energy Production and Use.* Energy Resources Conservation and Development Commission, Sacramento, Calif.

Wharton, C. H. (1970). The southern river swamp—a multiple-use environment. Bureau of Business & Economic Res., School Business Admin., Georgia State Univ., Atlanta, Ga.

White, L., Jr. (1967). The historical roots of our ecologic crisis. *Science* **155**:1203–1207.

Zeckhauser, R. (1975). Procedures for valuing lives. *Public Policy* **23**:419–464.

PART FOUR

PREDICTING IMPACTS: THE PHYSICAL ENVIRONMENT

6

LAND

A landscape segment is typically composed of patches that are discontinuous in some physical or biological sense. Topographic variation, for example, through its effect on air and water flow and sun angle, can induce differences in species composition and soil development at a continuum of scales of patchiness (Table 6.1). These landscape patches may be observed by using soil, vegetation, landform, or other attributes as indicators. Land planners and impact analysts have often used the distinguishing characteristics of a landscape to predict the effect of actions on landscape patches and ultimately, on the entire landscape.

Observation of a landscape at one point in time is much like observing a single frame of a movie. Processes of change become apparent only by observing the landscape periodically or by noting clues to the past that may be present (e.g., floodplain banks, tree growth rings, geological strata). Processes of landscape change occur in a continuum from high frequency or *continual* (e.g., surface soil, coastal erosion) to low frequency or *episodic* (e.g., flooding, earthquake, volcanic eruption, landslide, subsidence).

In this chapter we examine static attributes of landscapes which may be used both as indicators of landscape processes (dynamics) and of likely response to human action. For example, a soil type may indicate both vulnerability to the erosion process and suitability for agriculture development. We also consider how such landscape characteristic can be mapped, using field and remotely sensed data, computerized data storage and retrieval, and graphical presentation. Finally, we examine how the predicted landscape alterations from development proposals can be evaluated using economic, ecological, or aesthetic criteria.

LANDSCAPE CHARACTERISTICS AS INDICATORS OF LAND SUITABILITY AND VULNERABILITY

Vegetation, soil, landform, or combinations of these have been used as indicators of a larger suite of land characteristics in the different land evaluation systems in use around the world (see, e.g., Stewart 1968, McRae and Burnham 1981, McEntyre 1978, Whyte 1976). These initial pieces of land-

Table 6.1.   Terms Associated with Increasing Spatial Scales of Analysis of Biological Components of the Landscape

Spatial Unit of Habitat	Order of Magnitude of Typical Spatial Scale	Unit of Biological Assemblage	Distinguishing Characteristics of Biological Unit
Microhabitat	1–10 m^2	Microcommunity	Distinct species composition
Ecosystem	10^2–10^5 m^2	Community	Significant interaction among component species; self-contained flow of energy and materials
Region to subcontinent	10^6+ m^2	Biome	Distinct vegetative physiognomy (external form, e.g., height of canopy, number of strata of vegetation)

scape information may be used in predicting a variety of landscape responses. Thus soil attributes may indicate vulnerability of the landscape to impact from septic tank leachate, using soil porosity and texture as indicators. Soil porosity and texture may also serve to indicate potential of the soil for growing a crop. When considering the development potential of the land, it is useful to distinguish between suitability (immediate potential of the current state of the land), capability (full potential after development) and feasibility (likely potential, considering socioeconomic and political constraints on development) (Belknap and Furtado 1967). Thus a patch of marshland is currently suitable as a wildlife habitat, is capable of being developed for a marina, and may only be feasible for modest development as a recreation area for fishing and bird watching. The vulnerability of a landscape to impact will depend on the nature of the disturbance, the initial resistance of its ecosystems to change, and the rate and manner of recovery of the ecosystems following disturbance (see Chapter 12).

We consider here how landscape attributes can serve as indicators of response to purposeful development or as indicators of vulnerability to inadvertent impact.

Soils

Soils have been widely used as indicators of agricultural capability. Soils are first classified into types based on physical and chemical features. The areal extent of each soil type is mapped (Figure 6.1) using field sampling and some clues (surface color, topography, vegetation) from aerial photographs. The

Figure 6.1. Section of a soil survey map overlain upon an aerial photograph, showing soil types in an area near Santa Barbara, California. From Sheet 88 of Shipman (1972).

correlation of soil type attributes with potential crop growth is then determined from local agronomic experience, and the soil types are classified into one of several agricultural capability classes. By concentrating on soil attributes relevant to support of built structures (e.g., compaction, drainage, frost heave, expansion potential), soils have also been classified for construction capability in more recent U.S. soil surveys (see, e.g., Golden et al. 1979, Table 9-5). Table 6.2 shows eight capability classes used by the U.S. Soil Conservation Service.

Figure 6.1 shows soil types for particular portions of a Californian landscape as mapped by the U.S. Soil Conservation Service. Each soil type is classified into a capability classification in the accompanying soil survey document. For example, in Figure 6.1, soil type TdF in the lower middle of the photograph has been classified as capability unit VIIe. The accompanying soil survey document informs us that such soils are found on uplands and terrace escarpments, that they are somewhat excessively to moderately

Table 6.2. Soil Capability Classes for Agricultural Use According to the U.S. Soil Conservation Service

Class I: Soils that have few limitations that restrict their use. Suitable for cultivation.

    Unit I-4: Deep, well-drained, nearly level, upland soils.

    Unit I-6: Nearly level, well-drained, silty soils on floodplains and low terraces.

Class II: Soils that have some limitations that reduce the choice of plants or require moderate conservation practices. Suitable for cultivation.

    Subclass IIe:[a] Nearly level to gently sloping soils, subject to erosion if tilled.

    Subclass IIw:[b] Moderately wet soils.

Class III: Soils that have severe limitations that reduce the choice of plants, require special conservation practices, or both. Suitable for cultivation.

    Subclass IIIw: Wet soils that require artificial drainage if tilled.

    Subclass IIIs:[c] Soils that are severely limited by stoniness.

Class IV: Soils that have very severe limitations that restrict the choice of plants, require very careful management, or both. Marginal soils.

    Subclass IVe: Soils severely limited by risk of erosion if tilled.

    Subclass IVw: Soils severely limited for use as cropland because of excess water.

Class V: Soils that have little or no erosion hazard but have other limitations that are impractical to remove and that limit their use largely to pasture, woodland, or wildlife food and cover. Level but wet.

    Subclass Vw: Soils limited in use to grazing or woodland because of poor internal drainage.

Class VI: Soils that have severe limitations that make them generally unsuitable for cultivation and limited by steepness, drought, or moisture. Suitable for grazing and forestry uses.

Class VII: Soils with very severe limitations that restrict their use to pasture or trees.

    Subclass VIIe: Hilly, steep, erosive.

    Subclass VIIs: Stony, rolling, steep, shallow to bedrock.

Class VIII: Soils with no agricultural use, mountains.

---

[a] The letter "e" indicates the soil is erodible.

[b] The letter "w" indicates wet.

[c] The letter "s" indicates extreme stoniness.

well-drained sandy loams to silty clay loams, of 15–75% slopes, with 15–150 cm depth to bedrock, low to high fertility, moderate to very slow permeability, 3–20 cm of available water capacity, high to very high erosion potential and agricultural capability limited to controlled grazing.

The Canada Land Inventory uses a similar classification on a seven-point scale, primarily by combining classes VI and VII of the U.S. system. Soil surveys are available for many parts of the world primarily for regions with agricultural potential. Useful discussions of the land use capabilities and management problems associated with different soil types may be found in Foth and Schafer (1980) and Steila (1976).

Vegetation

Vegetation maps may record the current nature of the vegetation, indicating such features as dominant species, height of canopy, and extent of canopy closure as well as vegetation at various stages of succession and areas where the vegetation has been cleared. Alternatively, maps may present the *potential* vegetation of the area, that is, the climax vegetation likely to be present in the absence of human interference, given the climate, soil, and topography of the region (Figure 6.2). Potential-vegetation maps (see, e.g., Küchler 1964, USGS 1970) are necessarily more speculative and less accurate but do provide information on vegetation in relation to habitat which are useful as indicators of land capability. The U.S. Forest Service is in the process of developing a National Vegetation Classification system (see, e.g., Paysen et al. 1981, Driscoll et al. 1982), modeled after the proposed international system of vegetation classification (UNESCO 1973).

As with soils, vegetation may serve as an indicator for a wide range of landscape conditions and capabilities. The U.S. Forest Service, for example, uses native vegetation types, or vegetation and soils, as indicators of the potential for growth of commercial timber in plantations or by management (selective cutting) of uneven-aged natural stands (Figure 6.3).

A "site index" system for predicting forest growth capability based on the height of dominant and codominant trees of a specific age on a site is widely used by U.S. foresters and the Soil Conservation Service. Typically the site index is the height of trees at 50 years for shorter-lived species east of the Great Plains, 100 years for longer-lived species more common in the west (see Carmean 1975). For a description of other forest site quality indexes in use worldwide, see McRae and Burnham (1981, Ch. 8).

Ecosystems

A third tradition in land classification has been to use the combined information from various ecosystem components, such as soils, vegetation, land-

Figure 6.2. Differences between maps of existing and of potential vegetation. Shaded portions show areas of coastal sage scrub ("coastal sagebrush") mapped by the California Forest and Range Experiment Station (Wieslander 1945) from field observations in the 1930s and 1940s in southern California. The outlined areas show the regions potentially supporting coastal sage scrub as the climax vegetation type, as predicted by Küchler (1977).

form, and climate, to map ecological units. The rationale for this approach is that the ecological unit derives from a larger information base and should therefore be a more successful indicator of a range of land capabilities. A difficulty with the approach is that the natural boundaries for soil, vegetation, landform, and climatic differences do not always coincide. Some criteria must be used to establish boundaries. Since this involves judgment by the mapmaker, replication by other mapmakers is more difficult.

In Canada ecological land classification has been performed by a variety of federal provincial and university groups since the 1960s (Rubec 1979, Wiken 1980). The system uses a variety of biological and physical criteria for classification of land into units of increasing size (Table 6.3). Within a given climatic region, landform (including substrate) is often the major influence on vegetation and soil development. As a result at the level of an ecoregion it is possible to generalize about the relationship between soils, vegetation

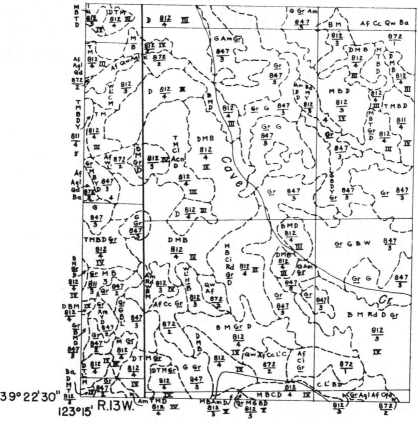

Figure 6.3. A section from a California vegetation-soil map, identifying native vegetation, soil types, and timber growth potential, produced by the U.S. Forest Service and state agencies (Cushman et al. 1948). Three sets of classification symbols are used. Letter groups, like Gr, AvCc, or RDTMCt symbolize names of dominant plant species and other land status elements. Numbers shown as fractions, e.g., 812/4, designate the soil series and average soil depth. Single numbers like 5 or IV rate the capability of the area for growing timber, based on vegetation and soil characteristics of the area.

(defined physiognomically), and topography. The series of predictable soil changes with landform, given homogeneous climate and parent material, is called a *catena* and the corresponding vegetation, a *toposequence*. The catena concept has been used both in the Canadian system (e.g., Rowe and Sheard 1981) and the Australian land survey system (Christian and Stewart 1968). A catena and toposequence for the low subarctic ecoregion of the Lockhart River area of Canada is shown in Figure 6.4, and a map of ecoregions and districts for the Lockhart River area in Figure 6.5. At present, ecological land classification in Canada has not been linked with guides to land capability, though the potential is there.

Table 6.3. Levels of Ecological Generalization Proposed for Use in Ecological Land Classification by the Canada Committee on Ecological (Bio-physical) Land Classification

Level of Generalization Common map scale[a]	Common Benchmarks for Recognition					
	Geomorphology	Soils[c]	Vegetation[d]	Climate	Water[e]	Fauna
Ecoregion 1:3,000,000 to 1:1,000,000	Regional landforms or assemblages of regional landforms	Great groups or associations thereof	Plant regions or assemblages of plant regions	Meso or small scale macro	Water regime	High species diversity; may correspond either to a widely distributed species (e.g., deer mouse), or to the habitat of individuals within a species
Ecodistrict 1:500,000 to 1:125,000	Regional landform or assemblages thereof	Subgroups or associations thereof	Plant districts or assemblages of plant districts	Meso or large scale micro	Drainage pattern; water quality	
Ecosection 1:250,000 to 1:50,000	Assemblages of local landforms or a local landform	Family or associations thereof	Plant associations or a plant association	Large scale micro to small scale micro	River reaches, lakes and shoreland	Less diverse species complement; habitat requirements of typical species more restricted (e.g., beaver, otters); may coincide with specialized areas of animal total habitat (e.g., wintering area, calving grounds)
Ecosite[b] 1:50,000 to 1:10,000	A local landform or portion thereof	Soil series or an association of series	Plant association or seral stage	Small scale micro	Subdivision of above	

Ecoelement 1 : 10,000 to 1 : 2,500	Portion of or a local land-form	Phases of soil series or a soil series	Parts of a plant as-soc. or subassociation	Small scale micro	Sections of small streams	Low species diversity; habitat of smaller mammals, reptiles, and amphibians etc.; specialized areas of some fauna's habitat requirements (e.g., denning areas, local wintering deer yards)

Source: Reprinted, with permission, from Wiken (1980).

Note: Definitions for the levels of generalization.

Ecoprovince—an area of the earth's surface characterized by major structural or surface forms, faunal realms, vegetation, hydrological, soil, and climatic zones.

Ecoregion—a part of an ecoprovince characterized by distinctive ecological responses to climate as expressed by vegetation, soils, water, fauna, etc.

Ecodistrict—a part of an ecoregion characterized by a distinctive pattern of relief, geology, geomorphology, vegetation, soils, water, and fauna.

Ecosection—a part of an ecodistrict throughout which there is a recurring pattern of terrain, soils, vegetation, waterbodies, and fauna.

Ecosite—a part of an ecosection having a relatively uniform parent material, soil and hydrology, and a chronosequence of vegetation.

Ecoelement—a part of an ecosite displaying uniform soil, topographical, vegetative, and hydrological characteristics.

[a] Map scales should not be taken too restrictively, as they will vary with the environment setting and objectives of the survey.
[b] This level is frequently subdivided into phases according to the stage of plant succession.
[c] Canadian System of soil classification, Agriculture Canada, 1979.
[d] These vegetative groupings are only suggested ones; agreement on a common system is yet to be achieved.
[e] See D. Welch, 1978. *Land/Water Classification.* ELC Series No. 5, Lands Directorate, Ottawa.

LOW SUBARCTIC

Landform-Soil-Vegetation Catenas

Noncalcareous granitic sandy loam and loamy sand glacial till	Noncalcareous sand and gravel; ice contact and fluvial	Deep peat; medium to high ice content
Soil Assoc. **Porter Lake**	**Odin Lake**	**Dymond Lake**
1. Eluviated Dystric Brunisol lithic phase	1. Eluviated Dystric Brunisol	
2. Eluviated Dystric Brunisol	2. Eluv. Dystric Brunisol	5. Fibric Organic Cryosol
3. Eluviated Dystric Brunisol	3. Eluv. Dystric Brunisol	
4. Gleyed Dystric Brunisol; Rego Gleysol, peaty phase	4. Gleyed Dystric Brunisol; Rego Gleysol, peaty phase	6. Mesic Organic Cryosol

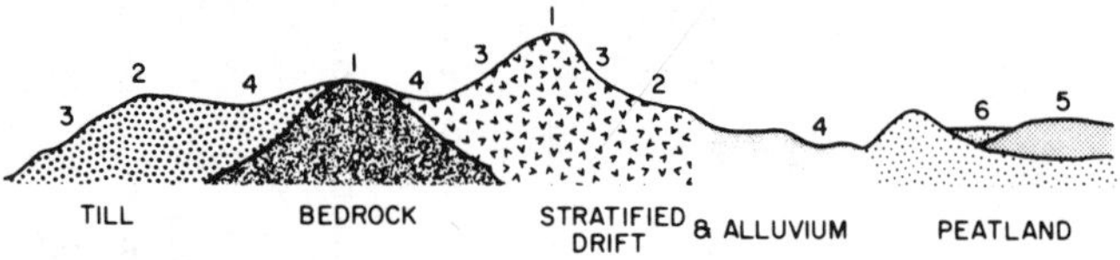

TILL BEDROCK STRATIFIED DRIFT & ALLUVIUM PEATLAND

Drainage Classes	Symbol	Vegetation
Excessively drained knolls	1	Rock Lichen & Rock-Lichen Woodland
Well-drained flats & convexities	2	Lichen Woodland and Heath-Lichen Woodland
Well-drained side slopes		
- south aspect	3	Shrub-Heath & Shrub-Heath Woodland
- north aspect		Moss-Lichen Woodland
Imperfectly drained		
- toe slopes		Moss Forest
- alluvium	4	Shrub-Herb Forest, Shrub Thicket
Poorly drained flats & concavities	5	Bog Woodland, Heath-Lichen Bog
Saturated lowlands		Sedge Fen, Shrub-Sedge Fen

Figure 6.4. The generalized relationships of soils, physiognomic vegetation types, and the topographic facets of typical landforms in the low subarctic region of the Lockhart River map area of Canada. Reprinted, with permission, from Rowe and Sheard (1981). Copyright Springer-Verlag, New York.

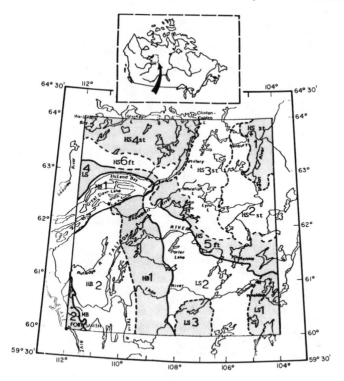

Figure 6.5. Ecological regions and districts of the Lockhart River map area, Northwest Territories, Canada. The regions, separated by solid lines, are designated by letter as MB (midboreal), HB (high boreal), LS (low subarctic), and HS (high subarctic). The latter region includes belts of forest tundra (HSft) and of shrub tundra (HSst). Numbers on the map refer to the districts that, within regions, are separated by broken lines. Reprinted, with permission, from Rowe and Sheard (1981). Copyright Springer-Verlag, New York.

In the Australian land survey system such catenae are being used to map the continent, under the auspices of the federal government (CSIRO, Division of Land Resources Management) (Stewart 1968). Figure 6.6 shows an example of the set of landform-soil units identified for a landscape in western Australia. The current land uses of each landform unit is noted by the surveyors (McArthur et al. 1977). Bennett et al. (1978) have used the units, along with rainfall data, to describe economically feasible land development opportunities for the region.

In the United States the ecoclass and ecoregion (Figure 6.7, Table 6.4) are relatively new classificatory proposals (Crowley 1967, Bailey 1976, 1978). Klopatek et al. (1981) used the ecoregions as one of the spatial scales at which to examine the variety of particular types of vegetation, birds, mammals, and endangered and threatened species. Betters and Rubingh (1978) used the system to classify aspen forest resources in the Central Rockies.

Ellis et al. (1977) summarize vegetation and land use classification systems used in the western United States. The landscape system used in the

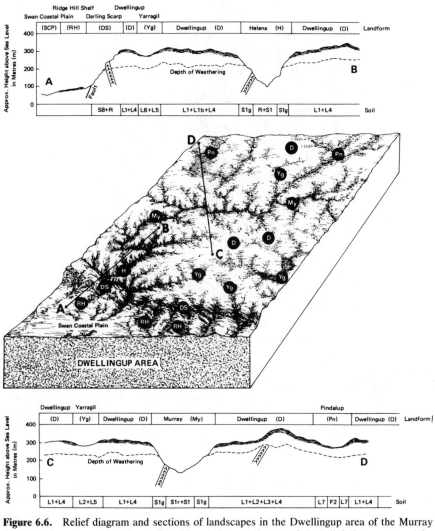

Figure 6.6. Relief diagram and sections of landscapes in the Dwellingup area of the Murray River catchment, western Australia. Abbreviations at the top of the relief diagrams stand for landform types; at bottom, for component soil types. Reprinted from McArthur et al. (1977).

Soviet Union is described by Isachenko (1973). Selman (1982) describes the use of land classification in the United Kingdom to examine the occurrence and value of wildlife for purposes of strategy planning.

Land Use Maps

In classifying land by biological and physical characteristics, one encounters the problem of how to classify land that has already been urbanized. One can

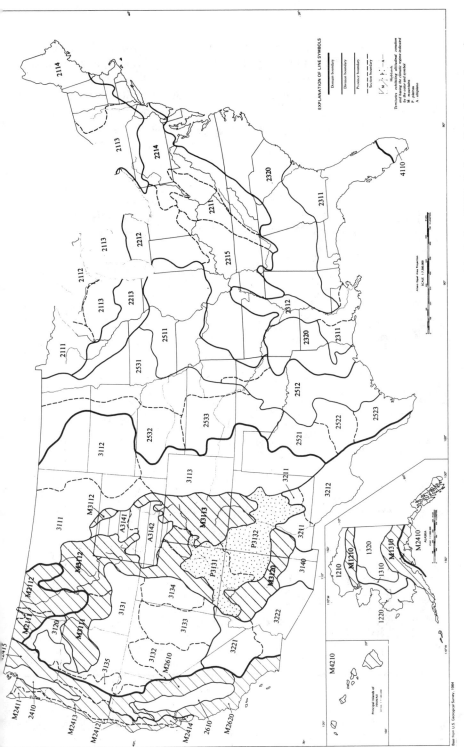

Figure 6.7. A map of the ecoregions of the U.S. based on vegetation, soil, and climate, produced by Bailey (1976). Refer to Table 6.4 for legend.

proceed, as with potential vegetation maps, to determine the ecological category to which the land belongs based on underlying soil, former vegetation and landform, and existing climate. Alternatively, one may limit land classification to relatively undeveloped areas. A third alternative is to map each parcel of land by its existing land use. This is no longer an ecological land classification but rather a record of existing land uses at one point in

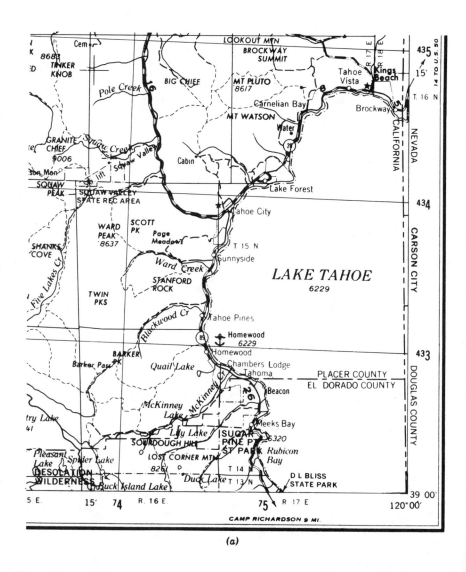

(a)

Figure 6.8. (a) A base map of geographic features in the Lake Tahoe area of California, 1 : 250,000 scale.

time. Land use maps exist for urbanized areas in many parts of the world. In the United States both public and private agencies have produced land use maps. Recently the U.S. Geologic Survey has begun to produce land use and land cover (vegetation) maps keyed to 1 : 250,000-scale feature maps. An example is shown in Figure 6.8.

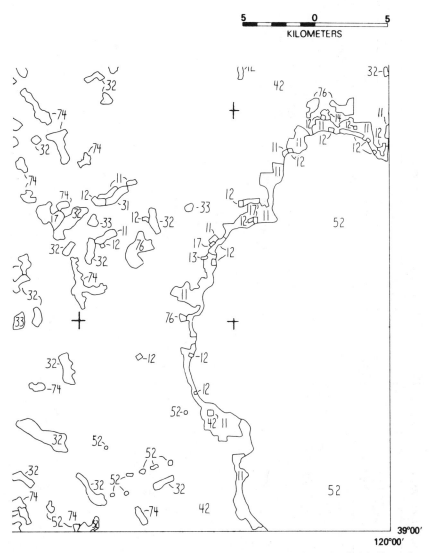

Figure 6.8. (*b*) Land use and land cover units corresponding to the base map in (*a*). Examples of key to units: 11, residential; 12, commercial and services; 32, shrub and brush rangeland; 42, evergreen forest land; 74, bare exposed rock. Section of map "Land Use and Land Cover, 1975–77, Chico, California–Nevada" produced by the U.S. Geological Survey (1979).

Table 6.4. Legend to Accompany the Map of Ecoregions of the United States

Domain	Division	Lowland Ecoregions		Highland Ecoregions	
		Province	Section	Province	Section
1000 Polar	1200 Tundra	1210 Arctic tundra 1220 Bering tundra		M1210 Brooks range	
	1300 Subarctic	1310 Yukon parkland 1320 Yukon forest		M1310 Alaska range	
2000 Humid temperate	2100 Warm continental	2110 Laurentian mixed forest	2111 Spruce-fir forest 2112 Northern hardwoods–fir forest 2113 Northern hardwoods forest 2114 Northern hardwoods–spruce forest	M2110 Columbia forest (dry summer)	M2111 Douglas-fir forest M2112 Cedar–Hemlock–Douglas-fir Forest
	2200 Hot continental	2210 Eastern deciduous forest	2211 Mixed mesophytic forest 2212 Beech–maple forest 2213 Maple–basswood forest + oak savanna 2214 Appalachian oak forest 2215 Oak–hickory forest		
	2300 Subtropical	2310 Outer coastal plain forest 2320 Southeastern mixed	2311 Beech–sweetgum–magnolia–pine–oak forest 2312 Southern floodplain forest		

Division	Province	Section	Mountain Province	Mountain Section
2400 Marine	2410 Willamette–Puget forest		M2410 Pacific forest	M2411 Sitka spruce–cedar–hemlock forest
				M2412 Redwood forest
				M2413 Cedar–hemlock–Douglas-fir forest
				M2414 California mixed evergreen forest
				M2415 Silver fir–Douglas-fir forest
2500 Prairie	2510 Prairie parkland	2511 Oak–hickory–bluestem parkland		
		2512 Oak + bluestem parkland		
	2520 Prairie brushland	2521 Mesquite–buffalo grass		
		2522 Juniper–oak–mesquite		
		2523 Mesquite–acacia		
	2530 Tall-grass prairie	2531 Bluestem prairie		
		2532 Wheatgrass–bluestem–needlegrass		
		2533 Bluestem–grama prairie		
2600 Mediterranean (dry-summer subtropical)	2610 California grassland		M2610 Sierran forest	
			M2620 California chaparral	
3000 Dry / 3100 Steppe	3110 Great Plains short-grass prairie	3111 Grama–needlegrass–wheatgrass	M3110 Rocky Mountain forest	M3111 Grand fir–Douglas-fir forest
		3112 Wheatgrass–needlegrass		M3112 Douglas-fir forest
		3113 Grama–buffalo grass		M3113 Ponderosa pine–Douglas-fir forest
	3120 Palouse grassland		M3120 Upper Gila Mountains forest	

(continued on next page)

Table 6.4. (Continued)

		Lowland Ecoregions		Highland Ecoregions	
Domain	Division	Province	Section	Province	Section
		3130 Intermountain sagebrush	3131 Sagebrush–wheatgrass 3132 Lahontan saltbush–greasewood 3133 Great Basin sagebrush 3134 Bonneville saltbush–greasewood 3135 Ponderosa shrub forest	P3130 Colorado Plateau	P3131 Juniper–Pinyon woodland + sagebrush–saltbush mosaic P3132 Grama–Galleta steppe + Juniper–Pinyon woodland mosaic
		3140 Mexican highlands shrub steppe		A3140 Wyoming Basin	A3141 Wheatgrass–needlegrass–sagebrush A3142 Sagebrush–wheatgrass
	3200 Desert	3210 Chihuahuan desert	3211 Grama–tobosa 3212 Tarbush–Creosote bush		
		3220 American desert (Mojave–Colorado–Sonoran)	3221 Creosote bush 3222 Creosote bush–Bur sage		
4000 Humid Tropical	4100 Savanna	4110 Everglades			
	4200 Rainforest			M4210 Hawaiian Islands	

Source: Bailey (1976).

Note: Key to letter symbols: M–mountains, P–plateau, A–altiplano.

LANDSCAPE CHARACTERISTICS AS INDICATORS OF LANDSCAPE PROCESSES

Continual Processes

Continual processes like soil erosion are highly correlated with underlying structural features of the landscape. This correlation can serve as a basis for prediction. For example, Wischmeier and Smith (1965) have suggested that average annual soil loss (A) (tonnes km^{-2} yr^{-1}) from agricultural soils in the United States is a function of six variables:

$$A = 88.27 (R \times K \times L \times S \times C \times P) \tag{6.1}$$

where

R = a measure of rainfall intensity; an index value related to the maximum 30 minute rainfall intensity per storm (in cm hr^{-1}), averaged over all storms in a given period (obtainable from Golden et al. 1979, Figures 9-4, 9-5; and from the U.S. Soil Conservation Service).

K = a measure of soil erodibility; an index from 0.001 (nonerodible) to 1 (erodible) based on soil texture, structure, organic matter content, permeability (available for U.S. soils from the Soil Conservation Service, or in Golden et al. 1979, Table 9.7).

$L \times S$ = effect of slope on erodibility; S is slope angle (% of 45°), L is slope length (m). The factor $L \times S$ is expressed as the ratio of erosion from the slope angle and length under consideration to that experienced on a slope of 9% and length 22 m. The latter data were obtained from extensive field trials on experimental plots (Wischmeier and Smith 1965). Ratios are obtainable from slope-effect charts (e.g., Golden et al. 1979, Fig. 9-6) for the agricultural soils studied, or by the following formula (Wischmeier and Smith 1965):

$$S \times L = \frac{(0.52 + 0.36s + 0.052s^2) \sqrt{L}}{30.862} \tag{6.2}$$

C = Plant cover and management factor; ranges from 0.001 for well-managed woodland to 1.0 for no cover. Values can be computed from procedures in Wischmeier and Smith (1965) or in Golden et al. (1979, Table 9-9).

P = Management practice factor; ranges from .001 for effective contour plowing, terracing and other erosion control for tilled land, to 1.0 for absence of erosion control factors on tilled land. Values obtainable from Golden et al. (1979, Table 9-10).

This equation is known as the universal soil loss equation. Each variable is an equally weighted scalar (see Chapter 4); the variables are multiplied together, rather than summed, to reflect their interdependence. The 88.27 in Eq. 6.1 is a factor to convert the soil loss (A) from tons acre^{-1} yr^{-1} to tonnes km^{-2} yr^{-1}. Although values were derived for U.S. agricultural soils, the equation has been applied to a wide variety of soils in the United States and elsewhere; Wischmeier (1976), however, cautions against undue extrapolation.

Miller et al. (1979) have written a flexible computer program to calculate soil loss from an area divided into grid units, using the universal soil loss equation. The effect of different management practices and cover values on predicted soil loss can readily be computed in this manner (see, e.g., Briggs and France 1982a; Figure 6.9). An analogous approach has been used to compute predicted soil erosion by wind (Briggs and France 1982b), using a five-variable wind erosion equation developed by Chepil and Woodruff (1963). The variables used are climate, soil erodibility, surface roughness, effective field length, and vegetation.

The likely rates of change for other continual processes, such as coastal erosion or groundwater movement, can also be predicted based on structural features of the landscape. Coastal erosion is dependent on both the geological structure of sea walls and the manner of exposure to wave action; groundwater movement depends on such factors as the depth and angle of

Figure 6.9. Soil erosion by rainfall in South Yorkshire, U.K., mapped on a 1 km² unit grid using the universal soil loss equation (Eq. 6.1). Reprinted with permission from D. J. Briggs and J. France (1982), Mapping soil erosion by rainfall for regional environmental planning. *Journal of Environmental Management* **14**:219–227. Copyright: Academic Press, Inc. (London) Ltd.

impermeable rock layers, water inputs and losses, and the porosity of the rock layers. Maps of average depth to groundwater are typically available from water resource or flood control agencies. For a discussion of these processes, readers may consult environmental geology texts such as Coates (1981), Griggs and Gilchrist (1983), Keller (1979), and Tank (1976).

Episodic Processes

Natural processes of landscape change are more difficult to predict the more infrequently they occur. This is partly because there have been fewer such events within a monitoring period from which to develop predictive regressions. It is partly also because low frequency, high impact events (major earthquakes, volcanic eruptions) tend to attract less sustained social concern, and support for research and management, than higher frequency, lower impact events (floods, fires) that may be of equal social risk (frequency × damage). Thus our ability to predict, manage, and prepare for fires, floods, and storms is greater than for avalanches and mudslides, and greater still than for earthquakes or volcanoes. The study of natural hazards, and possible social responses to them, is broadly reviewed in such recent books as Botkin and Keller (1982), Heathcote and Thom (1979), and Kates (1978), as well as in the environmental geology texts cited earlier.

Static landscape characteristics may be used as predictors of vulnerability to natural hazards of an episodic, catastrophic nature. We will consider wildfires and large earthquakes as examples of frequent and infrequent hazards, respectively.

Fire

Vegetative mass (fuel loads) and climate (wind, air temperature, relative humidity) are important predictors of fire hazards, and such factors as slope help predict rate and pattern of fire spread. Aspect also serves as a predictor because of its influence on fuel moisture. Based on earlier work of Rothermel (1972), Albini (1976) has developed a computer model (FIREMOD) to predict the intensity and rate of spread of fire in wildlands based on the vegetative and climatic parameters noted above as well as terrain slope (see Figure 6.10). The model is also available for use with a programmable calculator. The U.S. Forest Service is now working to incorporate FIREMOD into a larger set of computer models (FIRESCOPE; Albini and Anderson 1982) which will predict the probability of successful containment and control of a wildfire using a given level of fire suppression effort, and the expected fire perimeter location over a fire period. Kessell (Kessell 1979, Kessell and Catelino 1978, Kessell 1981) has developed computer programs to predict pattern of fire spread across major landscape segments which can be readily updated by incorporating information on recent fires. The model itself accounts for successional changes in fuel load and changes in vegeta-

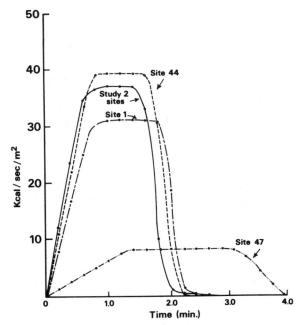

Figure 6.10. Fire intensity as a flame front passes through two coastal (Site 1, Study 2 sites) and two inland sites (Sites 44, 47) of California coastal sage scrub. The figure shows rate of heat release from dead fuels after they have been ignited, as calculated using Albini's (1976) computer model. Reprinted, with permission, from Westman et al. (1981). Copyright Dr. W. Junk, The Hague.

tive composition along environmental gradients. This "gradient modeling" system relies for its survey of vegetation and fuel load on gradient analysis techniques discussed in Chapter 10. For a general review of fire ecology and management techniques, see Wright and Bailey (1982). Green (1981) discusses prescribed burning techniques in greater detail.

A variety of systems have evolved to rate fire hazard. The U.S. Fire Danger Rating System (Deeming et al. 1972) is useful for tracking the changing probability of fire in wildlands as weather and fuel moisture changes over a season. Of greater use for long-range planning is the Australian Fire Hazard Mapping System (Morris and Barber 1980, Barber 1982) which allows mapping of fire hazard to built structures in rural areas on a map of scale 1 : 50,000. Fire "hazard" here reflects both the likelihood of fire occurrence and the extent of likely damage. Each of 10 factors (Table 6.5) is rated on a 1–5 ordinal scale with built-in weighting factor (i.e., nonlinear scalar), and the ratings summed and mapped. Because the ordinal ratings have been internally scaled to be equivalent between criteria, they can be considered interval scales and therefore summed.

A more quantitative approach to rating fire hazard was applied to the Angeles National Forest in southern California by Omi et al. (1979). Because

Table 6.5.   Australian Fire Hazard Rating System

Frequency of fire season
Length of fire season
Slope aspect
Slope steepness
Vegetation—ground cover
Vegetation—average annual driest state
Fire history
Amount of existing development or use
Egress from area
Fire services available

Source: Morris and Barber (1980).

Note: Factors rated on a 1–5 ordinal scale and summed to obtain a fire hazard rating of landscape units suitable for mapping.

the nature of the substrate and the intensity of storms makes this area prone to severe erosion and mudslides after vegetation is burned, the rating system includes criteria for sediment loss hazard as well. The several watersheds in this forest were divided into 71 land units of differing aspect or drainage. Each land unit was characterized by a range of landscape characteristics (Table 6.6). Urban and recreation potential, based on criteria established by the U.S. Forest Service (1972a), were included in the rating system as a guide to the human significance of fire and flood damage in different areas, even though this results in mixing analysis and evaluation functions in a single index.

The 71 units were next classified into groups based on similarity in the attributes listed (see Chapter 10 for discussion of multivariate classification methods). By this means, four classes of land with differing potential for fire and flood damage were recognized. Upon mapping, the classes occurred in distinct zones geographically and, with a few exceptions ("outliers"), were reclassified into these geographic zones for purposes of simplification (Figure 6.11). These zones serve as a basis for applying different management procedures to mitigate fire or erosion hazard. This system differs from the Australian one in requiring interval or ratio rather than ordinal data, and it does not use information on fire hazard due to vegetation amount and condition.

Earthquakes

By contrast to fire prediction, the static landscape indicators for earthquake prediction are less helpful, because the periodicity of earthquake activity is

Table 6.6.   An American Fire Hazard Assessment System Incorporating Hazard From Mudslides and Sediment Loss Following Fire

Attribute	Measured Variable
Location	Latitude and longitude (degrees)
Elevation-aspect	% of unit in each of five strata:
	1. Upper slopes with prevailing north exposure
	2. Lower slopes with prevailing north exposure
	3. Principal canyon bottoms
	4. Lower slopes with prevailing south exposures
	5. Upper slopes with prevailing south exposures
Available sediment area	Total area less reservoir area (ha)
Steepness	Relief ratio (Strahler 1957): gradient in elevation/ longest dimension
Major drainage density	Sum of drainage lengths/unit area (km/ha $\times$ 10^2)
Annual precipitation	Average and range (cm yr^{-1})
Geology and soil dispersion	Erosion hazard index based on soil erodibility and substrate type (1–5 ordinal scale)
Earthquake fault density	Sum of fault lengths/unit area (km/ha $\times$ 10^4)
Unimproved road density	Sum of road lengths/unit area (km/ha $\times$ 10^4)
Urban and recreational	% of unit area in the highest of seven resource potential classes established for this forest by the U.S. Forest Service

Sources: Information from Table 1 of Omi et al. (1979), An application of multivariate statistics to land-use planning: classifying land units into homogeneous zones. *Forest Science* **25**(3):399–414; adapted with permission of the Society of American Foresters.

Note: Each of 71 land units in the Angeles National Forest were characterized by each of the criteria listed, and these used to classify the units by fire erosion damage potential.

so much longer. Earthquake fault lines can be identified by topographic features in aerial photos (e.g., fault valleys, saddles, scarps, linear ridges, landslides, offset streams, sag ponds), from geologic features (juxtaposition of different rock types and ages, crushed and deformed rocks); and from dramatic vegetation changes (Griggs and Gilchrist 1983). Geologists attempt to distinguish between *active* faults, along which movement has occurred within "recent" times (the last 11,000 years), *potentially active* faults, in which evidence of movement is dated from 11,000–2.5 million years, and older and *inactive,* faults. The inactivity along fault lines in the last several hundred years is not a sufficient indication of its potential for further movement; records in Kansu and northern China regions, where historic records of earthquake activity exist for a 3000 year period, indicate that the region experienced an 800 year period without large shocks, preceded and followed by periods of major earthquakes (Allen 1975).

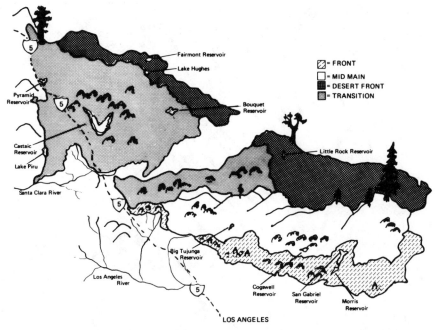

Figure 6.11. Fire damage potential zones, Angeles National Forest, California. Reprinted from Figure 3 of Omi et al. (1979), An application of multivariate statistics to land use planning: classifying land units into homogeneous zones, *Forest Science* 25(3):399–414, with permission of the Society of American Foresters.

Earthquake fault lines are more accurately considered fault *zones,* since the width of the region in which active ground shaking may occur is wide (on the order of a kilometer or more). The area affected by post-earthquake fires, rupture of water, sewerage, gas, and electrical lines, and damage to roads, bridges, and dams is of course much larger.

Apart from the location of earthquake fault zones, land use planners recognize that certain substrates reverberate in a way that increases damage to built structures during earthquakes. Thus buildings on wet, marshy, or unconsolidated ground suffer more damage than buildings on bedrock. Hence bayfill, sediments, landfill sites and cut-and-fill pads are particularly inappropriate places to build in earthquake-prone regions. Earthquake fault maps, of the type shown in Figure 6.12, can be very useful in land use planning.

Injection of wastes into deep wells can also induce earthquake activity by increasing strain of surrounding rock structures; this phenomenon has prompted the suggestion that purposeful deep-well injection could be used to alleviate earth strain and dissipate the strength of potentially large earthquakes, but too little is known to experiment with such technology in urbanized areas (see Griggs and Gilchrist 1983, Healy et al. 1968).

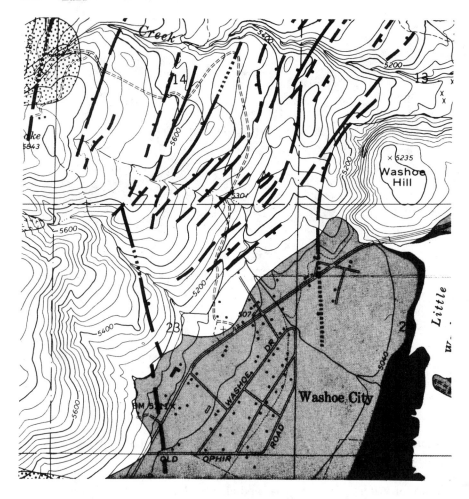

Figure 6.12. Section of the Natural Hazards Map of the U.S. Geological Survey, Washoe City, Nevada, 7½′ quadrangle (Tabor et al. 1978). Dashed lines show locations of earthquake faults; these lines are color coded on the map original to indicate how recently the last known movement along the fault occurred (six categories from <10,000 years to up to 12 million years ago). Shaded areas indicate zones subject to different severities of shaking during an earthquake (six categories mapped). Dotted area at upper left indicates maximum expectable inundation by rockfall avalanches and associated debris flows during an earthquake. Areas of potential landslide during earthquakes were also mapped (not shown).

Certain signs of impending major earthquake activity have been used as a basis for earthquake prediction. One may measure "creep" or small changes in earth position on either side of a fault trace, with theodolites, lasers, or wires strung between poles across the fault line. Small changes in earth rock angles can be measured with tiltmeters. Increases in electrical conductivity of the ground, apparently due to infiltration of water into pores and cracks (Hammond 1973), can be measured with a conductivity meter. Other indicators include patterns of change in the velocity of transmission of small seismic waves through the earth, changes in local magnetic fields, increased emission of radon gas into well water, and changes in behavior of ground-dwelling animals, including the increased appearance of snakes, jumpy behavior in dogs and cats, and the refusal of chickens, hogs, and cows to enter their pens (see Asada 1982, Griggs and Gilchrist 1983, Office of Earthquake Studies 1976, Tributsch 1982).

Earthquake prediction presents as yet unresolved social problems. Because predictions are statements of likelihood rather than certainty, they may cause an alarm in the public which later proves unwarranted. When a seismologist in Los Angeles predicted an earthquake in a portion of the county in 1976, local political officials talked of suing him for any loss of property tax revenue resulting from declines in property values in the region due to the prediction. On the other hand, failure to inform the public of imminent earthquake danger may also result in legal suits.

Long-term plans to reduce earthquake hazard include reinforcing of buildings, enforcing zoning restrictions in fault zones, using flexible piping for underground utility supply lines and developing emergency response capacities (see, e.g., Los Angeles, Department of City Planning 1975; California Office of Emergency Services 1975). Discussions of earthquake prediction and its problems include those of Griggs and Gilchrist (1983), National Research Council (1975), and Press (1975).

Wildfires and earthquakes are only two of the many natural hazards which land planners must consider, but these examples illustrate the fact that planning for natural hazards requires consideration of the complex interplay between physical features, built structures, and social attitudes toward natural phenomena.

MAPPING LANDSCAPE CHARACTERISTICS

Principles of Land Capability Mapping

A common set of landscape characteristics may be used, alone or in combination, to predict the suitability or vulnerability of the land for various uses. Land resource analysts have thus developed flexible systems for data storage and retrieval for use in a variety of land-planning tasks. Typically each landscape attribute is separately mapped, and relevant maps overlaid to determine land units that contain the combination of landscape attributes of interest for particular land uses. Since the 1960s such approaches have rapidly evolved from hand-drawn transparency maps suitable for overlay to elaborate, computerized mapping systems, sometimes attached to automatic systems of data input from satellite photos. Collections of spatial data on landscape attributes, organized for flexible automated use, are termed *geographic information systems* (see, e.g., Calkins and Tomlinson 1977).

Gestalt Method

Hopkins (1977) has summarized major approaches to land suitability mapping; the discussion that follows owes much to his lucid article. The two major mapping tasks are to identify homogeneous units of land with respect to the mapped attribute and to rate the suitability of land units for a particular use. The *gestalt* method of land suitability mapping determines homogeneous land units by direct field observation, implicitly using clues from such characteristics as topography, vegetation, and substrate to establish the distinct units. The homogeneous regions (e.g., valley floors, north-facing slopes) are then rated for suitability for a particular land use (e.g., suited for dense vs. sparse housing), and the units are mapped with a different color or symbol for each suitability class (Figure 6.13). Such an approach suffers from the implicit, hence subjective, nature of judgments regarding land homogeneity and suitability.

Parametric Systems

The gestalt method is also called the "landscape approach" (Fabos et al. 1978), since its first step is the identification of homogeneous landscape units whose suitability for particular land uses is to be judged. The alternative approach, which is now more widely used, is the *parametric* one, in which individual landscape parameters or attributes (soils, vegetation, landform, etc.) are separately mapped and rated for suitability, and these ratings are combined into a grand index of suitability.

Mathematical Combination of Factors

In the *ordinal combination* method, ordinal ratings for suitability for a particular land use (e.g., housing) assigned to each separate landscape characteristic (soil, vegetation, etc) are summed to produce ratings on a composite land use suitability map (Figure 6.13). This approach was used by Ian McHarg (1969) in the Richmond Parkway study and elsewhere. As Hopkins (1977) notes, because an ordinal scale is used, ratings cannot meaningfully be summed between maps (see Chapter 4). The intervals between rating scores may not be the same for different landscape attributes (e.g., a "3" rating on soils may be equivalently suitable to a "2.3" for vegetation). Summation is also inappropriate because the landscape attributes are often not independent in their effect on land use suitability. Thus a particular vegetation may indicate high suitability for housing, and so may its underlying soil type, but the vegetation may be growing there only because the appropriate soil is there. To sum these two suitability ratings therefore results in "double counting."

The factors may not only be dependent (vegetation on soils, in the preceding example) but interdependent and multiplicative in interrelation. For example a 25% slope on well-drained soil over clay may result in a mudslide, but a 25% slope with well-drained soil over granite could be quite suitable for housing, as could well-drained soil over clay on a 5% slope (Hopkins 1977). Hence overlaying ordinally rated suitability maps (e.g., slope, soil drainage, subsoil type) is inappropriate both because of the nonadditivity of ordinal scales and the interdependence of landscape factors contributing to suitability.

An ordinal scale can be converted to a common interval scale (in which addition of ratings is meaningful) by transforming the ordinal scale so that the intervals between units are equal to those of the second scale with which it is being compared. If vegetation is rated on a 1–10 scale of suitability in which a "10" is equally suitable to a "5" on a 1–5 scale used for soil suitability and an "8" to a "4", the soil scale might be made equal to the vegetation scale by multiplying all soil ratings by a factor of 2. This transformation between the two scales is a nonlinear function. Hence we would need curvilinear scalars for weighting the different ordinal suitability scales (Figure 6.14).

The Battelle scalar EES system described in Chapter 4 is an example of such an approach, in which weighted scores for separate impact (rather than suitability) categories are added together, after scalar transformation of raw data, to obtain interval ratings. While the EES system uses nonlinear transformation based on expert judgment or empirical data, other planning systems assume linear transformations in the absence of additional data on which to base the scalars.

An example of a linear weighting system is that described by Lyle and von Wodtke (1974) for use in land suitability mapping in San Diego County, California. In Table 6.7 the effect of using land for citrus production on various environmental processes (including productivity) are assigned relative weights totaling 100 points (bottom row). Each relative weight is then subdivided among vegetation, climate, and land variables according to the role of these landscape or climatic attributes in causing the ecological effect (columns). Thus as the table shows, 10% of the total effect of citrus production is assigned to its effect on sediment transport (by expert judgment). Of these 10 points, 5 are ascribed to slope factors, 2 to the effect of rivers, channels, and other water features, and 3 to the soil runoff potential. Each landscape attribute (e.g., slope class; eight classes in Table 6.8) is then rated as to its relative suitability for inducing all of the ecological effects, up to the maximum weight for the attribute (e.g., total of 19 points for slope; Table 6.7). With this information each land unit is scored for its suitability to citrus production based on vegetation type, slope class, and so on, and these weighted ratings are summed. In the final mapping stage the suitability scores are classed into three categories: low, moderate, or high suitability (Figure 6.15).

In this process the assumption has been made that 40 points for plant productivity are equal in significance for land use suitability to 10 points for erosion or 20 points for pesticide transport. This "weighting" is an evalua-

Landscape approach:
Gestalt Method

Step 1.
Identification of
homogeneous units
for each landscape
attribute

Land use

Land type	Housing	Parkland	• • •

Step 2.
Rating of
land types
by suitability

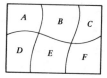

Code:

Well
suited

Moderately
suited

Poorly
suited

Step 3.
Map land types by suitability, one map for each land use

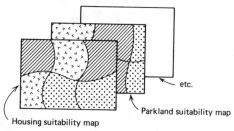

etc.

Parkland suitability map

Housing suitability map

Figure 6.13. Diagrammatic view of alternative land suitability mapping systems. Adapted with permission from Figures 1 and 3 of Hopkins (1977), *Journal of the American Institute of Planners* **43**(4), October issue. Copyright 1977 by the American Institute of Planners (now the American Planning Association). 1313 E. 60th St., Chicago, IL 60637. See next page.

Parametric approach:
Ordinal Combination Method

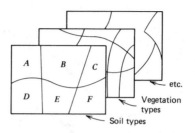

Step 1.
Identification
of homogeneous
land units

A B C

D E F

← etc.

← Vegetation
types

← Soil types

		Land use	
	Soil type	Housing	Parkland	• • •
	A	2	1	• • •
	B	1	2	• • •
	C	3	2	• • •
	D	3	3	• • •
	E	2	1	• • •
	F	1	1	• • •

Step 2.
Rating of types in
landscape attributes
for suitability

Code
1 = well suited
2 = mod. suited
3 = poorly suited

	Land use		
Vegetation type	Housing	Parkland	• • •
A	1	2	• • •
B	2	1	• • •
C	2	3	• • •
•	•	•	
•	•	•	
F	3	2	• • •

Other land types
·
·
·

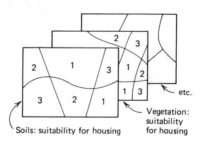

Step 3.
Map types
by suitability,
one map for
each
landscape
attribute

2   3

2   1   3   1
                2

3   2   1   1   3

← etc.

← Vegetation:
suitability
for housing

Soils: suitability for housing

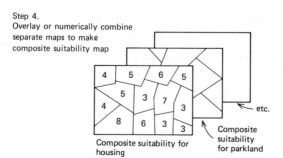

Step 4.
Overlay or numerically combine
separate maps to make
composite suitability map

4   5   6   5

5   3   7
4   3

8   6   3   3

← etc.

Composite
suitability
for parkland

Composite suitability for
housing

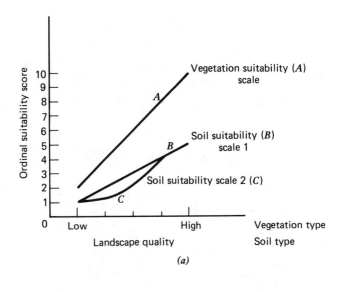

(a)

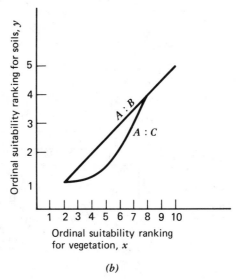

(b)

Figure 6.14. Examples of scalars to convert ordinal to interval scales. (*a*) The relationship between landscape quality and the ordinal suitability score. *A*, a linear relationship between vegetation quality and the suitability scale. *B*, a linear relationship for the soil scale. *C*, an alternative, nonlinear relationship for the soil scale. (*b*) Line *AB* shows a linear relationship between the vegetation and soil scales, in which the soil scale can be weighted by a constant (0.5, the slope of line *A* of form $y = 0.5x$) to convert it to the vegetation scale. Line *C* bears a linear relationship to *A* only between 8 and 10 on the vegetation scale; its scalar for interconversion to *A* is consequently nonlinear.

Table 6.7. Weights Assigned to the Relative Importance of Each Ecological Effect and Landscape Attribute in Influencing the Suitability of Land for Citrus Production in Southern California

Citrus Suitability	Ecological Effect							Total Weight
	Sediment Transport	Nutrient Transport	Pesticide Transport	Erosion	Productivity	Disruption of Wildlife Habitats		
Vegetation types						5	=	5
Slope	5	4	5	5			=	19
Soil					30		=	30
Water features	2	3	5	2			=	12
Rainfall					10		=	10
Runoff potential	3	8	10	3			=	24
Total Weight	10	15	20	10	40	5	=	100

Source: Reprinted from Lyle and von Wodtke (1974, Fig. 8), with permission of the American Planning Association, 1313 E. 60th St., Chicago IL 60637.

Note: The weights are assigned subjectively by expert judgment.

Table 6.8.   Each Landscape Attribute as Divided into Numbered Classes ("Attribute Code") and Assigned a Score ("Model Value") on the Extent of Its Suitability for Promoting Plant Productivity or Reducing Negative Ecological Effects if Land is Used for Citrus Production

Vegetation types	Attribute code	0	2	3	4	5	6	7	8	9	10	11	12	13	14
	Model value	0	1	0	2	3	2	3	3	5	5	0	1	4	4
Slope	Attribute code	0	1	2	3	4	5	6	7	8					
	Model value	0	1	3	5	8	11	13	15	19					
Soil	Attribute code	1	2	3											
	Model value	1	10	30											
Water features	Attribute code	0	2	3	4	5	6	7	8	9	10				
	Model value	0	6	12	6	12	12	6	12	12	12				
Rainfall	Attribute code	1	2	3	4	5	6								
	Model value	10	10	10	5	5	1								
Runoff potential	Attribute code	1	2	3	4										
	Model value	1	8	16	24										

Source: Reprinted from Lyle and von Wodtke (1974, Fig. 8), with permission of the American Planning Association, 1313 E. 60th St., Chicago, IL 60637.

Note: The maximum possible score is equal to the total weight assigned to the landscape attribute (right-hand column of Table 6.7).

tive and arbitrary process. It is a form of linear scaling since points within these weights are assumed of equal value (i.e., are on an interval scale); hence the weighted scores can be added. In the Lyle and von Wodtke example all weightings were assigned by subjective "expert judgment." Thus there is no guarantee that these are truly interval scales. Roberts et al. (1979) have constructed a land suitability model (WIRES) that permits the user to modify the weights in a user-interactive computer system. This feature not only makes such a system more flexible but permits a sensitivity analysis of the choice of weights.

A second problem concerns interdependence of landscape attributes. The EES and Lyle and von Wodtke (1974) weighting schemes (so-called "linear combination" methods) convert different attributes to linear scales for addition into a grand index, hence assuming independence of landscape factors (e.g., slope effects are independent of vegetation effects). As noted earlier, this is usually not a valid assumption. Other systems attempt to represent the interdependence between landscape factors by multiplying, rather than adding, the separate attributes ("nonlinear combination methods"). The universal soil loss equation exemplifies this multiplicative combination of separate landscape attributes. The main problem with the multiplicative approach is that the interdependencies may not be simple multiplicative functions. The dependence of vegetation on slope may be a sine, log, or other complex mathematical function of which simple multiplication is a better approximation than addition, though still not necessarily accurate. Other examples of multiplicative indexes are Storie's (1978) index of soil factors for agricultural production in California and the FAO index of soil

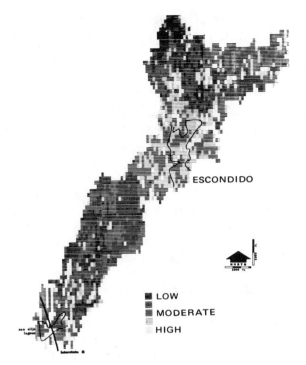

Figure 6.15. A unitized map indicating suitability of land units for citrus production in a portion of San Diego County, California, based on landscape attributes and ecological effects as weighted in Tables 6.7 and 6.8. Reprinted, with permission, from Lyle and von Wodtke (1974), *Journal of the American Institute of Planners* **40**(6), November issue. Copyright 1974 by the American Institute of Planners (now the American Planning Association), 1313 E. 60th St., Chicago, IL 60637.

productivity (Riquier et al. 1970). Rarely is the ecological information available with which to determine the exact nonlinear relationship between parameters. For further discussion on additive, multiplicative, and more complex systems, see McRae and Burnham (1981, Ch. 6).

A third problem, of an ecological rather than statistical nature, exists with all "unitized" mapping procedures which break watersheds or other coherent landscape areas into small grid units for rating and mapping. The ecological interactions between subunits is ignored. Thus in the citrus suitability example each mapped unit was rated separately for effect of slope on pesticide transport, ignoring the transport of pesticide *between* mapped units. The holistic nature of ecosystems is easily lost sight of when interacting landscape segments or ecosystems are subdivided (Westman 1975). This is a problem of the *interdependence of spatial units* and is additional to the problem of interdependence of landscape factors, or the requirement that each mapping unit be homogeneous in relation to some landscape attribute. The problem of ignoring the interdependence of spatial units exists with all unitized mapping approaches. Ecologists have recently become interested in the dynamics of movement of species and materials between "homogeneous" landscape patches. The study of this flow of species, material and energy between landscape patches is part of the emerging discipline of landscape ecology, discussed in Chapter 11.

Identification of Homogeneous Regions

Returning to the problem of interdependence of landscape factors, several other approaches have been used to deal with the problem. One is to identify "homogeneous" land units by overlaying maps for all the separate landscape factors and regarding the resulting landscape units on the composite map as irreducible, homogeneous landscape segments. Such units are then each assigned suitabilities by expert judgment. This procedure ("factor combination" method) differs from the gestalt method in that the process of identifying homogeneous units has been made explicit, though the assignment of suitability rankings remains implicit (Hopkins 1977). In addition to the subjectivity of the ranking process, a major problem with this approach is that the number of landscape units needing to be ranked can be enormous. If, say each of 10 landscape attributes occurred with 10 types in the landscape, 10^{10} or 10 billion separate landscape units would have to be rated.

A possible remedy for this problem is to apply a numerical classification procedure ("cluster analysis") to the units to group them into a manageable number of classes (see Chapter 10). The rating of units of the Angeles National Forest for fire hazard by Omi et al. (1979) cited earlier exemplifies this approach. There factor analysis was used to reduce the initial number of landscape attributes into a smaller number of highly correlated landscape factor combinations. Then land units with similar factor-combination scores were grouped by cluster analysis (Sneath and Sokal 1973), and this classification was further refined by discriminant function analysis (Cooley and Lohnes 1971). The final "homogeneous" fire hazard areas were then mapped (Figure 6.11).

Combination of Ecologically Related Factors

Still another approach for dealing with interdependence of landscape factors is the "rules of combination" method (Hopkins 1977). The combination of landscape attributes (soil, vegetation, etc.) occurring in a landscape unit are considered, and different units are given the same suitability rating if they have certain landscape attribute levels in common. Thus in setting rules for suitability for a golf course, all flat areas with good drainage may be suitable, regardless of vegetation type, since the existing vegetation will be cleared. Thus the units with different vegetation types do not have to be separately considered for ranking. In other words, topography and drainage are interdependent variables, with a certain combination of them acceptable for the land use purpose, whereas vegetation is an independent variable which in this case is insignificant for the land use purpose. Table 6.9 provides another example of rules of combination.

Table 6.9.   Illustration of the Nonhierarchical Rules of Combination Approach for Assigning Suitability Ratings to Landscape Units

Landscape Attributes

A.	Slope, %	Surface Soil Drainage	Presence of Clay Subsurface Horizon
Class 1	>30%	Good	Yes
Class 2	<30%	Poor	No

Rules of Combination

B.	Nonhierarchical (Excluding Consideration of Vegetation). High Suitability
Rule 1.	<30% slope (2, −, −)
Rule 2.	>30% slope, poor surface drainage (1, 2, −)
Rule 3.	>30% slope, subsurface nonclay (1, −, 2)

C.		Slope	Drainage	Claypan	Suitability
Factor combinations in the		1	1	2	High, rule 3
landscape (number indi-		1	2	2	High, rule 2 or 3
cates class in part A)		1	2	1	High, rule 2
		1	1	1	Low
		2	2	2	⎤
		2	1	1	⎟ High, rule 1
		2	1	2	⎟
		2	2	1	⎦

Landscape Attributes

D.	Hierarchical Combinations. Suitability Based on Slope, Drainage-Claypan	Presence of Soil-Binding Vegetation
Class 1	High	Yes
Class 2	Low	No

E.	Rules of Combination. High Suitability
	Rule 1. All highly suitable units with soil-binding vegetation (1, 1)

F.	Factor Combinations	Previous Suitability	Vegetation	Suitability
		1	1	High, rule 1
		1	2	Low
		2	1	Low
		2	2	Low

Note: Suitability for housing development (two classes) is rated and varies with the combination of landscape attributes present. The main concern in this example is mudslide hazard. In the hierarchical example, classes established in the nonhierarchical rating are now rated considering the vegetation factor.

Once a group of interdependent attributes that contribute to high or low suitability is identified, it may be considered a single factor combination in the next round of suitability ratings when new landscape attributes are considered. In this way suitability ratings may be accomplished in a hierarchical fashion, with new landscape attributes added to an interaction matrix after smaller combinations of landscape attributes have been grouped into a single suitability class. This "hierarchical rules of combination" approach is also illustrated in Table 6.9. The advantage of the hierarchical approach is that fewer factor combinations ultimately need to be ranked.

Comparison of Methods

Table 6.10 compares the different approaches to land suitability classification. Because of the problems of subjectivity with the gestalt method, and mathematical invalidity with ordinal combination methods, Hopkins (1977) suggests starting with a linear or nonlinear combination method and using rules of combination to deal with interdependence of factors. With the increasing trend for automation of data, the use of multivariate analysis (cluster analysis, hierarchical classification) to reduce the data to a manageable number of units for ranking is likely to increase. An additional advantage of the cluster analysis method is that, by applying appropriate rules of combination, one can begin to deal with the problem of spatial interdependence of units. Thus by screening adjacent cells for certain levels of flow of energy, materials, or species, one can group adjacent units in which this spatial interaction is important.

Uses of Remote Sensing

Remote sensing of landscapes, both from airplanes and earth resources satellites (LANDSAT, SKYLAB), has enabled the generation and transmission of spatial data to computers in ways that are revolutionizing land suitability mapping. Whereas the interpretation of aerial photos often still involves a human interpreter of the photograph, who may then encode the information in cells for input to a computer cartographic (map-making) system, it is now common for airplane and most satellite data to be digitized and "interpreted" directly by computers. These machines interpret data from computer-compatible magnetic tapes which record digitized information from a sensor such as a multispectral scanner in the airplane or satellite. A "multispectral scanner" senses light in a series of bands of wavelengths from the visible through the thermal infrared radiating from the earth. Because different earth surfaces (e.g., different crops, air or water of different qualities) reflect or reradiate slightly different amounts of radiation in each of these wavelengths, the sensing of these differences and their interpretation as ground features become possible. Much as a television image can be broken into small units, the information transmitted by radio wave through

Table 6.10. Comparison of Methods for Land Suitability Mapping

Methods	Handles Interdependence of Landscape Factors	Rates Units for Suitability by Explicit Process	Comments	Examples
Gestalt	Yes	No	Has no explicit process for identifying homogeneous landscape units	Hills (1961)
Mathematical Combination of Factors				
Ordinal combination	No	Yes	Involves invalid mathematical operations	McHarg (1969, pp. 31–41)
Linear combination	No	Yes	Makes often untested assumptions of colinearity of scales used	Ward and Grant (1971); Lyle and von Wodtke (1974)
Nonlinear combination	Yes	Yes	Requires functional relationships generally not known	Voelker (1976, pp. 49ff.)
Identification of Homogeneous Regions				
Factor combination	Yes	No	Requires very large number of evaluative judgments	Wallace-McHarg (1964)
Cluster analysis	Yes	No	Can be used to deal with spatial interdependence of units	Rice Center (1974); Omi et al. (1979)
Combination of Ecologically Related Factors				
Rules of combination	Yes	Yes	Requires much time and ecological expertise in establishing rules	Kiefer (1965)

(continued on next page)

Table 6.10.   (*Continued*)

Methods	Handles Interdependence of Landscape Factors	Rates Units for Suitability by Explicit Process	Comments	Examples
Hierarchical combination	Yes	Yes	May save time by reducing number of separate evaluations	Murray et al. (1971, pp. 131–174)

Source: Modified from Hopkins (1977) with permission from the American Institute of Planners (now the American Planning Association), 1313 E. 60th St., Chicago, IL 60637.

the air, and reconstituted electronically as cells of an image on a TV screen, so it is with satellite photographs. Antennae on the ground can receive satellite radio waves and record the impulses on magnetic tape. This information can then be used directly for computer-aided mapping.

Because satellites orbit the earth many times a year, temporal changes in land surface features can readily be followed: crop growth, weather features, oil spill movements, progression of disease, insect plagues or air pollution damage in vegetation, coastal erosion, to name a few. Indeed, the amount of information generated is so large that considerable filtering of information is necessary. At the same time, airplane photographs (black and white, color, infrared, multispectral) are useful for smaller-scale studies which can identify individual plants or buildings—a level of resolution not normally possible from satellites.

Remotely sensed data are regularly used in forestry, agriculture, and urban land use studies; mineral prospecting, natural hazard study, and other earth science concerns; studies of movement of water, sediment, and pollutants in lakes, rivers, and the ocean; climatology and weather forecasting; geomorphology, and many other areas. Reviews of principles and applications of remote sensing to environmental sciences include those of Christenson (1979), Lintz and Simonett (1976), Richason (1982), Sabins (1978), and Schanda (1976). Computer-generated maps derived from satellite photographs are shown in Figure 6.16. This example shows how changes in rangeland cover could be followed over time by satellite.

Examples of Computer-Aided Land Resource Analysis: Geographic Information Systems

Three examples of the application of computerized systems to the study of land resources and their utilization are reviewed here to illustrate the range of applications of the techniques discussed in this chapter.

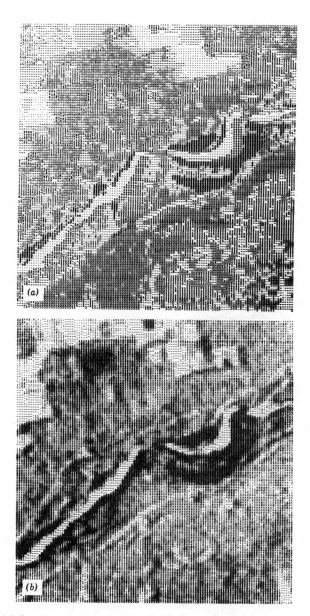

Figure 6.16. (*a*) Land cover classification of Cimarron National Grassland, derived from LANDSAT satellite photo. Each character represents 1.1 acres. Key: *G*, range grasses, >75% cover, no sage; *g*, range grasses, 50–75% cover, no sage; *, range grasses, >75% cover with open tree canopy near river, little or no sage; #, woodland with grass understory, >80% cover, no sage; ~, soil with sparse stubble or weed cover; ·, bare soil; *S*, sagebrush, >75% cover; *s*, sagebrush and yucca, 50–75% cover; -, scattered sagebrush and yucca with grass (<50% sage/yucca, 50–75% ground cover). (*b*) Estimate of vegetative biomass of Cimarron National Grassland, derived from LANDSAT satellite photo. This is a seven-step gray shade print of the LANDSAT multispectral scanner Band 5/Band 7 ratio. Darkest tone (−) represents greatest biomass, lightest tone (−) the least. Each character represents 1.1 acres. Figures *a* and *b* reproduced with permission from "Inventory and evaluation of rangeland in the Cimarron National Grassland," by J. W. Merchant and E. A. Roth, Pecora VII Symposium. Copyright 1982 by the American Society of Photogrammetry.

The METLAND Study

A team of researchers at the University of Massachusetts developed the Metropolitan Landscape Planning Model (METLAND), using the Boston metropolitan region as an example (Fabos and Caswell 1977, Fabos et al. 1978). METLAND is a computer-interactive parametric approach to landscape assessment. Most of the parametric approaches in Table 6.10 are used. In Phase I (Figure 6.17) information on various physical features of the landscape are coded and mapped. Three types of "suitability" judgments are made for each landscape attribute: (1) the value of the landscape attribute as a resource for human use, such as source of water, minerals, or recreational value; (2) degree of hazard: air pollution, noise, or flooding potential; (3) suitability for development (housing, etc.). In order to combine these three suitability ratings into a grand index (the "landscape value") by linear combination, the suitability ratings are converted into dollars (or calories of energy, in a second valuation approach) (Figure 6.18).

"Ecological compatibility" of land units is assessed in two ways. First, landscape units are grouped into five classes based on their plant productivity or on the extent of urban development. The authors intended to use existing biomass and the production (P)/respiration (R) ratio as strict criteria for land classification, based on the suggestions of Odum (1969; see Chapter 8), in which $P/R > 1$ in successional communities, and $P/R = 1$ at climax. Data of this type were scanty for the 104 land use types, so that in practice a combination of available data, extrapolation, and expert judgment were used, and the resulting land use classes were quite conventional (Figure 6.17). Second, "biological potential" is calculated by using the eight crop capability classes derived from U.S. Soil Conservation Service maps, combined with solar radiation input to the land (in three classes) to form a "crop potential index." The U.S. Soil Conservation Service's forest site index is also combined with solar data to form the "forest potential index." These two ratings are then combined in a nonlinear, but nearly additive, way to form a "biological potential" index. An analogous procedure is used to generate "denudation potential" of soil and slope based on Soil Conservation Service's soil-mapping data. Nonlinear scalars are used to convert slope classes into erosion- and runoff-potential interval scales. After the various erosion runoff maps are overlaid, the resulting land units are rated (factor combination approach). The "biological potential" index and the "denudation potential" index are then combined, using nonhierarchical rules of combination, to produce a "substrate profile" index. Finally, the ecological land classes (rated by productivity) and the substrate profile index are combined by nonhierarchical rules of combination to form an "ecological compatibility index."

In a third part of Phase I the potential of the landscape to provide public services (sewerage, water supply, recreation, police and fire protection) is also determined and mapped (not shown in Figure 6.17). For example, for

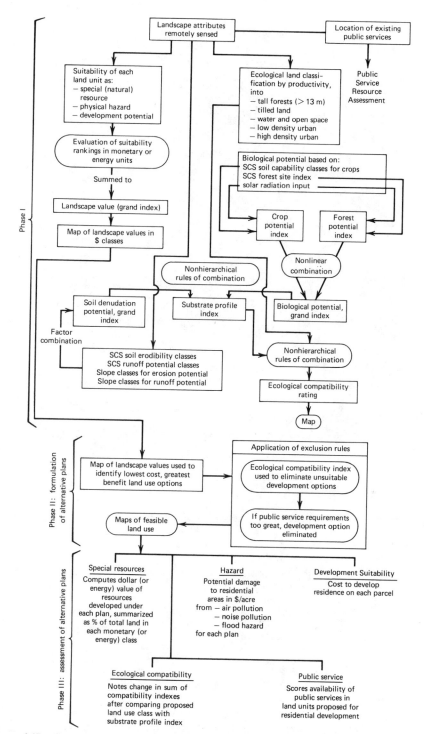

Figure 6.17. Steps in the three phases of the METLAND approach to composite landscape assessment (from information in Fabos et al. 1978).

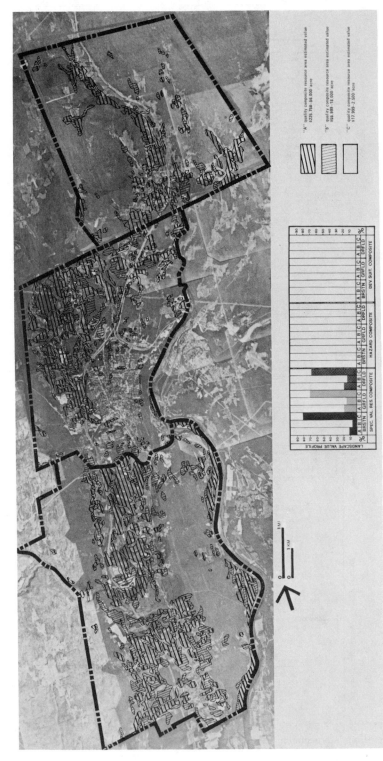

Figure 6.18. Map of composite resource values for an area along Interstate 91 in Massachusetts, evaluated by the METLAND study. Values are combined worth of agricultural, forest, and wildlife productivity, groundwater, sand and gravel, and visual amenity values. Reprinted from Figure 7.4.1.3 of Fabos et al. (1978) with permission of the author and the Massachusetts Agricultural Experiment Station.

water supply land is classified by distance from the water main and available water pressure.

Because the various landscape indexes are computer encoded, it is possible to generate alternative plan scenarios (Phase II) by computer. For example, land units of high development potential based on landscape value are identified; these are screened for "ecological compatibility," and remaining developable areas analyzed in terms of existing public services. By changing weights assigned to different components of the model, different land use plans will be generated. Community preferences may be incorporated into assignment of weights.

In Phase III alternative plan scenarios are evaluated by their potential for achieving three distinct community goals (landscape values, ecological compatibility, provision of public services; see goals-achievement matrix, Chapter 5). Landscape values are evaluated for net dollar benefit or loss from development of the land. Potentially exploitable natural resources are expressed in dollars per acre, hazard zones are evaluated by potential damage in dollars per acre, and development potential by dollar value of developed resource (e.g., value of house on hill with view) and cost of overcoming physical limitations (e.g., cost of building house, including cut and fill, draining high water table). Ecological compatibility is evaluated by summing the loss in ecological compatibility index scores (ordinal ranking) due to development of land parcels. For example, a house may change the ecological compatibility score from $+3$ to -3 on a particular site and from $+2$ to 0 on another site. Public service values of land are ranked on a three-point ordinal scale, and changes in these values similarly noted in the evaluation phase.

While the potential for automating the entire task of land use planning and evaluation is present in the METLAND model, numerous limitations with the model exist. First, data necessary for the mapping tasks are rarely complete, so that much "soft" information is put into the data base. Because we have no way to evaluate statistically our confidence in the accuracy of such data, the resulting output is without bands of confidence. Furthermore the data are compounded with other data in ordinal scales and then summed or combined nonlinearly. None of these mathematical operations are based on empirical relationships or later validated with empirical data. Additional sources of error included errors in mapping boundaries, identifying homogeneous units, and combining overlay boundaries accurately. Also, although many ecological criteria have been included, others have not (e.g., potential for dispersion of pollutants in air, water, or soil); the same could be said for the social concerns. The evaluation of results in dollar terms involves numerous assumptions, especially in relation to nonmarketed goods and services (see Chapter 5), which remain implicit in the valuation procedure. Indeed, despite the opportunity for users to change some weights in the model, so many decisions have been made by the model-builders, and the model calculations themselves are so numerous, that the exact derivation of the final output is not clear to decision makers, the public, or even to any one planner contributing to the model. Although this may be an inevitable feature of any detailed planning process, the use of more and more complex computer models for land use planning does have the effect of alienating the public still further from the assumptions built into the planning process.

The METLAND model was built with the intention that it could be "reparameterized" with new data for use in other parts of the world. Building the model required the work of 40 people for seven years, and hundreds of thousands of dollars.

To date, the model has been reparameterized for use in the Geelong Region, Victoria, Australia; the upper Hussatonic River basin, Massachusetts, and the city of Durban, South Africa, and copies sent to parties in Germany, the Netherlands, and Canada as well as the United States (J. E. Fabos, personal communication, 1983).

The Australian CSIRO South Coast Study

The CSIRO Land Use Research Division, an Australian Federal research organization, developed a computer-aided land use planning program (SIRO-PLAN) during the same period that METLAND was developed (Austin and Cocks 1978). The CSIRO team applied its methodology (Figure 6.19) to the Eurobadalla Shire and environs, on the south coast of New South Wales. The CSIRO methodology also started with an inventory of landscape attributes and identified 3900 homogeneous land units by computerized map overlay. Rather than classify existing land uses by biological productivity, SIRO-PLAN classified land by land tenure (public, private), and current development status (cleared, farmed, forested). Whereas METLAND rated each land parcel for biological and soil denudation potential and rated the compatibility of these with five land use classes, the SIRO-PLAN system coded "raw" information on geology, vegetation, landform, and soils and determined the compatibility of these with eight land uses by a set of subjective "exclusion rules."

An "exclusion rule" states which landscape attributes are incompatible with particular land uses. The eight land uses were agriculture, forestry, urbanization, recreation, beekeeping, conservation, and residue assimilation (landfills, septic tanks). An exclusion rule for forestry took account of slope and potential log volume (see Figure 6.20); for urbanization, one rule excluded land with median slopes >20°.

Application of exclusion rules generates a composite map showing all nonexcluded land uses possible on each land parcel. Whereas METLAND makes greater use of indexes and uses dollars or energy units to rate suitability of land for development, SIRO-PLAN uses a larger number of exclusion rules, in which ecological or economic judgments remain more implicit. The SIRO-PLAN does attempt to deal with the problem of spatial interdependence of land units by establishing exclusion rules for land units that are "incompatible" with uses on adjoining units. In this regard SIRO-PLAN is more sensitive to this problem than METLAND, and research on this aspect continues (Baird 1981).

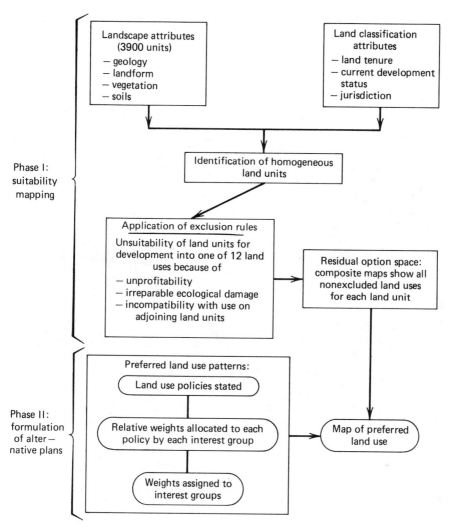

Phase I: suitability mapping

Phase II: formulation of alternative plans

Figure 6.19.   Steps in the CSIRO approach to land use planning (SIRO-PLAN), as applied to the south coast of New South Wales, Australia (from information in Austin and Cocks 1978).

In Phase II (formulation of alternative plans) SIRO-PLAN uses explicit policy statements to establish land use priorities. All policy statements are given weights (relativized to 100) by representatives of five public interest groups (agriculture, conservation, forestry, recreation, urbanization). Each interest group is also weighted in importance. The application of the weighted goals to the residual land use option space generates a preferred land use plan by optimization through linear programming (see Chapter 3). For example, policy statements give preference to a use on land most ecologically suited to it, to agriculture on private land, to recreation near cities, and to existing land uses.

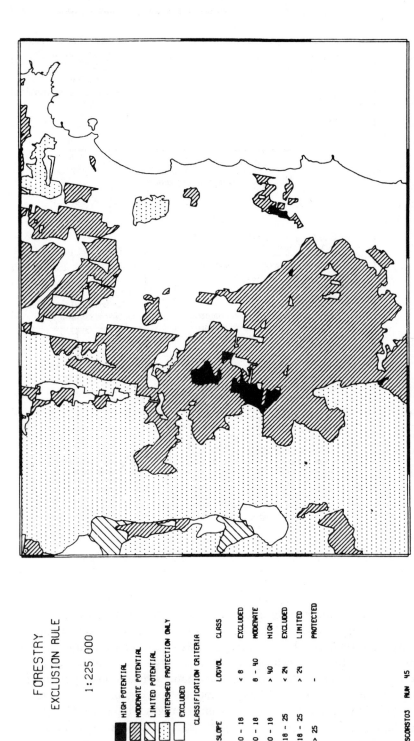

Figure 6.20. Classes of forest growth capability based on landscape attributes, generated by SIRO-PLAN. Reprinted from Austin and Cocks (1977) with permission of the Division of Water and Land Resources, CSIRO.

By contrast, METLAND proposes three plan formulation approaches. The first (Figure 6.17, Phase II) uses an economic criterion: choose land use options with lowest development cost and greatest economic benefit. These possibilities are then screened for ecological compatibility and feasibility of providing public services. This approach has been modeled only for residential development to date. A range of possible plans is thus generated by METLAND. The plans may also be generated by extrapolating existing trends in development, given existing zoning and master plan restrictions, or by inserting community group priorities. The main differences in plan formulation, then, are that SIRO-PLAN provides greater initial flexibility in choosing guiding policies and does not use provision of public services as a suitability criterion.

Although geographic information systems provide the capacity to generate a multitude of alternative land use plans, the choice of a preferred scenario must occur by reference to explicit planning goals. The METLAND approach to plan selection incorporates ecological, economic, and social goals in the initial indexes for land units (Phase I), uses these to derive a small number of land use plans (Phase II), and evaluates each plan by reference to the final grand index scores in dollars or noneconomic units (Phase III), leaving final choice to decision makers. The CSIRO approach uses ecological and economic criteria to filter possibilities to a small number of land use plans (Phase I) and then applies enough policy goals to filter choices to a single option by linear optimization (Phase II). The latter approach more completely specifies the nature of the trade-offs to be made between economic, ecological, and other goals, although by doing so with a computer optimization program, the exact nature of the trade-offs remains obscured. In the CSIRO system "unprofitability" of a land unit for development is made as a qualitative judgment via an exclusion rule, rather than as a quantitative cost/benefit ratio. Such exclusion rules have equal weight to those based on ecological criteria. The METLAND approach to evaluation is more disaggregated, with quantitative economic and semiquantitative index scores not summed; it therefore permits the final trade-offs to be judged more explicitly by the decision maker. Despite their differences in detail the two models are quite similar in broad approach, and many of the same strengths and weaknesses occur.

The CSIRO model took 6 years and 31 professional staff to build at a cost in excess of $500,000. To aid the reparameterization of SIRO-PLAN for use in other parts of Australia, investigators at CSIRO have taken two steps. First, they have designed a somewhat simplified version of SIRO-PLAN, called LUPLAN, for use on microcomputers and line printers owned by most local government planning agencies (Ive 1980). Second, they have compiled a check list of policies or guidelines for use, as a starting point, by local planners (Cocks et al. 1980, Compagnoni and Cocks 1981) and have developed a computerized data bank (ARIS—Australian Resources Information System; Cocks and Walker 1980) containing raw spatial data on resources, bibliographic information, and a mapping capability. With these developments SIRO-PLAN has begun to be used more widely by local agencies (M. Austin, personal communication, 1983). It was also used to prepare a management plan for the Cairns section of the Great Barrier Reef Marine Park (Cocks et al. 1982).

Corridor Siting

A subclass of land use planning problems to which computerized techniques have been applied involves selection of an optimum corridor for such purposes as roads, transmission lines, pipelines, or tanker routes. End points on a map are specified, and search procedures among land units identify contiguous parcels that satisfy specified criteria such as minimum length, cost, or optimization of conflicting economic and environmental goals.

Rasmussen et al. (1980) illustrate a computerized technique for the selection of road paths through forest which would keep construction costs low and scenic quality high. The costs of road building were calculated based on the cost of clearing forest from land units of particular slope and stand density. The economic value of the scenic quality of forest stands was estimated by use of a panel-of-experts technique (Daniel and Boster 1976). Because data on slope, forest density, and scenic quality were available on a coarser grid-cell density than was to be used for path selection, a computer program called SYMAP (Dougenik and Sheehan 1975) was used to interpolate values between available data points, to fill in values for each cell of the map grid. Different weights were applied to the importance of scenic quality or construction costs, and different road routes were generated. Once a desirable corridor is determined by these criteria, other computer programs, such as OPTLOC (Bennett 1973) or FSRDS (George 1975), can be used to locate and align the road within the corridor.

General Comments on Geographic Information Systems

From these three examples we see that geographic information systems offer the potential for finding optimum one- or two-dimensional spaces for particular development purposes based on a range of landscape attributes. Following the mapping of quantitative or economic indexes of landscape value, contiguous cells are chosen that optimize particular weighted objectives or remain as "option spaces" following the elimination of unsuitable areas. In the process of constructing indexes much information is lost, information is extrapolated beyond the empirical data base, and sources of statistical error are compounded.

A question needing further research in unitized land capability mapping is how to account better for the spatial interdependence of landscape units. One approach is to combine individual units into larger, ecologically homogeneous areas using cluster analysis. Typically the landscape attributes used in such a cluster analysis reflect structural rather than dynamic features of ecosystems. One way to isolate areas that act as a functional unit for flow of materials, energy, or species is to start by identifying land classes at an ecosystem, ecoregion, or larger level and devising regional plans that account for these larger units of integration. For example, Cooper and Zedler (1980) mapped natural areas as regional units in southern California, ranked these on a four-point "ecological sensitivity" scale, and recommended avoidance of the most sensitive areas when planning developments such as rights-of-way.

A major issue is how to identify the natural boundaries of the "regional ecological units." As illustrated diagrammatically in Figure 6.21, there is a continuum in the areal extent of the flow of materials (including pollutants), energy, and species in the landscape, punctuated by significant declines in the rate of exchange between landscape units at certain boundaries of scale. Thus as the edge of a forest stand is reached, a number of species reach the boundaries of their likely migration, though the dispersal probability is itself a continuous function within and between species. Some individuals of minimally dispersing species will escape the stand border, just as some species are more likely to emigrate than others (e.g., deer vs. burrowing rodents). Waterborne pollutants are confined within watershed boundaries in certain

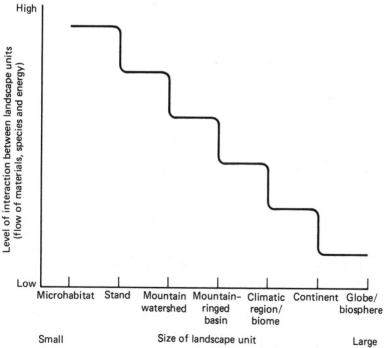

Figure 6.21. Flux of materials and energy between landscape units of increasing scale. The exchange of materials (including pollutants), energy, and species between landscape units is continual across areas of increasing extent, but certain ecological boundaries cause a decline in exchange of these, resulting in a step curve.

directions, but can escape this boundary via river or groundwater flow. Large quantities of pollutants are often confined within "airsheds" defined by basin topography (mountain ranges surrounding flatter areas), but a certain proportion of pollutants will disperse beyond such a basin. For effective land capability mapping, these natural ecological boundaries should be used, but at the same time planners should recognize that the appropriate ecological boundary to use may differ depending on whether one is focusing on species dispersal, water or air pollutant movement, or some other feature. Hence dispersal of air or water pollutants are typically modeled separately (see Chapter 7), as is movement of species between landscape units (see Chapter 11).

EVALUATING ALTERNATIVE LAND USE PLANS

The geographic information systems discussed in the previous section incorporated economic, ecological, and aesthetic criteria into the process of scenario generation. Because final plans are a compromise between conflicting criteria, one may evaluate the final plans by examining how close each comes to achieving specific economic, ecological, or aesthetic goals. The methods of evaluation in relation to economic goals were discussed in Chapter 5. Here we will briefly review methods of evaluation in relation to ecological goals and refer readers to additional work on aesthetic evaluation.

Ecological Criteria

Carrying Capacity Approaches

One approach to evaluating a proposed land use plan in relation to ecological goals is to examine whether the proposed level of resource use will exceed the natural "carrying capacity": the ability of the natural ecosystem to support such levels of use without adverse ecological effect.

The carrying capacity concept derives from the study of population growth in ecology. In the presence of a limitless supply of resources essential for growth, the rate of population growth (the rate of change in number of individuals N present at time t vs. $t + dt$) will be proportional to the initial population size (N_0), and the intrinsic rate of natural increase (r), which is a function of generation time and reproductive biology of the species. Hence the J-shaped exponential growth which results (Figure 6.22a) can be expressed in the equation,

$$\frac{dN}{dt} = rN_0 \tag{6.2}$$

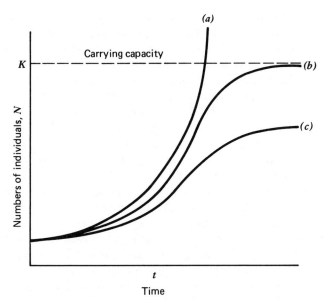

Figure 6.22. Population growth and carrying capacity. (*a*) The exponential curve of population growth; (*b*) the sigmoidal or logistic curve of population growth of a species in the presence of a limiting resource. *K* is the maximum sustainable population, or "carrying capacity," on this resource base. (*c*) A possible growth curve of a species in the presence of competition from another species for the limiting resource.

In the presence of a limited supply of some essential resource, the population growth will be slowed due to negative feedback (e.g., competition for the scarce resource), and eventually net population growth will cease at a level equal to the maximum sustainable by the essential resource in least supply (the limiting resource). This maximum sustainable level (*K* in Figure 6.22*b*) is the carrying capacity of the resource base for the species. The growth curve is S-shaped, sigmoidal or "logistic," and can be modeled as

$$\frac{dN}{dt} = rN \left(\frac{K - N}{K} \right) \tag{6.3}$$

The population growth rate of an organism is further affected by competition with other species, so that the equilibrium population size of a species in the presence of competitors will normally be lower than in the absence of competitors (Figure 6.22*c*). For discussions of population ecology, see such texts as Krebs (1972) and Whittaker (1975).

When the concept of carrying capacity is extended beyond population studies, it quickly becomes evident that defining the natural carrying capacity involves subjective judgment regarding what constitutes "adverse" ecological effects. Whether the trampling of a meadow in a national park, and creating a hiking trail, exceeds the carrying capacity of that environment will depend on the evaluation criteria applied. A preservationist concerned with maintaining the physical environment may argue that such trampling has changed the natural function of that portion of the ecosystem and is therefore a breach of the carrying capacity. A hiker concerned with wilderness appreciation may feel that this level of damage has not ruined his or her ability to experience the meadow, and indeed has enhanced it by improving access. A park official may feel that the park's ability to accommodate the increased number of visitors to the area (parking, sanitation, etc.), resulting from the increased access, has been exceeded. Carrying capacity has thus been judged in at least three different ways: environmental, perceptual, and institutional. Godschalk and Parker (1975) note that the carrying capacity of a region for urban growth is often set by all three of these perspectives. Local water supply may initially be limiting to growth (environmental); as rivers are dammed and channelized to increase water supply, a new limit may be reached as citizens feel the environment is becoming too unnatural (perceptual); if this threshold is passed, the ability of institutions to raise the taxes to supply abundant, unpolluted water to the growing population may finally be exceeded (institutional). In many Mediterranean- and arid-climate urban regions precisely this progression has occurred, and institutions struggle to find socially acceptable means to increase water supply.

Frissell et al. (1980) tried to use a carrying capacity concept to develop a land use plan for Yosemite National Park. Their technique involved mapping the park by level of scenic and biotic value and allocating acceptable levels of visitor use to each zone. This was done first by determining land areas physically unsuitable for extensive campground and facility development due either to natural hazard, susceptibility to soil erosion or compaction, dust buildup, or presence of sensitive wetlands. The amount of remaining area judged suitable for development was 26% less than the area currently developed. The existing and developed land area was then reduced by 26% to obtain a plan considered within carrying capacity limits. Such an approach assumes that the existing ratio of number of visitors to developed area is acceptable and simply reduces both to match the level of development considered ecologically acceptable. In fact, however, there was no empirical evidence that the existing ratio of people to developed land was ecologically, perceptually, or institutionally within acceptable limits.

Gilliland and Clark (1981) also explored use of carrying capacity concepts in planning the future of Lake Tahoe, a recreational lake surrounded by cabins and hotels in the Sierra Nevada bordering California and Nevada. Table 6.11 shows sediment and phosphorus loading rates for the lake under estimated "natural" (predevelopment) and present conditions, as well as under three alternative land use plans or "environmental threshold standards." The urban land use in Table 6.11 is divided into areas adjacent to water ("stream environment zones") and landlocked areas. The "natural conditions" column in Table 6.11 takes a preservationist view and assumes that *any* urban development will exceed the carrying capacity. This assumption does not take into account any natural ecosystem resistance (see "inertia," Chapter 12) or assimilative capacity (Chapter 7), so that any empirical attempt to determine the ecological carrying capacity is forgone. The three alternatives A, B, and C simply attempt to establish levels of use given different perceptual and institutional (cost) constraints. Thus in this example there has been no attempt to determine the true ecological carrying capacity of the lake.

Starting in 1974, the U.S. Fish and Wildlife Service attempted to use detailed ecological data to establish carrying capacity for wildlife habitats. The U.S. Fish and Wildlife Service (1980a, b; 1981) methodology attempts to evaluate changes in the carrying capacity of a habitat for a particular species of wildlife. For each species of interest, ecological field and laboratory studies are conducted or reviewed in an effort to determine the optimum conditions for survival and reproduction of the species. The optimum value for each habitat variable is given a habitat suitability index (HSI) value of 1.0, and habitat values less than (or in some cases, greater than) the optimum are scaled linearly from 1.0 to 0.0 on the HSI scale (see Figure 6.23a). In the absence of more detailed information, a linear relationship is assumed between the abundance of a species and the habitat variable, and between HSI and the carrying capacity (modal peak). There is a good theoretical basis for questioning this assumption, since most species show a Gaussian response curve (Figure 6.23b) to the habitat variable controlling it most strongly and more complex nonlinear responses to less significant habitat factors (Austin 1976; Westman 1980, Ch. 10).

Once a suitability index scalar is obtained for each habitat variable affecting the species, existing habitat conditions are rated on the scalar as a proportion of the optimum or carrying capacity condition (e.g., HSI = 0.4 for 25% canopy cover in Figure 6.23). The theoretical weakness with this procedure is that if single-species experimental data are used, it assumes that the "optimum" habitat for a species in the absence of competition is equal to its optimum in the field. Due to niche differentiation in a multispecies community, however, habitat optima are typically different for a species in the community setting than in isolation (see, e.g., Whittaker 1975, pp. 77–82).

To combine separate HSI values for different ecological parameters into a single grand index of habitat suitability for the species, any of several methods are proposed (U.S. Fish and Wildlife Service 1981). If the parameters are such that a low suitability in one parameter (e.g., herbaceous cover in which hawk prey lives) is compensated for by a high value of another variable (e.g., high cover of tall, isolated trees used as posts to look out for prey), the two (or more) parameters are to be averaged arithmetically or geometrically. If the relationship between habitat suitability variables is cumulative (e.g., herb cover and tree cover both encourage hawk prey species), then the variables are simply added. If, however, the sum exceeds 1.0,

Table 6.11. Three Alternative Environmental Threshold Standards and Their Carrying Capacity Implications for the Lake Tahoe Basin, California

| | Natural Conditions | Present Conditions | Alternatives | | |
			A	B	C
			Desired Environmental Quality		
Water quality of Lake Tahoe	Equilibrium or undetectable degradation Primary productivity[c] 40 gC/m²/yr	Exponential deterioration Primary productivity was 80 gC/m²/yr in 1978; the increase averages 5%/yr.	Maintain existing quality by reversing current trend of exponential deterioration with a margin of safety.	Maintain existing quality by reversing current trend of exponential deterioration.	Allows the present exponential deterioration to continue.
	Clarity[c] 29 m (Secchi Disk depth annual average)	Clarity was 26 m in 1978; decline averages 1%/yr.	(It is not possible to determine precisely what reduction in nutrient and sediment loads is required to reverse the current trend toward eutrophication.)		

Environmental Threshold Standards[a] (Metric Tons/Yr as Runoff)

Sediment loading rate	3100	61,000	36,000	38,000	82,000
Total nitrogen loading rate					
Dissolved	10	142	84	89	191
Particulate	16	242	143	150	325
Total	26	384	227	239	516
Total phosphorus loading rate					
Dissolved	5	77	45	48	103
Particulate	32	530	313	330	713
Total	37	607	358	378	816
Mitigation cost (million 1979 dollars)	N/A	N/A	95	95	0
Carrying Capacity					
Urban land use[a] (hectares)					
Stream environment zones	0	1,740	1,740	1,740	3,764
Total	0	9,543	9,543	9,543	14,211
Population[b] (summer peak)					
Residents	0	73,200	73,200	84,400	106,600
Total	0	223,200	223,200	257,500	325,000

Source: Reprinted from M. W. Gilliland and B. D. Clark (1981) The Lake Tahoe Basin: a systems analysis of its characteristics and human carrying capacity, *Environ. Manage.* **5**:397–407, with permission of Springer-Verlag, New York.
[a] Based on data from the California State Water Resource Control Board (1980), natural sediment, nitrogen, and phosphorus loading rates are somewhat controversial; land use represents subdivided land.
[b] Assumes that the land use to people ratio under each alternative is the same as in 1978.
[c] Primary productivity measurements began in 1960; clarity measurements in 1968.

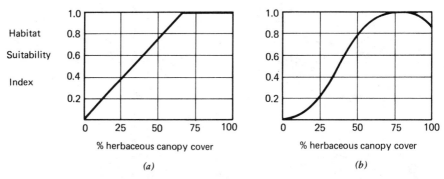

Figure 6.23. U.S. Fish and Wildlife Service HSI scalars. (*a*) Habitat Suitability Index (HSI) in relation to herbaceous cover for the red-tailed hawk, from Figure 3.12 of U.S. Fish and Wildlife Service (1981). (*b*) A Gaussian curve fitted to the same modal optimum as in part *a*. On ecological grounds this is a more likely shape for scalars.

1.0 is taken as the maximum value for the HSI grand index. This introduces an undesirable nonlinearity into the relationship between component HSIs and the grand index. If a single habitat parameter is considered a limiting factor to the welfare of the species, its value is suggested to be used alone for the grand index. (The Fish and Wildlife Service's method actually recommends taking the value of the HSI parameter with the lowest score as the "limiting factor." This procedure has no basis in ecological theory since the limiting factor will not necessarily be the one with the lowest HSI index.)

A major problem with any of these approaches to aggregation is that they require extensive empirical study before the true nature of the relationships between variables can be ascertained. Further, of the many ecological variables being rated by an HSI index, it is unlikely that the relationships between all of them will be cumulative, or all compensatory. Hence to compile a grand index, a more complex formula—involving addition of some variables, averaging of others, and so forth—would have to be derived. The interrelations between sets of variables (e.g., cumulative with compensatory variables) would also have to be determined. Such requirements exceed the current capacity of ecological science. In the U.S. Fish and Wildlife Service system all variables are assumed to be cumulative, compensatory, or part of a "limiting factor" complex. No empirical evidence is used in this determination. The grand index of HSI values is finally multiplied by the area thus assessed (e.g., 10 km^2) to obtain habitat units (HU) in areal units (km^2). Habitat units are later summed for different subareas of a species range of concern. The evaluation of the economic significance of the habitat suitability changes is treated in an additional phase (U.S. Fish and Wildlife Service 1980c).

Although the carrying capacity notion is used in the habitat evaluation method, it is applied with disregard to several fundamental ecological concepts, and the mathematical manipulations are not empirically justified. The overall method is therefore theoretically flawed at present. Some features of the method, however, do hold promise for use once refined. The U.S. Fish and Wildlife Service (Fort Collins, Colo.) had prepared HSI models for 77 species of American fish, birds, and mammals by the end of 1983.

The notion of carrying capacity is an important one, but its application in assessment techniques to date has been troubled either by a lack of use of detailed ecological data or by the neglect of ecological theory in the application of those data. The application of the carrying capacity concept is clearly ripe for further research.

Environmental Performance Standards and Impact Zoning

By disaggregating an ecosystem into specific components whose natural functions can be exceeded, one has a somewhat more manageable approach to the carrying capacity concept in ecological land use planning. The "environmental performance standards" discussed in Chapter 2 exemplify this approach. For example, development may be permitted if natural runoff rates do not increase or if they decrease by less than 10% (Rahenkamp et al. 1977). Such a performance standard limits the degree of change from the natural condition, which is assumed to represent a natural carrying capacity. Of course, if a natural site is in a stage of successional change to some other condition, taking the background level as an unchanging standard is of dubious merit.

Nevertheless, a given performance standard can be combined with land suitability maps to indicate the development constraints necessary on a particular land unit to achieve the environmental performance standard. Thus Figure 6.24 shows the percentage of impervious cover (e.g., concrete roofs, roads) allowable on a land unit of given soil, slope, and vegetation, so that a given level of runoff will not be exceeded. Expected runoff was computed using U.S. Soil Conservation Service's equations for calculating runoff from

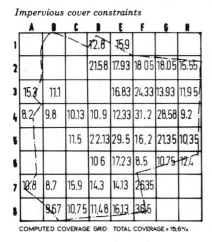

Impervious cover constraints

COMPUTED COVERAGE GRID: TOTAL COVERAGE = 15.6%

Figure 6.24. Impact zoning. Map shows the percent impervious cover that can be added to each land unit and still comply with a particular environmental performance standard. In this case the standard was that the direct runoff shall not exceed that created by a land use of one single-family unit per acre. Landscapes with greater natural infiltration capacity can accommodate a larger % of impervious cover and still meet the standard. Reprinted from Figure 10 of Rahenkamp et al. (1977), with permission of *Plan Canada*, Journal of the Canadian Institute of Planners.

climate, slope, vegetation, and soil porosity data (U.S. Soil Conservation Service 1972), modified by the extent of impervious surface. The resulting map is not unlike a SIRO-PLAN map in which a particular performance standard is applied as a policy.

In the present example the map was derived from a Canadian geographic information system (Rahenkamp et al. 1977). The use of environmental performance standards, combined with land suitability maps, to produce acceptable levels of environmental change on particular land units, is termed *impact zoning*. It has been used in two communities in Massachusetts for several years (Kelly 1975, Lynch and Herr 1973).

Although the combining of performance standards with capability maps permits a somewhat more manageable approach to establishing ecological carrying capacity, the problem of the spatial interdependence of units remains. A true ecosystem carrying capacity derives from the interaction of ecological elements (land, air, water, species) in space and time. The unitized "impact zoning" approach does not integrate across these ecosystem elements, or spatially across land units. Perhaps factor analysis of ecological elements and cluster analysis of spatial units can help to reassemble the holistic notion of carrying capacity, starting with impact zoning as a basis. In so doing, we are reassembling parts of a geographic information system. The METLAND or SIRO-PLAN systems have built-in (inflexible) or user-applied performance standards, respectively.

Aesthetic Criteria

In addition to economic or ecological criteria for evaluating the effects of development on a land resource, aesthetic criteria may also be applied. Assessment of the visual or scenic qualities of a landscape is sometimes termed "landscape evaluation." Some analysts have attempted to evaluate the scenic qualities of a landscape by dissecting it into "universally valued" landscape elements (mountains, waterfalls, long vistas, etc), or design elements (color, texture, line, contrast). Others (e.g., Jacques 1980) have declared the process of landscape evaluation totally subjective and have sought, for example, to compare public opinion on vistas before and after changes, using photographic comparisons and other techniques. Dearden (1981) suggests that landscape appreciation derives from some mix of "intrinsic beauty" and individualistic pleasurable responses. He notes that landscapes of superlative beauty are more likely to be judged similarly by a wide variety of groups than landscapes of inferior quality. The field of landscape evaluation is young, but active. While a treatment of issues in this field is beyond the scope of this book, readers may consult comprehensive literature reviews by Dalzell (1978), Daniel and Boster (1976), Dearden (1980), Krutilla and Fisher (1975), Lang and Armour (1980), McAllister (1980, Ch. 11), Moss and Nickling (1980), Penning-Rowsell (1980), and books by U.S. Forest Service (1972b, 1973–75), and Zube et al. (1975).

CONCLUDING REMARKS

The use of landscape characteristics as indicators of land suitability or vulnerability, and the recognition of homogeneous land units in relation to these characteristics, are two of the key steps in computerized land use planning. Many such systems have further assumed that the landscape characteristics are independent both in effect on land suitability (factor independence) and in interaction between spatial units (spatial independence). In both cases interdependence is usually the more realistic assumption. While attempts to deal with factor interdependence are reasonably well developed, attempts to deal with spatial interdependence are less so.

Many geographic information systems tend to combine indexes of suitability or vulnerability in additive, multiplicative, or weighted fashion, without empirical evidence to characterize the true form of interrelationship. The use of nonlinear scalars holds promise in providing more realistic forms of interrelation, but the empirical data needed to develop these are time-consuming and difficult to collect. Data collected on single species under laboratory conditions, furthermore, are not likely to characterize the performance of the same species in a community setting in the field. Yet the use of field data often limits application of the scalar to a particular physical locale. In the short term the solution to these problems would appear to lie in accepting generalization and extrapolation from available data. In the longer term the science of ecology may clarify the attributes of species, communities, and landscapes that serve as the best predictors of ecosystem response to stress and hence permit more effective prediction from a limited data base. The progress in this area is reviewed in Part V.

REFERENCES

Albini, F. A. (1976). Computer-based models of wildland fire behavior: a user's manual. U.S. Forest Service Intermtn. For. Range Expt. Sta., Ogden, Utah.

Albini, F. A., and Anderson, E. B. (1982). Predicting fire behavior in U.S. Mediterranean ecosystems. In Dynamics and management of Mediterranean-type ecosystems. U.S. Forest Service, Gen. Tech. Rep. PSW-58. Berkeley, Calif., pp. 483–489.

Allen, C. R. (1975). Geological criteria for evaluating seismicity. *Geol. Soc. Amer. Bull.* **86**:1041–1057.

Asada, T., ed. (1982). *Earthquake Prediction Techniques: Their Applications in Japan.* Transl. M Ohmiki. Univ. Tokyo Press, Tokyo.

Austin, M. P. (1976). On non-linear species response models in ordination. *Vegetatio* **33**:33–41.

Austin, M. P., and Cocks, K. D. (1977). Introduction to the South Coast project. Tech. Memo. 77/18. CSIRO Div. Land Use Research, Canberra.

Austin, M. P., and Cocks, K. D., eds. (1978). *Land Use on the South Coast of New South Wales: A Study in Methods of Acquiring and Using Information to Analyse Regional Land Use Options.* 4 vols. CSIRO, Melbourne.

Bailey, R. G. (1976). Ecoregions of the United States. U.S. Forest Service, Ogden, Utah.

Bailey, R. G. (1978). Description of the ecoregions of the United States. U.S. Forest Service, Ogden, Utah.

Baird, I. A. (1981). The application of SIRO-PLAN to rural planning—recent research and development activity of the Land Use Planning Group. Tech. Memo. 81/31. CSIRO Div. Land Use Research, Canberra.

Barber, J. R. (1982). Research and development for improved fire prevention and suppression in rural Victoria. In Dynamics and management of Mediterranean-type ecosystems. U.S. Forest Service, Gen. Tech. Rep. PSW-58. Berkeley, Calif., pp. 490–495.

Belknap, R. K., and Furtado, J. G. (1967). Three approaches to environmental resource analysis. Grad. School Design, Harvard Univ., Cambridge, Mass.

Bennett, D. W. (1973). OPTLOC II general description and user's manual. Univ. Melbourne, Melbourne.

Bennett, D., Thomas, J., and Havel, J. J., eds. (1978). The Murray River land use study. CSIRO Div. Land Resources Management Series, Canberra.

Betters, D. R., and Rubingh, J. L. (1978). Suitability analysis and wildland classification: an approach. *J. Environ. Manage.* 7:59–72.

Botkin, D. B., and Keller, E. A. (1982). *Environmental Studies: The Earth as a Living Planet.* Merrill, Columbus, Ohio.

Briggs, D. J., and France, J. (1982a). Mapping soil erosion by rainfall for regional environmental planning. *J. Environ. Manage.* 14:219–227.

Briggs, D. J., and France, J. (1982b). Mapping soil erosion by wind for regional environmental planning. *J. Environ. Manage.* 15:159–168.

California Office of Emergency Services (1975). Los Angeles and Orange Counties earthquake planning project. NOAA Rep. Synopsis. Fed. Disaster Assistance Admin. and Calif. Off. Emerg. Serv., Sacramento, Calif.

Calkins, H. W., and Tomlinson, R. F. (1977). Geographic information systems: methods and equipment for land use planning. U.S. Geological Survey, Reston, Va.

Carmean, W. H. (1975). Forest site quality evaluation in the United States. *Adv. Agron.* 27:175–269.

Chepil, W. S., and Woodruff, N. P. (1963). The physics of wind erosion and its control. *Adv. Agron.* 15:211–302.

Christenson, J. W. (1979). Environmental assessment using remotely sensed data. GPO, Washington, D.C.

Christian, C. S., and Stewart, G. A. (1968). Methodology of integrated surveys. In Aerial surveys and integrated studies. UNESCO, Paris, pp 233–280.

Coates, D. R. (1981). *Environmental Geology.* Wiley, New York.

Cocks, K. D., Baird, I. A., and Anderson, J. R. (1982). Application of the SIRO-PLAN method to the Cairns section of the Great Barrier Reef Marine Park. Div. Rep. 82/2. CSIRO Div. Water and Land Resources, Canberra.

Cocks, K. D., McConnell, G., and Walker, P. A. (1980). Matters for concern—tomorrow's land use issues. Div. Rep. 80/1. CSIRO Div. Water and Land Resources, Canberra.

Cocks, K. D., and Walker, P. A. (1980). An introduction to the Australian Resources Information System. Tech. Memo. 80/19. CSIRO Div. Land Use Research, Canberra.

Compagnoni, P., and Cocks, K. D. (1981). Operational policies for shire and region land use planning. Tech. Memo. 81/10. CSIRO Div. Land Use Research, Canberra.

Cooley, W. W., and Lohnes, P. R. (1971). *Multivariate Data Analysis*. Wiley, New York.

Cooper, C. F., and Zedler, P. H. (1980). Ecological assessment for regional development. *J. Environ. Manage.* **10**:285–296.

Crowley, J. M. (1967). Biogeography (in Canada). *Canadian Geogr.* **11**:312–326.

Cushman, M., Knight, E., and McGlynn, C. (1948). Vegetation-soil map example. In The timber stand and vegetation-soil maps of California. U.S. Forest Service, Calif. For. Range Expt. Sta., Calif. Dept. Natural Resources and Div. Forestry, Berkeley and Sacramento, Calif., p. 2.

Dalzell, L. (1978). Environmental aesthetics preferences and assessments. Council of Planning Libraries, Monticello, Va.

Daniel, T. C., and Boster, R. S. (1976). Measuring landscape esthetics: the scenic beauty estimation methods. U.S. Forest Service Res. Paper RM-167, Fort Collins, Colo.

Davidson, D. A. (1980). *Soil and Land Use Planning*. Longman, New York.

Dearden, P. (1980). A statistical technique for the evaluation of the visual quality of the landscape for land-use planning purposes. *J. Environ. Manage.* **10**:51–68.

Dearden, P. (1981). Landscape evaluation: the case for a multi-dimensional approach. *J. Environ. Manage.* **13**:95–105.

Deeming, J. E., Lancaster, J. W., Fosberg, M. A., Furman, R. W., and Schroeder, M. F. (1972). National fire danger rating system. U.S. Forest Service, Res. Paper RM-84, Fort Collins, Colo.

Dougenik, J. A., and Sheehan, D. E. (1975). SYMAP user's reference manual. Lab. for Computer Graphics and Spatial Analysis, Grad. School Design, Harvard Univ., Cambridge, Mass.

Driscoll, R. S., Russell, J. W., and Meier, M. C. (1982). Recommended national land classification system for renewable resource assessments. U.S. Forest Service, Rocky Mtn. For. Range Expt. Sta, Ft. Collins, Colo.

Ellis, S. L., Fallat, C., Reece, N., and Riordan, C. (1977). Guide to land cover and use classification systems employed by western government agencies. U.S. Fish and Wildlife Service. Fort Collins, Colo.

Fabos, J. G., and Caswell, S. J. (1977). Composite landscape assessment. Mass. Agric. Expt. Sta. Res. Bull. 637. Amherst, Mass.

Fabos, J. G., Green, C. M., and Joyner, S. A., Jr. (1978). The METLAND landscape planning process: composite landscape assessment, Alternative plan formulation and plan evaluation. Part 3, Metropolitan landscape planning model. Mass. Agric. Expt. Sta. Res. Bull. 653. Amherst, Mass.

Foth, H. D., and Schafer, J. W. (1980). *Soil Geography and Land Use*. Wiley, New York.

Frissell, S. S., Lee, R. G., Stankey, G. H., and Zube, E. H. (1980). A framework for estimating the consequences of alternative carrying capacity levels in Yosemite Valley, California. *Landscape Planning* **7**:151–170.

George, T. A. (1975). The Forest Service's computer-aided road design system. *Transp. Res. Bd. Spec. Rep.* **160**:75–81.

Gilliland, M. W., and Clark, B. D. (1981). The Lake Tahoe Basin: a systems analysis of its characteristics and human carrying capacity. *Environ. Manage.* **5**:397–407.

Godschalk, D. R., and Parker, F. H. (1975). Carrying capacity, a key to environmental planning? *J. Soil Water Conserv.* **30**:160–165.

Golden, J., Ouellette, R. P., Saari, S., and Cheremisinoff, P. N. (1979). *Environmental Impact Data Book.* Ann Arbor Sci., Ann Arbor, Mich.

Green, L. R. (1981). Burning by prescription in chaparral. U.S. Forest Service Gen. Tech. Rep. PSW-51. Berkeley, Calif.

Griggs, G. B., and Gilchrist, J. A. (1983). *Geologic Hazards, Resources, and Environmental Planning.* 2nd ed. Wadsworth, Belmont, Calif.

Hammond, A. L. (1973). Earthquake prediction: breakthrough in theoretical insight. *Science* **180**:851–853.

Healy, J. H., Robey, W. E., Griggs, D. T., and Raleigh, C. B. (1968). The Denver earthquakes. *Science* **161**:1301–1310.

Heathcote, R. L., Thom, B. G., eds. (1979). *Natural Hazards in Australia.* Australian Acad. Sci., Canberra.

Hills, G. A. (1961). The ecological basis for land use planning. Res. Rep. 46. Ontario Dept. Lands and Forests, Ontario.

Hopkins, L. D. (1977). Methods for generating land suitability maps: a comparative evaluation. *J. Amer. Inst. Planners.* **43**:386–400.

Isachenko, A. G. (1973). *Principles of Landscape Science and Physical-Geographic Regionalization.* Melbourne Univ. Press, Melbourne.

Ive, J. R. (1980). LUPLAN—a preliminary explanation and description of the features of a land use planning package. Proc. URPIS 8. Australian Urban and Regional Info. Systems Assn., Surfer's Paradise, Queensland.

Jacques, D. L. (1980). Landscape appraisal: the case for a subjective theory. *J. Environ. Manage.* **10**:107–113.

Kates, R. W. (1978). *Risk Assessment of Environmental Hazard.* SCOPE 8. Wiley, New York.

Keller, E. A. (1979). *Environmental Geology.* Merrill, Columbus, Ohio.

Kelly, E. D. (1975). Impact zoning: concept for growth management. *Colorado Municipalities* **51**(5):142–145.

Kessell, S. R., and Cattelino, P. J. (1978). Evaluation of a fire behavior information integration system for southern California chaparral wildlands. *Environ. Manage.* **2**:135–159.

Kessell, S. R. (1979). *Gradient Modeling: Resource and Fire Management.* Springer-Verlag, New York.

Kessell, S. R. (1981). Application of gradient analysis concepts to resource management modeling. *Proc. Ecol. Soc. Aust.* **11**:163–173.

Kiefer, R. W. (1965). Land evaluation for land use planning. *Building Sciences* **1**:105–126.

Klopatek, J. M., Kitchings, J. T., Olson, R. J., Kumar, K. D., and Mann, L. K. (1981). A hierarchical system for evaluating regional ecological resources. *Biol. Conserv.* **20**:271–290.

Krebs, C. J. (1972). *Ecology: The Experimental Analysis of Distribution and Abundance.* Harper & Row, New York.

Krutilla, J. V., and Fisher, A. C. (1975). *The Economics of Natural Environments: Studies in the Valuation of Commodity and Amenity Resources.* Johns Hopkins Univ. Press, Baltimore, Md.

Küchler, A. W. (1964). Manual to accompany map: Potential natural vegetation of the conterminous United States. Amer. Geogr. Soc. Spec. Pub. 36.

Küchler, A. W. (1977). The map of the natural vegetation of California. In M. J. Barbour and J. Major, eds. *Terrestrial Vegetation of California.* Wiley, New York, pp. 909–938.

Lang, R., and Armour, A. (1980). Sec. 3.5. Scenic Areas. In *Environmental Planning Resourcebook.* Lands Directorate, Environment Canada, Ottowa, pp 151–156.

Lintz, J., Jr., and Simonett, D. S., eds. (1976). *Remote Sensing of Environment.* Addison-Wesley, Reading, Mass.

Los Angeles Department of City Planning (1975). Seismic safety plan. Los Angeles, Calif.

Lyle, J., and von Wodtke, M. (1974). An information system for environmental planning. *J. Amer. Inst. Planners* **40**:394–413.

Lynch, K., and Herr, P. B. (1973). Performance zoning: the small town of Gay Head, Massachusetts, tries it. *Planners Notebook* **3**(5).

McAllister, D. M. (1980). *Evaluation in Environmental Planning. Assessing Environmental, Social, Economic, and Political Trade-offs.* The MIT Press, Cambridge, Mass.

McArthur, W. M., Churchward, H. M., and Hick, P. T. (1977). Landforms and soils of the Murray River catchment area of western Australia. *CSIRO Div. Land Resources Manage. Ser.* **3**:1–23.

McEntyre, J. G. (1978). *Land Survey Systems.* Wiley, New York.

McHarg, I. (1969). *Design with Nature.* Natural History, New York.

McRae, S. G., and Burnham, C. P. (1981). *Land Evaluation. Monographs on Soil Survey.* Clarendon, New York.

Merchant, J. W., and Roth, E. A. (1982). Inventory and evaluation of rangeland in Cimarron National Grassland, Kansas. In B. F. Richason, Jr., ed. *Remote Sensing—An Input to Geographic Information Systems in the 1980's.* Pecora Symp. VII Proc., Amer. Soc. Photogrammetry, Falls Church, Va, pp. 104–113.

Miller, B. A., Daniel, T. C., and Berkowitz, S. J. (1979). Computer programs for calculating soil loss on a watershed basis. *Environ. Manage.* **3**:237–270.

Morris, W., and Barber, J. R. (1980). Fire hazard mapping. Operations Suppl. 96. *The Fireman* **35**(3).

Moss, M. R., and Nickling, W. G. (1980). Landscape evaluation in environmental assessment and land use planning. *Environ. Manage.* **4**:57–72.

Murray, T., Rogers, P., Sinton, D., Steinitz, C., Toth, R., and Way, D. (1971). Honey Hill: a systems analysis for planning the multiple use of controlled water areas. Reps. AD 736 343, AD 736 344, NTIS. Graduate School of Design, Harvard Univ., Cambridge, Mass.

National Research Council (1975). *Earthquake Prediction and Public Policy*. Natl. Acad. Sci., Washington, D.C.

Odum, E. P. (1969). The strategy of ecosystem development. *Science* **164**:262–270.

Office of Earthquake Studies (1976). Abnormal animal behavior prior to earthquakes, I. Proc. Earthquake Hazard Reduction Program Conf., U.S. Geological Survey, Menlo Park, Calif.

Omi, P. N., Wensel, L. C., and Murphy, J. L. (1979). An application of multivariate statistics to land-use planning: classifying land units into homogeneous zones. *For. Sci.* **25**:399–414.

Paysen, T. E., Derby, J. A., Black, H., Bleich, V. C., and Mincks, J. W. (1981). A vegetation classification system applied to southern California. U.S. Forest Service, Gen. Tech. Rep. PSW-45. Berkeley, Calif.

Penning-Rowsell, E. C. (1980). Fluctuating fortunes in guaging landscape value. *Prog. Human Geogr.* **5**:25–41.

Press, F. (1975). Earthquake prediction. *Sci. Amer.* **232**:14–23.

Rahenkamp, J., Ditmer, R. W., and Ruggles, D. (1977). Impact zoning: a technique for responsible land use management. *Plan Canada* **17**:48–58.

Rasmussen, W. O., Weisz, R. N., Ffolliott, P. F., and Carder, D. R. (1980). Planning for forest roads—a computer-assisted procedure for selection of alternative corridors. *J. Environ. Manage.* **11**:93–104.

Rice Center for Community Design and Research (1974). Environmental analysis for development planning, Chambers County, Texas. Tech. Rep. Vol. 1: An approach to natural environmental analysis. Houston, Tex.

Richason, B. F., Jr., ed. (1982). *Remote Sensing—An Input to Geographic Information Systems in the 1980's*. Pecora VII Symp. Proc., Amer. Soc. Photogrammetry, Falls Church, Va.

Riquier, J., Bramao, D. L., and Cornet, J. P. (1970). A new system of soil appraisal in terms of actual and potential productivity. AGL/TESR/70/6. Food and Agriculture Organization, Rome.

Roberts, M. C., Randolph, J. C., and Chiesa, J. R. (1979). A land suitability model for the evaluation of land-use change. *Environ. Manage.* **3**:339–352.

Rothermel, R. C. (1972). A mathematical model for predicting fire spread in wildland fuels. U.S. Forest Service Res. Paper INT-115, Ogden, Utah.

Rowe, J. S., and Sheard, J. W. (1981). Ecological land classification: a survey approach. *Environ. Manage.* **5**:451–464.

Rubec, C. D. A. (1979). Applications of ecological (biophysical) land classification in Canada. Ecological Land Classification Ser. 7, Lands Directorate, Environment Canada, Ottowa.

Sabins, F. F. (1978). *Remote Sensing: Principles and Interpretation*. Freeman, San Franciso, Calif.

Schanda, E., ed. (1976). *Remote Sensing for Environmental Sciences: Ecological Studies, Analysis and Synthesis*. Vol. 18. Springer-Verlag, Berlin.

Selman, P. H. (1982). The use of ecological evaluations by local planning authorities. *J. Environ. Manage.* **15**:1–13.

Shipman, G. E. (1972). Soil Survey of Northern Santa Barbara Area, California. U.S. Soil Conservation Service. GPO, Washington, D.C.

Sneath, P. H. A., and Sokal, R. R. (1973). *Numerical Taxonomy*. Freeman, San Francisco.

Steila, D. (1976). *The Geography of Soils: Formation, Distribution and Management*. Prentice-Hall, Englewood Cliffs, N.J.

Stewart, G. A., ed. (1968). *Land Evaluation*. Macmillan, New York.

Storie, R. E. (1978). Storie Index Soil Rating. Univ. Calif. Div. Agric. Sci., Spec. Pub. 3203.

Strahler, A. N. (1957). Quantitative analysis of watershed geomorphology. *Trans. Am. Geophys. Union* **38**:913–920.

Tabor, R. W., Ellen, S., and Clark, M. M. (1978). Washoe City Folio. Geologic. Hazards Map. Scale 1 : 24,000. Nevada Bur. Mines & Geol., Univ. Navada, Reno.

Tank, R. W., ed. (1976). *Focus on Environmental Geology*. 2nd ed. Oxford Univ. Press, New York.

Tributsch, H. (1982). *When Snakes Awake: Animals and Earthquake Prediction*. The MIT Press, Cambridge, Mass.

UNESCO (1973). International classification and mapping of vegetation. Ecol. & Conserv. Serv. 6, Paris.

U.S. Fish and Wildlife Service (1980a). Habitat as a basis for environmental assessment. 101 ESM. Div. Ecol. Serv., Washington, D.C.

U.S. Fish and Wildlife Service (1980b). Habitat evaluation procedures (HEP). 102 ESM. Div. Ecol. Serv., Washington, D.C.

U.S. Fish and Wildlife Service (1980c). Human use and economic evaluation. 104 ESM. Div. Ecol. Serv., Washington, D.C.

U.S. Fish and Wildlife Service (1981). Standards for the development of habitat suitability index models. 103 ESM. Div. Ecol. Serv., Washington, D.C.

U.S. Forest Service (1972a). National fire planning instructions. Angeles National Forest, Pasadena, Calif.

U.S. Forest Service (1972b). Forest Landscape Management. Northern Region, La Grande, Oreg.

U.S. Forest Service (1973–75). National Forest Landscape Management. Vols. 1, 2. GPO, Washington, D.C.

U.S. Geological Survey (1970). Potential natural vegetation of the United States (map). In *The National Atlas of the United States of America*. Washington, D.C., pp. 89–92.

U.S. Geological Survey (1979). Land Use and Land Cover, 1975–77, Chico, California–Nevada. Scale 1 : 250,000. Open file 79-1582-1. Land Use Series, Reston, Va.

U.S. Soil Conservation Service (1972). National engineering handbook. Sec. 4, Hydrology. GPO, Washington, D.C., Ch. 10.

Voelker, A. H. (1976). Indices: a technique for using large spatial data bases. Rep. ORNL/RUS-15. Oak Ridge National Lab, Oak Ridge, Tenn.

Wallace-McHarg Associates (1964). Technical report on the plan for the valleys. 2 vols. Philadelphia, Pa.

Ward, W. S., and Grant D. P. (1971). A computer-aided space allocation technique.

In M. Kennedy, ed. Proc. Kentucky Workshop Computer Applications to Environmental Design. Coll. Architecture, Univ. Kentucky, Lexington, Ky.

Westman, W. E. (1975). Letter. *J. Amer. Inst. Planners* **41**:213.

Westman, W. E. (1980). Gaussian analysis: identifying environmental factors influencing bell-shaped species distributions. *Ecology* **61**:733–739.

Westman, W. E., O'Leary, J. F., and Malanson, G. P. (1981). The effects of fire intensity, aspect and substrate on post-fire growth of Californian coastal sage scrub. In N. S. Margaris, and H. A. Mooney, eds. *Components of Productivity of Mediterranean-Climate Regions—Basic and Applied Aspects*. Junk, The Hague, pp. 151–179.

Whittaker, R. H. (1975). *Communities and Ecosystems*. 2nd ed. Macmillan, New York.

Whyte, R. O. (1976). *Land and Land Appraisal*. Junk, The Hague.

Wieslander, A. E. (1945). Forest survey maps of California. U.S. Forest Service, Calif. For. Range. Expt. Sta., Berkeley, Calif.

Wiken, E. B. (1980). Rationale and methods of ecological land surveys: an overview of Canadian approaches. In D. G. Taylor, ed. Land/Wildlife Integration. Ecological Land Classification Ser. 11. Lands Directorate, Environment Canada, Ottowa, pp. 11–19.

Wischmeier, W. H. (1976). Use and misuse of the universal soil loss equation. *J. Soil Water Conserv.* **31**:5–9.

Wischmeier, W. H., and Smith, D. D. (1965). Predicting rainfall-erosion losses from cropland east of the Rocky Mountains. Agric. Handbook 282, USDA/ARS, Washington, D.C.

Wright, H. A., and Bailey, A. W. (1982). *Fire Ecology: United States and Southern Canada*. Wiley-Interscience, New York.

Zube, E., Brush, R. O., and Fabos, J. G. (1975). *Landscape Assessment: Values, Perceptions and Resources*. Dowden, Hutchinson & Ross, Stroudsburg, Pa.

7

AIR AND WATER

To predict the ecological effects of a pollutant, one should know the rate of pollutant release, its pattern of dispersion, and the pollutant-sensitivity of receptor species. One must also consider the background level of pollutant, the spatial and temporal variability of ambient concentrations, the variation in susceptibility to pollution within a species population and between species, and the reactions in air, water, soil, and biota that can serve to adsorb, absorb, or transform the pollutant to a different chemical form. This chapter describes methods for predicting pollutant dispersion in air and water as well as ecological processes which influence pollutant effects. Methods for assessing the biological effects of pollutants are discussed more fully in Part V, especially Chapter 9.

AIR

Calculating Pollutant Emissions

How might we predict the effect of air pollutants from burning coal? Dvorak and Lewis (1978) discuss an example for a 350 megawatt coal-fired power plant burning coal from the western United States. Coal typically varies in composition, and is expressed in Table 7.1 as the average and worst grade values so that average and worst-case scenarios of emission may be calculated. Dvorak and Lewis (1978) present a more complete elemental analysis for a range of coal types.

To convert concentrations of elements in coal [as % or parts per million (ppm) of coal dry weight] to rates of emission of elements or oxidized compounds from the stack, one must know the rate of utilization of the coal ("usage rate") and the amount of each pollutant released to the stack for each unit of coal burned ("emission factor"). The latter depends on the design of the particular furnace or boiler structure, the combustion temperature, and other technical matters. As coal burns, different solid elements segregate differentially and remain with ash in the furnace ("bottom ash") or travel with ash in the stack ("fly ash"). Average emission factors for particu-

Table 7.1. Average Concentration of Selected Elements in Coal

	Coal Grade	
	Average	Worst
Heating value, J/kg[a]	1.91×10^7	2.23×10^7
Sulfur, %	0.48	1.2
Ash, %	6.0	12.2
Moisture, %	29.7	36.9
Uranium, ppm	0.8	23.8
Thorium, ppm	2.0	34.8
Mercury, ppm	0.05	
Cadmium, ppm	0.46	

Source: Data from Dvorak and Lewis (1978).
Note: Values are for coal from Wyoming, except for radionuclide values, which are averages of coal from five western states.
[a] To convert J/kg to Btu/lb, divide by 2.324×10^3.

lar types of technology are periodically tabulated (e.g., U.S. Environmental Protection Agency 1977c). Clearly the emission rate for a particular power plant can vary from these average performance levels. Table 7.2 provides typical values for solid and gaseous elements following combustion of western coal, using average emission factors for this coal type, multiplied by the average rate of coal in a 350 MWe (megawatts of electricity) plant.

The daily coal requirement for a plant of given capacity (in our example, 350 MWe) can be calculated as follows:

$$\text{Daily coal requirement} = \frac{\text{plant capacity} \times \text{capacity factor} \times \text{hours in operation/day}}{\text{plant efficiency} \times \text{heat value of coal}} \tag{7.1}$$

The "capacity factor" is the percentage of maximum fuel use at which the plant actually operates over a year. We assume a typical value of 70% of capacity as the capacity factor for this example. "Plant efficiency" is the percent of electrical energy output per kilowatt hour (kWh) of coal energy input. We assume 38% efficiency. There are 3412 Btu's in a kWh. In our example, then,

$$
\begin{aligned}
\text{Daily coal requirement} \atop \text{(English tons/day)} \quad = \quad & \frac{\begin{aligned}70\% \times (350 \text{ MWe}) \times 10^3 \text{ kW/MW} \\ \times 24 \text{ hr/day} \times 3412 \text{ Btu/kWh}\end{aligned}}{38\% \times 8200 \text{ Btu/lb} \times 200 \text{ lb/English ton}} \quad (7.2) \\
= \quad & 3219 \text{ English tons/day (or 2920 metric} \\
& \text{tonnes/day)}
\end{aligned}
$$

To convert from English tons to metric tonnes, we multiply by 0.9068.

Table 7.2.   Average Rates of Emission of Selected Elements and Combustion Products from Burning of Pulverized Western U.S. Coal in a Plant Generating 350 MWe

	Average Weight (metric tonnes/day)	Radiation (Ci/yr)	Maximum Allowable Emissions, (metric tonnes/day[a])
Collected Ash			
Bottom ash	9.9	—	—
Fly ash[b]	140	—	—
Released to Atmosphere			
Fly ash	0.7	—	2.4
Sulfur dioxide[c]	28.0	—	28.7
Nitrogen oxides[e]	8.6	—	16.8
Carbon monoxide[d]	0.8	—	—
Hydrocarbons[d]	0.2	—	—
Aldehydes[d]	0.004	—	—
Cadmium	5×10^{-5}	—	—
Mercury	22×10^{-5}	—	—
Thorium	1.68×10^{-5}	6.5×10^{-4}	—
Uranium	0.90×10^{-5}	1.1×10^{-3}	—

Source: Data obtained or calculated from Dvorak and Lewis (1978).

Note: Also shown are allowable rates of emission from new coal-fired steam generators set by the U.S. Environmental Protection Agency (1971a, 1979).

[a] Although the standard has been converted here to metric tonnes/day for a 350 MWe plant, the published standard is expressed as lb of pollutant/10^6 Btu of fuel, monthly averaging time.

[b] Assumes operation of an electrostatic precipitator in the stack with 99.5% collection efficiency.

[c] Assumes no removal of SO_2 by emission-control devices.

[d] Calculated by ratio to SO_2 released from bituminous coal (Dvorak and Lewis 1978, Table 62); hence this is a crude estimate.

[e] Extrapolated linearly from a plant of 1000 MWe capacity.

A next step is to compare emission rates to allowable levels of emission (Table 7.2). The EPA standard (1979) requires 70–90% removal (monthly average) of SO_2, depending on the sulfur content of the coal. Thus even though in our example the stack emissions meet the SO_2 stack emissions requirement, they do not meet the technological standard for further SO_2 removal, and some form of flue gas desulfurization technology (e.g., lime or limestone scrubbing, double alkali scrubbing; see Dvorak and Lewis 1978) must be installed. The emission of the trace and radioactive elements is regulated in the United States by national emission standards for hazardous pollutants. So far, of the elements listed, only standards for mercury have been promulgated (U.S. Environmental Protection Agency 1977a).

Predicting Ambient Concentrations

Additional standards in the United States apply to ambient concentrations in the region, including requirements to prevent significant deterioration of air quality (Chapter 2). To determine whether emissions from the coal-fired plant will result in acceptable ambient concentrations, one must predict ground-level concentration from a knowledge of existing air pollution levels and a model of dispersion of the new emissions into the region.

Rollback Models

A crude way to estimate changes in ambient concentrations with emissions is to assume a linear relationship between emissions and ambient concentrations in a regional airshed, using an existing inventory of emissions sources and ambient levels to determine a proportionality factor between the two. Thus if existing ambient concentrations of SO_2 are 50 μg m^{-3} and emissions in the region total 10,000 tonnes/day, the addition of 28 tonnes/day would be calculated to add 0.14 μg m^{-3} ((28 mt/day ÷ 10,000 mT/day) × 50 μg m^{-3} = 0.14 μg m^{-3}) to ambient concentrations. Such a procedure has been termed a "rollback" model because it was originally used to calculate the degree of emission reduction needed to achieve ambient standards in a noncompliance area.

Such models do not work well for secondary pollutants, such as ozone, which arise from the photochemical reaction of the primary pollutants (reactive hydrocarbons and nitrogen oxides) and in which the rate of formation of ozone is nonlinearly related to the reactants (see Eldon et al. 1978, Conn 1975). Furthermore the concept of a closed airshed whose contents arise wholly from emissions internal to the airshed is not realistic since long-distance transport of pollutants occurs between airsheds. Also the method can be only reasonably applied over a large area, where the variations in concentrations induced by individual sources can be considered to average out and homogeneous mixing can be assumed. "Rollback" models must be replaced by more precise atmospheric dispersion models when local concentrations near major point sources are to be modeled or the complexities of pollution movement introduced.

Dispersion Models

At its simplest, the process of modeling atmospheric dispersion of pollutants considers release and diffusion of pollutants at a constant rate from a single, ground-level point source over flat unvegetated terrain during stable atmospheric conditions. Many additional factors can be added to such a model in an attempt to make it more realistic (Table 7.3). Even so, the results of atmospheric dispersion models typically provide estimates of ambient concentrations that may be accurate only within 1–2 orders of magnitude. Turner (1970) considers that estimates of centerline (direct-downwind) concentrations from ground-level sources may be accurate within a factor of 3. Dvorak and Lewis (1978) estimate that annual average concentrations modeled for releases from a stack, using climatological input data, are only accurate within a factor of 10–20. Models that incorporate effects of complex terrain or intermittent emissions provide predictions which may be off by a factor of 20–50 (Van der Hoven 1976). Regional urban air pollution models, which average results from many sources, may be accurate to within a factor of 2 (Hameed 1974). Although current work is improving the accuracy of model predictions, it is clear that estimates should be evaluated according to the confidence intervals calculated from previous validations. Useful discussions of atmospheric dispersion modeling include those of Samuelson (1980) and U.S. Environmental Protection Agency (1974).

Box Model. Simple dispersion models assume emission of pollutants into an open-ended, rectangular volume of air ("box model"). Wind blows through the volume at fixed, constant speed, mixing the pollutant homogeneously (Figure 7.1a). Thus the concentration (c_j) of pollutant at any point throughout the length of the box is

$$\text{Concentration (g m}^{-3}) = \frac{\substack{\text{Emission rate} \\ \text{of pollutant (g sec}^{-1})}}{\substack{\text{Wind velocity (m sec}^{-1}) \\ \times \text{ width (m)} \times \text{ depth (m)}}} \quad (7.3)$$

Box models give crude estimates of ambient concentrations based on the volume of air within which mixing is occurring and a unidirectional, uniform wind speed. The assumption of instantaneous, homogeneous mixing of the pollutant within the volume is highly unrealistic.

Gaussian Model. A second model type is the Gaussian model, illustrated in Figure 7.1b for an elevated point source. This model type typically assumes a uniform rate of emission of pollutant. Concentrations of pollutant decrease, at a rate described by a Gaussian curve, in two orthogonal directions (y, z) along a line downwind of the source, thus forming a bell-shaped

Table 7.3.   Some Factors Influencing the Ambient Concentration of an Air Pollutant Following Emission from a Stack

Influencing Factor	Processes Particularly Affected	Influencing Factor	Processes Particularly Affected
Meteorological Conditions		*Atmospheric Chemistry*	
Wind speed	Transport, dilution	Concentrations of other gaseous pollutants	Rate of chemical transformation
Wind direction	Plume direction and shape		
Precipitation	Scavenging of pollutants	Rate of chemical interaction with other pollutants	
Duration of calm conditions	Dry deposition, diffusion		
Temperature changes with atmospheric elevation	Plume rise and shape; diffusion	Solar radiation	Catalysis of photochemical reactions
Humidity and dew	Transport and solution of gaseous pollutants	Concentration of particulates	Adsorption of gaseous pollutants
Cloud cover	Thermal convection at ground level	*Terrain*	
Atmospheric pressure	Plume rise	Water bodies	Absorption of pollutants; surface heating
Technological Design Factors		Vegetation and soil	Absorption or adsorption of pollutants; surface heating
Stack height	Long-distance transport, dispersion		
Exit temperature	Plume buoyancy	Mountains, valleys	Trapping, channeling of pollutants; effects on wind speed and direction
Exit velocity	Height of plume rise		
Stack internal diameter	Concentration of gases		
Rate of gas emission			

plume. A Gaussian model for calculating the ground-level concentration from a stack, incorporating average data on atmospheric turbulence, takes the following form (Hesketh 1972, p. 50):

$$C_{x,y,0} = \frac{Q}{\bar{U}\pi\,\sigma_y\sigma_z} \exp -\left(\frac{H^2}{2\sigma_z^2} + \frac{Y^2}{2\sigma_y^2}\right) \tag{7.4}$$

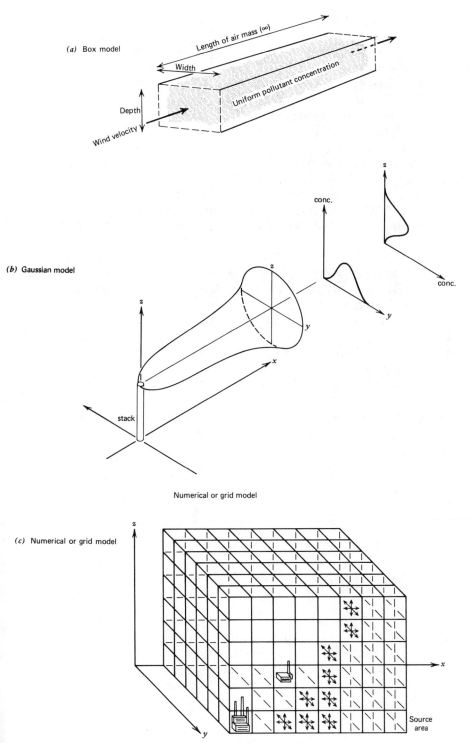

Figure 7.1. Conceptual diagrams of air pollutant dispersion for three model types.

where

C = ground level concentration, $\mu g\ m^{-3}$
Q = release rate from stack, $\mu g\ sec^{-1}$
σ_y = crosswind standard deviation, obtained from Figure 7.3, for a given distance downwind
σ_z = vertical standard deviation, obtained from Figure 7.2, for a given distance downwind.
$\bar{U}$ = mean wind speed, $m\ sec^{-1}$
H = effective stack height (stack height and plume rise), m
x, y = downwind and crosswind distances, respectively, m
exp means $e(= 2.732)$, raised to the power indicated in the expression

In Eq. 7.4 the pollutants reaching the ground are assumed to be absorbed by the ground, although the equation may be adjusted to assume reflection of pollutants. A major advantage of the Gaussian model over a box model is that the dispersion of pollutants due to turbulence in the air can be incorporated, using empirical meteorological data to obtain constants which modify the rate of diffusion of the pollutant.

To calculate ground level concentrations at a given distance downwind of a stack using Eq. 7.4, one first determines the atmospheric conditions at the time. For a worst-case scenario we might consider a clear night of low wind speed (2–3 m sec^{-1}). According to Table 7.4, this is atmospheric stability category F. At a distance of 10 km downwind the dispersion coefficients are σ_y = 300 m, σ_z = 60 m obtained from Figures 7.2 and 7.3. In our earlier example the release rate of SO_2 from the stack of a 350 MWe plant was 28 tonnes/day ($= 3.24 \times 10^8\ \mu g\ sec^{-1}$). If wind speed is 2.5 m sec^{-1} and effective stack height is 40 m, then, using Eq. 7.4 the concentration at 10 km downwind is

$$C_{10,000,\,0,0} = \frac{3.24 \times 10^8\ \mu g\ sec^{-1}}{(3.14)\,(300\ m)\,(60\ m)\,(2.5\ m\ sec^{-1})}\ \exp - \left[\frac{(40\ m)^2}{2(60\ m)^2} + 0 \right]$$

$$= 1834\ \mu g\ m^{-3}\ SO_2 \qquad (7.5)$$

The maximum ground-level concentration downwind will occur approximately where $\sigma_z = H/\sqrt{2}$ under given atmospheric conditions (Hesketh 1972). In our example, σ_z = 40 m/$\sqrt{2}$ = 28 m; from Figure 7.3 we obtain a distance of approximately 2.5 km downwind, using curve F. The maximum ground-level concentration itself may be calculated from the following equation:

$$C_{x,\,0,0_{max}} = \frac{0.117\ Q}{\bar{U}\sigma_y\sigma_z} \qquad (7.6)$$

Table 7.4. Atmospheric Instability Classes for Use with Figures 7.2 and 7.3

Surface Wind Speed at 10 m Height (m sec^{-1})	Insolation Stability Classes[a]				
	Day			Night	
	Strong[b]	Moderate[c]	Slight[d]	Thinly Overcast or $> \frac{1}{2}$ Cloud[e]	Clear to $< \frac{1}{2}$ Cloud
>2 (4.5 mi hr^{-1})	A[f]	A–B	—	—	
2–3 (4.5–6.7)	A–B	B	C	E	F
3–5 (6.7–11)	B	B–C	C	D	E
5–6 (11–13.5)	C	C–D	D	D	D
>6 (>13.5 mi hr^{-1})	C	D	D	D	D

Source: H. E. Hesketh (1972), *Understanding and Controlling Air Pollution,* Ann Arbor Sci., Ann Arbor, Mich.

[a] Insolation, amount of sunshine.
[b] Sun >60° above horizontal; sunny summer afternoon; very convective.
[c] Summer day with few broken clouds.
[d] Sunny fall afternoon; summer day with broken low clouds; or summer day with sun from 15 to 35° with clear sky.
[e] Winter day.
[f] Class A indicates greatest amount of spreading and most unstable atmospheric conditions, and class F indicates least spreading and most stable atmospheric conditions.

In our example, $\sigma_z = 28$, $\sigma_y = 70$ (Figures 7.2 and 7.3), and the maximum ground-level concentration is

$$C_{x,0,0_{max}} = \frac{(0.117)\ (3.24 \times 10^8\ \mu g\ sec^{-1})}{(2.5\ m\ sec^{-1})\ (70\ m)\ (28\ m)} = 7736\ \mu g\ m^{-3}$$

This concentration is clearly toxic, implying that alterations to the power plant design or siting must be considered.

Although the model as presented here does not account for such factors as absorption by vegetation and soil, and chemical transformations, terms for these factors can be readily added to the model. For example, the amount of SO_2 absorbed by vegetation is considered to be a fixed proportion of the ground-level concentration. This fact can be expressed by multiplying the right side of Eq. 7.4 by $(1 - x)$, where x is the proportion absorbed, as determined empirically for the vegetation type in question.

Grid Model. The third type of dispersion model is the numerical, or grid model. Here a three-dimensional imaginary rectangular grid is overlaid on a source region (Figure 7.1c), and the movement of pollutants is calculated by

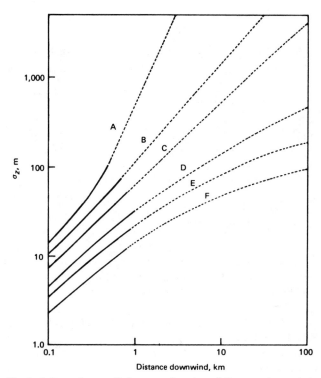

Figure 7.2. Vertical dispersion coefficient as a function of distance downwind of the source, from Turner (1970).

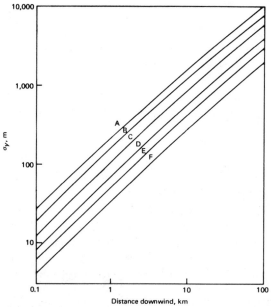

Figure 7.3. Horizontal dispersion coefficient as a function of distance downwind of the source, from Turner (1970).

278

considering the inputs and outputs of pollutant to each cell, the convective (wind) transport of pollutant in any direction (x, y, z axes) through the cell, and the turbulent mixing coefficient (analogous to the dispersion coefficients, σ_n, in the Gaussian model) in any direction (x, y, or z). In other words, for each grid cell the pollutant concentration is calculated by mass balance. The amount of pollutant entering the system from ground sources, and by convective transport from outside the source region (long-range transport), can be considered as initial inputs to grid cells at the margin of the airshed volume. Input to other cells occurs by convective transport and dispersion, under conditions of atmospheric turbulence, from adjacent cells.

The operation of such a model clearly requires a large amount of information, including a complete emissions inventory, site-specific knowledge about meteorological conditions, information on additions and losses of pollutants to the airshed, and site-specific information on pollutant absorption and transformation. How each of these parameters changes with time of day and year is also important. Such numerical models are the most complex computationally of the three dispersion models discussed, although they have the greatest ability to incorporate information about spatial and temporal variation in pollutant dispersion conditions.

Meteorology

Some of the metereological information needed for modeling can be summarized in a wind rose (Figure 7.4). Each radius in the 16 sectors of the circle is scaled to show the percentage of all records for the year in which the wind blew from that sector direction at the particular wind speed indicated in the legend. The proportional occurrence of calm periods is indicated to scale at the center of the figure. Although Figure 7.4 summarizes wind records on an annual basis, separate wind roses may be constructed for each month or season, or for diurnal versus nocturnal periods. In some areas (e.g., California, Department of Transportation; see Samuelson 1980) computer programs are available to calculate wind roses, and the polarity of occurrence of each of the six atmospheric stability classes (Table 7.4) in a particular region, based on data from the National Climatic Center.

The temperature profile above ground-level is also important to an understanding of stack plume behavior and the volume of air available for mixing of air pollutants emitted at ground level. Figure 7.5 shows the different ways a stack plume will expand downwind, depending on the actual temperature profile, and the rate of cooling of an air mass as it expands under decreased pressure at higher elevations (adiabatic lapse rate). Under "trapping" conditions one can note that a warm air layer sits above the cooler ground air mass, inverting the expected trend based on the adiabatic curve. This "inversion layer" serves to trap pollutants beneath it, lowering the mixing height and hence decreasing the volume of air within which pollutants can mix. In such a situation even a simple box model will predict the rise in ground-level pollutant concentration, although often the pollutants are not

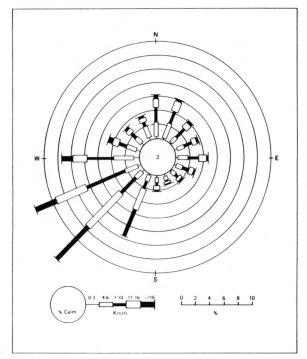

Figure 7.4. Wind rose for Casper, Wyoming (1967–1972), annual average conditions. From Woodward-Clyde Consultants (1976), reprinted in Dvorak and Lewis (1978, p. 36).

homogeneously mixed during inversion conditions. Rather they will tend to increase with elevation up to the height of the inversion layer. Many other aspects of atmospheric meteorology are important to a sophisticated modeling of pollutant dispersion (see, e.g., Samuelsen, 1980; Stern 1976).

Comparing Ambient Concentrations to Standards

Returning to the results of the Gaussian model calculation of SO_2 downwind of a 350 MWe plant, we found (Eq. 7.6) that the maximum ground-level concentration under worst-case atmospheric conditions will be 7736 μg m^{-3}, 2.5 km downwind. Assuming that this atmospheric condition occurs for more than one day of the year, such a concentration exceeds both a primary (365 μg m^{-3}, 24 hr) and secondary standard (1300 μg m^{-3}, 3 hr; see Table 2.2). It now becomes important to know how often in the year the worst-case atmospheric condition occurs, since it may occur often enough to exceed also the 80 μg m^{-3} annual average (primary standard). A several-year meteorological record is needed to determine the probability of occurrence of stability class F.

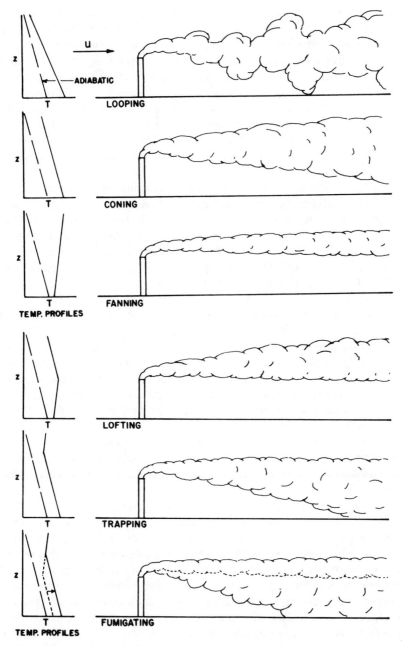

Figure 7.5. Plume behavior under various atmospheric temperature profiles. z = elevation; T = air temperature; dashed line = standard adiabatic rate of decrease in temperature with elevation; solid line = actual temperature profile; u = wind direction. Reprinted from Strom (1976), with permission of Academic Press, New York, and the author.

Suppose we were to install best available technology: a flue gas desulfurization device to remove 90% of the SO_2 from the emissions. Assuming a linear rollback in maximum ambient concentration at ground level to 774 μg m^{-3}, we find that atmospheric stability class F must last less than 12 consecutive hours in any 24 hour period over a year (except for one day) in order not to exceed the 365 μg m^{-3} standard. This assumes that SO_2 concentrations will drop to near zero for the other 12 hours of the day, which is unlikely unless the plant is shut down. It also assumes that background levels of SO_2 are zero. Since SO_2 levels are likely to be somewhat higher during the remainder of the day, a further reduction in allowable hours at 774 μg m^{-3} must occur.

The actual calculation would refer to the probability of occurrence of each stability class, the computed maximum ground-level concentration under each meteorological condition, and the effect of existing background levels. The combinations of stability classes that would result in exceeding the standard could be calculated, and their likelihood of occurrence within more than one 24-hour period per year determined. A similar set of calculations would have to be performed to determine the likelihood of exceeding the annual average (80 μg m^{-3}). Even if all of these standards are likely to be met, one must further consider (under the U.S. Clean Air Act 1970, 1977) the PSD (prevention of significant deterioration) requirements (Table 2.4). In a class II (noncompliance) area, for example, the annual arithmetic mean ambient concentration cannot rise more than 20 μg m^{-3} due to this source.

If one or more of these standards is predicted to be exceeded, several options remain: (1) An offset agreement is negotiated whereby sources contributing to background levels in the region are further controlled; this will only be useful if the plant does not already exceed standards before background levels are considered. (2) The plant agrees to monitor ambient concentrations in the region and cut back or shut down operations during adverse atmospheric conditions. (3) The stack height is raised; this may reduce ground-level concentrations to acceptable levels, but it will contribute to long-distance transport, and such regional problems as acid rain and sulfate particulates, which are not directly regulated from point sources in the United States presently. (4) The capacity of the plant is reduced from 350 MWe to a size compatible with regional air pollution loads. (5) The plant applies for a variance, exempting it from compliance with standards. (6) The plant is not built in the region but perhaps at a site where background levels are lower and atmospheric conditions more favorable, and electricity is transmitted over a longer distance. Alternatively, a less polluting energy source (e.g., solar), or measures to conserve energy, may be substituted.

Long-Distance Transport and Transformation of Pollutants

A certain fraction of emitted pollutants is lofted to upper levels of the troposphere and eventually to the stratosphere, where major wind currents can carry the pollutants over regional, national and intercontinental distances (e.g., Barnes 1979). These processes result, for example, in the transnational acid rain deposition noted in Chapter 2. The analysis of air pollution dispersion discussed in the previous section did not consider the contributions of SO_2 to the regional and global sulfur cycle (Figure 7.6), yet clearly the cumulative effect of these releases has been to add substantially to the global mass of sulfur in the atmosphere.

Once released to the atmosphere, SO_2 is subject to chemical oxidation to SO_3 (sulfur trioxide) and H_2SO_4 (sulfuric acid), which after washout, or release to the soil from plants, can be reduced to H_2S or organic sulfur compounds (e.g., carbonyl sulfide) by microbes and rereleased to the atmosphere as reduced gases to be oxidized again. Thus the sulfur, once released from fossil fuels, has been added to the more active sulfur cycle of the biosphere, with major repositories in air and water. Sulfur is not unique in this behavior. Carbon released as CO_2 (carbon dioxide) and CO (carbon monoxide) and nitrogen as NO, NO_2, and N_2O (nitrogen oxides), for example, similarly enter and accumulate in active cycles of the biosphere once unlocked from fossil fuels. Organisms, especially microorganisms, play critical roles in many of these chemical transformations of pollutants. Other chemical transformations are strictly physical, as for example, the oxidation of SO_2 to SO_3 and the photochemical reaction of reactive hydrocarbons and nitrogen oxides to form oxidants.

At the present time the impacts of large-scale movement of air pollutants are given little attention in ecological assessments of new point sources, in part because there is no legislative requirement to address them besides broad mandates for environmental protection. This situation is changing,

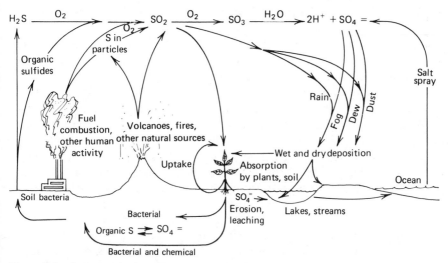

Figure 7.6. Pathways in the global sulfur cycle. Sources of information: Hill (1973), Smith (1981), Unsworth et al. (1985).

however. As noted in Chapter 2, proposals are currently pending in the United States to establish interstate superbubble regions within which total SO_2 emission reduction is to be achieved. An ecological assessment of SO_2 emissions from a new source would then consider not only the effect of SO_2 emission on ambient concentrations and standards but would consider the effect of the additional emission on the total SO_2 emission inventory for the superbubble (interstate) region, and increases would be allowed only if surplus emission credits were available. This mechanism differs from that of the existing prevention-of-significant-deterioration standards, in that focus is on emissions rather than ambient concentrations and credits are bankable or salable. Both tactics are examples of emerging legislative attempts to deal with regional, or long-range transport, air pollution problems.

Predicting Ecological Responses to Air Pollutants

Species Tolerances

Principles for examining the effects of pollutants on plants, animals (including people), soils, and materials can be viewed as common to pollutants in all media (air, water, soil). Such principles of ecotoxicology are discussed in Chapter 9. We will note briefly here some of the particular effects of air pollutants on ecosystems to illustrate some concepts of importance to Part V.

Table 7.5 summarizes some of the ecological effects of major air pollutants. Effects of air pollutants on materials are not included in the table, although rusting or oxidation (by ozone, other oxidants), soiling (by particulates), and acid corrosion (by sulfate, SO_2, NO_x) of metals, rocks, plastic, and other surfaces can be major problems.

In addition to synergisms that can occur between gaseous pollutants, and between gases and particulates (particularly fine particulates <0.5 μ diameter) to which they adhere, the effect of an air pollutant on an organism is much affected by environmental conditions. Thus humidity and temperature can act to alter respiratory rates in animals and stomatal openings and metabolism in plants, hence influencing rates of intake and accumulation of air pollutants. Soil moisture and nutrients also interact with plant susceptibility to air pollutants (Smith 1981, Westman et al. 1985). The pH of soil and water affects solubilities of other ions (especially metal cations) and can result in toxic concentrations of the latter (Smith 1981, Cronan and Schofield 1979) in soil or water, due either to pH lowering or raising. A lowering of pH internal to plant cells, for example, due to acid rain or SO_2 absorption, can in turn solubilize toxic metals in the cell such as iron or aluminum. Animal species typically have higher respiratory rates as body size decreases; respiratory rates in turn influence rate of pollutant uptake and are another factor influencing pollutant susceptibility.

Ecosystem-Level Effects

At the ecosystem level the effects of gaseous air pollutants are multiple and interactive, but a few generalizations may be noted: (1) Because pollution tolerances differ between species, pollution stress will produce a gradient of response downwind of a point source (e.g., Gordon and Gorham 1963; Figure 7.7) and a patchwork of differential response by species to a diffuse

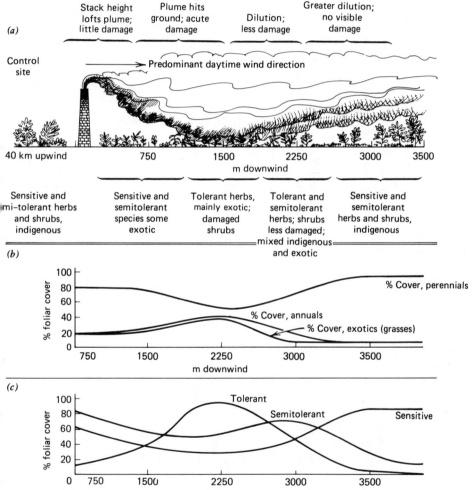

Figure 7.7. Ecosystem level effects of chronic SO$_2$ exposure. (*a*) Diagrammatic view of response of a California coastal sage scrub community to 25 years of chronic SO$_2$ exposure (0.09 ppm annual average maximum ground level concentration at 1600 m). Dominant shrub was black sage (*Salvia mellifera*). (*b*) Changes in foliar cover at the site of perennials (both shrubs and herbs), annuals and exotics (all grasses). The major exotic was an SO$_2$-tolerant strain of annual bromegrass (*Bromus rubens*) (Westman et al. 1985). All curves fit polynomial regression with $r^2 = 0.82$, $p < 0.001$. (*c*) Changes in cover of the 21 most widespread species downwind. Curves fit by polynomial regression. Data for figures (*b*) and (*c*) used with permission of K. P. Preston.

Table 7.5. Some Ecological Effects of Major Air Pollutants[a]

			Effects on		
Pollutant	Humans [Wildlife Effects in Brackets][b]	Plants	Soil and Microorganisms	Climate/Atmosphere	Known Synergisms with
Total suspended particulates (TSP)	Asthma, other respiratory or cardio-respiratory problems; increases cough and chest discomfort; increases mortality; synergistic interaction: gases and metal ions adhere to particle surfaces, are drawn deeper into lungs[c] Fine particulates ($<0.5 \mu$) are more damaging	Clogs stomates, reducing photosynthesis and growth	Deposits attached heavy metals and radionuclides in soil; clogs soil pores; provides possible substrate for decomposers	Forms clouds; reduces incoming radiation, visibility; can result in atmospheric cooling	Animals: SO_2, radionuclides, heavy metals
Sulfur dioxide (SO_2)	Aggravates respiratory diseases, reduces lung function; irritates eyes and respiratory tract	Injures leaves, reduces growth of tops and roots; increases mortality	Can inhibit nitrogen fixation in symbiotic bacteria, blue-green algae; is oxidized by microorganisms to SO_4^{2-}; will be	Will oxidize to SO_3, then H_2SO_4 in atmosphere (see SO_4^{2-} below)	Plants: NO_2, O_3, H_2O vapor Animals: SO_2, H_2O

Pollutant	Effects on humans and animals	Effects on plants	Effects on soil		Atmospheric effects	Reactions
Sulfate particulates (SO_4^{2-})	Effects on humans similar to SO_2 [in fishes, pH 4.5–6 can inhibit reproduction and survival of young, pH 4–5 interferes with ion balance, pH 3 coagulates mucous on gills; amphibian reproduction impaired]	With moisture, corrodes epidermis and leaf tissue; leaches nutrients from terrestrial plants; increased acidity of water kills diatoms, then other algae	Leaches nutrients, changes solubility of ions under lowered pH, particularly increasing concentration of toxic metal ions; inhibits symbiotic nitrogen-fixing bacteria	adsorbed by soil and absorbed by soil microorganisms; lichens very susceptible to damage	Will combine with atmospheric moisture to acidify rain, snow, fog, mist, and dew	*Plants and animals:* H_2O, heavy metals
Oxidants (O_3, PAN, others)	Aggravate respiratory and cardiovascular illnesses; irritate eyes, respiratory tract, impair cardiopulmonary functions; increase mortality	Injure leaves; reduce growth; cause premature leaf and fruit drop; lower resistance to disease	Inhibit mycorrhizae, symbiotic nitrogen-fixing bacteria; lichens susceptible		Reduce visibility; oxidize other atmospheric constituents; absorb UV radiation	*Plants:* SO_2
Nitrogen dioxide (NO_2)	Aggravates respiratory and cardiovascular illness and chronic nephritis	Reduces growth; causes premature leaf drop; in moisture, forms strong acid (see SO_4^{2-})	In moisture, leaches nutrients; lowers soil pH; releases metal ions		Colors atmosphere brown; can form nitric acid in atmospheric moisture; quenches ozone	NO_2/SO_2; $NO_x + HC \rightarrow O_x$; NO_x/H_2O

(continued on next page)

287

Table 7.5. *(Continued)*

	Effects on				
Pollutant	Humans [Wildlife Effects in Brackets][b]	Plants	Soil and Microorganisms	Climate/Atmosphere	Known Synergisms with
Hydrocarbons	Some may cause cancer; injure respiratory system, irritate eyes	Ethylene can cause leaf injury and premature leaf drop	Absorbed by microorganisms	Reduce visibility	uv and NO_x (to form oxidants)
Fluorides	Hydrogen fluoride irritates respiratory system, mucous membranes; burns eyes and skin [inorganic fluoride accumulated by insects; bees sensitive]	Hydrogen fluoride causes leaf injury; reduces plant growth	Little known	Organic chlorofluorocarbons reduce stratospheric ozone	Little known
Radionuclides (e.g., Co-60, I-137, Sr-90)	Cause cancer, leukemia and genetic mutations; can bioaccumulate to hazardous levels through food chains	Injure leaves; reduce plant growth	Can remain in soil for long periods and reenter active biogeochemical cycles	Capable of global transport and deposition	Particulates

				Particulates
Heavy metals (e.g., lead, mercury, arsenic, beryllium, cadmium, chromium)	Different metals can cause cancer, respiratory impairment, damage to nervous system [some toxic to raptors, other top carnivores; most heavy metals can bioaccumulate through food chains]	Mercury, cadmium, others known to cause leaf injury, reduce growth	Adsorbed onto soil particles; depress decomposition rates by microorganisms; may also inhibit nitrification; some (e.g., mercury) can be converted to toxic organic forms by microorganisms	Capable of long-distance transport and deposition	
Carbon monoxide	Induces fatigue; impairs alertness; inhibits fetal development; aggravates cardiovascular diseases	Can produce acute injury at relatively high levels (>100 ppm).	Absorbed by microorganisms; inhibits N-fixing symbionts		
Carbon dioxide	High levels induce fatigue, interfere with normal respiration	Stimulates photosynthesis up to a point; can become toxic at very high levels	Little known	Can reradiate heat, inducing atmospheric warming	SO_2

Sources: Smith (1981), Westman and Conn (1976), Winner et al. (1985), Dvorak and Lewis (1978), Council on Environmental Quality (1976).

[a] Additional major air pollutants include asbestos, chlorine, hydrogen sulfide (H_2S), ammonia (NH_3), pesticides, alcohols, aldehydes, organic acids, ketones, mercaptans, N-nitroso compounds, vinyl chloride, heat, and others.

[b] Much less is known about air pollution effects on wildlife than on humans. For additional references, see Dvorak and Lewis (1978); for arthropods, see Hay (1977) and Smith (1981).

[c] Natusch et al. (1974).

pollution source (e.g., Westman 1979). (2) Species populations with short generation times (annuals, short-lived perennials) are more likely to be resistant to air pollution stress than those with longer generation times (e.g., trees). This occurs for at least two reasons. First, individuals with greater genetic resistance to the pollutant will increase more rapidly in a stressed population, the shorter the interval between generations (see, e.g., development of resistant strains of grasses under SO_2 stress in Bell and Clough 1973, Bell and Mudd 1976, Horsman et al. 1978, Westman et al. 1985). Second, species with short generation times are more likely to reproduce before succumbing to pollution-induced death; thus a population of annuals may maintain itself at a polluted site by a "pollution-evading" strategy that is not open to a long-lived perennial. The annual population replenishes itself with fresh, unstressed organisms each year, and avoids pollution stress during part of the year by seed dormancy. (3) The weakening or death of sensitive species in an ecosystem is likely to be followed by an increase in tolerant indigenous species or an invasion by tolerant nonindigenous ("exotic") species (e.g., Winner and Bewley 1978; Figure 7.7) due to competition.

In Chapter 9 various attributes of species that impart resistance to pollution are discussed. It is important to recognize, however, that relative resistance to pollution exhibited in the laboratory setting may not be a good predictor of relative resistance of these same species in the field. Neither interspecies competition nor the influence of differing generation times are effectively reproduced in laboratory tests.

In predicting effects of chronic air pollutant stress on terresterial ecosystems, then, one might initially rank indigenous species by rapidity of growth to sexual maturity as well as known sensitivity to the pollutant in single-species tests. Species with rapid growth rates and low pollutant sensitivity are likely to increase in cover and be joined initially by tolerant exotic annual and short-lived perennial species. Over time, resistant strains in the indigenous population may be selected for, evident first in the herbaceous community with shortest generation times. Further discussions of ecological effects of air pollutants may be found in such books as Butler (1978), Dvorak and Lewis (1978), Guderian (1977), Mansfield (1976), Miller (1980), Mudd and Kozlowski (1975), Smith (1981), and Winner et al. (1985).

WATER

Categorization of Pollutants

The way in which we categorize air or water pollutants influences how useful the information will be for ecological assessment purposes. The U.S. Clean Air Act, for example, chose to focus on the effects of atmospheric substances on human health and welfare. Seven pollutants were identified, whose ambient concentrations were to remain within healthful levels. Five of these were categorized by specific or general chemical formula (carbon monoxide, sulfur dioxide, oxides of nitrogen, lead, hydrocarbons), one by physical property (suspended particulates), and one by chemical reactivity (oxidants, later changed to a chemical formula basis, singling out ozone). In addition substances hazardous to human health or welfare that are emitted from industrial point sources were designated for control, and a list of these were promulgated by chemical formula. Pollutants that did not fall clearly into one of these categories tended to be neglected, such as chemicals affecting climate (CO_2), the ozone layer (freons), or the acidity of precipitation (sulfate) and unhealthful substances released by private individuals (e.g., pesticides, cleaning fluids, tobacco smoke) or from natural sources (e.g., allergens).

In the case of water pollutants the 1965 U.S. Water Quality Act focused concern on interference of pollutants with a variety of human uses of water (agricultural irrigation, industrial cooling, maintenance of aquatic life, swimming and boating, drinking). As a result a variety of criteria for categorizing pollutants were developed. Table 7.6 illustrates the categorization of water pollutants by three such criteria. For direct human contact the effect of water pollutants on the human senses (Table 7.6C) is important, but this is not true if the water is to be used, for example, for irrigation or industrial cooling. The interaction of pollutants with wastewater treatment technology (Table 7.6B, Figure 7.8) is the oldest and most widespread means for classifying pollutants but clearly of little aid in assessing the quality of water for other purposes such as protection of human health or influence on aquatic life. "Dissolved solids," for example, are a mixture of biostimulatory, toxic, and relatively inert chemical substances from the point of view of biological effect (Table 7.6A). Parameters of water quality to be monitored are frequently chosen using several such classificatory criteria. As a result a person seeking to assess the influence of water constituents on any one target, such as aquatic ecosystems, may find the list of water quality parameters incomplete and inappropriate to the assessment task. For analysis of effects on

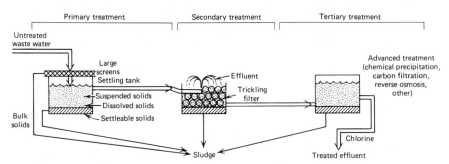

Figure 7.8. A conventional wastewater treatment system. Chlorination (and sometimes dechlorination) occurs at the last stage before effluent discharge.

Table 7.6.   A Classification of Water Quality Parameters by Three Criteria of Concern

	Criterion A. Effects on Organisms	
Category	*Components*	*Primary Biological Effects*
Physical	Heat	Affects metabolic rates; can denature proteins
	Turbulence	Entrainment in current, disturbance to mobility; suspension of particles, decrease in visibility, affecting predation, feeding, mating, etc.
Chemical	Nutrients including:	Affect rate and absolute growth of species populations
	Oxygen	Respiration
	Carbon	Photosynthesis
	Nitrogen ⎫ Phosphorus ⎬ Silicon ⎭	⎰ Common limiting elements to ⎱ growth in aquatic ecosystems
	Hydrogen, magnesium, calcium, sodium, chlorine, copper, zinc, manganese, molybdenum, potassium, iron, sulfur ⎬	⎱ Other essential elements to ⎰ growth
	Toxic substances including: Pesticides (DDT); other synthetic (PCBs) or natural organic compounds (phenol), heavy metals (mercury, cadmium); essential elements in toxic concentrations; radioactive substances	Interfere with metabolism at low concentrations, leading to physiological disruption, mutagenesis and/or death of organisms or their offspring
	Other (neither immediately biostimulatory nor toxic) including: Decomposable (oxygen-demanding) substances, slowly decomposable (relatively inert) substances	Smother sediment-surface organisms, decrease visibility; during oxidation (decomposition) will rapidly decrease dissolved oxygen and release nutrients or toxic substances
Biological	Pathogens including: Bacteria, fungi, viruses	Infection and disease to aquatic organisms or those that drink the water

Table 7.6. (*Continued*)

Criterion B. Interaction with Treatment Technology

Category	Components	Interaction with Technology
Bulk solids	Large objects	Removable by coarse screens
Settleable solids	Large particles passing through coarse screens	Settle quickly (minutes) to bottom of still tanks as sludge
Suspended solids	Finer particles	Remain suspended for hours or days in still tanks
Dissolved solids	Ions	Remain in solution in effluent
Coliform bacteria	Readily recognizable rod-shaped bacteria found in feces	Killed by chlorination

Criterion C. Interaction with Human Senses

Category	Components	Interaction with Sense
Sight	Ions, particles	Color clarity, appearance of slicks, foam
Smell	Volatile substances	Odor
Taste	Inorganic or organic chemicals	Flavor
Touch	Inorganic or organic chemicals	Consistency, oiliness

ecosystems, classification by biological effect (Table 7.6A) is most often appropriate.

Table 7.7 lists typical levels of removal of selected pollutants by wastewater treatment technologies. Notice that because the technologies were designed primarily to reduce suspended matter, oxygen-demanding substances, and bacteria the percentage of removal is greater for these items at primary and secondary stages of treatment than for other biologically important substances (nutrients, toxic substances, viruses). Were technologies redesigned to remove primarily substances of ecological effect, quite different treatment processes might well be used.

Water quality and water quantity are intimately related. The volume of water flow helps determine the concentration at which a pollutant will occur and its pattern of dispersion. The discussion of water quality in the next subsection will note some interactions with water quantity and will be followed by a brief discussion of hydrology and water movement.

Pollution Dispersion

Approaches to pollutant dispersion in water can be illustrated for a biologically degraded pollutant by considering the case of dissolved organic carbon

Table 7.7. Efficiency of Wastewater Treatment Methods in Removing Pollutants Defined by Criteria A and B Presented in Table 7.6

| | Suspended Matter | Oxygen-Demanding Substances (BOD) | Pathogens | | Nutrients | | Toxic Substances, Heavy Metals |
			Bacteria	Viruses	Phosphorus	Nitrogen	
Primary treatment							
Fine filters (screens)	5–20	5–10	10–20	0			
Settling tank	40–70	25–40	25–75	0–10	0–10	0–?	0
Secondary treatment							
Trickling filters	65–92	65–95	80–95	0–75?	10–70[b]	30–50	0–10
Testiary treatment							
Advanced chemical							
treatment	80–95 (98[a])	65–95 (98[a])	90–98 (99[a])	0–(90?)[b]	60–98	85	99

Sources: Weinberger (1971), Sosewitz (1971), Klein (1966), Havens and Emerson (1971), and adapted from Westman (1972a) with permission of Sigma Xi.

Note: Figures are percentage removed from effluent.

[a] EPA estimates (1971).

[b] Sosewitz (1971).

(DOC) in streams. The relationship between the concentration of dissolved organic carbon in the River Aare (Switzerland) and the flow rate of the river (Q) is illustrated in Figure 7.9a. As would be expected given a constant rate of discharge of organic wastes into the river from human sources, the concentration of organic carbon decreases (though not linearly) as the flow rate of the river increases. The concentration (C_t) of dissolved organic carbon at any time, in milligrams of carbon per liter (mgC liter^{-1}), multiplied by the rate of flow of the river (Q) in cubic meters per second (m^3 sec^{-1}) should give the amount of carbon flowing through a cubic meter of river each second, or the "mass flux" of carbon, in units of gC m^{-3} sec^{-1} (since 10^3 liter = 1 m^3). This mass flux of carbon ($C \times Q$) is dependent on the volume of waste carbon initially discharged per unit time (C_0/t) plus the flow rate (Q) of the river, times the level of carbon (b) eroded from the natural riverine ecosystem (Stumm and Morgan 1981) so that

$$C_t \times Q = \frac{C_0}{t} + bQ \qquad (7.7)$$

or

$$C_t = \frac{C_0}{tQ} + b \qquad (7.8)$$

In Figure 7.9b the data of Figure 7.9a have been replotted with $1/Q$ rather than Q on the abscissa, so that the straight-line relationship of Eq. 7.8 becomes apparent. By fitting a linear regression through the data, one can calculate the slope A as 460 gC sec^{-1}. This is the average flux of dissolved organic carbon due to waste discharge. The y-intercept, b, is 2.06 mgC liter^{-1}, which is the background level of dissolved organic carbon removed from the river segment.

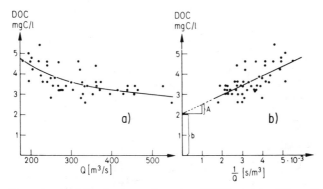

Figure 7.9. Dependence of dissolved organic carbon (DOC) on rate of flow, Q, in the River Aare, Switzerland. From Figure 2 of Zobrist et al. (1977). Reprinted with permission of Schweizerischer Verein von Gas- und Wasserfaches, Zurich, Switzerland.

Equation 7.8 assumes that the "background" amount of carbon (b) removed is due to erosive forces and hence is flow-rate dependent. It also assumes that the organic carbon discharge is simply being diluted and is not being assimilated by organisms, sedimenting out or being decomposed by bacteria. We know that all three of these processes act on carbon in an aquatic system, so that Eq. 7.8 is only an oversimplified starting point for modeling pollution dilution, best applied to substances that remain suspended in water for long periods without removal or decomposition.

In the case of dissolved organic carbon in water, as with terrestrial litter, the rate of decomposition is generally found to be a function of the concentration present at a given time (C_t) (Stumm and Morgam 1981) so that

$$\frac{-dC_t}{dt} = kC_t \tag{7.9}$$

where k is a constant expressing the rate of decomposition, and the expression dC_t/dt refers to the rate of change in the instantaneous concentration of carbon (C_t) over time. This is a mathematical way of expressing an exponential rate of decay. Given an initial mass C_0 of decomposable organic matter, the amount left at time t (C_t) will be

$$C_t = C_0 e^{-kt} \tag{7.10}$$

Equation 7.10 is the integral of Eq. 7.9. The reason the decay of organic carbon in ecosystems is a negative exponential (i.e., a function of the concentration at time t; see Figure 7.10) is that decomposition rates of organic matter slow down as the more readily decomposed carbohydrates are metabolized and the more refractory, or resistant, substances remain. If the rate of input of fresh organic matter (L for effluent discharge and natural erosion, or

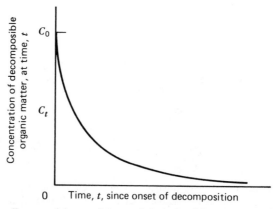

Figure 7.10.   Exponential rate of decay of decomposible organic matter over time.

litterfall in the case of terrestrial ecosystems) is in equilibrium with the decomposition of organic matter, and its loss to the atmosphere or water as CO_2 and H_2O, then the amount of organic matter (C_s) left in steady state in the river (water plus sediment) or forest floor (in the case of litter) is

$$C_s = \frac{L}{k} \tag{7.11}$$

If we start with a stream, lake, or landscape containing no organic matter, the rate of buildup of organic matter will be

$$C_t = C_s (1 - e^{-kt}) \tag{7.12}$$

The "half-life" ($t_\frac{1}{2}$) for decaying of an initial mass of organic matter can be calculated by substituting $C_t = 0.5\,C_0$ in Eq. 7.10 once k is known and solving for t (Eq. 7.10–7.12 adapted from Whittaker 1975). Table 7.8 shows some average quantities of organic matter in litter or aquatic sediments and their rates of decay. Decay rates for leaves in water are slower than on land because of the more anaerobic conditions.

Knowing the rate of discharge of organic carbon waste, C_0/t, into a river, one would need to know what proportion sediments out per unit time (k_1), what proportion is decomposed in suspension (k_2), and what proportion is assimilated by detritivores (k_3). One simple way to begin modeling the rate of change in organic matter in a river at any time would be to assume that the total change in organic matter concentration per time will be a function of the initial rate of input of organic matter minus losses due to the partitioning of this discharged mass into sediment (k_1), instream decomposition (k_2), and assimilation (k_3) (Figure 7.11). Thus

$$-\frac{dC_t}{dt} = \frac{C_0}{t} - k_1 \frac{(C_0)}{t} - k_2 \frac{(C_0)}{t} - k_3 \frac{(C_0)}{t} \tag{7.13}$$

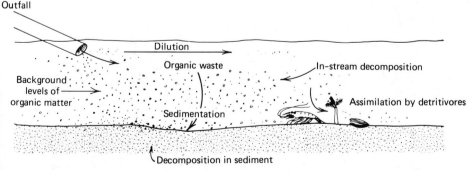

Figure 7.11.   Fates of organic matter discharged into a stream as discussed in the model in the text.

Table 7.8. Average Rates of Decay and Steady-State Values of Organic Matter in Terrestrial Ecosystems (Litter Layer) and Aquatic Ecosystems (Sediment)

Terrestrial	Steady-State Organic Mass	Rate of Organic Input L	Decay Rate	Decay Half-Life
	Forest Floor (g m^{-2})	Litterfall (g m^{-2} yr^{-1})	k (yr^{-1})	t$_{\frac{1}{2}}$ (yr)
Tropical rain forest	200	1200	6.0	0.12
Temperate deciduous forest	1200	800	0.67	1.0
Boreal conifer forest	6000	600	0.10	7.0
Temperate grassland	2000	500	0.25	2.8

Aquatic	Sediment (g m^{-2})	Decay Rate, k (yr^{-1})		Decay Half-Life, t$_{\frac{1}{2}}$ (yr)	
		Coarse particulate organic matter	Fine particulate organic matter	Coarse particulate organic matter	Fine particulate organic matter
Boreal conifer forest (Quebec)					
First Choice Creek	610	0.014	0.065	74	15
Moisie River	71	0.011	0.144	97	7
Temperate deciduous forest (New River, Va.)					
Acer negundo leaves		0.018			
Platanus occidentalis leaves		0.003			

Sources: Terrestrial values from Whittaker (1975). Aquatic values for boreal conifer forest from Naiman (1983); for deciduous forest from Paul et al. (1983).

The gradual process of building a conceptual model for degradation of organic matter in a river starts with such simplified conceptual beginnings, and modifies or adds terms as more realistic relationships are determined. Notice, for example, that to take account of changes in concentration of organic carbon due to variations in river flow rate, we could return to Eq. 7.8. Thus C_0/t could be replaced by $Q (C_t - b)$ wherever it occurs in Eq. 7.13, in order to incorporate dilution due to the flow rate, Q.

Even as we progress to make such a model more elaborate, we need to note problems with the assumptions made along the way. For example, organic matter decomposition does not always proceed by a neat first-order exponential decay rate (see, e.g., Birk and Simpson 1980 for a discussion of departures from this model in eucalypt forest litter). Further the decay rate calculated in a stream containing only organic matter inputs will differ from one where other pollutants, such as heavy metals, may be present in the sediments or adhering to the organic matter, thus killing some of the decomposer population. Additional refinements needed in the model include ways to characterize exhaustion of the dissolved oxygen supply in water or sediment and rates of reoxygenation by incorporation of surface air, fluctuations in decay rates due to water and sediment temperature, variations in assimilation of detrital carbon by organisms due to growth in these populations, and other items. As with any modeling process the possible degree of refinement of the model to increase its realism is virtually infinite.

The modeling of organic matter decomposition in aquatic ecosystems is usually approached as part of the larger problem of considering all substances that use up oxygen in the water as they are respired by decomposers, or oxidized chemically, as well as processes of reaeration. The amount of oxygen used up in the respiration of organic matter by decomposers is termed the *biological oxygen demand* (BOD) of the substance and, for chemical oxidation, *chemical oxygen demand* (COD). The two processes together are referred to as "biochemical oxygen demand." In a laboratory BOD is usually measured over a fixed period (five days) at a standard water temperature (20°C) with a known concentration of organic waste in water seeded with decomposing bacteria, expressed as $BOD_{5d,20°}$. Models of BOD and COD removal in streams and lakes are discussed in Krenkel and Novotny (1980).

Typically models of dissolved oxygen (DO) predict a "sag" in DO levels downstream of the organic pollution source due to bacterial respiration of the organic matter, followed by recovery of DO levels downstream as the organic matter is degraded and diluted. Often tube worms (Tubificidae) and midge larvae (*Chironomus*) which can tolerate low oxygen conditions thrive immediately downstream, along with certain fungi and protozoa. As oxygen levels recover, algae bloom on the released nutrients; eventually a full complement of species returns. A simplified illustration of these changes is shown in Figure 7.12.

We have used relatively simple formulae to express relationships. The

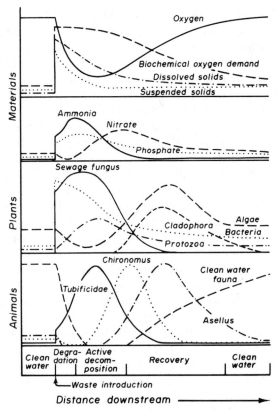

Figure 7.12. Simplified diagram of some chemical and biotic changes downstream following introduction of organic sewage to a stream. Reprinted, with permission, from Figure 16 (p. 94) in H. B. N. Hynes (1960), *The Biology of Polluted Waters*, Liverpool University Press.

actual empirical data on which they are based show variation around the modeled trend, however. Thus in Figure 7.9*b* the coefficient of correlation, *r*, for the regression was 0.65. Ideally one should keep track of these terms of variation so that they may be used to calculate theoretical confidence intervals around final model estimates. In complex computer models, however, calculating such "error terms" is usually not possible because of the fact that not all relationships were empirically derived and because the contributions of "errors" or variations from each component of the model vary in complex ways. Nevertheless, it is important to remember that simulation models are built using empirical relationships associated with such natural variation, and that the resulting estimates have compounded a series of such error terms in the process of computation.

Temperature

Some major effects of temperature on aquatic ecosystems are summarized in Table 7.9. In Figure 7.13 some of the physical effects of heated water dis-

Table 7.9.   Some Ecological Effects of Thermal Discharges into Aquatic Ecosystems

Physical

1. Discharge upsets existing thermal stratification. Depending on depth of discharge, heat can expand the warm layer, warm the cooler waters, induce thermal currents or mixing, and can change time of year when lakes turn over.
2. Fishes and other mobile organisms will avoid a thermal plume, or zone of water of different temperature. If the plume extends across a river, it will be a barrier to migration. If the plume is narrow but deep, it may also be a barrier to migration, as many organisms can change depth or habitat only within a narrow range.

Chemical

1. Higher temperature decreases the solubility of gases in water. Thus concentrations of dissolved oxygen and carbon dioxide in water decrease as temperature rises.
2. The effect of heat addition on the solubility of gases is greater in salt water than fresh water because (a) gases are less soluble in salt water and (b) the temperature of salt water is raised more per unit addition of heat.
3. Higher temperatures increase the rate of chemical oxidation of substances, further decreasing dissolved oxygen content of water.
4. Salts become more soluble, releasing increasing concentrations of ions that may be toxic into solution.

Biological

1. As temperature rises, tolerances of aquatic organisms can be exceeded, (lethal level typically 35°C, with exceptions), causing denaturation of enzymes, other physiological disruptions, and death.
2. Because different species have different temperature optima for growth, as temperature changes different species are favored, changing the composition of the aquatic community.
3. Alteration of natural temperature regime can adversely affect reproduction, survival of offspring, predator-prey behavior, resistance to disease of organisms. Generally a change of 3°C is the maximum tolerable shift in ambient temperature for species.
4. Higher temperatures increase rate of assimilation of substances, enhancing body concentrations of heavy metals, pesticides, cyanide, and other substances in water; nontoxic concentrations of pollutants can become toxic in warmer water.
5. Higher temperatures increase rate of respiration by organisms, including decomposers, leading to greater biological oxygen demand, and further drop in dissolved oxygen levels.
6. Higher temperatures speed rate of growth of organisms, leading to an unwanted "bloom" of phytoplankton and macroalgae, which can in turn deplete dissolved oxygen supplies.

(continued on next page)

Table 7.9. *(Continued)*

Biological

7. A sudden change in temperature, either by onset or cessation of thermal discharge, can cause thermal shock to organisms, which can be fatal. Some fishes have been shown to be more susceptible to damage from sudden decreases, than to sudden increases in temperature (Speakman and Krenkel 1972).

Sources: Cole (1969), Westman (1972b); see also Gibbons and Sharitz (1974), U.S. Environmental Protection Agency (1977b).

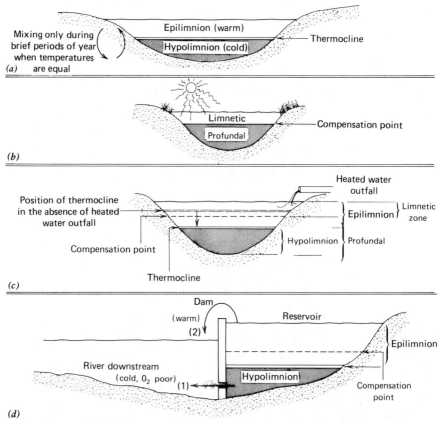

Figure 7.13. Examples of stratification in quiet water bodies. (*a*) Thermal stratification; (*b*) light stratification; (*c*) effects of thermal discharges on relative positions of layers; (*d*) effects of water withdrawals from different layers of a dam on the receiving water.

charges are illustrated. Figure 7.13a illustrates the natural layering of warm water over cold in a relatively still water body (lake, reservoir, open ocean) during warm seasons. The thermocline marks the sharp transition from warm, surface waters (epilimnion) to cold, deepwaters (hypolimnion). In temperate regions mixing between these two layers occurs only during the brief period when the surface waters have cooled (fall) or melted (spring) to the same temperature as the deep waters. In warmer climates, where waters do not freeze, only a single period of mixing occurs during the cooler season. In the tropics complete mixing is typically absent, and only weak currents exist. A result of this thermal stratification is that profundal layers are cut off from reaeration by surface or air contact, while experiencing continual oxygen demand from biological and chemical oxidation of sunken detritus. Dissolved oxygen concentrations may drop to levels inadequate to support most aerobic life (<5 ppm).

Figure 7.13b illustrates a second type of layering that occurs simultaneously in water bodies, due to the progressive absorption of solar radiation as it penetrates the water. A point is reached where there is insufficient light for plants to carry on net photosynthesis (gross photosynthesis − respiration). This point is called the *compensation point*. Below this level of light in a water body, plants can no longer survive. The level above this point is the limnetic zone; below it, the profundal zone.

Figure 7.13c considers a situation where the thermocline has been moved below the limnetic zone due to heated water discharges to a lake. (Such stratification can also occur naturally.) In the absence of the heated water discharge a portion of the limnetic zone is in the hypolimnion. Oxygen produced by the net photosynthesis of phytoplankton can help replenish oxygen supplies in the hypolimnion. The heated water discharge, by lowering the thermocline below the limnetic zone, cuts off this source of oxygen to the hypolimnion, increasing the potential for reductions of dissolved oxygen in the hypolimnion to levels lethal to benthic (bottom) and demersal (bottom-swimming) organisms.

Figure 7.13d considers the situation where a dam has created a still reservoir that has developed thermal stratification, although the downstream water is sufficiently shallow and turbulent to remain well mixed and oxygenated. If water from low enough in the dam wall to be drawing from the hypolimnion (point 1 in Figure 7.13d) is released to the river, the river will receive a slug of oxygen-poor (and frigid) water that may be damaging to riverine aquatic life. By the same token, if water is introduced from the top of the spillway (point 2 in Figure 7.13d), this warm surface water may be a thermal shock to the cooler, well-mixed river water. Ways to mitigate these problems include designing the dam with several depths of water outlet, so that water of downstream temperature can be drawn, and introducing artificial mixing (hydrofoils, deep propellers) into the dam to reoxygenate the hypolimnion. In the case of heat discharges to a stream, plant operations may be adjusted to keep the size of plume within acceptable limits. Cooling towers may be used to dissipate heat to the atmosphere, avoiding the need for heat discharge to water.

A mixing zone is the region below an outfall where pollutant levels in a stream remain above ecologically acceptable levels (Figure 7.14). It is possible to model heat dispersion from a knowledge of the river flow rate and the rate and volume of heated water input. Such models are discussed in a variety of engineering texts, such as Krenkel and Novotny (1980) and Threlkeld (1970). A detailed text on mixing of water and pollutants in water bodies is that of Fischer et al. (1979).

Among the methods used to trace thermal plumes are infrared aerial photography (remote sensing). The heated waters, by releasing more infrared radiation, develop a lighter color on IR-sensitive film, as opposed to a dark background (Figure 7.15). The use of remote sensing to trace differences in sediment load and in dissolved oxygen contents of water bodies is also well developed. Another method, applicable to any effluent, is to introduce a colored or radioactive tracer into the effluent and follow its path of dispersal visually, with a Geiger counter, or by analyzing aliquots of water downstream for color or fluorescence. More traditional methods involve releasing a series of flagged buoys into a dispersing current (Fischer et al. 1979).

The chemical effect of thermal pollution on the solubility of dissolved oxygen is particularly significant (Table 7.9). Because of the higher specific heat and lower solubility of gases in salt water, the discharge of heated waters into estuaries may be unusually damaging ecologically. The effects of heat and salt on oxygen in the open ocean are countered somewhat by the greater volume of water available for mixing. The mixing of chlorine or antiscaling agents with heated discharges may prove more damaging to phytoplankton than the heat itself (Brook and Baker 1972).

One of the benefits of thermal discharges sometimes cited is the increase in abundance of fish life noted around heated outfall pipes. The fish may well

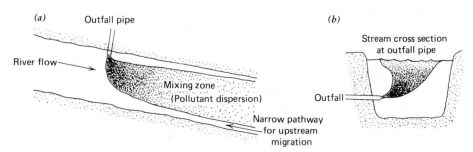

Figure 7.14. Mixing zones in streams. (*a*) Longitudinal section of a dispersing plume of effluent in a stream. (*b*) Cross-section of a buoyant plume from a below-surface outfall pipe. Downstream the plume will have become more evenly mixed across the stream. Modified, with permission, from H. B. Fischer, E. J. List, R. C. Y. Koh, J. Imberger, and N. H. Brooks (1979), *Mixing in Inland and Coastal Waters*. Copyright (c) 1979 by Academic Press, New York.

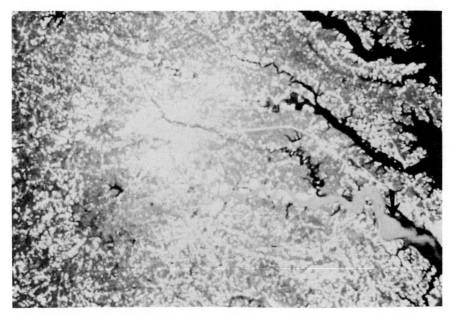

Figure 7.15. Pollution in the Potomac River, Washington, D.C. (arrow at right) shows up as lighter shade on infrared-sensitive film. LANDSAT photo courtesy of NASA.

be attracted to the rich supply of prey growing more rapidly in the warmer waters and feeding off any nutrients in the discharge. The increased local abundance of fish is probably a result both of the increased local food supply, and the migration of fish from surrounding regions, so that total increases in fish productivity may be smaller than would appear from a sample limited to the outfall region. Furthermore fish feeding on prey around the outfall are also more prone to adverse effects from any chemical constituents in the effluent and the adverse effects of heat. Thus, for example, while fish abundance around the marine sewage outfall from the Hyperion plant in Los Angeles is increased, the incidence of fin-rot disease is also higher in these fish (Cross 1982).

By contrast, under managed conditions the benefits of heated water in stimulating growth can be put to good use (e.g., mariculture, aquaculture; hydroponic growth of terrestrial plants; heating of greenhouses).

Nutrients

Because a living cell has a well-defined structure, requiring a given set of proteinaceous and carbohydrate building blocks, there is a strong tendency for plants to assimilate essential elements from the environment in fixed proportions. To be sure, some elements can be accumulated in excess amounts ("luxury accumulation"), and species differ in the ratios with which they need the elements, but generalizations are nevertheless possible.

Stumm and Morgan (1981) provide the following equation for uptake of essential elements by phytoplankton on the average:

$$106\ CO_2 = 16\ NO_3^- + HPO_4^{2-} + 122\ H_2O + 18H^+\ (+\ \text{Trace elements})$$

$$
\begin{array}{cc}
\text{Photosynthesis} & \Big\uparrow \text{Respiration} \\
\text{(in light)} \Big\downarrow & \\
\end{array}
$$

(7.14)

$$C_{106}\ H_{263}\ O_{110}\ N_{16}\ P_1 + 138\ O_2$$
$$\text{Algal Protoplasm}$$

Justus von Liebig formulated a hypothesis regarding growth of organisms which has since been called "Liebig's Law of the Minimum": the total amount of growth in a population is limited by that essential growth factor present in least supply. To be sure, the levels of such environmental variables as temperature, moisture, and light will influence the *rate* of growth, but should an absolute scarcity of one essential growth factor develop, further growth will depend on the supply of that factor. Provided that environmental factors affecting rate of growth are within thresholds of tolerance, it is most likely that an absolute scarcity will develop first in some essential element, vitamin, or chemical cofactor. From empirical study ecologists have found that phosphorus is most often the limiting element to growth of primary producers (primarily phytoplankton) in freshwater ecosystems, and nitrogen in estuarine and marine ecosystems (see, e.g., Likens 1972, Ryther and Dunstan 1971, Schindler 1974, 1977, Stumm and Morgan 1981).

One crude way to determine which of the macronutrients (C, H, O, N, or P) might be in least supply in a given water body is by analyzing the water chemically for the concentrations of these substances in soluble form. Suppose we found that a lake under study had the following ratio of the five elements dissolved in its water:

$$
\begin{array}{c}
C\ :\ H\ :\ O\ :\ N:P \\
250:550:300:50:2
\end{array}
$$

(7.15)

whereas we know from Eq. 7.14 that algae are more likely to need these elements in the following ratios:

$$
\begin{array}{c}
C\ :\ H\ :\ O\ :\ N:P \\
106:263:110:16:1
\end{array}
$$

(7.16)

We can see, by doubling all the figures in Eq. 7.16 and comparing them to Eq. 7.15, that algal growth will have consumed all the available phosphorus before the supply of any other element is exhausted. Thus, if there is no other even more limiting factor that we have failed to consider, once algal biomass has grown to consume all the available phosphorus, further increases in biomass of the plants will occur in a 1 : 1 ratio with the addition of external phosphorus (from soil erosion, fertilizer runoff, injection of P-laden sewage, or other effluent).

A more direct test of the limiting element in water can be achieved by an algal bioassay (e.g., U.S. Environmental Protection Agency 1971b, 1978; Weber 1973). Here an algal species is introduced to an illuminated test tube containing sterilized or filtered lake water and adequate supplies of CO_2 (bubbled in if necessary). The test organism is typically one of three unicellular algae: *Selenastrum capricornutum, Anabaena flos-aquae, Microcystis aeruginosa.* The algal culture will quickly grow to consume the nutrients in the lake water sample; at that point algal growth will occur only from release of nutrients by the death of other algal cells, and population numbers will be in steady state. To some of these algal cultures is added a known concentration of an element such as P or N. If a sudden increase in growth occurs from the addition of one of these elements, it is identified as the limiting nutrient. Sometimes other elements can become limiting once the initial limiting element is supplied in excess amounts.

For example, Figure 7.16 shows that nitrogen was the primary limiting element to the growth of *Selenastrum capricornutum* in Ganzert Lake, Texas, in December, 1980. EDTA is a chelating agent that keeps relatively insoluble ions, like iron, in suspension. Algal biomass can be determined readily by measuring the optical density of the culture with a colorimeter or spectrophotometer and calibrating measurements to a dry weight basis on a

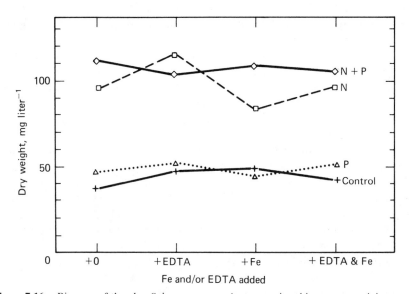

Figure 7.16. Biomass of the alga *Selenastrum capricornutum* in a bioassay containing water collected in December 1980 from Ganzert Lake, Texas. Graph shows growth of the algae upon addition of N, P, Fe, and EDTA in various combinations. From Hoban et al. (1981).

sample of the algal solution (Weber 1973). For a description of a computer model of eutrophication in a lake, see Krenkel and Novotny (1980, Ch. 13); for an account of trial fertilization of a lake with phosphorus, see Schindler (1974, 1977).

Some rules of thumb regarding eutrophication have appeared in the literature, but these must be regarded with extreme skepticism since they are averages of a limited sample; particular water bodies may differ dramatically from these generalizations. Ryther and Dunstan (1971) report that in marine waters N/P ratios > 15 : 1 usually imply P limitation, < 15 : 1, N limitation, with the actual threshold varying from 5 : 1 to 20 : 1 (see also Krenkel and Novotny 1980, Sawyer 1947, and Vollenweider 1968, 1975).

Toxic Substances: Heavy Metals, Radionuclides and Synthetic Organic Compounds

Heavy metals (metals generally in the first two columns of the periodic chart), radionuclides (radioactive (unstable) isotopes in ionic form), and synthetic organics (organic compounds that do not occur naturally) are chemically quite different. Yet their movement through the aquatic portion of biogeochemical cycles has certain similarities. All three tend to be adsorbed onto organic particulates and hence are prone to be assimilated by organisms that eat detritus (dead organic matter particles). These so-called detritivores excrete detritus depleted of energy-rich compounds relative to these substances. As the detritus is consumed, excreted, and reconsumed, the heavy metals, radionuclides, or synthetic organics become concentrated in the detritus on a dry-weight basis. In addition to a detritus food chain, aquatic ecosystems contain a grazing food chain composed of plants and animals that eat living organisms.

Several additional factors contribute to the increase in concentration of toxic substances as they pass from one consumer to the next in the detritus and/or the grazing food chain in aquatic ecosystems (see Figure 7.17). First, the organisms themselves, be they detritivores (e.g., filter feeders: shrimp, barnacles, etc), grazers (e.g., most fish), or even bacteria (Patrick and Loutit 1976, McLerran and Holmes 1974), respire the large quantity of organic matter they ingest and release it as CO_2 and water. The heavy metals and radionuclides, like nutrients, are elemental and must be excreted as such rather than respired. They therefore have a different, often longer, residence time in the body than the carbon, oxygen, and hydrogen in respirable organic compounds. Many synthetic organic compounds, because their structure is novel to the catabolic (breakdown) processes of organisms, are refractory— that is, slowly degraded or not degraded at all. They therefore also must be excreted, rather than respired, to leave the body. In the case of phytoplankton and other plants that absorb these substances with the water stream, differential membrane permeability can act to retain certain ions and organic compounds relative to water and dissolved gases. Thus some algae can concentrate cadmium 140,000 times relative to the ambient water envi-

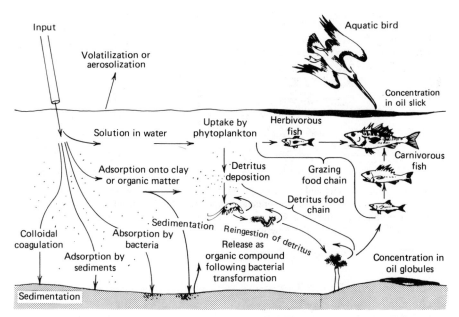

Figure 7.17. Generalized diagram of possible routes of transport or transformation of heavy metals, radionuclides, and synthetic organic compounds in aquatic ecosystems. Main sinks for these substances are sediments, oil slicks, and lipids of top carnivores.

ronment, for example. Subsequent ingestion of the algae by a planktonic grazer will result in respiration and release of the carbohydrate as CO_2 and H_2O and retention of much of the cadmium.

A second factor influencing bioconcentration is the water solubility of the compounds. The more highly water-soluble substances are likely to be more readily lost from the body by excretion, secretion, or simply water passage across cell membranes (in the case of unicellular organisms). Organic compounds, being nonpolar, are relatively water insoluble and tend to dissolve in body fats (lipids) or other organic compounds in the body. In the case of higher animals, the organics pass through the lipoproteins in the gut lining much more readily than do water-soluble compounds (Freed & Chiou 1981).

Figure 7.18 shows the differential partitioning of a variety of pesticides and other organic compounds in n-octanol (an organic compound) relative to water. Figure 7.18 shows that those substances that have the highest preference for dissolving in the n-octanol also are those that bioaccumulate to greatest relative concentrations in rainbow trout. Thus the degree of affinity to body fats ("lipophilicity") of a compound is one indication of its potential for bioconcentration. The molecular composition and lipophilicity (including n-octanol/water partition coefficients) of organic compounds can readily be found in chemical handbooks (e.g., CRC 1981, Hansch and Leo 1979).

In addition to affinity to organic solvents, other chemical affinities of a compound may be used as a basis for predicting where it will accumulate in

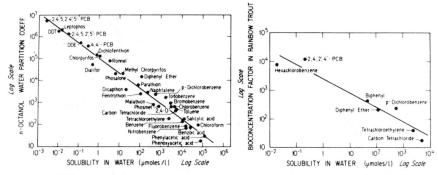

Figure 7.18. Lipophilicity and bioaccumulation. Both the *n*-octanol–water partition coefficient and the bioconcentration factor of the organic compound in rainbow trout are inversely proportional to the solubility of the compound in water. Reprinted, with permission, from Chiou et al. (1977). Copyright (c) 1977 by the American Chemical Society.

the environment. Figure 7.19 shows such predictions based on preferences of chemicals for water, organic solvents, sediments or air.

Gossett et al. (1982) rated the potential toxicity of a range of organic pollutants in sewage effluent by dividing the *n*-octanol/water partition coefficient by the lethal dose of the chemical required to kill 50% of a population of rats in 96 hours ($LD_{50,96}$; see Chapter 9). Chemicals with the highest value of this index are presumed to have both the greatest toxicity and the greatest potential for bioaccumulation. Toxicity data on rats, rather than aquatic organisms, were used because such data are available for a wider range of compounds (Lew and Tatken 1979). This index was then multiplied by the concentration of the compounds in the effluent discharge being studied, in order to rank pollutants of greatest hazard to aquatic life in that particular effluent. The differential tendency of such compounds to be adsorbed onto organic matter and the levels already present in the ecosystem are not taken into account in such an approach, but the resultant ranking may serve as a crude guide to potential hazard for proposed discharges.

Heavy metals and radionuclides in inorganic form are often relatively water soluble compared to organic compounds. However, they enter detritus food chains adhering to organic particulates from sewage or other natural or human sources. In addition a number of the inorganic substances are converted to organic form by aquatic microorganisms. For example, certain bacteria are capable of transforming inorganic mercury compounds into methyl and dimethyl mercury (McEntire and Neufeld 1975), and inorganic selenium to dimethyl selenide (Chau et al. 1976). The latter are subject to bioconcentration due to lipophilicity as well as the other mechanisms discussed. Microbial conversions from inorganic to organic form are also known for lead (Wong et al. 1975), arsenic (Wood 1973, 1974), tin, platinum, gold, and thallium (Agnes et al. 1971, Huey et al. 1974).

		Water Solubility Testing																															
		+																-															
		Partition Coefficient (Octanol/H_2O) Testing																															
		+								-								+								-							
		Adsorption Testing																															
		+				-				+				-				+				-				+				-			
		Desorption (Leachability) Testing																															
		+		-		+		-		+		-		+		-		+		-		+		-		+		-		+		-	
		Volatility Testing [Hi-(+), Lo-(-)]																															
		+	-	+	-	+	-	+	-	+	-	+	-	+	-	+	-	+	-	+	-	+	-	+	-	+	-	+	-	+	-	+	-
Probable Sites of Distribution	Air	X		X		X		X		X		X		X		X		X		X		X		X		X		X		X		X	
	H_2O - Aqueous Phase	X	X	O	O	X	X	X	X	X	X	O	O	X	X	O	O																
	H O - Sediment	O	O	X	X			X	X	O	O	X	X	O	O	X	X	O	O	X	X	O	O	O	O	O	O	O	O	O	O	O	O
	Soil	X	X	X	X			X	X			X	X			X	X	X	X	X	X												
	Animals - (Bioaccumulation)	X	X	X	X	X	X	X	X									X	X	X	X	X	X	X	X								

X - Primary Sites of Distribution
O - Distribution Sites of Secondary Importance

Figure 7.19. Likely sinks for compounds based on their chemical affinities for air, water, organic solvents, and sediment. From Stern and Walker (1978). Copyright American Society for Testing and Materials, 1916 Race Street, Philadelphia, PA 19103. Reprinted with permission.

The chemical forms that heavy metals or radionuclides can take in water are actually many more than the simple free metal ions and the organic compounds formed biologically. Additional forms include inorganic ion pairs and complexes, organic chelates or complexes, humic acid polymers, lipids, and polysaccharides. Heavy metals or radionuclides bound to other inorganic ions or to clay may form relatively large clusters bound together by ionic charge, forming colloidal-sized particles that stay in suspension but are large enough to reflect light. Finally, the heavy metals or radionuclides may be absorbed by bacteria or adsorbed onto the surfaces of soil particles or detritus and precipitate to the sediment, or move with sediment as it is transported downstream. Indeed, the major mechanism of transport of heavy metals in rivers is typically as organometallic particulates (see, e.g., Gibbs 1973, Trefry and Presley 1976). Other water quality parameters such as pH, redox potential, salinity, temperature, hardness, and CO_2 concentrations will ultimately affect the partitioning, transport, and bioaccumulation of toxic substances. For example, in acid waters ($\leq$pH 5) most heavy metals are more soluble and enter organisms through water uptake. Under more basic pH conditions these metals tend to form less soluble hydroxides and other ion complexes and are more likely to be ingested with detritus (Merlini and Pozzi 1977, Tsai et al. 1975).

The extent of concentration of these substances in the body of higher organisms will depend on many factors, including the age, size, metabolic rate, gender, and stage of life cycle of the organism, as well as its feeding pattern. Substances will differentially lodge in certain organs, often the liver, brain, egg-shell-laying organs, or skinfat in the case of pesticides and other organic compounds, the hair or feathers in the case of heavy metals and radionuclides in inorganic form. Hence in sampling organisms for concentrations of a toxic substance it is important to know which organs are the major repository for the substance in the particular species at hand.

Because of these complexities it has proved more difficult to model the fate of heavy metals, radionuclides, or even pesticides in aquatic systems than to model dissolved carbon, oxygen, or nutrients. If one assumes that a majority of the toxic pollutant will be transported with organic particulates, one can use conventional approaches that model particle transport with bottom sediment (see Krenkel and Novotky 1980, Ch. 10). The pollutant will meanwhile be subject to microbial transformation and degradation and biological uptake. In addition organic compounds, including organometallic complexes, pesticides, and pheromones (chemical messages), will differentially partition into any oil in the water, present either as surface slicks or as sunken oil on sediments (Seba and Corcoran 1969). The modeling of pollutants transported on particles is clearly only a very crude approximation to the true fate of such pollutants.

The complexity of ecological processes has inhibited the development of effective computer models of bioaccumulation. Studying uptake in microcosms, or sampling receiving water organisms for levels of these materials in their tissues, would seem to offer more suitable means for such estimation at present. It is tempting to extrapolate from concentration factors observed in previous studies (see, e.g., Table 7.10). However, because of the large number of factors influencing bioconcentration, the use of these figures for ecosystems other than those in which they were collected would be ill-advised. Large differences in behavior of these metals occur between water bodies, and between freshwater, estuarine and marine environments. Whereas bioaccumulation of certain pesticides continues monotonically to the highest levels of the food chain (e.g., DDT; Woodwell et al. 1967), heavy metal levels may be highest in detritivores and decrease in concentration in grazers at higher trophic levels (Dvorak and Lewis 1978; see also Table 7.10). Table 7.10 does emphasize, however, that concentrations of these substances in water can magnify in organisms by 1–5 orders of magnitude. Concentrations of the element in water that are sometimes below the level of detection can therefore rise to levels in organisms that may prove lethal or sublethal.

Species Tolerances

In addition to computerized sources of aquatic toxicity literature and data (see Table 9.1), various countries have produced compendia summarizing information on species tolerances to toxic substances as criteria documents (e.g., U.S. Environmental Protection Agency 1977b, Hart 1974); for a critique of the EPA document, see Thurston et al. (1979). These documents

Table 7.10. Example of Some Heavy Metal and Radionuclide Concentration Factors Relative to Ambient Water

Element	Concentration (ppm) in Water for *Lemna* Study Only	*Lemna minor* (Duckweed)	Freshwater Invertebrates	Freshwater Fish	Marine Invertebrates	Marine Fish
Cadmium	0.0038	600	2000	200	250,000	3000
Cobalt	0.0150	450	200	20	1000	100
Copper	0.0350	950	1000	200	1700	700
Lead	0.0157	850	100	300	1000	300
Manganese	0.1485	11,000	40,000	100	10,000	600
Mercury	—	—	100,000	1000	33,000	1700
Nickel	0.0270	400	100	100	250	100
Radium	—	—	250	50	100	50
Selenium	—	—	150	150	1000	4000
Silver	0.0004	82	800	2	3300	3300
Zinc	0.0290	1100	10,000	1000	100,000	2000

Source: Data for plants from Hutchinson and Czyrska (1975); for animals (edible portions only) from Vaughan et al. (1975), in Dvorak and Lewis (1978).

Note: These figures are from particular ecosystems and cannot be extrapolated with accuracy to other systems.

contain information on species' responses to pollutant concentrations primarily from single-species bioassays and to a lesser extent from field studies. Recommended standards are based on a compromise figure weighing the many complexities of pollutant transport, transformation, and the toxicity of various chemical forms to various species. Some of the shortcomings of this approach are discussed in Chapter 9. The recommended "standard" is a panel-of-experts judgment based on available literature. Because appropriate levels will differ with species, site, other water quality constituents, and other factors, these "standards" are obviously only crude guides to acceptable levels.

Water Quantity

Changes in land use affect the amount of water running off the land, in turn influencing both water quantity and quality. A discussion of hydrology and its interaction with land use is treated in many excellent texts (e.g., Dunne and Leopold 1978, Griggs and Gilchrist 1983, Marsh and Dozier 1981). Here we note some approaches used to study interactions between land and water quantity.

Analysis of Historical Data

Predictions of flood frequency and extent derive from long-term measurements of river volume. To calculate the volume of water discharged, the cross-sectional area of the stream at its mouth must be known, as well as the velocity of flow at various points in this cross section. Discharge is computed as area × velocity for each subsection of the river cross section. Because stream discharge varies continually, some simpler measurement that is correlated with stream discharge is desirable. Stream height, as measured by a recording gauge, is such a measure. A *rating curve* relates estimated discharges to measured stream heights. Stream heights (or estimated discharge rates) can be plotted over time in a *hydrograph*. From such hydrographs, preferably averaged over many years, a plot of the frequency of each stream height can be drawn. Another way to express river discharges is to compare the highest discharge for the year to the average of such figures over a several year period (the "mean annual flood"). The frequency of given stream heights can then be related to discharge volumes using a rating curve, and the frequency of recurrence of a given discharge (expressed as a ratio to the "mean annual flood" discharge value) plotted. Such a curve is called a *flood frequency curve*. Using information from such graphs, one can plot the boundaries of flooding along a river course for a given recurrence interval. Figure 7.20 shows the estimated extent of inundation by water, debris, and boulders, from floods with recurrence intervals of 10, 25, 50, or 100 years in an area of Nevada.

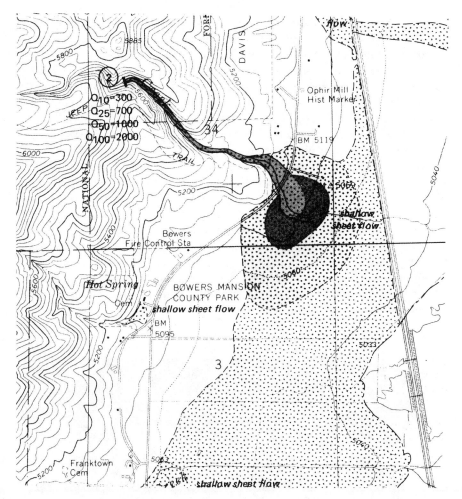

Figure 7.20.   Flood inundation and debris flow under four floods of increasing magnitude, Washoe City, Nevada. $Q_{10} = 300$ refers to a flood peak of 300 cubic feet per second (8.5 m³ sec⁻¹), with a recurrence interval of once every 10 years, or a 10% chance of occurring every year. The increasing boundaries of flooding and debris flow in the creek are shown for 10-, 25-, 50-, and 100-year floods on the background of a 1 : 24,000 topographic map. From Glancey and Katzer (1977).

Physical Models

The development of such historical data, and its application to topographic maps, is most readily accomplished for single gauged streams draining a relatively small watershed of stable land use mixes. In more complex water systems involving lakes or impoundments, the flooding patterns are more difficult to predict, as are the impacts of new land uses. An alternative approach to prediction of water volumes has been to build scaled-down physical models of the river or bay system and to follow water movement under the influence of various artificial alterations of channels, water input volumes, and so forth. In the United States such large hydraulic models have been built for San Francisco Bay and the Sacramento–San Joaquin Delta (Sausolito, California), Chesapeake Bay (Baltimore, Maryland), and portions of the Mississippi River (Waterways Experiment Station, Vicksburg, Mississippi). In Australia a scale model of a coastal river system in the Noosa–Caloundra area of Queensland was built to test effects of upstream development on marinas ("canal estates") at the river mouth.

The San Francisco Bay Model (Figure 7.21) occupies 47,000 square feet, the size of a large warehouse or two football fields. It has been used to study effects of ship channel dredging (deepening) on water flow, effects of barriers, effects of fill or increased freshwater flows on water quality, the dispersion of pollutants and sediment, and methods to alleviate salt water intrusion into the Delta area. Much smaller models (e.g., flumes) have been built to study sediment transport and other phenomena on a more general level. Analogous scale models of air flow around structures (e.g., buildings) have also been constructed, using wind tunnels.

Figure 7.21.   A physical model used to study effects of human alterations on water flow in San Francisco Bay and the Sacramento–San Joaquin Delta. Strips of metal in the model floor help guide water movement and adjust flow rate. Photo courtesy U.S. Army Corps of Engineers, Sausolito, Calif.

Computer Simulation Models

Computer simulation models of water flow and pollutant dispersion are another alternative for hydraulic modeling (e.g., Krenkel and Novotny 1980, Shen, 1979). Computer models have the advantage that they do not require a large physical space, they can be transferred quickly to other parts of the world, and they can be modified readily. Typically computer models incorporate fewer details of streambed morphology and channel contours; such models are therefore simple and potentially less accurate. However, the level of detail obtainable from a physical model may not be necessary for many assessment purposes. Often a cruder determination of worst-case events is sought, for which a computer model is more appropriate. For a discussion of further considerations in planning a river basin water quality and quantity assessment, see, for example, Goldman et al. 1973, Griggs and Gilchrist 1983, U.S. Geological Survey 1975–79, Warner et al. 1974, White 1977.

Table 7.11. Some Computerized Bibliographies and Data Banks on Air, Water, and Noise Pollution Emissions and Ambient Levels in the United States

Name and Location of File	Contents
Bibliographic Files	
Air	
APTIC (Air Pollution Technical Info. Center), U.S. EPA, Research Triangle Park, N.C.	Citations and abstracts on all aspects of air pollution
ENVIRON (Environmental Information Retrieval ON-line), U.S. EPA, Washington, D.C.	Ongoing R & D project information, published literature, and monitoring information
Water	
RECON, University of Arizona Tucson, Ariz.	Citations and abstracts on all aspects of water pollution
WRSIC (Water Resources Scientific Information Center), U.S. DOI, Washington, D.C.	Literature on water resources
Noise	
ENVIRON	See above.
NOISE (Noise Information Retrieval System), U.S. EPA, Washington, D.C.	Worldwide literature on environmental noise

(continued on next page)

Table 7.11. (*Continued*)

Name and Location of File	Contents

Data Banks

Air

AEROS (Aerometric and Emissions Reporting System), U.S. EPA, Washington, D.C.

Data on emissions from U.S. point and area sources; contains National Emissions Data System (NEDS) and ambient air quality Monitoring data (SAROAD: Storage and Retrieval of Aerometric Data); information on monitoring accuracy (QAMIS: Quality of Aerometric Data); point and source monitoring (SOTDAT: Source Test Data Storage); emissions of hazardous and trace substances (HATREMS: Hazardous and Trace Substance Inventory System); texts of state implementation plans (SIPS: State Implementation Plans); fuel use, emissions, and ambient air quality around energy facilities (EDS: Energy Data System)

Water

NEI (Natural Estuarine Inventory), U.S. EPA Washington, D.C.

Users of, and pollution status of, U.S. estuaries and coastal waters

Water Quality Standards, U.S. EPA, Washington, D.C.

All state water quality standards for all stream segments

NWDS (National Water Data System), U.S. Geological Survey, Water Resources Division, Reston, VA

Surface water flows and levels; ground-water levels and geologic data; flood frequency and inundation mapping; chemical quality of waters

ENDEX (Environmental Data Index). U.S. NOAA, Washington, D.C.

Locations of marine and coastal water environmental pollution data; what parameters are monitored; when program started

National Oceanographic Data Center, U.S. NOAA, Washington, D.C.

Physical, chemical, and biological data on the world's oceans and estuaries

Lake Survey Center, U.S. NOAA, Washington, D.C.

Charts, hydrologic, and limnologic data on the Great Lakes and their outflow rivers

STORET (Storage and Retrieval of Water Quality Data), U.S. EPA, Washington, D.C.

Data on contents of pollutant discharges and ambient water quality concentrations in 150,000+ U.S. locations; also information on plans and permits for water quality control

Table 7.11. (*Continued*)

Name and Location of File	Contents
GPSF (General Point Source File) U.S. EPA, Washington, D.C.	Enables recall of data on specific point sources or types of discharge from STORET

Source: Fennelly et al. (1976); see also Golden et al. (1979).
Note: For data banks on biological effects, see Table 9.1.

USE OF COMPUTER DATA BANKS FOR AIR AND WATER POLLUTION STUDIES

Information on emissions, monitored ambient levels of air and water pollution, and literature citations, are often retrievable from computerized data bases (Table 7.11). In addition the U.S. Department of Energy makes available a computer program (MERES) that calculates emissions of air and water pollutants from a variety of energy sources upon input of specifications on the size and design of the power plant.

Throughout this chapter we have seen that assessment of ecological effects of pollutants has involved as a first step the prediction of pollutant dispersion. Approaches to predicting dispersion have included computer (mathematical) simulation, the building of small-scale physical models, and the use of remote-sensing and field monitoring to track pollutant dispersion. The use of estimated ambient concentrations in predicting ecological effects involves an understanding of the tolerances of individual species to pollutants, the transfer and transformation of pollutants in ecological systems, and the effect of toxification of one part of the ecosystem on remaining portions. In Part V we will consider assessment of effects on the biota in greater detail.

REFERENCES

Agnes, G., Hill, H. A. O., Pratt, J. M., Ridsdale, S. C., Kennedy, F. S., and Williams, R. J. P. (1971). Methyl transfer from methyl vitamin B-12. *Biochem. Biophys. Acta* **252**:207–212.

Barnes, R. A. (1979). The long range transport of air pollution. Review of European experience. *J. Air. Poll. Control Assn.* **29**:1219–1235.

Bell, J. N. B., and Clough, W. S. (1973). Depression of yield in ryegrass exposed to sulfur dioxide. *Nature* **241**:47–49.

Bell, J. N. B., and Mudd, C. H. (1976). Sulphur dioxide resistance in plants: a case study of *Lolium perenne*. In T. A. Mansfield, ed. *Effects of Air Pollutants on Plants*. Cambridge Univ. Press, Cambridge, pp. 87–103.

Birk, E. M., and Simpson, R. W. (1980). Steady state and the continuous input model of litter accumulation and decomposition in Australian eucalypt forests. *Ecology* **61**:481–485.

Brook, A. J., and Baker, A. L. (1972). Chlorination at power plants: impact on phytoplankton productivity. *Science* **176**:1414–1415.

Butler, G. C. (1978). *Principles of Ecotoxicology*. SCOPE 12. Wiley, New York.

Chau, Y. K., Wong, P. T. S., Silverberg, B. A., Luxon, P. L., and Bengert, G. A. (1976). Methylation of selenium in the aquatic environment. *Science* **192**:1130–1131.

Chiou, C. T., Freed, V. H., Schmedding, D. W., and Kohnert, R. L. (1977). Partition coefficient and bioaccumulation of selected organic chemicals. *Environ. Sci. Tech.* **11**:475–478.

Cole, L. C. (1969). Thermal pollution. *BioScience* **19**:989–992.

Conn, W. D. (1975). The difficulty of forecasting ambient air quality—a weak link in pollution control. *J. Amer. Inst. Planners* **41**:334–346.

Council on Environmental Quality (1976). *Environmental Quality*. Seventh Ann. Rep. Washington, D.C.

CRC (1981). *Handbook of Chemistry and Physics. A Ready-Reference Book of Chemical and Physical Data*. 62nd ed. CRC, Boca Raton, Fla.

Cronan, C. S., and Schofield, C. L. (1979). Aluminum leaching response to acid precipitation effects on high-elevation watersheds in the Northeast. *Science* **204**:304–306.

Cross, J. N. (1982). Trends in fin erosion among fishes on the Palos Verdes shelf. In W. Bascom, ed. *Biennial Report 1981–1982*. Southern Calif. Coastal Water Res. Proj., Long Beach, Calif., pp. 99–110.

Dunne, T., and Leopold, L. B. (1978). *Water in Environmental Planning*. Freeman, San Francisco.

Dvorak, A. J., and Lewis, B. G. (1978). Impacts of coal-fired power plants on fish, wildlife, and their habitats. Rep. FWS/OBS-78/29. Off. Biol. Serv. and Environ. Contaminants Eval., U.S. Fish and Wildlife Service, Ann Arbor, Mich.

Eldon, J., Trijonis, J., and Yuan, K. (1978). Statistical oxidant relationships for the Los Angeles region. Contract A5-020-87. Calif. Air Resources Board, Sacramento, Calif.

Fennelly, P. F., et al. (1976). Environmental Assessment Perspectives. Rep. PB-257-911. GCA Corp, Bedford, Mass.

Fischer, H. B., List, E. J., Koh, R. C. Y., Imberger, J., and Brooks, N. H. (1979). *Mixing in Inland and Coastal Waters*. Academic Press, New York.

Freed, V. H., and Chiou, C. T. (1981). Physiochemical factors in routes and rates of human exposure to chemicals. In J. D. McKinney, ed. *Environmental Health Chemistry*. Ann Arbor Sci., Ann Arbor, Mich., pp. 59–74.

Gibbons, J. W., and Sharitz, R. R., eds. (1974). *Symp. on Thermal Ecology*. CONF-730505, AEC Symp. Ser. 31. Tech Info. Center, Oak Ridge, Tenn.

Gibbs, R. J. (1973). Mechanisms of trace metal transport in rivers. *Science* **189**:71–73.

Glancey, P. A., and Katzer, T. L. (1977). Washoe City Folio. Flood and related debris flow hazards map. Scale 1:24,000. Nevada Bur. Mines and Geology, Univ. Nevada, Reno.

Golden, J., Ouellette, R. P., Saari, S., and Cheremisinoff, P. N. (1979). *Environmental Impact Data Book.* Ann Arbor Sci., Ann Arbor, Mich.

Goldman, C. R., McEvoy, J., III, and Richerson, P. J., eds. (1973). *Environmental Quality and Water Development.* Freeman, San Francisco.

Gordon, A. G., and Gorham, E. (1963). Ecological aspects of air pollution from an iron-sintering plant at Wawa, Ontario. *Can. J. Bot.* **41**:1063–1078.

Gossett, R. W., Brown, D. A., and Young, D. R. (1982). Predicting the bioaccumulation and toxicity of organic compounds. In W. Bascom, ed. *Biennial Report 1981–1982.* Southern Calif. Coastal Water Res. Proj., Long Beach, Calif., pp. 149–156.

Griggs, G. B., and Gilchrist, J. A. (1983). *Geological Hazards, Resources, and Environmental Planning.* 2nd ed. Wadsworth, Belmont, Calif.

Guderian, R. (1977). *Air Pollution. Phytotoxicity of Acidic Gases and Its Significance in Air Pollution Control.* Springer-Verlag, New York.

Hameed, S. (1974). Modeling urban air pollution. *Atmos. Environ.* **8**:555–561.

Hansch, C., and Leo, A. (1979). *Substituent Constants for Correlation Analysis in Chemistry and Biology.* Wiley-Interscience, New York.

Hart, B., ed. (1974). *A Compilation of Australian Water Quality Criteria.* Aust. Water Resources Council Tech. Paper 7. Canberra.

Havens & Emerson Ltd. (1971). Feasibility study for wastewater management program. A report to the Buffalo District, U.S. Army Corps of Engineers.

Hay, C. J. (1977). Bibliography on arthropoda and air pollution. U.S. Forest Service, Gen. Tech. Rep. NE-34. Upper Darby, Pa.

Hesketh, H. E. (1972). *Understanding and Controlling Air Pollution.* Ann Arbor Sci., Ann Arbor, Mich.

Hill, F. B. (1973). Atmospheric sulfur and its link to the biota. In G. M. Woodwell and E. V. Pecan, eds. *Carbon and the Biosphere.* AEC Tech. Info. Center, Brookhaven, New York, pp. 159–181.

Hoban, M. A., Covar, A. P., and Lippe, J. C. (1981). Nutrient limitation study of Ganzert Lake, Williamson County, Texas. Radian Corp, Austin Tex.

Horsman, D. C., Roberts, T. M., and Bradshaw, A. D. (1978). Evolution of sulphur dioxide tolerance in perennial ryegrass. *Nature* **276**:493–494.

Huey, C. et al. (1974). The role of tin in the bacterial methylation of mercury. Proc. Int. Conf. Transport of Persistent Chemicals in Aquatic Ecosystems, II-73, Ottowa (as cited in *J. Water Poll. Control Fd.* **48**:1459–1486).

Hutchinson, T. C., and Czyrska, H. (1975). Heavy metal toxicity and synergism to floating aquatic weeds. *Verh. int. Ver. Limnol.* **19**:2102–2111.

Hynes, H. B. N. (1960). *The Biology of Polluted Waters.* Liverpool Univ. Press, Liverpool.

Krenkel, P. A., and Novotny, V. (1980). *Water Quality Management.* Academic Press, New York.

Klein, L. (1966). *River Pollution. Vol. 3. Control.* Butterworths, London.

Lew, R. J., Jr, and Tatken, R. L., eds. (1979). *Registry of Toxic Effects of Chemical Substances.* 2 vols. U.S. Dept. Health and Human Services, Natl. Inst. Occup. Safety and Health, Cincinnati, Ohio.

Likens, G. E., ed. (1972). Nutrients and eutrophication: the limiting-nutrient controversy. Amer. Soc. Limnol and Oceanogr. Spec. Symp. 1.

Mansfield, T. A., ed. (1976). *Effects of Air Pollutants on Plants.* Cambridge Univ. Press, Cambridge.

Marsh, W. M., and Dozier, J. (1981). *Landscape: An Introduction to Physical Geography.* Addison-Wesley, Reading, Mass.

McEntire, F. E., and Neufeld, R. D. (1975). Microbial methylation of mercury: a survey. *Water Poll. Control* **74**:465–470.

McLerran, C. J., and Holmes, C. W. (1974). Deposition of zinc and cadmium by marine bacteria in estuarine sediments. *Limnol. Oceanogr.* **19**:998–1001.

Merlini, M., and Pozzi, G. (1977). Lead and freshwater fishes. 1. Lead accumulation and water pH. *Environ. Poll.* **12**:167–172.

Miller, P. R., ed. (1980). *Effects of Air Pollutants on Mediterranean and Temperate Forest Ecosystems.* U.S. Forest Service, Gen. Tech. Rep. PSW-43. Pacific SW For. Range Expt. Sta., Berkeley, Calif.

Mudd, J. B., and Kozlowski, T. T., eds. (1975). *Responses of Plants to Air Pollution.* Academic Press, New York.

Naiman, R. J. (1983). The annual pattern and spatial distribution of aquatic oxygen metabolism in boreal forest watersheds. *Ecol. Monogr.* **53**:73–94.

Natusch, D. F. S., Wallace, J. R., and Evans, C. A., Jr. (1974). Toxic trace elements: preferential concentration in respirable particles. *Science* **183**:202–204.

Patrick, F. M., and Loutit, M. (1966). Passage of metals in effluents, through bacteria to higher organisms. *Water Res.* **10**:333–335.

Paul, R. W., Jr., Benfield, E. F., and Cairns, J., Jr. (1983). Dynamics of leaf processing in a medium-sized river. In T. D. Fontaine, III, and S. M. Bartell, eds. *Dynamics of Lotic Ecosystems.* Ann Arbor Sci., Ann Arbor Mich., pp. 403–423.

Ryther, J. H., and Dunstan, W. M. (1971). Nitrogen, phosphorus, and eutrophication in the coastal marine environment. *Science* **171**:1008–1013.

Samuelsen, G. R. (1980). Air quality impact analysis. In J. G. Rau and D. C. Wooten, eds. *Environmental Impact Analysis Handbook.* McGraw-Hill, New York, Ch. 3.

Sawyer, C. N. (1947). Fertilization of lakes by agricultural and urban drainages. *J. New Engl. Water Works Assn.* **51**:109–127.

Schindler, D. W. (1974). Eutrophication and recovery in experimental lakes: implications for lake management. *Science* **184**:897–899.

Schindler, D. W. (1977). Evolution of phosphorus limitation in lakes. *Science* **195**:260–262.

Seba, D. B. and Corcoran, E. F. (1969). Surface slicks as concentrators of pesticides in the marine environment. *Pesticides Monitoring J.* **3**:190–193.

Shen, H. W., ed. (1979). *Modeling of Rivers*. Wiley, New York.

Smith, W. H. (1981). *Air Pollution and Forests: Interactions between Air Contaminants and Forest Ecosystems*. Springer-Verlag, New York.

Sosewitz, B. (1971). Statement to the Subcommittee on Air and Water Pollution, Comm. Public Works, U.S. Senate. Fed. Water Poll. Control Hearings, Part 8, 3687–3716; 3812–4047. Ser. 92-H18. GPO, Washington, D.C.

Speakman, J. N., and Krenkel, P. A. (1972). Qualification of the effects of rate of temperature change on aquatic biota. *Water Res.* **6**:1283–1290.

Stern, A. C. (1976). *Air Pollution*. 3rd ed. Academic Press, New York.

Stern, A. M., and Walker, C. R. (1978). Hazard assessment of toxic substances: environmental fate testing of organic chemicals and ecological effects testing. In J. Cairns, Jr., K. L. Dickson, and A. W. Maki, eds. *Estimating the Hazard of Chemical Substances to Aquatic Life*. STP 657. Amer. Soc. Testing Materials, Philadelphia, Pa., pp. 81–131.

Strom, G. H. (1976). Transport and diffusion of stack effluents. In A. C. Stern, ed. *Air Pollution*. Academic Press, New York, pp. 401–501.

Stumm, W., and Morgan, J. J. (1981). *Aquatic Chemistry: An Introduction Emphasizing Chemical Equilibria in Natural Waters*. 2nd ed. Wiley-Interscience, New York.

Threlkeld, J. L. (1970). *Thermal Environmental Engineering*. 2nd ed. Prentice-Hall, Englewood Cliffs, N.J.

Thurston, R. V., Russo, R. C., Fetterolf, C. M., Jr., Edsall, T. A., and Barber, Y. M., Jr., eds. (1979). *A Review of the EPA Red Book: Quality Criteria for Water*. Water Quality Section, Amer. Fisheries Soc., Bethesda, Md.

Trefry, J. H., and Presley, B. J. (1976). Heavy Metal Transport from the Mississippi River to the Gulf of Mexico. In H. L. Windom and R. A. Duce, eds. *Marine Pollutant Transfer*. Lexington Books, Lexington, Mass., pp. 39–76.

Tsai, S. C., Boush, G. M., and Matsumura, F. (1975). Importance of water pH in accumulation of inorganic mercury in fish. *Bull. Environ. Contam. Toxicol.* **13**:188–194.

Turner, D. B. (1970). Workbook of Atmospheric Dispersion Estimates. Rev. ed. AP-26, U.S. Environmental Protection Agency, Research Triangle Park, N.C.

Unsworth, M. H., Crawford, D. V., Gregson, S., and Rowlatt, S. M. (1985). Pathways for sulfur from the atmosphere to plants and soil. In W. E. Winner, H. A. Mooney, and R. Goldstein, eds. *Sulfur Dioxide and Vegetation. Physiology, Ecology, and Policy Issues*. Stanford Univ. Press, Stanford.

U.S. Environmental Protection Agency (1971a). Performance standards for new coal-fired steam-generating plants. Fed. Reg. 36:24876 (1971), revised Fed. Reg. **40**:46250 (1975).

U.S. Environmental Protection Agency (1971b). Algal assay procedure: bottle test. Natl. Eutrophication Research Prog., Corvallis, Oreg.

U.S. Environmental Protection Agency (1974). Guidelines for air quality maintenance planning and analysis. Vol 12. Applying atmospheric simulation models to air quality maintenance areas. Off. Air Qual. Plann. Stds., Research Triangle Park, N.C.

U.S. Environmental Protection Agency (1977a). National emission standards for mercury. Code Fed. Reg. 40, Part 61, 66.

U.S. Environmental Protection Agency (1977b). Quality Criteria for Water. Off. Water Hazardous Materials, Washington, D.C.

U.S. Environmental Protection Agency (1977c). Compilation of Air Pollutant Emission Factors. 2nd ed. Rep. AP-42. Washington, D.C.

U.S. Environmental Protection Agency (1978). The *Selenastrum capricornutum* Prinz algal assay bottle test. EPA Rep. 600/9-78-018. Washington, D.C.

U.S. Environmental Protection Agency (1979). Performance standards for new coal-fired steam-generating plants. Fed. Reg. **44**:33580.

U.S. Geological Survey (1975–79). River-quality Assessment of the Willamette River Basin, Oregon. USGS Circular 715.

Van der Hoven, I. (1976). A survey of field measurements of atmospheric diffusion under low-wind-speed inversion conditions. *Nuclear Safety* **17**:223–230.

Vaughan, B. E., Abel, K. H., Cataldo, D. A., Hales, J. M., Hane, C. E., Rancitelli, L. A., Routson, R. D., Wildung, R. E., and Wolf, E. G. (1975). Review of potential impact on health and environmental quality from metals entering the environment as a result of coal utilization. Battelle Mem. Inst., Pacific NW Labs., Richland, Wash.

Vollenweider, R. A. (1968). The Scientific Basis of Lake and Stream Eutrophication. Tech. Rep. DAS/CSI/68. OECD, Paris.

Vollenweider, R. A. (1975). Input-output models, with special reference to the phosphorus loading concept in limnology. *Schweiz. z. Hydrol.* **37**:53–84.

Warner, M. L., Moore, J. L., Chatterjee, S., Cooper, D. C., Ifeadi, C., Lawhon, W. T., and Reimers, R. S. (1974). An assessment methodology for the environmental impact of water resource projects. EPA Rep. 600/5-74-016. Washington, D.C.

Weber, C. I., ed. (1973). Biological field and laboratory methods for measuring the quality of surface waters and effluents. Natl. Environ. Res. Center, U.S. Environmental Protection Agency, Cincinnati, Ohio.

Weinberger, L. (1971). Statement to the Subcommittee on Air and Water Pollution, Comm. Public Works, U.S. Senate. Fed. Water Poll. Control Hearings, Part 8, 3646-3652. Ser. 92-H18. GPO, Washington, D.C.

Westman, W. E. (1972a). Some basic issues in water pollution control legislation. *Amer. Sci.* **60**:767–773.

Westman, W. E. (1972b). Environmental problems of the future: development offshore. *Ecology Today* **2**(2):11–13, 48–50.

Westman, W. E. (1979). Oxidant effects on Californian coastal sage scrub. *Science* **205**:1001–1003.

Westman, W. E., and Conn, D. W. (1976). *Quantifying Benefits of Pollution Control: Benefits of Controlling Air and Water Pollution from Energy Production and Use.* Calif. Energy Comm., Sacramento, Calif.

Westman, W. E., Preston, K. P., and Weeks, L. B. (1985). Sulfur dioxide effects on the growth of native plants. In W. E. Winner, H. A. Mooney, and R. Goldstein, eds. *Sulfur Dioxide and Vegetation. Physiology, Ecology, and Policy Issues.* Stanford Univ. Press, Stanford.

White, G. F., ed. (1977). *Environmental Effects of Complex River Development.* Westview, Boulder, Colo.

Whittaker, R. H. (1975). *Communities and Ecosystems.* 2nd ed. Macmillan, New York.

Winner, W. E., Mooney, H. A., and Goldstein, R., eds. (1985). *Sulfur Dioxide and Vegetation. Physiology, Ecology, and Policy Issues.* Stanford Univ. Press, Stanford.

Wong, P. T. S., Chau, Y. K., and Luxon, P. L. (1975). Methylation of lead in the environment. *Nature* **253**:263–264.

Wood, J. M. (1973). Metabolic cycles for toxic elements in aqueous systems. *Rev. Int. Oceanogr. Med.* **31–32**:7–16.

Wood, J. M. (1974). Biological cycles for toxic elements in the environment. *Science* **183**:1049–1052.

Woodward-Clyde Consultants (1976). Environmental Report, Sweetwater Uranium Project, Sweetwater County, Wyoming. Prepared for Minerals Exploration Co., San Francisco.

Woodwell, G. M., Wurster, C. F., Jr., and Isaacson, P. A. (1967). DDT residues in an East Coast estuary: a case of biological concentration of a persistent pesticide. *Science* **156**:821–824.

Zobrist, J., Davis, J., and Hegi, H.-R. (1977). Charakterisierung des chemischen Zustandes von Fliessgewässern. *Gas-Wasser-Abwasser* **57**:402–415.

PART FIVE

PREDICTING IMPACTS: THE BIOTA

8

STRUCTURE AND FUNCTION
OF BIOLOGICAL
COMMUNITIES

In designing an ecosystem management or mitigation plan, one must decide which attributes of ecosystems are to be the focus of restoration or preservation. Several alternative goals are possible. One may seek to create a landscape that optimizes yield of a particular natural resource, to restore or preserve to the extent possible the natural functions (processes) of ecosystems, or to restore or preserve the integrity and diversity of biological structure (molecular, population or community levels). One may seek to restore original structure and function fully (restored ecosystems), partially (rehabilitated ecosystems), or not at all (altered ecosystems) (see Cairns 1983). Manipulating ecosystems for these different goals implies the need for different management strategies. Some of these implications are explored in this chapter for three ecosystem processes: energy flow, nutrient flux, and population regulation.

ALTERNATIVE GOALS FOR MANAGEMENT OF ECOSYSTEMS

1. Preservation of ecosystem processes. A landscape plan may be designed to optimize restoration or preservation of one or more ecosystem functions, such as soil binding, hydrological balance, radiation or gas exchange, energy fixation and flow, nutrient assimilation and release, or population fluctuation and regulation. Indeed, these functions occur simultaneously and are interrelated. Nevertheless, if the main aim is to preserve maximal soil binding, this may be accomplished with exotic species and management practices (fertilization, irrigation, weeding) that will not necessarily achieve the goal of, say, minimal nutrient loss from the system.

Awareness of the minimum scale at which an ecosystem process occurs is important to its preservation. The binding of soil by a tree involves retaining at least enough area to encompass its entire root system and to permit offspring to grow and replace the tree upon its death. Controlling larger processes of soil mass movement (e.g., slumping) will require preserving larger patches of vegetation. The movement of nutrients in ground or surface water may at a minimum encompass an entire watershed. The minimum acreage of salt marsh or mangrove forest necessary to supply sufficient detritus to sustain a minimum breeding population of marine fish species is much more difficult to determine, and possibly no threshold exists below which habitat destruction will fail to impact the dependent marine populations (see, e.g., Westman 1975). Finally, some processes, such as gas exchange and radiation balance, are cumulative, without threshold, from the scale of the organism to that of the biosphere. Management plans should take into account the scale needed for preserving the particular ecosystem feature.

Botkin et al. (1982) suggest the concept of "partial" or "incomplete" ecosystems in which a landscape or seascape segment is too small to be managed as an independent unit because some vital aspect (e.g., breeding ground, source of nutrients) is excluded from its boundaries. The converse of this notion, however, that there is some spatial scale short of the biosphere at which an ecosystem is complete, does not follow, since some processes show no spatial threshold. As a result there is not likely to be one minimum area that clearly bounds all the processes of an ecosystem. This is an inevitable result of the interconnectedness and open-system nature of the world's ecosystems.

2. Preservation of integrity of biological communities. Sometimes the aim of "wilderness preservation" is to restore and preserve ecosystems in some self-regulating condition that is assumed to approximate their state before major disturbance by human society and technology. An ecosystem that "self-regulates" does not necessarily return to the same equilibrium after natural disturbance. Nevertheless, a *tendency toward* restoring balance (homeostasis or resilience) is widely observed in ecosystems (see, e.g., Bormann and Likens 1979; Chapter 12).

It is common for the focus of preservation to be not on ecosystem processes but on preservation of ecosystem structure, with the assumption that if structure is preserved, function will be also. At the landscape level efforts have been made to preserve entire patches of natural vegetation in a mosaic that is believed to represent an optimal mix for the preservation of species of concern. No one mosaic pattern will preserve all species equally. A landscape pattern with more open fields may favor ungulate browsers (e.g., deer), whereas a landscape of closed forest may favor certain carnivores (e.g., bear). Thus even in seeking to preserve structural integrity of a wilderness, choices about what and how much of each type to preserve must be made.

The balance of habitat types in a landscape mosaic that will maintain the integral functioning of its components is not completely arbitrary, however. A lake or river is affected in its trophic status and associated biota by the amount and quality of sediment, water, and nutrient input it receives from adjacent upland areas. The diversity of species in a forest is determined by its size and degree of isolation. Thus a quantitative knowledge of the flow of material, energy, and species between landscape components is desirable to the design and maintenance of a chosen mix of habitat types (see Chapter 11).

Another approach to preserving biological structure is to focus on preserving species threatened with extinction. Managing habitat to favor these species will inevitably hamper growth of certain more abundant species, but if the criterion is to preserve the full mix of species regardless of relative abundance, this approach is appropriate, even if management intensive.

The goal of preserving the structural *integrity* of ecosystems is sometimes confused with preserving species or habitat *diversity*. As discussed in Chapter 11, "diversity" has several specific meanings in ecology; it is a concept quite distinct from "integrity." Biological "integrity" refers to the ability of species to interact and maintain their structure and function in some self-regulating homeostatic fashion. A natural ecosystem has evolved over the millennia to sustain a group of species that can coexist in balance (with temporal fluctuations). The addition or extinction of species occurs at a very slow, probably intermittent, rate usually of the order of thousands or hundreds of thousands of years, in the absence of major human intervention. The self-regulating interactions among a coevolved suite of species in an ecosystem is what is sought to be preserved in a balanced or integrated ecosystem. The addition of an exotic species to an ecosystem could destabilize it very quickly, even though species diversity is initially increased.

3. Management of ecosystems for human purposes. A natural ecosystem may be modified to maximize provision of some service to people: yield of a commercial product, recreational enjoyment, visual beauty. Because of the tendency of natural processes to return the system to its evolutionary form, such management will require sustained and costly intervention (weeding, stocking, pruning, fertilizing, watering, hunting, etc.).

Ecosystems composed of many noncoevolved species will take on a structure and function that is virtually impossible to predict in the absence of prior examples. Whether this new ecosystem will contain properties as desirable to people as the natural ecosystem it replaced is also not possible to predict in advance with precision. Planned communities (landscaped gardens, cropland) can clearly yield desirable resources and functions whose benefits exceed their management costs. Adventive communities with a large proportion of nonnative species (e.g., forests of exotics on Hawaii; an old field of native and exotic herbs on abandoned farmland) can also be judged desirable. The management of altered ecosystems for particular human purposes is indeed the most common goal applicable to landscape management. The assumption underlying the biological integrity goal, however, is that given natural selection, the natural ecosystem is the system *most likely* to achieve and maintain an internal balance rapidly, without further human intervention. If the properties of such systems are considered desirable, we may choose to maintain the system that provides them. The goal of preserving natural ecosystem integrity involves choices about what to preserve, just as does implementing the goal of managing a drastically altered system. The two goals exist on a continuum that varies in the extent to which human intervention is required.

ECOSYSTEM PROCESSES AND IMPACT ASSESSMENT

Energy Fixation and Flow

As plants (primary producers) in an ecosystem are grazed by herbivorous animals and these in turn by carnivores, the energy of the sun fixed into the chemical bonds of molecules in plants is transferred to higher trophic levels in the food web. (Each level in the hierarchy of feeding relationships is termed a trophic level.) Typically, however, at least 50% of the energy fixed by plants in photosynthesis is released by them during respiration (plant metabolism).

Another 40% of the plant's energy typically remains in the uneaten portion of the plant and falls as litter onto the forest floor or lake or ocean bottom, there to be decomposed. Consequently only about 10% of the energy fixed by plants is ingested by herbivores. Animals in turn assimilate only a fraction of what they eat (typically 20–60% for herbivores, 50–90% for carnivores; Whittaker 1975) and again respire at least half of this. The gross growth efficiency (ratio of new protoplasm in growth and reproduction divided by food consumed) is on average about 10%, though it may be up to 20% or more in higher carnivores.

An inevitable result of this loss of energy as heat of respiration during food transfer in the food web is that each higher trophic level has a lower level of annual net production (gross production of biomass − respiration averaged over time). In Figure 8.1a the declining net production at higher levels of the food chain is illustrated as a pyramid of productivity. In this particular example the standing crop of each trophic level (biomass per unit area at an instant in time, Figure 8.1b), also happens to decline at higher trophic levels, but this need not always be the case. If higher carnivores (e.g., fish) have a longer life (slower turnover time) than primary producers (e.g., phytoplankton), the biomass present at an instant could be higher at the carnivore level, since the fish biomass represents an accumulation of net production over a longer period than does phytoplankton biomass. Similarly the pyramid of numbers (Figure 8.1c) may be inverted (e.g., many ants feed on a single tree). The pyramid of productivity is never inverted, however, since the total amount of energy available to produce mass is greatest as it is captured from the sun by plants. At higher levels much of this energy is dissipated to unusable forms as heat.

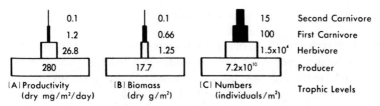

0.1		0.1		15	Second Carnivore
1.2		0.66		100	First Carnivore
26.8		1.25		1.5×10^4	Herbivore
280		17.7		7.2×10^{10}	Producer
(A) Productivity (dry mg/m²/day)		(B) Biomass (dry g/m²)		(C) Numbers (individuals/m²)	Trophic Levels

Figure 8.1. Pyramids of productivity, biomass and numbers for an experimental pond, using data of Whittaker (1961). Reproduced from R. H. Whittaker (1975), *Communities and Ecosystems,* 2nd ed., with permission of Macmillan Publishers, New York.

Key Trophic Levels

The pyramid of productivity holds important implications for impact assessment. First, it is clear that the productivity of primary producers in the ecosystem puts an upper bound on the size of animal populations in the system. Any reduction in the size of the plant community will have adverse repercussions on wildlife population sizes, so that plants are an obvious major focus for study even when an animal species is the direct concern. Second, the animal species at or near the top of the food web will always be the smallest in productivity, and generally its population will be relatively small in biomass and numbers. As a result such species will be most vulnerable to extinction, since any stress causing even small fluctuations may drive the population size to zero. For this reason upper-level carnivores such as mountain lions, Bengal tigers, condors, and Tasmanian devils have long ago been brought to the edge of extinction by human manipulations of their habitat (as well as hunting) that have left species at lower trophic levels still abundant. In general, the higher on the food web an animal is, the less dense its prey and the larger its feeding territory. A single California condor, for example, hunts over a territory of hundreds of square miles. The likelihood of habitat disturbance by human activity over such a large area is increased, and the vulnerability to extinction thus further enhanced. Consequently top carnivores become a second key focus of concern when seeking to preserve a food web intact.

Community Energetics and Ratios

Ecologists speak of net primary production (NPP) as the gross amount of energy fixed into chemical form by plants (gross primary production, GPP), less the energy respired (R). The net primary production is plant growth, available for ingestion by animals and for litterfall and decay:

$$NPP = GPP - R \qquad (8.1)$$

When all tropic levels are considered, ecologists speak of net ecosystem production (NEP) where R_c is community respiration by all plants, animals, and saprobes:

$$NEP = GPP - R_c \tag{8.2}$$

The net primary production of a plant community may be entirely consumed by animals and saprobes so that no additional biomass accumulates over time. In such a situation net ecosystem production is zero, and the community is in a climax or steady-state condition from an energetic point of view. Figure 8.2 illustrates the general tendency for NEP to be positive during early stages of succession in forest (Kira and Shidei 1967) or algal test-tube microcosm (Cooke 1967) and to approach zero at the equilibrium or climax phase (Odum 1969).

O'Neill (1976) has attempted to examine whether the ratio of primary production to biomass (rather than respiration) can be related to ecosystem resistance to stress or recovery ability. He built an energy model of ecosystems with three compartments: photosynthesizing organisms or "autotrophs," P; nonphotosynthesizing organisms or "heterotrophs," that is, animals and saprobes, R; and detritus (detached pieces of organic matter from living things), S (Figure 8.3).

O'Neill simulated energy flow through the ecosystem compartments and tested the relative ability of six different biomes to recover from disturbance, using field data on organic mass (P, R, S) and rates of transfer of energy between compartments. He simulated a disturbance in each ecosystem by decreasing plant mass by 10% initially, noting how much P, R, and S subsequently deviated from their equilibrium values ("malleability"; Chapter 12). He used simulated data at 1-year intervals for 25 model years. The recovery index computed the sum of squares of deviations of each variable over the period of analysis.

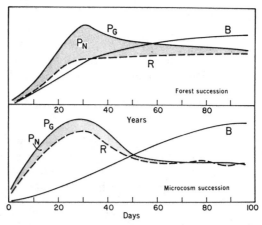

Figure 8.2. Comparison of the energetics of succession in a forest and a laboratory microcosm. P_G, gross production; P_N, net ecosystem production; R, total community respiration; B, total biomass. These are broadly generalized curves based on the studies of Kira and Shidei (1976) and Cooke (1967). Reproduced with permission from Odum (1969), *Science* **164**:262–270. Copyright 1969 by the American Association for the Advancement of Science.

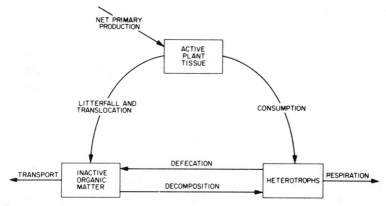

Figure 8.3. A model of the ecosystem as used by O'Neill (1976). The boxes represent biomass ("state variables"), and the arrows represent transfers of energy between components. From "Ecosystem persistence and heterotrophic regulation" by R. V. O'Neill, *Ecology,* 1976, **57,** 1244–1253. Copyright (c) 1976 by the Ecological Society of America. Reprinted by permission.

Each biome showed different extents of average deviation from the equilibrium values over the 25-year period. The recovery index for each biome was plotted against the ratio of net primary production (NPP) to biomass $(P + R)$, a ratio that Odum and Pinkerton (1955) termed the "power parameter." The power parameter reflects the rate of energy fixation per unit of standing crop of plants and animals. Figure 8.4 shows that the biomes with

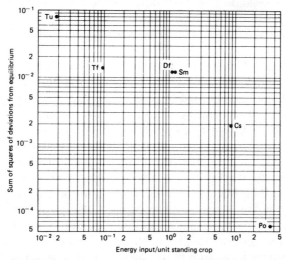

Figure 8.4. The recovery index (a measure of average malleability over a 25-year period) on the ordinate versus the power parameter (ratio of net primary production to biomass) on the abscissa. The recovery index was computed by simulation modeling of changes in biomass for the state variables shown in Figure 8.3. *Tu,* tundra; *Tf,* tropical forest; *Df,* deciduous forest; *Sm,* salt marsh; *Cs,* cold spring; *Po,* pond. From "Ecosystem persistence and heterotrophic regulation" by R. V. O'Neill, *Ecology,* 1976, **57,** 1244–1253. Copyright (c) 1976 by the Ecological Society of America. Reprinted by permission.

the highest ratio of net primary production to biomass (pond, cold spring) show the least average deviation from equilibrium. Salt marshes and deciduous and tropical forests showed intermediate values of both ratios, and the tundra, with the lowest net growth rate per standing crop, showed greatest average change following disturbance (i.e., highest average "malleability," Chapter 12). These results suggest that even in climax communities from which these data were drawn (except for the temperate forest), high rates of energy flowthrough relative to standing biomass may be advantageous to the resilience of ecosystems. Clearly much more empirical data will have to be examined before these relationships can be confirmed.

Based on his model, O'Neill (1976) also suggested that heterotrophs may play an important role in regulating the rate of primary production in an ecosystem, a conclusion also suggested by a number of other ecologists (Golley 1973, Lee and Inman 1975, O'Neill et al. 1975). In particular, herbivorous insects have been suggested to be key regulators of plant production, through their role in accelerating nutrient movement from canopy to forest floor in insect frass and in responding to increased plant production by rapid increase in insect numbers (Mattson and Addy 1975).

Diversity and Stability in Food Webs

The resistance of an ecosystem to change under stress is termed "inertia" in this text. The rate and manner of recovery of an ecosystem following disturbance is here termed "resilience" (Chapter 12). Ecological studies of inertia and resilience in relation to energy ratios generalize at a level that ignores the number or relative abundance of species in a food web. A second line of investigation has sought to examine whether inertia and resilience can be related to the number of species in a food web.

The British animal ecologist C. S. Elton (1927) first highlighted the specific feeding relationships in a biotic community in terms of food webs. Lindeman (1942) noted that food webs tended to have a certain hierarchical structure at each level of which organisms had a distinctive feeding relationship. Thus within a given trophic level organisms tended to feed on creatures at the level below them and were predated upon, in turn, by creatures in the level above. Lindeman recognized that this trophic structure oversimplifies the true complexity of a food web, in which creatures often feed on organisms from more than one other level. Indeed, in the case of detritivores, organisms may be feeding on residues of organisms at their own level.

R. H. MacArthur (1955) first suggested that the larger the number of species in a food web, the more "stable" the community should be in response to disturbance, because of the greater number of alternate pathways for energy flow. Thus, if one species were decimated by stress, a predator had more alternative feeding options in a species-rich food web than in a species-poor one. MacArthur used the word "stability" to refer both to the inertia (resistance to change) and resilience (recovery from change; particularly, *rate* of recovery, termed "elasticity"; Chapter 12) of the web. He characterized changes in response to stress primarily as changes in the number of individuals in a species population. MacArthur's concept become rather overgeneralized in the succeeding 15 years, to the notion that the more "diverse" in species a community is, the more "stable" it will be. A number of ecologists in the 1970s showed that this was indeed an overgeneralization. They further noted that both "diversity" and "stability" required more precise definition. Ecologists working with computer models followed a rather different approach to this question than field ecologists.

Gardner and Ashby (1970) and May (1972, 1973), working with computer-simulated food webs in which links were assigned at random, showed that the "stability" of these webs actually decreased as species were added. Systems were defined as more stable if they were more elastic; that is, population sizes more rapidly returned to equilibrium values following disturbance. Simulating food webs permitted investigators to vary several attributes: the number of species in the web, the number of connections per species, and the intensity of interaction between members.

In the model of May (1972, 1973) an ecosystem was modeled with m species in it. The feeding interaction between each species and every other was represented by coefficients in the off-diagonals of an $m \times m$ matrix. Each interaction coefficient was assigned from a distribution of random numbers with mean value zero, and standard deviation value, s. The value for s determines the average strength of interactions between species, since the lower the standard deviation, the more the feeding of a given species is concentrated on a few other species rather than dispersed among a large number. The proportion of interaction coefficients that is not zero determines the number of links in the web and is called the connectance, C.

May (1973) found that his model ecosystem moved sharply from stable to unstable whenever m, s, or C exceeded a critical value. Only when $s < \sqrt{mC}$ was the system highly likely to be stable. In other words, only when the number of feeding links of any one species in the web was relatively small, was the food web likely to be stable (elastic). May (1972, 1973) thus suggested that the concentration of feeding behavior within small blocks of species or compartments may stabilize food webs in nature.

To use a simple analogy, May's model was not unlike a spider web with some broken links. The more links that were added to the web, the more likely that a juncture of threads in the web would vibrate from an insect flying into the web somewhere else, since the interconnected threads reverberate throughout the web due to a disturbance to any part of it. In nature it would seem that something acts to dampen the reverberations, not unlike droplets of dew that settle on some of the web junctures at dusk, acting as shock absorbers to hamper the spread of reverberations (Figure 8.5). The dew-encased subsets of web, or "modules," are characterized by strong interactions within the web patch and weak interactions beyond the patch.

Field ecologists have attempted to test the hypothesis that food webs in nature contain compartments or modules. McNaughton (1978) tried to test

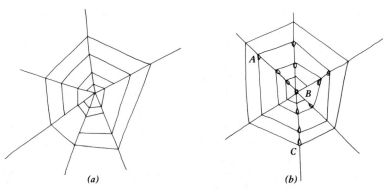

Figure 8.5. A spider web with some missing links, like R. M. May's model of randomly connected species in food webs, has no mechanisms to dampen reverberations. As the web becomes larger, the number of reverberations passing through a juncture increases. (*b*) A spider web with dew at some junctures creates subsets or "modules" of the web matrix (e.g., $\overline{ABC}$) within which reverberations are carried freely, but between which they are dampened. This second version seems to be a more realistic model of food webs in nature.

for guilds of species that might correspond to the species in a web compartment, by examining tendencies for nearest neighbors among plants in the Serengeti grasslands (Tanzania) to be more similar than random chance would dictate. Although his initial analysis indicated that guilds of four or five plants existed, Harris (1979) and Lawton and Rallison (1979) recalculated the data using more valid statistical methods and found no evidence for guilds. Rejmánek and Starý (1979) did claim evidence for guilds in a plant-aphid-parasitoid food web in the canopy of an oak-pine-birch forest in Central Europe: they noted that as the number of species (*m*) in real food webs increased, the average connectance (*C*) decreased, so that $C \times m$ was a constant of the order of 2–6.

Pimm and Lawton (1980), however, noted that decreased connectance does not necessarily result in the formation of isolated compartments; webs of equal connectance may be compartmented or not (Figure 8.6). In their review of an oak leaf gall system (which includes several levels of parasitoids) they noted the existence of modules but determined that the number observed could have arisen by chance. These authors did, however, find evidence for compartments corresponding to habitat boundaries when they analyzed a large landscape patch consisting of terrestrial, freshwater, and marine habitats on an arctic island (data of Summerhayes and Elton 1923). They also found evidence for compartments separating primary consumers (e.g., fish) from primary producers (e.g., phytoplankton) and detritivores (e.g., filter feeders) in aquatic ecosystems and to some extent separating plants and the litter layer in forests.

Pimm (1979) has constructed a simulated food web in which compartmentalization does not increase stability. Pimm and Lawton (1980) therefore

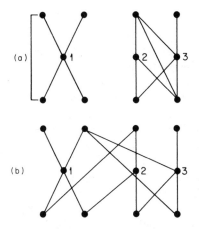

Figure 8.6. Webs of equal connectance but different compartmentalization. The two webs have identical numbers of species, connectance, species on which no other species feed (top-predators), species that feed on nothing else within the web (basal species), and species feeding on more than one trophic level (omnivores). System (*a*) is a compartmented system, (*b*) is not. Reprinted, with permission, from Pimm and Lawton (1980). Copyright (c) 1980, Blackwell Scientific Publications Ltd.

argue that the relatively weak evidence they have found for compartmentalization in nature to date is not necessarily contrary to the theoretical predictions from computer models. The tests of compartmentalization applied by Pimm and Lawton (1980) are severe because they treat food web links between species as present or absent, rather than allowing for distinction between strong and weak links. Pimm and Lawton (1980) point out that the existence of compartments is an important question in ecosystem management, since without compartments a disturbance to one species in an ecosystem would be expected to reverberate in effect much more widely throughout the food web (see Figure 8.5*a*). For recent discussions of food web structure, see Briand (1983), de Angelis et al. (1983), and Pimm (1982).

Viewing food web links as strong or weak, rather than simply as present or absent, is crucial to a realistic understanding of food webs in nature. When field evidence on the strength of interactions is obtained, the existence of compartments or "modules" and guilds in nature emerges quite clearly (Paine 1980).

Keystone Species and Modules in Nature

The biological community occupying the marine rocky intertidal zone of Pacific North America is characterized not by a reticulate food web with links of equal strength but by a web with some strong, and some weak, links. Through experimental removal of all the individuals of a particular starfish species (*Pisaster*), R. T. Paine (1966, 1974, 1980) has shown that this species plays a key role in regulating the composition of the associated community. At Mukkaw Bay, Washington, Paine (1966) found that when the starfish *Pisaster* was removed, the community composition went through a series of changes and was ultimately dominated by *Mytilus californianus,* a mussel. Benthic algae, chitons, and large limpets disappeared. Indeed the species richness of the community was reduced from 15 to 8 species. Thus *Pisaster* acted as a *keystone* species, the crucial unit supporting the arch of the intertidal community. Its removal led to major changes, resulting in widespread repercussions in this module.

The results obtain because *Pisaster* eats most mussel size categories. As long as mussels do not grow too large, they can also be eaten by other predators, and their populations remain small. Upon removal of the starfish, the mussels grow too big to be eaten by other predators, and achieve dominance, affecting other species by their own, quite different, feeding preferences. Other keystone species can be identified within the rocky intertidal community. For example, experimental removal of sea urchins (*Strongylocentrotus* spp.) produces a lush algal community where none had existed previously; reintroduction of the urchins causes reversion to a relatively barren, herbivore-resistant flora (Paine 1977, Paine and Vadas 1969). Removal of other algal herbivores does not have the same effect (Paine 1980).

Probably the most dramatic example of a keystone species in the Pacific nearshore community is the sea otter. The left side of Figure 8.7 illustrates that the sea otter is a top carnivore in this community (excluding humans). Its removal has massive repercussions for the community. Estes and Palmisano (1974) compared the nearshore communities of two sets of Aleutian islands, one (Rat Island) in which sea otters were present and another (Near Island group) in which they had been hunted to local extinction ("extirpation"). On Rat Island the otters had fed heavily on sea urchins, permitting the extensive growth of macroalgae (particularly kelp). Barnacles, mussels, sea urchins, and chitons were relatively scarce and small; rock greenling (a fish), harbor seals, and bald eagles were abundant. By contrast, on the Near Island group where sea otters were absent, sea urchins were large and abundant, kelp beds were much reduced in size and heavily grazed, there were extensive beds of mussels and barnacles, and rock greenling, harbor seals, and bald eagles were scarce or absent.

Simenstad et al. (1978) have more recently excavated shell middens on another Aleutian island (Amchitka I). They contrasted the composition of bones and shells found in middens of the Aleuts (2500 B.C. to recent) and pre-Aleut human inhabitants. Evidence from the middens indicated that the Aleuts gradually overexploited the sea otter populations, causing a shift in community structure from that on the left of Figure 8.7 to that on the right. The food web in Figure 8.7 is oversimplified, as it excludes many other species present in both communities, but the change in magnitude of the dominant species is dramatic enough to show the influence of human extirpation of an upper-trophic-level keystone species (the sea otter) on community composition. The Aleuts became the new top carnivore and, by overhunting a keystone species, dramatically changed the composition of the entire natural community. The impact of reintroduction of sea otters can also be predicted with accuracy (Duggins 1980).

Clearly if impact analysts can identify keystone species in a food web, studies of the effect of the proposed impact can be more sharply focused on these species, with considerable gain in research efficiency. Furthermore the

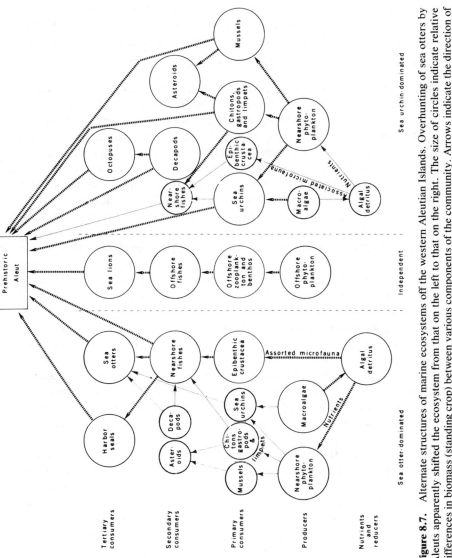

Figure 8.7. Alternate structures of marine ecosystems off the western Aleutian Islands. Overhunting of sea otters by Aleuts apparently shifted the ecosystem from that on the left to that on the right. The size of circles indicate relative differences in biomass (standing crop) between various components of the community. Arrows indicate the direction of biomass or energy flow; heavy arrows indicate importance or magnitude of an interaction compared with the alternate community. Reprinted from Simenstad et al. (1978), *Science* **200**:403–411, with permission. Copyright 1978 by the American Association for the Advancement of Science.

identification of modules would be of potential benefit, in that studies could be focused on keystone or at least dominant species within these modules, without the necessity for examining responses of all associated species.

Use of the Guild Concept in Impact Analysis

The identification of keystone species in the previous examples involved experimental or historical removal and/or replacement of a species that turned out to have a critical role in ecosystem structure and function. Clearly such studies are time-consuming. Indeed, the marine intertidal environment may demonstrate these changes more rapidly than terrestrial environments (Paine 1980). The knowledge of modular structure and evidence for existence of such modules in terrestrial communities within a habitat type is at present scarcer. Risch and Carroll (1982) have demonstrated that removal of the fire ant by insecticide (Mirex) application to corn-squash plots in southern Mexico increased the diversity of other arthropods on the plot relative to controls. Shugart and West (1977) explored the long-term effect of removal of the native American chestnut (*Castanea americana*) by chestnut blight on associated oak species in temperature deciduous forests, using a simulation model. Further study is needed in each of these cases to establish modular structure.

Ecologists have long recognized that the introduction of an exotic species into a habitat can create major changes in the communities invaded over a short time (e.g., Bennett 1976, Elton 1958, Zaret and Paine 1974). The prickly pear (*Opuntia stricta*), for example, escaped from cultivation in Australia in the mid-1800s and by 1920 had densely infested 60 million acres of grazing land. It was brought under control by the introduction of one of its native herbivores, the moth *Cactoblastis cactorum,* which lays its eggs in the cactus pads (Krebs 1972). Such exotic weeds as *Opuntia* exhibit strong interactions with the community into which they are introduced, usually resulting from superior competition for sources. Such exotic species may thus act as "artificial keystone" species in some cases, but it should be clearly recognized that these interactions are not the result of coevolution.

Ideally the identification of modules for use in impact analyses would proceed from a knowledge of where the strong links in the food web lie. In the absence of that information, simply identifying the group of species that exploits the same class of resources in a similar way ("guild" as defined by Root 1967) may be a first approach toward identifying modules. Paine (1980, p. 668) has proposed to restrict the term "module" to a group of species with "strong trophic interactions [which] can produce predictable, persistent patterns in the resource guilds. Species that seem dependent on these resources, give evidence for evolved modification for use of, or association with, these resources, and that disappear upon the removal of a strongly interacting species (or appear with its addition) will be said to belong to a module." Thus not all guilds will form modules, but it is likely that modules will be composed of guilds.

Severinghaus (1981) has suggested a basis for using the guild concept in impact assessment. His suggestion is that "actions that affect environmental resources will similarly affect the members of the guilds using those resources. Once the impact on any one species in a guild is determined, the impact on every other species in that guild is known" (p. 187). He presents proposed guild classifications for mammals and birds in the contiguous United States (Figures 8.8 and 8.9). Johnson (1981) has made a similar

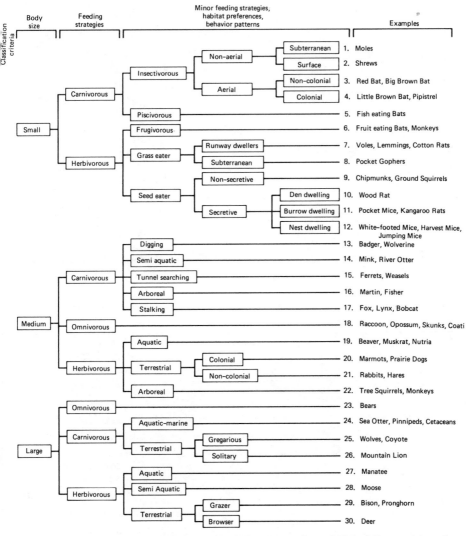

Figure 8.8. Proposed guilds of mammals inhabiting the continental United States. Adapted with permission from Figure 1 of Severinghaus (1981). Copyright (c) 1981, Springer-Verlag, New York.

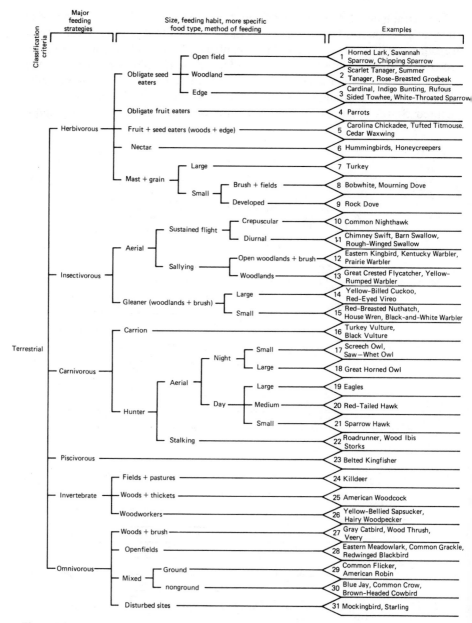

Figure 8.9. Proposed guilds of predominantly land-dwelling birds inhabiting the contiguous United States. Adapted from Figure 2 of Severinghaus (1981), with permission. Copyright (c) 1981, Springer-Verlag, New York.

proposal for classifying terrestrial plants into "guilds" (Figure 8.10). John-son's classification was designed to classify plant species based on their likely similarities in use of bare ground as a resource for recolonization following disturbance (mining, clearing, etc.). The same set of plants might be assigned to quite different "guilds" by some other classification criterion such as sensitivity to SO_2 pollution (use of SO_2 as a "resource").

Unless the resources chosen for classification of species into guilds are in some sense critical or limiting to growth and function of the species, it is

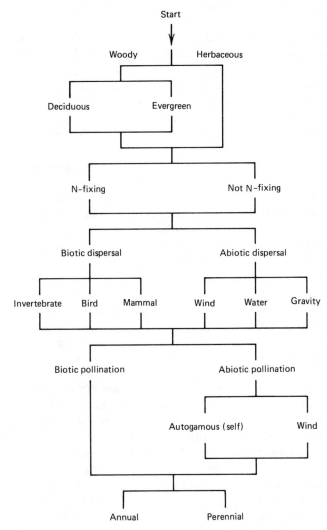

Figure 8.10. Sequential steps in the classification of terrestrial plants into one of 95 "guilds" (structural/functional types) in relation to recolonization of bare ground. Adapted with permission from Johnson (1981). Copyright: Academic Press, Inc. (London) Ltd.

unlikely that the guilds will correspond to modules that have coevolved evolutionarily. In the marine intertidal communities, space on the rocks was a limiting resource, although this fact was not used to identify the guilds. In terrestial communities moisture is often the key limiting factor to plant growth, but some secondary stressor, such as an air pollutant, may impose additional limits to growth. Frequently the growth-limiting factors change over the course of a day, season, or year, or interact to cause multiple inhibition. For example, from morning to early afternoon a plant like *Salvia mellifera* (a shrub of the coastal sage scrub of California) will experience increasing drought stress as the sun's radiation and evapotranspirative stress increase. Stomates will close and leaves may wilt or turn over (Gill and Mahall 1981) at the peak of drought stress. Under low SO_2 exposure (0.1–0.2 ppm), *S. mellifera* will open its stomates (Winner 1981). If this were to happen in the peak of the day, the leaves would likely dry up, and death to parts or all of the plant would occur. At higher levels of SO_2 exposure (0.5 ppm) stomates of this species close even when moisture is abundant (Winner 1981). Thus depending on the combination and levels of environmental stressors, one factor may override, or enhance, the effect of another in limiting plant growth.

In order to select criteria appropriate for classifying species in terms of likely response to stress, therefore, it would be desirable to understand the likely role of this stress factor relative to other stressors in the environment. In view of the inevitable uncertainties regarding the role of these relative stress factors, it seems preferable to consider such classifications of species *structural* or *functional* classifications relative to a particular stressor, rather than to use the word "guild" for all such classifications.

Howell (1981) has proposed such a structural/functional classification of plant species from the point of view of how solitary campers and hikers interact with the plants. She lists several criteria on which plant species could be classified in this regard (Table 8.1). She further suggests that quantitative measures of the abundance of the classified species could be used to rate each land unit. Such scores would reflect the extent to which the existing vegetation enhances the recreational experience, or provides resistance to adverse impact by trampling or vandalism from campers (Table 8.1). Such a classification of species no longer bears any relation to the evolutionary interrelationships of the species. It is very useful for the third criterion of landscape design noted in the introduction to this chapter—management of ecosystems for human purposes—as is Johnson's classification for land rehabilitation. The guilds of Figures 8.8 and 8.9 are closer to the search for modules, and are more appropriate for use in pursuing the second criterion of landscape design: preservation of integrity of biological communities.

The U.S. Army used the guild classification in Figure 8.8 to assess the impact of Army track-making vehicles (e.g., jeeps, tanks) and training maneuvers on small mammals of the area. Figure 8.11 illustrates the impact of short-term use (woodland clearing, brief period of training) versus long-term

Table 8.1.   Structural/Functional Classification of Vegetation in Relation to Solitary Camping

Structural/Functional Classification of Plants	Possible Measures for Sampling and Rating
Enhancement Features	
Shade	% canopy (line intercept)
	60–80 excellent
	40–59; 81–100 good
	<40 poor
Screening	% shrub cover (line intercept at 2 m height)
	61–80 excellent
	30–60 good
	<30; >80 poor
Noxious species	% frequency (quadrats)
	≤30 excellent
	31–60 good
	61–100 poor
Visual diversity	Life form diversity (quadrats)
	≥8 excellent
	5–7 good
	≤4 poor
Impact Features	
Resistant life forms	% frequency (quadrats)
	61–100 excellent
	31–60 good
	≤30 poor
Barrier species	% frequency
	61–100 excellent
	31–60 good
	≤30 poor
Largest species	% frequency (quadrats)
	≤30 excellent
	31–60 good
	60–100 poor

Source: Reprinted from Table 4 of Howell (1981), with permission of Springer-Verlag, New York.

Note: Left column classifies plants by features that enhance camping experience or inhibit adverse impact by campers. Right column suggests methods of sampling and rating plant abundances in relation to these features.

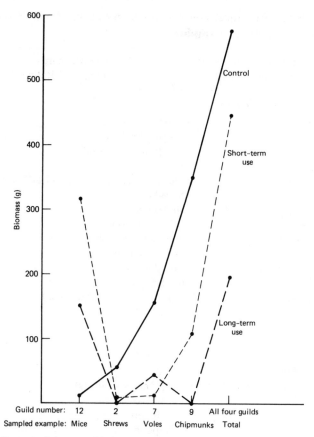

Figure 8.11. Impact of short- and long-term (35-year) use of an area for Army track-making vehicles and training maneuvers on four small mammal guilds. Guild number corresponds to that in Figure 8.8. Data of Severinghaus (1981).

use (35 years of training) on biomass of some small mammal guilds in the area. Shrews and voles (guilds 2,7) exclusively use surface runways and were impacted severely. Chipmunks (guild 9) also prefer the ground but will climb trees occasionally; they were impacted less severely in the short term. Mice (guild 12) were favored because of their ability to climb trees and their effective utilization of food resources made available by the decline of other small mammals (Severinghaus 1981). A food web of the small mammal community would have been helpful in determining whether these interactions might be resulting from a modular structure, and in further predicting the changes due to track-vehicle impact. Nevertheless, a knowledge of animal guilds appears to be useful in interpreting the data and making future predictions. Brenner et al. (1982) apply a similar approach to analyzing changes in small mammal communities on reclaimed strip-mined land over a four-year period (see also Brenner and Kelley 1981 for bird changes in the same area).

Short (1983) has used extensive information on wildlife guilds in Arizona desert habitats to predict the impacts of land-use changes on wildlife in westcentral Arizona.

Nutrient Assimilation and Release

The potentially critical role of nutrients in the ecosystem dynamics of aquatic ecosystems was discussed briefly in Chapter 7. Here we will examine how nutrient flux in terrestrial ecosystems can be used to elucidate ecosystem functioning and response to disturbance. We will further examine what indexes of the complex nutrient fluxes in ecosystems might be used for more rapid assessment needs.

Nutrient Budgets

The construction of a "nutrient budget" for a terrestrial ecosystem involves quantifying standing stocks and rates of transfer of essential nutrients between ecosystem compartments and the inputs and losses of nutrients to the system. The determination of such nutrient budgets permits the ecologist to understand where the major repositories of particular nutrients are and what the key regulators of nutrient movement are. This understanding can permit predictions of the effect of disturbances on subsequent nutrient retention or loss by the system, which in turn will affect the ability of the system to maintain its biomass and prevent major losses to surrounding water bodies.

The construction of a nutrient budget even for a small stand of vegetation is a complex and time-consuming task, requiring several years of measurement and analysis, though the results of such a study can be very illuminating. As an example, the nutrient budget of a forest may permit an understanding of the problems involved in restoring the natural vegetation following clearing of a forest for strip mining. Just such a situation arose on islands off the coast of southeast Queensland, Australia, where giant dune forests were being bulldozed to mine zirconium and titanium buried in the dune sands (Figure 8.12).

To obtain a nutrient budget for a 50 × 50 m sample of the 15-m tall *Eucalyptus*-dominated forest that grew on these North Stradbroke Island high dunes, Westman and Rogers (1977a) first quantified the biomass of the forest. This involved cutting down and excavating 10 trees of each of the three dominant species on a nearby area slated for mining and obtaining dry weights of each plant part, including roots (Figure 8.13). These measurements were then related to a more easily measured plant part (diameter of the trees at breast height: DBH), using logarithmic regressions (Whittaker and Woodwell 1971). These regressions could then be applied to the DBH measurements for all trees on the 50 × 50 m intact study plot to estimate total biomass of trees.

Biomass of the 45 understory species were estimated by relating foliar

Figure 8.12. Surface mining of sand dunes on North Stradbroke Island, Australia. Foreground, recently bulldozed; midground, rehabilitated vegetation after approximately three years; background, undisturbed forest.

canopy cover values obtained by line transect (Chapter 10) to above- and below-ground weights of sample individuals of known canopy size. The nutrient contents of plant tissues (leaves of various ages, stem wood and bark, root wood and bark) were determined for each major species and multiplied by biomass values to calculate nutrient stocks for the forest (Westman and Rogers 1977b). The amount of nutrients transferred from canopy to forest floor by litterfall was quantified by catching and analyzing litter fall at six-week intervals for two and a half years (Rogers and Westman 1977). Decomposition rates of litter, studied by periodic sampling of litter in bags or marked piles, permitted determination of rates of release of nutrients from litter (Rogers and Westman 1977).

The rate of growth of each plant tissue for each species was obtained by measuring changes in stem and root girths, elongations of twigs, and additions and losses of leaf tissue (including losses to herbivory) every six weeks on a sample of trees (Rogers and Westman 1981; Figure 8.14). Multiplying

Figure 8.13. A partially excavated root system of *Eucalyptus umbra*. Trees were excavated to obtain biomass measurements in the study of Westman and Rogers (1977a).

nutrient contents of plant parts by productivity measurements, assimilation rates of nutrients at different seasons of the year could be calculated (Westman 1978). The concentrations of nutrients in soil horizons (Westman and Rogers 1977b), the inputs of nutrients in rainfall, the partitioning of this as precipitation falling through the canopy (throughfall) or running down tree trunks (stemflow), and the loss of nutrients from the dunes by groundwater drainage were all measured (Westman 1978). Finally, these measurements were assembled into nutrient budgets for each of eleven elements in the forest. An example of the potassium budget is shown in Figure 8.15.

Some useful findings resulted. First, it was apparent that for potassium and phosphorus the net annual uptake demand of the forest would exhaust the entire soil stores in 35–70 years, even when inputs from precipitation and external recycling by litter fall were accounted for. This implied first of all that the forests were dependent on the periodic fires that swept through their

Figure 8.14. Measuring leaf and twig growth on intact eucalypt trees to obtain production estimates every six weeks over two and a half years for the study reported in Rogers and Westman (1981).

canopies every 5–10 years for the release of enough potassium and phosphorus from leaves and twigs to renew available soil supplies. Second, because potassium in particular is highly soluble and easily lost to drainage, it implied that these ecosystems were leaky in an essential element initially in short supply in the sands and that the forest biomass might eventually shrink over the millennia due to nutrient shortages in this element (at least). Indeed, biomass was observed to be less on older dunes (500,000 years old) than younger ones (20,000 years old) on the island (Westman 1978).

Third, the existing mass of the forest was seen to be "fossil" in character, since the present biomass resulted from accumulation of nutrients from rainfall, salt spray, and soil weathering over many centuries and could not regrow from today's soil with anything like its present productivity. This meant that to replace the cleared forest after mining would also take many

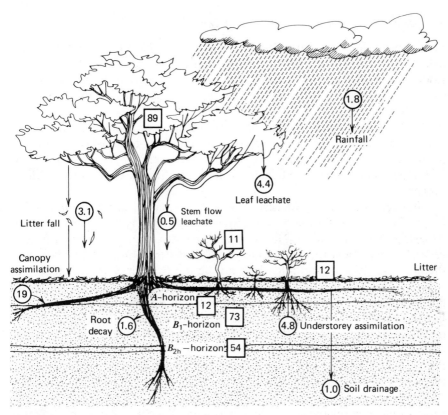

Figure 8.15. Potassium budget for a mixed *Eucalyptus* forest, North Stradbroke Island, Australia. Figures show stocks (boxes) and flows (circles) of potassium as a percentage of the stock in live vegetation (192 kg ha^{-1}). Data from Westman (1978).

centuries, unless supplemented with artificial fertilizer. But fertilizer experiments on similar species in South Australia showed that certain native heath species important in the mature community cannot tolerate high fertility levels in the seedling stage and fail to replace themselves on fertilized plots. (Specht 1963, Heddle and Specht 1975). Experiments with fertilizing the strip-mined dunes on North Stradbroke Island showed poor establishment of dominant tree seedlings in the first 2–5 years (Thatcher and Westman 1975), but it was too early to say whether fertilizer would prove inhibitory here. These findings were important, since the mining companies had been given leases to strip-mine the forests on these highly prized recreational and wilderness areas with the understanding from the mining court that the original vegetation would be replaced in 10 years, yet from experimental and nutrient budget studies it was clear that rehabilitation would require 100–250 years at a minimum (Thatcher and Westman 1975, Westman 1978).

Nutrient Flux-Related Indicators of Impact

Similar or even larger-scale nutrient budget studies have been undertaken in a number of vegetation types around the world (see, e.g., Howell et al. 1975, Rodin and Bazilevich 1967), and in certain cases experimental manipulation (timber-cutting; herbicide application) has allowed study of the effects of forest practices on nutrient flux (e.g., Bormann and Likens 1979, Likens et al. 1977, Swank and Douglass 1977). Informative as such nutrient budget studies can be, they are clearly only practical for long-term research studies. A critical question for impact assessment is whether more readily measured features of nutrient flux can be used as indexes of the larger nutrient cycling process.

Vitousek et al. (1981) suggest that the mechanisms of retention and release of nitrate (NO_3^-) ion by ecosystems may be an appropriate focus for terrestrial nutrient studies. Nitrogen is frequently a limiting element in terrestrial ecosystems; the nitrate ion is highly water soluble and mobile and therefore easily lost, and release of nitrate ion from litter is dependent on several steps of microbial decomposition in the litter layer. For all these reasons it may be a particularly appropriate ion for study. The accelerated loss of nitrate ion following clear-cutting of the Hubbard Brook temperature deciduous forest in New Hampshire was one of the most striking early indicators of impact to this forest (Bormann and Likens 1979), and losses of nitrate have similarly proved to be more sensitive indicators of disturbance than other ions in most forest studies reported (Vitousek and Melillo 1979).

Vitousek et al. (1979, 1981) compared nitrate losses from different forest ecosystems by a relatively simple experiment. They dug a trench below the rooting zone entirely surrounding small forest patches. This "trenching" technique, which has been used by ecologists for decades, cuts off the major supplies of water and nutrient to the trees by severing their fine, active roots. Vitousek et al. (1981) then observed nitrate losses from the trenched plots by collecting soil water below the rooting zone in porous-cup lysimeters (long probes with porous tips that collect water from the soil interstices) and analyzing this soil leachate for nitrate ion. Figure 8.16 shows that different forest ecosystems showed different levels of inertia (resistance to change, measured as absolute level of NO_3^--loss) and elasticity (measured as rate of recovery to pre-impact levels of NO_3^--loss) by this technique. This method therefore seems able to illustrate some major features of forest regulation of nutrient flux rather readily.

Still another sensitive indicator of stress related to nutrient cycling is the change in the rate or nutrient content of litter fall. Litter fall is readily collected and analyzed by litter fall traps placed below vegetative canopies, over the major season of leaf drop. Toxic air pollutants typically accelerate litter fall by causing premature death and abscission of leaves. An example of this phenomenon is illustrated in Figure 8.17 for the coastal sage scrub community downwind of the SO_2 source illustrated in Figure 7.7. Changes in

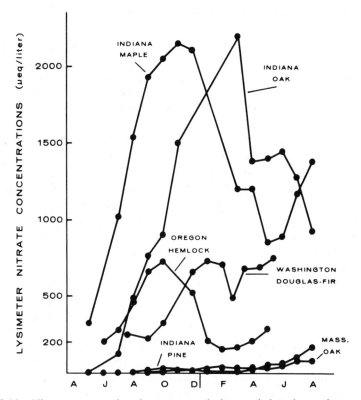

Figure 8.16. Nitrate concentrations in porous-cup lysimeters below the rooting zone within trenched plots in six forest ecosystems from April 1977 to August 1978. Each point represents the mean of between 6 and 10 lysimeters. The nitrate concentrations in control-plot lysimeters averaged less than 20 μeq liter^{-1} in all sites at most times. Reprinted with permission from Vitousek et al. (1981). Copyright: John Wiley and Sons, New York.

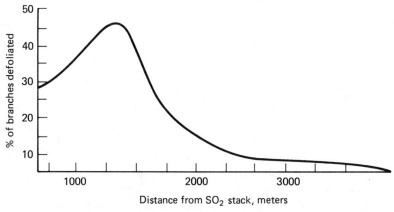

Figure 8.17. Percent of branches of *Salvia mellifera* defoliated at increasing distances downwind of an SO_2 point source. Ground-level concentrations are greatest where the plume hits the ground at 1600 m. Curve fitted to a polynomial regression, $R^2 = 0.81$, $P < 0.001$. Control, percent defoliation = 0.0%. Data used with permission of K. P. Preston (Cf. Figure 7.7).

the nutrient contents of leaves of impacted species relative to the control could give some quantitative indication of the acceleration of nutrient loss from the system and would be a much better indicator of changes in a key aspect of nutrient flux than mere observation of changes in nutrient contents of living leaves (see Legge 1980, Wiersma and Brown 1980).

These examples do not exhaust the list of ways in which nutrient flux studies may assist in predicting impacts on ecosystems. They are nevertheless illustrative of the kinds of insights that may arise. It was through the study of the nutrient budget for Hubbard Brook, for example, that ecologists first recognized the problem of acid rain in the northeastern United States (see, e.g., Johnson et al. 1972). That discovery has since proved to be of widespread significance, since increasing acidity of rainfall from pollution sources may be inflicting billions of dollars in damages to crops, materials, and wilderness in the United States and Canada.

The fact that release of nitrate following clear-cutting is a result of stimulation of nitrifying bacteria in soil and litter by the warmer surface temperature of the cleared ground (Likens et al. 1969) further emphasizes the critical role of decomposers in nutrient cycling. Although microbes and litter invertebrates are rarely included in species lists or lists of threatened or endangered species, from the point of view of the integrity of ecosystem functioning, they are "critical-link" species. Critical-link species are those which play a vital role in ecosystem functioning, regardless of their biomass, place in a food web, or possible role as a keystone species.

Population Dynamics and Genetic Diversity

Loss of Genetic Diversity in Small, Isolated Populations

The problem of preserving adequate genetic diversity in a population that has been drastically reduced in size was noted briefly in Chapter 2. Processes that can reduce genetic variability in small populations include inbreeding, genetic drift, and "bottleneck" and "founder" effects (Miller 1979). For every gene location (locus), higher organisms contain two genes—one from each parent. These two genes for a particular characteristic are called *alleles*. If the alleles are identical, the individual is *homozygous* for the characteristic; if the alleles are different, the individual is *heterozygous,* and only the dominant gene character will be expressed as an observable trait (*phenotype*). A typical individual may be heterozygous at 15% or more of its gene loci (Soulé et al. (1982), though this figure varies enormously.

Inbreeding (the mating of genetically similar individuals) can result in loss of genetic variability in the population since, following the expression of homozygous recessive genes in the population, selection will often act to eliminate these individuals because of the typically maladaptive character of expressed recessive genes (Dobzhansky 1970). In other words, a loss of "fitness" occurs in inbred offspring.

Experience of animal and plant breeders suggests that for every 10% increase in the amount of homozygosity due to inbreeding, there is as much as a 25% loss in fitness of the inbred line in fecundity, vigor, and longevity (Soulé et al. 1982).

The experience of animal breeders suggests that a minimum population of 50–100 individuals is necessary to prevent severe inbreeding in the short term (20–30 generations), and 500 individuals in the long term (Soulé et al. 1982). As Soulé (1980) has pointed out, however, the actual minimum size for wild species is likely to be substantially larger for at least two reasons. First, not all individuals in a population breed. It is common for a dominant male in a hierarchical breeding structure to suppress the breeding of other males in the population (e.g., moose, hunting dogs). Thus the effective breeding population relevant to genetic mixing may be 3–4 times less than the population size. Second, overlapping generations in a population further decrease the genetic heterogeneity of a population, since many closely related individuals may mate.

Soulé (1980, p. 163) provides an interesting example of the application of these constraints in the case of the wolf. Suppose that (1) the density of wolves is about one adult per 20 km^2 (Rutter and Pimlott 1968), (2) only about one-third of the adults actually breed, and (3) the population fluctuates. Soulé assumes that the population reaches a minimum of 10% of carrying-capacity level once every 10 years and increases by 50% per year following a population crash. If one assumes initially a minimum necessary breeding size of 100, based on animal inbreeding experience, one would need to double this to 200 to account for overlapping generations and to triple it to 600 to account for nonbreeding individuals. Based on wolf density, to maintain a population of 600 wolves would require a preserve of 12,000 km^2, larger than Yellowstone National Park. This still does not account for loss of genetic variability due to fluctuations in population numbers (assumption 3).

Furthermore the minimum breeding size of 100 is extremely conservative. Franklin (1980) estimates 500 as a lower starting limit necessary for native species, which would imply a wolf park of 60,000 km^2, or 1% of the area of the continental United States. Clearly if this reasoning is valid, the size of the average critical habitats set aside to preserve endangered species is much too small to preserve the genetic diversity of the population over a series of generations. Much more experience with genetic variation in native populations is needed, but the maintenance of genetic variability should clearly be considered in deriving effective areas of habitat to maintain species.

Genetic drift is the change in gene frequency that occurs upon random sampling of gene characters (alleles) in each generation. A random drift of gene frequencies from generation to generation typically results. In small populations the variation in frequencies of alleles becomes larger (Miller 1979), so that as the process continues, it is more likely that some alleles will disappear from the population altogether.

Soulé et al. (1982) have pointed out that the sex ratio of a breeding population has an important influence on whether loss of variability due to genetic drift will occur. The "effective breeding population" is the actual number of organisms in a population whose genes are passed on to the next generation. In a harem-forming species such as caribou or zebra, a herd may consist, for example, of one male and nine females. All the offspring will therefore have either one or both of the same parents. Thus the genetic variability of the offspring will be much less than if the same number of offspring derived from five different male-female pairs. The chance of an allele being lost by genetic drift is much greater in a herd-forming species.

Soulé et al. (1982) present the following formula for calculating the size of the effective breeding population (N_e) based on the number of males (N_m) and females (N_f) in the population:

$$N_e = \frac{4N_m N_f}{N_m + N_f} \tag{8.3}$$

In the case of a zebra herd with one male and nine females, the effective size of the breeding population is 3.6, rather than 10. Thus to avoid short-term inbreeding depression (loss of fitness), instead of a population of 50 individuals, one would need (10/3.6) × 50, or 139 individuals with a 1 : 9 sex ratio.

A *bottleneck effect* occurs when a population becomes reduced to a fraction of its former size and proceeds to repopulate on the basis of this reduced gene pool. A *founder effect* refers to the reduced gene pool resulting from establishment of a population by a few individuals in a new area.

The effective size of a breeding population (N_e) whose size is fluctuating over time is the harmonic mean of the effective number in each generation (N_t). This average effective size can be calculated by the following formula (Soulé et al. 1982):

$$\frac{1}{N_e} = \frac{1}{t} \left(\frac{1}{N_1} + \frac{1}{N_2} + \cdots + \frac{1}{N_t} \right) \tag{8.4}$$

Suppose the effective size of a population of zebras were as follows over five generations, in which hunting plus drought reduced the effective breeding size of the second generation drastically: 150, 20, 70, 100, and 150. The effective size of the population over the five generations would be 28.5.

Among the steps which may be taken to improve genetic mixing in a managed population are translocating individuals from other populations to add to the effective breeding population and encouraging a larger number of different male-female pairs to mate (Soulé et al. 1982). Additional discussions of problems and strategies for maintaining genetic diversity in managed populations include those of Frankel and Soulé (1981), Miller (1979), and Shaffer (1981).

Principles of population dynamics are also widely applied to the determination of optimum yield in the harvest of fisheries, forests, and wildlife populations. The problem is well researched but beyond the scope of this book. Interested readers may refer to such works as Clark (1976), Cushing (1975), Goh (1980), Krebs (1972), Walter and Hilborn (1978), and Watt (1968).

CONCLUDING REMARKS

We have briefly noted in this chapter three of the many processes of ecosystems: energy flow, nutrient flow, and the dynamics of population genetics. The flux of radiation, gases, and water and the binding of soil in the ecosystem are equally vital to the maintenance of ecosystem integrity. The preservation of ecosystem processes is as important a goal for conservation as is the preservation of ecosystem structure (genetic or taxonomic diversity). Although both of these goals can be modified in designing a landscape for optimal human use, no human design for the landscape can afford to ignore ecosystem processes and structures, which interact with human impacts continually. The science of ecology has elucidated ecosystem processes to the extent that some management principles are evident, yet much research on ecosystem structure and functioning is still needed. In Chapter 9 we shall examine how to predict the response of ecosystem structures and processes to toxic pollutants, using biological characteristics of the component species as indexes.

REFERENCES

Bennett, C. M. (1976). *Man and Earth's Ecosystems: An Introduction to the Geography of Human Modification of the Earth*. Wiley, New York.

Bormann, F. H., and Likens, G. E. (1979). *Pattern and Process in a Forested Ecosystem*. Springer-Verlag, New York.

Botkin, D. B., et al. (1982). Ecological characteristics of ecosystems. In W. R. Siegfried and B. R. Davies, eds. *Conservation of Ecosystems: Theory and Practice*. South African Natl. Sci. Prog. Rep. 61. CSIR, Pretoria.

Brenner, F. J., and Kelly, J. (1981). Characteristics of bird communities on surface mine lands in Pennsylvania. *Environ. Manage.* 5:441–449.

Brenner, F. J., Kelly, R. B., and Kelly, J. (1982). Mammalian community characteristics on surface mine lands in Pennsylvania. *Environ. Manage.* 6:241–249.

Briand, F. (1983). Environmental control of food web structure. *Ecology* 64:253–263.

Cairns, J., Jr. (1983). Management options for the rehabilitation and enhancement of surface mined ecosystems. *Minerals Environ.* 5:32–38.

Clark, C. W. (1976). *Mathematical Bioeconomics: The Optimal Management of Renewable Resources*. Wiley-Interscience, New York.

Cooke, G. D. (1967). The patterns of autotrophic succession in laboratory microcosms. *BioScience* **17**:717–721.

Cushing, D. H. (1975). *Fisheries Resources of the Sea and their Management.* Oxford Univ. Press, Oxford.

De Angelis, D. L., Post, W. M., and Sugihara, G., eds. (1983). *Current Trends in Food Web Theory.* Rep. ORNL/TM 8643. Oak Ridge Natl. Lab, Oak Ridge, Tenn.

Dobzhansky, Th. (1970). *Genetics of the Evolutionary Process.* Columbia Univ. Press, New York.

Duggins, D. O. (1980). Kelp beds and sea otters: an experimental approach. *Ecology* **61**:447–453.

Elton, C. S. (1927). *Animal Ecology.* Sidgwick & Jackson, London.

Elton, C. S. (1958). *The Ecology of Invasions by Plants and Animals.* Methuen, London.

Estes, J. A., and Palmisano, J. F. (1974). Sea otters: their role in structuring nearshore communities. *Science* **185**:1058–1060.

Frankel, O. H., and Soulé, M. E. (1981). *Conservation and Evolution.* Cambridge Univ. Press, Cambridge.

Franklin, I. R. (1980). Evolutionary change in small populations. In M. E. Soulé and B. A. Wilcox, eds. *Conservation Biology: An Evolutionary-Ecological Perspective.* Sinauer, Sunderland, Mass., pp. 135–149.

Gardner, M. R., and Ashby, W. R. (1970). Connectance of large dynamical (cybernetic) systems: critical values for stability. *Nature* **228**:784.

Gill, D. S., and Mahall, B. E. (1981). Leaf curling and recovery with reference to water relations in *Salvia mellifera,* a Mediterranean-climate shrub. *Bull. Ecol. Soc. Amer.* **62**:66.

Goh, B. S. (1980). *Management and Analysis of Biological Populations.* Elsevier, Amsterdam.

Golley, F. B. (1973). Impact of small mammals on primary production. In J. A. Gessman, ed. *Ecological Energetics of Homeotherms.* Utah State Univ. Press, Logan, pp. 142–147.

Heddle, E. M., and Specht, R. L. (1975). Dark Island Heath (Ninety-Mile Plain, South Australia). VIII. The effects of fertilizers on composition and growth, 1950–72. *Aust. J. Bot.* **23**:151–164.

Howell, E. A. (1981). Landscape design, planning, and management: an approach to the analysis of vegetation. *Environ. Manag.* **5**:207–212.

Howell, F. G., Gentry, J. B., and Smith, M. H., eds. (1975). *Mineral Cycling in Southeastern Ecosystems.* ERDA Symp. Ser. CONF. 740513. NTIS, Springfield, Va.

Johnson, N. M., Reynolds, R. C., and Likens, G. E. (1972). Atmospheric sulfur: its effect on the chemical weathering of New England. *Science* **177**:514–516.

Johnson, R. A. (1981). Application of the guild concept to environmental impact analysis of terrestrial vegetation. *J. Environ. Manage.* **13**:205–222.

Kira, T., and Shidei, T. (1967). Primary production and turnover of organic matter in different forest ecosystems of the western Pacific. *Jap. J. Ecol.* **17**:70–87.

Krebs, C. J. (1972). *Ecology: The Experimental Analysis of Distribution and Abundance*. Harper & Row, New York.

Lee, J. J., and Inman, D. L. (1975). The ecological role of consumers—an aggregated systems view. *Ecology* **56**:1455–1458.

Legge, A. H. (1980). Primary productivity, sulfur dioxide, and the forest ecosystem: an overview of a case study. In P. R. Miller, ed. *Effects of Air Pollutants on Mediterranean and Temperate Forest Ecosystems*. U.S. Forest Service, Gen. Tech. Rep. PSW-43. Berkeley, Calif., pp. 51–62.

Likens, G. E., Bormann, F. H., and Johnson, N. M. (1969). Nitrification: importance to nutrient losses from a cut-over forested ecosystem. *Science* **163**:1205–1206.

Likens, G. E., Bormann, F. H., Pierce, R. S., Eaton, J. S., and Johnson, N. M. (1977). *Biogeochemistry of a Forested Ecosystem*. Springer-Verlag, New York.

Lindeman, R. L. (1942). The trophic-dynamic aspect of ecology. *Ecology* **23**:399–418.

MacArthur, R. H. (1955). Fluctuations of animal populations, and a measure of community stability. *Ecology* **36**:533–536.

Mattson, W. J., and Addy, N. D. (1975). Phytophagous insects as regulators of forest primary production. *Science* **190**:515–522.

May, R. M. (1972). Will a large complex system be stable? *Nature* **238**:413–414.

May, R. M. (1973). *Stability and Complexity in Model Ecosystems*. Princeton Univ. Press, Princeton.

Miller, R. I. (1979). Conserving the genetic identity of faunal populations and communities. *Environ. Conserv.* **6**:297–304.

Odum, E. P. (1969). The strategy of ecosystem development. *Science* **164**:262–270.

Odum, H. T., and Pinkerton, R. C. (1955). Times speed regulator, the optimum efficiency for maximum output in physical and biological systems. *Amer. Sci.* **43**:331–343.

O'Neill, R. V. (1976). Ecosystem persistence and heterotrophic regulation. *Ecology* **57**:1244–1253.

O'Neill, R. V., Harris, W. F., Ausmus, B. S., and Reichle, D. E. (1975). Theoretical basis for ecosystem analysis with particular reference to element cycling. In F. G. Howell, J. B. Gentry, and M. H. Smith, eds. *Mineral Cycling in Southeastern Ecosystems*. ERDA Symp. Ser. CONF. 740513, NTIS, Springfield, Va., pp. 28–40.

Paine, R. T. (1966). Food web complexity and species diversity. *Amer. Nat.* **100**:65–75.

Paine, R. T. (1974). Intertidal community structure: experimental studies on the relationship between a dominant competitor and its principal predator. *Oecologia* **15**:93–120.

Paine, R. T. (1977). Controlled manipulations in the marine intertidal zone and their contributions to ecological theory. *Acad. Nat. Sci. Phila. Spec. Pub.* **12**:245–270.

Paine, R. T. (1980). Food webs: linkage interaction strength and community infrastructure. *J. Anim. Ecol.* **49**:667–685.

Paine, R. T., and Vadas, R. L. (1969). The effects of grazing by sea urchins, *Strongylocentrotus* spp., on benthic algal populations. *Limnol. Oceanogr.* **14**:710–719.

Pimm, S. L. (1979). The structure of food webs. *Theor. Popn. Biol.* **16**:144–158.

Pimm, S. L. (1982). *Food Webs.* Chapman & Hall, London.

Pimm, S. L., and Lawton, J. H. (1980). Are food webs divided into compartments? *J. Anim. Ecol.* **49**:879–898.

Risch, S. J., and Carroll, C.R . (1982). Effect of a keystone predaceous ant, *Solenopsis germinata,* on arthropods in a tropical agroecosystem. *Ecology* **63**:1979–1983.

Rodin, L. E., and Bazilevich, N. I. (1967). *Production and Mineral Cycling in Terrestrial Vegetation.* Oliver & Boyd, Edinburgh.

Rogers, R. W., and Westman, W. E. (1977). Seasonal nutrient dynamics of litter in a subtropical eucalypt forest, North Stradbroke Island. *Aust. J. Bot.* **25**:47–58.

Rogers, R. W., and Westman, W. E. (1981). Growth rhythms and productivity of a coastal subtropical eucalypt forest. *Aust. J. Ecol.* **6**:85–98.

Root, R. B. (1967). The niche exploration pattern of a blue-gray gnatcatcher. *Ecol. Monogr.* **37**:317–350.

Rutter, R. J., and Pimlott, D. H. (1968). *The World of the Wolf.* Lippincott, Philadelphia, Pa.

Severinghaus, W. D. (1981). Guild theory development as a mechanism for assessing environmental impact. *Environ. Manage.* **5**:187–190.

Shaffer, M. L. (1981). Minimum population sizes for species conservation. *BioScience* **31**:131–134.

Short, H. L. (1983). *Wildlife Guilds in Arizona Desert Habitats.* U.S. Dept. of Interior, Bureau of Land Management, Tech. Note 362. Publ. No. BLM-YA-PT-83-005-4350, Washington, D.C.

Shugart, H. H., and West, D. C. (1977). Development of an Appalachian deciduous forest succession model and its application to assessment of the impact of the chestnut blight. *J. Environ. Manage.* **5**:161–179.

Simenstad, C. A., Estes, J. A., and Kenyon, K. W. (1978). Aleuts, sea otters and alternate stable-state communities. *Science* **200**:403–411.

Soulé, M. E. (1980). Thresholds for survival: maintaining fitness and evolutionary potential. In M. E. Soulé and B. A. Wilcox, eds. *Conservation Biology: An Evolutionary-Ecological Perspective.* Sinauer, Sunderland, Mass., pp. 151–169.

Soulé, M. E. et al. (1982). Genetic aspects of ecosystem conservation. In W. R. Siegfried and B. R. Davies, eds. *Conservation of Ecosystems: Theory and Practice.* South African Natl. Sci. Prog. Rep. 61. CSIR, Pretoria, pp. 34–45.

Specht, R. L. (1963). Dark Island Heath (Ninety-Mile Plain, South Australia) VII. The effect of fertilizers on composition and growth, 1950–1960. *Aust. J. Bot.* **11**:67–94.

Summerhayes, V. S., and Elton, C. S. (1923). Contributions to the ecology of Spitsbergen and Bear Island. *J. Ecol.* **11**:214–286.

Swank, W. T., and Douglass, J. E. (1977). Nutrient budgets for undisturbed and manipulated hardwood forest ecosystems in the mountains of North Carolina. In D. L. Correll, ed. *Watershed Research in Eastern North America. Vol 1.* Chesapeake Environ. Center, Edgewater, Md., pp. 343–364.

Thatcher, A. C., and Westman, W. E. (1975). Succession following mining on high dunes of coastal southeast Queensland. *Proc. Ecol. Soc. Aust.* **9**:17–33.

Vitousek, P. M., Gosz, J. R., Grier, C. C., Melillo, J. M., Reiners, W. A., and Todd, R. L. (1979). Nitrate losses from disturbed ecosystems. *Science* **204**:469–474.

Vitousek, P. M., and Melillo, J. M. (1979). Nitrate losses from disturbed ecosystems: patterns and mechanisms. *For. Sci.* **25**:605–619.

Vitousek, P. M., Reiners, W. A., Melillo, J. M., Grier, C. C., and Gosz, J. R. (1981). Nitrogen cycling and loss following forest perturbation: the components of response. In G. W. Barrett and R. Rosenberg, eds. *Stress Effects on Natural Ecosystems.* Wiley, New York, pp. 115–127.

Walters, C. J., and Hilborn, R. (1978). Ecological optimization and adaptive management. *Ann. Rev. Ecol. Syst.* **9**:157–188.

Watt, K. E. F. (1968). *Ecology and Resource Management.* McGraw-Hill, New York.

Westman, W. E. (1975). Ecology of canal estates. *Search* **6**:491–497.

Westman, W. E. (1978). Inputs and cycling of mineral nutrients in a coastal subtropical eucalypt forest. *J. Ecol.* **66**:513–531.

Westman, W. E., and Rogers, R. W. (1977a). Biomass and structure of a subtropical eucalypt forest, North Stradbroke Island. *Aust. J. Bot.* **25**:171–191.

Westman, W. E., and Rogers, R. W. (1977b). Nutrient stocks in a subtropical eucalypt forest, North Stradbroke Island. *Aust. J. Ecol.* **2**:447–460.

Whittaker, R. H. (1961). Experiments with radiophosphorus tracer in aquarium microcosms. *Ecol. Monogr.* **31**:157–188.

Whittaker, R. H. (1975). *Communities and Ecosystems.* 2nd ed. Macmillan, New York.

Whittaker, R. H., and Woodwell, G. M. (1971). Measurement of net primary production of forests. In P. Duvigneaud, ed. *Productivity of Forest Ecosystems.* UNESCO, Paris, pp. 159–175.

Wiersma, G. B., and Brown, K. W. (1980). Background levels of trace elements in forest ecosystems. In P. R. Miller, ed. *Effects of Air Pollutants on Mediterranean and Temperate Forest Ecosystems.* U.S. Forest Service, Gen. Tech. Rep. PSW-43. Berkeley, Calif., pp. 31–37.

Winner, W. E. (1981). The effect of SO_2 on photosynthesis and stomatal behavior of mediterranean-climate shrubs. In N. S. Margaris and H. A. Mooney, eds. *Components of Productivity of Mediterranean Regions—Basic and Applied Aspects.* Junk, The Hague, pp. 91–103.

Zaret, T. M., and Paine, R. T. (1974). Species introduction into a tropical lake. *Science* **182**:449–455.

9

ECOTOXICOLOGY: ASSESSING IMPACTS OF POLLUTANTS ON BIOTA

In studying the effects of toxic chemicals on ecosystems it is useful to consider both effects on a species population and the interactive effects on the community of organisms in an ecosystem. Gleason (1926) and Ramensky (1924) laid the foundation for the *principle of species individuality:* "each species is distributed in its own way, according to its own genetic, physiological, and life-cycle characteristics and its way of relating to both physical environment and interactions with other species; hence no two species are alike in distribution" in relation to an environmental factor (Whittaker 1975, p. 115). An implication of this principle for impact studies is that species will exhibit different sensitivities to a given dose of toxic substance. At the community level the result of this differential sensitivity will be the usurpation of relinquished resources by the more pollution-tolerant species and, as a result, increased dominance by the tolerant species.

Toxicology involves the experimental study of the response of a species population to known doses of a toxic chemical under controlled conditions; *autecology* is the comparable branch of ecology concerned with the interaction of a single species population with its environment and other species. *Epidemiology* is the statistical study of the response of one or more species populations to environmental stressors within the context of their natural environment; *synecology* is the analogous branch of ecology concerned with multispecies interactions with each other and their environment. In this chapter we shall first review some principles and procedures in toxicology as applied to nonhuman species: "ecotoxicology." We next examine autecological characteristics of species that permit prediction of the likely effect of a toxic substance to a species population or a community. We also examine ways in which species are used as indicators of the environmental conditions in which they live. We then review some approaches, including epidemiological ones, for assessing the effects of pollutants on whole ecosystems. The use of synecological characteristics to describe or predict response to stress is discussed primarily in Chapters 11 and 12. Finally, we note some advances in technology for monitoring pollutant concentrations and their effects on organisms.

ECOTOXICOLOGY

The effect of a toxic chemical on a species population is a function of both the *concentration* of the chemical in the immediate environment of the organism and the *duration* of exposure of the organism to that concentration. The two factors interact in their toxic effect in a multiplicative way; hence the *dose* of a substance received by an organism is defined as

$$Dose = (Concentration of chemical)$$
$$\times (Duration of exposure at that concentration).$$

The total dose will be the sum of the separate doses received at different concentrations of exposure, if concentration varied over the period of exposure. A high concentration over a short period of exposure (*acute* dose) may therefore result in the same dosage as a lower concentration over a longer period (*chronic* dose).

In theory, the biological effect of a toxic substance is a function of dose and will be the same at any combination of concentration and exposure that results in a certain dose level. In practice, the factors (concentration and exposure time) may interact differently on the organism's physiology, so that acute and chronic doses may not register the same biological effect. Garsed and Rutter (1982) found that the rank order of sensitivity of five conifer species and varieties to SO_2 treatment was exactly reversed depending on whether the plants were given an acute or a chronic dose of SO_2. The species that were most resistant to a low concentration of SO_2 for a long time (200 $\mu g \ m^{-3}$, 11 months) were most sensitive to a high concentration for a short time (8000 $\mu g \ m^{-3}$, 6 hours), and vice versa. Thus in measuring response of a species to a toxic substance, it is best to obtain measurements under both acute and chronic conditions of exposure.

Further evidence for the notion that injury due to acute or chronic dose may involve different physiological mechanisms is found in the work of Bell et al. (1982), who reported no correlation between those British grass species that were able to evolve a tolerance to acute doses of SO_2 and those evolving tolerance to chronic doses. Toxicologists commonly consider an acute dose one that is administered at high concentrations over a short period of up to 96 hours, and a *chronic dose* one administered at lower concentrations over longer periods.

The terms "acute" and "chronic" take on a different meaning when applied to the nature of the biological response to a toxic substance. An *acute response* (sometimes also called "acute toxicity") refers to rapid damage to the organism, leading quickly to death (Sprague 1969); a *chronic response* generally refers to sublethal effects such as interference with ability to grow, reproduce, or behave normally but not such as to be an immediate or direct cause of death (Warren 1971).

The testing of the response of plants, animals, or microorganisms to known doses of chemical substances is a well-developed field. An excellent introduction to the subject for aquatic organisms is provided by Warren (1971). Criteria data books, which provide information on toxic levels of chemicals to organisms, were briefly described in Chapter 7. Table 9.1 lists selected bibliographic and data-base sources available by computer in the United States on toxicological effects of chemical compounds on organisms.

Table 9.1. Computerized Bibliographies and Data Banks on the Biological Effects of Pollutants, Including Toxic Substances

Name and Location of File	Contents
Bibliographic Files	
APIBE (Biological Effects of Air Pollutants on Plants, Animals, and Microorganisms); Oregon State University Library and Computer Center, Corvallis, Oreg.	10,000+ literature references on biological effects of air pollutants
Data Banks	
Geoecology Data Base, Oak Ridge National Laboratory, Oak Ridge, Tenn.	County-level information on air quality, agriculture, climate, vegetation, forestry, natural areas, human population, water, terrain, and wildlife; statistical manipulations of data and mapping of geographic distribution of occurrences possible.
UPGRADE with SYMAP, Council on Environmental Quality, Washington, D.C.	Contains census-tract information on population density, ambient air, or water quality and public health; with input of pollutant dose-response curves, can calculate and map distribution of expected number of cases of pollution-related health problems in a region (city, state, country)
National Park Flora Data Base, U.S. National Park Service, Washington, D.C.	Plant species lists for class I parks are retrievable at any taxonomic level; with input on air pollution sensitivity of species, data base will indicate identity, location, and number of plant species sensitive to the pollutant in a park or parks

Table 9.1   (*Continued*)

Name and Location of File	Contents
Data Banks (continued)	
Environmental Information System office, Oak Ridge National Laboratory, Oak Ridge, Tenn.	Provides literature citations and unpublished data on (1) heavy metals and toxic organic compounds in media and organisms near industrial facilities: Toxic Materials Information Center; (2) environmental mutagens: Environmental Mutagen Information Center; (3) drugs, food additives, industrial chemicals: Toxicology Information Response Center; (4) radiation, radionuclides; Ecological Sciences Information Center
National Institute of Occupational Safety and Health, Cincinnati, Ohio	20,000 literature references on aspects of occupational safety and health: Technical Information Services Branch
On-line Computerized Bibliographies	
CHEMLINE	Literature on 60,000 substances can be searched by name or formula
MEDLINE	Medical references, including articles on effects of toxic compounds on people and domesticated animals
TOXLINE	325,000 references on human and animal toxicity studies

Sources: Bennett 1985, CEQ 1977, 1978; Fennelly et al. 1976. For additional computerized sources of information on air, water, and noise pollution, see Table 7.11.

Bioassays

A common approach to developing toxicological information on organisms is to expose a population of a standard test organism whose physiology is considered similar to the organisms of direct concern, to known levels of a toxic compound in an environmentally controlled chamber. This technique is known as a *bioassay*.

Standard bioassay methods for water pollutants have been developed for fish, invertebrates, and algae (e.g.; Sprague 1969, 1971, Katz 1971, APHA 1971, Cairns and Dickson 1973, Buikema and Cairns 1980).

Lethal-Dose tests

A typical bioassay procedure is that followed by Trama (1954) in determining the toxicity of copper ion (Cu^{2+}) to fish. A population of the common bluegill, one of 19 standard fish species used in bioassays in the United States, (Cairns 1969), was first acclimated to tank conditions and temperature. Deaths resulting from exposure of replicated tanks of bluegill populations to each of a range of concentrations for 96 hours are plotted in Figure 9.1. Two copper salts gave slightly different results, and these plus the average curve are plotted. Note that a semilog plot is used in Figure 9.1. The resulting sigmoidal curve approximately fits a lognormal distribution. The concentration at which 50% of the population dies after 96 hours of exposure is interpolated as 0.74 mg liter^{-1}, and is abbreviated $LD_{50,96hr}$, LD standing for "lethal dose."

Government agencies seeking to convert such information into recommended standards of safe concentrations for long-term exposure have applied the concept of a "margin of safety" very crudely by multiplying the

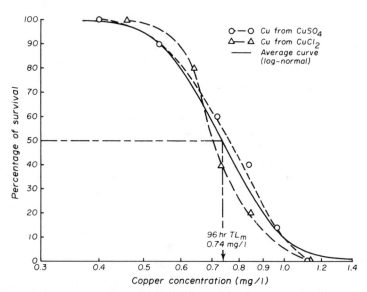

Figure 9.1. Percentage survival of bluegills following 96 hours of exposure to various concentrations of copper ion (with sulfate or chloride). The average of results is fitted to a lognormal curve, from which the LD_{50} is estimated. Data of Trama (1954) as redrawn by Warren (1971, Figure 13.1, p. 197). Reproduced with permission of the Academy of Natural Sciences, Philadelphia, and Saunders.

LD_{50} by some "application factor" such as 0.1 or 0.01.

In the forgoing example the standard for safe level of copper with an 0.1 application factor would be 0.074 mg liter^{-1}. ORSANCO, an interstate water pollution agency, recommended 0.1 of the 48-hour LD_{50} as an indication of safe levels for nonpersistent chemicals and pesticides, and this was widely adopted for many years.

The use of "application factors" and bioassay results for setting standards has at least the following limitations:

1. The bioassay is based on 48 or 96 hours of exposure. There is no basis for knowing how mortality will increase with longer-term exposure.

2. The bioassay is conducted on a single species. In a multispecies context the chemical can be transferred through food chains so that levels of exposure by ingestion of food (rather than uptake from water) may result in much higher levels of mortality from a given concentration in water. Furthermore competition, predation, and other multispecies interactions can increase the stress on the organism, resulting in increased sensitivity to a given concentration.

3. The potential for enhanced toxicity (positive synergisms) to other chemical stressors in the environment is ignored.

4. Only one stage in the life cycle is tested. The sensitivity of the species in juvenile stage, larval, or egg stage is frequently different (usually greater) from that in the adult stage. Ability to mate and reproduce is untested.

5. The sensitivity of different organisms to the pollutant will vary. A "safe level" for bluegill fish may not be safe for the other organisms in the biological community of which the bluegill is a part. One result may be that the bluegill will perish in nature anyway, due to toxification of some other species on which it depends.

6. The bioassay is not sensitive to sublethal effects which could prove lethal upon longer periods of exposure. Fishes that do not die in the bioassay are considered completely healthy.

Because of these problems the past two decades have seen increased emphasis on tests of species in multispecies microcosms and field settings—a practice referred to as "biological monitoring," discussed later. Several improvements to the bioassay technique have also been developed.

Sublethal Observations

One improvement in the bioassay technique that permits enhanced sensitivity to sublethal effects involves observing physiological and behavioral responses of a species population to a low concentration of pollutant. Abnormal behavior is a sublethal effect serving as an "early warning" of possible lethal effects at higher concentrations or longer exposure times.

One example involves automated observation of the activity pattern and breathing rates of fish in tanks. Figure 9.2 shows a portion of the system developed by Cairns et al. (1973). One test fish is placed in each of a series of tanks. Breathing rates of the test fish are sensed by stainless steel electrodes located at the ends of the aquaria, which register an electric signal with each flap of the gills. Activity patterns (swimming movements) are registered by three light beams passing through the aquarium. As the fish passes through and interrupts a beam, a photocell at the other end of the tank picks up the interruption of signal and records results in a computer. Both activity patterns and breathing rates are subject to normal diurnal fluctuation. These background fluctuations under clean water conditions are first observed and recorded on computer. Then the signals obtained from fish exposed to pollutants flowing through the tank are compared by computer to expected background patterns. A warning signal is registered (a bell, a message on video screen, or a telephone call) when a deviation of >5% from expected behavior occurs in two or more test fish and not more than one control fish in a given hour.

Cairns et al. (1973) suggest that such systems could be used to test for toxic concentrations either in industrial effluent (by flowing a diluted form of effluent through the aquaria) or in receiving waters (by flowing stream water from various locations up and downstream of the outfall through the aquaria). A variant of this system, using 12 fish and measuring breathing only, is operating on line at the Radford Army Ammunition Plant in Virginia

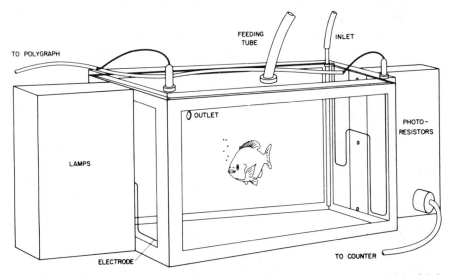

Figure 9.2. An automated biological monitoring system. The swimming activity of the fish is monitored by the break in light beams traversing the tank to the photoresistor; its breathing is recorded by impulses to the electrodes in the tank. Reprinted from Figure 5 of Cairns et al. (1973, p. 13) with permission of the Virginia Water Resources Research Center.

(Cairns and Gruber 1979). Several fish activity monitors developed at the Stevenage Laboratory in England are being used in British rivers (Cairns and van der Schalie 1980). Many other flow-through systems for continual monitoring of water quality by test organisms, including bacteria and vertebrates, using respiratory rates (oxygen consumption), nitrification rates, phototaxis (attraction to light), and other physiological/behavioral cues have been developed and are reviewed by Cairns and van der Schalie (1980). For a discussion of aquatic invertebrate bioassays and biological monitoring, see Buikema and Cairns (1980); for attached organisms, see Weitzel (1979) and Cairns (1982); for a range of aquatic organisms, including fish, see Cairns et al. (1977a). A suggested field- and laboratory-based protocol for testing the effects of chemicals on aquatic life is described by Cairns and Dickson (1978).

ATTRIBUTES FOR PREDICTING SPECIES' RESPONSE TO POLLUTION STRESS

One approach to predicting the sensitivity of organisms to pollutants is to examine particular features of their behavior, life cycle, growth form, physiology, or biochemistry that make them susceptible to pollutant damage. A few examples of the use of these attributes in prediction of species response to stress follow.

Physiognomy and Growth Forms

A number of morphological features of plants help them to exclude uptake of pollutants. For example, leaves with a large, smooth surface area and undissected leaf margins will cause a relatively thick layer of still air ("boundary layer") to remain between the layer of fast-flowing air and the leaf surface itself. Thicker boundary layers can act to reduce the rate at which air pollutants reach the pores of the leaf (stomates) which absorb gases, and hence reduce air pollutant damage (Taylor 1978). Resinous leaf surfaces, glands, and hairs will increase the degree to which air pollutants adhere to the leaf surface and, by acting as a filter, may ultimately decrease the amount of pollutant internally absorbed (see, e.g., Elkiey and Ormrod 1980, Sharma 1975, Sharma and Butler 1975). Elkiey and Ormrod (1980) found that varieties of petunia with larger and more numerous leaf hairs ("trichomes") were more resistant to SO_2. However, such hairs may also aid in entrapment of dew, fog, and particulates, so that sulfur (as SO_2 or SO_4) dissolved in fog or dew, or present as sulfate particulates, may actually be differentially collected by hairy leaves. It becomes clear that any one morphological feature, such as leaf hairs, cannot serve as a general indicator of susceptibility in the absence of knowledge of compounding variables (form and toxicity of pollutant collected, presence and nature of moisture in the environment, degree to which pollutants absorbed on leaf surfaces are later absorbed internally, etc.)

Cuticular wax, which coats the leaf surface, can also act to exclude pollutants. The greater susceptibility of lichens to air pollution damage relative to higher plants (Hawksworth and Ferry, 1973) may indeed be due to their lack of a waxy cuticle. Bystrom et al. (1968) found increased pollutant damage in younger leaves of beet (*Beta vulgaris*) which they attributed to the incomplete nature of the cuticle in young leaves. As leaves proceed beyond maturity to senescence they may also develop cracks in the cuticle (Chamel and Garrec 1977) which increase their susceptibility to pollutant absorption. One might expect middle-aged, fully grown leaves with intact cuticles to be most resistant to air pollutants, but in broad-leaved trees this is generally not the case (Linzon 1978). This again reminds us that single factor predictors are unlikely to account for much of the variation in pollutant susceptibility. Research in this field is as yet scant, but the notion that complexes of morphological features may serve as indexes of pollutant resistance is worth further exploration.

The notion that the form of growth of major plant organs may impart resistance to stress is an old idea. One of the best-tested generalizations is that fruticose lichens (with branched structures well above the surface) are more susceptible to SO_2 damage than foliose lichens (whose leaflike thallus lies nearly flat on surfaces) and that both in turn are more susceptible than crustose lichens (which embed their tissue in the cracks of bark, soil, or rocks) (Figure 9.3). The use of morphological lichen types as indicators of air pollution concentrations is now well developed (see, e.g., Skye 1968, Hawksworth 1976, de Wit 1976, Ferry et al. 1973). Bryophytes (especially mosses) have also been used as air pollution indicators (Gilbert 1968, Taoda 1972).

Whether woody plants can be considered more or less resistant to pollution damage than herbaceous ones is unresolved. Woodwell (1970) provided a series of examples in which woody species were less resistant to a variety of stresses (pollution, ionizing radiation, herbicides, fire) and noted that the high proportion of total mass in nonphotosynthetic tissue (wood, bark) puts a strain on larger woody organisms whose leaf areas are reduced in any way. In the study of SO_2 effects on boreal forests in Canada woody plants were more susceptible to pollutant damage than herbaceous ones (Gordon and Gorham 1963, Winner and Bewley 1978). In the case of oxidants Harkov and Brennan (1982) hypothesize that herbaceous plants will be more susceptible to ozone damage than woody ones, but data of Westman (1979) suggest that oxidant-damaged shrubs are in fact replaced by resistant herbs in polluted Californian coastal shrubland communities. This is consistent with the shorter generation times of herbs mentioned in Chapter 7. A much better understanding of the interaction among resistance factors in relation to particular pollutants is needed before generalizations can be considered firm.

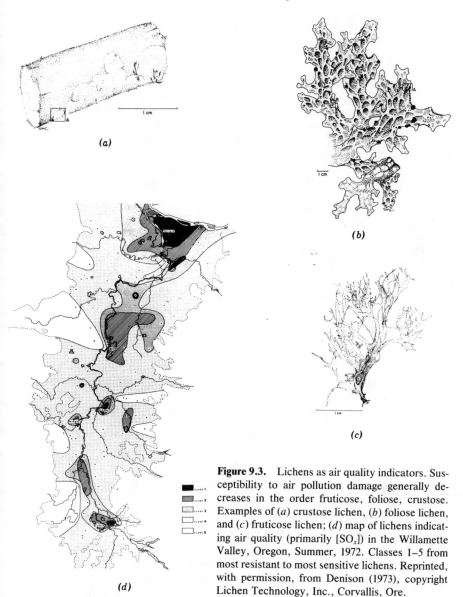

(a)

(b)

(c)

Figure 9.3.   Lichens as air quality indicators. Susceptibility to air pollution damage generally decreases in the order fruticose, foliose, crustose. Examples of (a) crustose lichen, (b) foliose lichen, and (c) fruticose lichen; (d) map of lichens indicating air quality (primarily [SO_x]) in the Willamette Valley, Oregon, Summer, 1972. Classes 1–5 from most resistant to most sensitive lichens. Reprinted, with permission, from Denison (1973), copyright Lichen Technology, Inc., Corvallis, Ore.

(d)

Raunkiaer (1934) developed a growth-form classification for plants based on the height of the growing bud and postulated that a plant community will have a larger number of plants with growing tips well exposed to the elements in climates of least moisture and temperature stress (Figure 9.4). He called his growth forms "life forms." Raunkiaer's generalization has proved broadly applicable (see, e.g., Cain 1950; Whittaker 1975, Table 3.2).

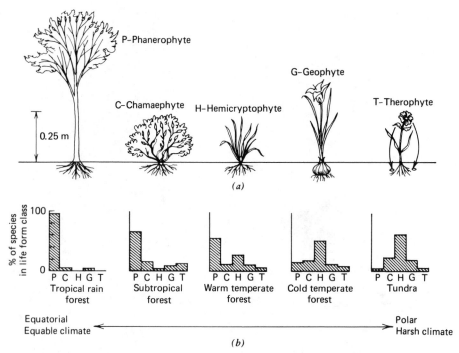

Figure 9.4. Life-forms. (*a*) Phanerophytes (trees or tall shrubs) have buds >0.25 m above ground. Chamaephytes are low shrubs with buds <0.25 m above ground. Hemicryptophytes (e.g., grasses) are perennial herbs with buds at ground surface. Geophytes (e.g., lily) are perennial herbs with bulb or other bud-containing organ below ground. Therophytes are annuals storing all their buds in seeds during part of the year. (*b*) Raunkiaer (1934) life-form spectra. There is a tendency for life forms with increasingly protected growing buds to predominate in harsher climates. Data of Cain and Castro (1959) and Whittaker (1960, 1965, 1975).

Many other growth form classification systems exist, based on a larger range of physiognomic characteristic of plants (see, e.g., Mueller-Dombois and Ellenberg 1974, Orshan 1983). These growth forms, for evolutionary reasons, are considered to reflect adaptation to natural rather than technologically induced stresses. Yet many such adaptations are capable of responding to novel stresses that mimic those for which they evolved. Thus the capacity of Mediterranean-climate shrubs in Californian coastal sage scrub to resprout twigs and leaves from root crowns following fire also permits them to resprout following defoliation by frost (Mooney 1977), herbivore attack, bulldozing or air pollution damage (Malanson and Westman 1984 and personal observation). *Eucalyptus* in Australia has a similar ability to respond to a variety of natural and novel stresses by the sprouting of buds beneath its bark surface. The root crown and epicormic buds are growth-form features that indicate a plant's resilience following stress, whereas the cuticle, leaf hairs, and other growth forms mentioned earlier are features indicating a plant's inertia (resistance to stress).

Life-Cycle Types

A number of ecologists have noted that as the intensity and periodicity of stress in the environment change, different patterns of longevity and reproduction will be favored. For example, MacArthur and Wilson (1967) (see also Pianka 1970) examined what attributes of a species would be most adaptive in a near-equilibrium climax community in which competition for resources becomes fierce as the carrying capacity (K) is approached. They reasoned that under such circumstances species will be favored which have long life expectancies and low proportions of energy devoted to reproduction, freeing energy to enhance mechanisms of competition for resources ("K-selection"). By contrast, in nonequilibrium, early successional habitats, species with short generation times and large reproductive effort (high intrinsic rate of reproduction "r"; see Chapter 6) will be selected for, since the need to disperse and colonize over large areas rapidly is the key to resource domination in such temporary habitats ("r-selection").

Successional changes in the chaparral and coastal sage scrub of California exemplify this trend. In the first year following a widlfire, such shrubland sites are dominated by a profusion of annual and short-lived perennial species which produce large numbers of seeds (Hanes 1971, Westman 1981, Westman et al. 1981). Within 2–5 years, however, the sites are shaded by resprouting shrubs of 20–30+ year longevity (Malanson and Westman 1984, Hanes 1971) which reproduce by seed only scantily (Malanson and O'Leary 1982); the annuals and short-lived perennials meanwhile decline in abundance drastically. The cycle repeats at each successive fire (Figure 9.5). While the distinction between r- and K-selected species is a useful one conceptually, many species in nature in fact fall between the extremes of the r–K axis.

Grime (1977, 1979) has suggested a somewhat more complex classification of plant types into competitors, stress-tolerant types and ruderals. Grime (1979) uses the term "stress" to refer to growth-limitations by habitat features such as light, water, nutrients, or suboptimal temperature and "disturbance" to refer to biomass-destroying activities such as herbivory, frost, fire, trampling, and cutting. The three plant strategies then are defined as those adapted to three different combinations of stress and disturbance in the habitat (Table 9.2).

The ruderal strategy exploits conditions where growth is favorable (low stress) but disturbance is frequent. Ruderals are typically herbaceous annuals or short-lived perennials with rapid vegetative growth and flowering and prolific production of seeds (r-selected). Stress-tolerant species occur in conditions that are severely growth limited but in which disturbance is infrequent. Such species may be of any growth form (lichen, herb, shrub, tree). Grime (1979) suggests that they are typically evergreen with small or leathery or needlelike leaves, long lived, slow growing, intermittently flowering, and producing a low mass of seeds (K-selected). Competitors, which are

Figure 9.5. (*a*) Coastal sage scrub in coastal Los Angeles County, five months after fire. Many *r*-selected annual and short-lived perennials have already grown to flowering. (*b*) The same site three years later. The site is now dominated by *K*-selected perennial shrubs, and herbaceous cover is much diminished.

Table 9.2. Habitat Conditions Conducive to the Evolution of Three Plant Strategies

Intensity of Disturbance	Intensity of Stress	
	Low	High
Low	Competitors	Stress tolerators
High	Ruderals	No viable strategy

Source: From Table 1 of J. P. Grime, *Plant Strategies and Vegetation Processes* (1979). Reprinted by permission of John Wiley & Sons, Ltd.

adapted to habitats of little stress or disturbance, may be herbs, shrubs, or trees, often with a broad, dense canopy, extensive root system, periodic and regular leaf and flower production, rapid growth in response to pulses of increased resource availability, and an intermediate strategy of seed production on the $r–K$ spectrum. Combinations of these strategies may of course also occur.

In Figure 9.6 it is seen that different combinations of these three strategies characterize the more traditional growth forms. Grime (1974) has shown that different clusters of strategies characterize the species found in different disturbed habitats, such as pastures, mown roadsides, soil heaps at building sites, and sewage sludge heaps (Figure 9.7). The axes of the triangles in Figure 9.7 are intended to correspond to the three strategies. The actual measurements used for each axis were relative growth rate (lowest for stress tolerators), and a "morphology index" based on height of canopy, lateral spread of canopy and roots, and litter mass. Each of the characters in the

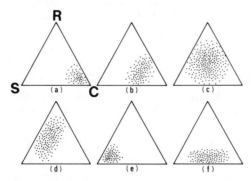

Figure 9.6. Distribution of species of particular growth form within the range of strategies *C* (competitors), *S* (stress tolerant) and *R* (ruderal). (*a*) Annual herbs; (*b*) biennial herbs; (*c*) perennial herbs and ferns; (*d*) trees and shrubs; (*e*) lichens; (*f*) bryophytes. Adapted from J. P. Grime in *American Naturalist 111*, by permission of the University of Chicago Press. (c) 1977. The University of Chicago Press.

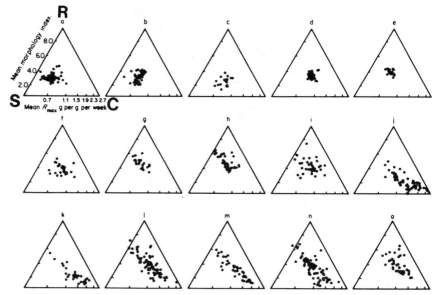

Figure 9.7. Position of square meter samples of herbaceous vegetation from 15 habitats, in relation to axes intended to measure three plant strategies: competitor (*C*), stress tolerant (*S*), and ruderal (*R*). Each sample is weighted by the relative frequency of species in it (Grime 1974). (*a*) Unenclosed sheep pastures on acidic strata; (*b*) unenclosed sheep pastures on limestone; (*c*) limestone outcrops; (*d*) meadows; (*e*) road margins, mown frequently; (*f*) enclosed pastures; (*g*) road margins, mown infrequently; (*h*) hedge bottoms; (*i*) derelict banks of rivers, ponds, and ditches; (*j*) paths; (*k*) fallow arable; (*l*) heaps of mineral soil, such as building sites; (*m*) demolition sites, brick and mortar rubble; (*n*) cinders (dumps and railway ballast); (*o*) manure heaps and sewage sludge. Adapted by permission from *Nature* **250**:26–31. Copyright (c) 1974 Macmillan Journals Limited.

morphology index is highest in competitors and lowest in ruderals. Grime (1979) has elaborated somewhat on this scheme, but it remains in need of much further testing.

Stearns (1976, 1977) has reviewed some notions of adaptive life history traits, especially for animals. Most investigators agree that early sucessional enviroments will favor *r*-selected species with short generation times and high reproductive effort, producing large numbers of small offspring which are given no parental care. "Stable" climax communities, by contrast, will favor *K*-selected species with organisms that produce offspring later in life, have multiple broods, and concentrate their energies on rearing a few, large young. While these generalizations are borne out by some studies with birds and plants (e.g., Cody 1971, Gadgil and Solbrig 1972, Abrahamson and Gadgil 1973), there are occasions when they do not hold (e.g., Menge 1974). One complicating feature is that the occurrence of stress or disturbance in a habitat is often intermittent, and the optimal reproductive strategy is likely to vary depending on the length of the stress-free period and the predictability (i.e., regularity) of the recurrence of stress episodes relative to generation time or life span.

Stearns (1976) offers a set of hypotheses regarding the optimal reproductive strategies likely to be favored under varying levels of predictability of stress (Table 9.3). In general, organisms are likely to evolve more flexible, reversible reproductive strategies the more unpredictable the interval and intensity of stress. Although these generalizations require much further testing, they may provide a basis for predicting which organisms are likely to be preadapted to, and hence more likely to survive, particular kinds of stresses. Thus by this criterion herbaceous annuals could be predicted to increase under chronic air pollution stress (environmental condition I, Table 9.3), while microorganisms could be predicted to recover best from randomly timed oil spills (environmental condition II, Table 9.3).

Physiological and Biochemical Attributes

A wide range of physiological and biochemical attributes may impart resistance to stress in organisms. Of interest to impact analysts is whether some readily measured feature of an organism can provide an indication of its likely resistance to stress. One such physiological example involves the rate at which water vapor escapes through the stomates of leaves (so-called "leaf conductance"). This feature of plants varies both with environmental conditions and with the structural/physiological makeup of the organism and is readily measured in the field with an electronic "diffusive resistance" meter and attachment that clamps onto a leaf, measuring changes in humidity in the chamber above the leaf.

Winner et al. (1982) found that when moisture and other environmental conditions were carefully controlled and nonlimiting to growth, broad-leaved trees and shrubs of coastal California exhibited characteristic differences in leaf conductance which correlated with their sensitivities to short-term SO_2 exposure (as measured by changes in photosynthetic rate). Deciduous-leaved shrubs characteristically had higher conductance rates in SO_2-free air, and a greater suppression of photosynthesis when exposed to SO_2, than evergreen, sclerophyllous (leather-leaved) species. If this trend in only 10 species proves more generally applicable, it could provide a basis for predicting SO_2-sensitivity of species based on their leaf conductance values under nonlimiting growing conditions. Other factors may confound this generalization (see, e.g., Winner et al. 1982, Westman et al. 1985), but it represents one example of the incipient development of a physiological predictor of stress response .

At the biochemical level Winner and Mooney (1980) contrasted the effects of short-term SO_2 exposure to two annual species of *Atriplex* (a genus of mainly arid-zone shrubs) that differed in their biochemical pathways of photosynthesis. *A. triangularis* has a C_3-pathway, so-called because the first

Table 9.3. Adaptive Strategies for Survival in Fluctuating Environments

Environmental Situation	Traits Selected	Examples	References
I. Cyclic, fixed period, period $\gg$ life span	1. Reproduction early in cycle 2. Large clutches 3. Parthenogenesis 4. Diapause forms	Aphids, clado-cerans, pelagic tunicates, multivoline insects	
II. Cyclic, period fixed, period $<$ life span			
A. Start of cycle and conditions during cycle predictably favorable	1. Synchronization of breeding time at optimal point in cycle 2. Synchronization of release of young to swamp preda-tors 3. Separation of vegetative (so-matic) and reproductive effort in time	Many ungulates, many trees, intertidal organ-isms, periodic cicadas, sea turtles	Janzen (1971b)
B. Start of cycle predictable only within limits, cycle predictably favorable	Spread risk by de-veloping within-clutch variance in hatching or ger-mination date to match the histori-cal probability distribution of the optimum	Univoltine insects, annual plants near center of their range	Cohen (1966), Palmblad (1969), Marshall and Jain (1970)
C. Start of cycle predictable but condi-tions dur-ing cycle unpredict-able; no in-formation on future conditions available at start of cycle	1. Iteroparity and long life span 2. Larger variance in diapause length 3. Intermingling of vegetative and reproductive growth during the season; several clutches per season	Univoltine insects and annual plants near limits of their range; desert plants, wild oats, annual fish	Cohen (1968, 1971), Mountford (1971), Marshall and Jain (1968, 1970), Wourms (1972)

380

Table 9.3   (*Continued*)

Environmental Situation	Traits Selected	Examples	References
D. Start of cycle predictable, conditions during the cycle unpredictable but some information available on the future	1. Ability to resorb reproductive tissue 2. Flexible timing of reproduction 3. Ability to skip a season entirely if it looks bad (e.g., mast years)	Condors, albatrosses, the red kangaroo, trees	Short (1972), Smith (1970), Janzen (1971b)
III. Not cyclic, but distributed as a random variable in time	1. Rapid development, large reproductive commitment 2. Ability to resorb reproductive tissue if a mistake has been made 3. Ability to enter a resting stage as an adult	Microorganisms?	

Source: Reprinted from Stearns (1976, Table 6) with permission of the Quarterly Review of Biology.

major carbohydrate formed in the photosynthetic pathway is a 3-carbon compound. *A. sabulosa* has a C_4-pathway, quite different in biochemical steps. The two biochemical pathways are typically associated with readily observed differences in leaf anatomy: in C_4-plants, chloroplasts are concentrated around vascular bundle sheaths ("veins"), whereas in C_3-plants, chloroplasts are more uniformly distributed throughout the internal leaf cells. Winner and Mooney (1980) found that the C_3-plant was more sensitive to SO_2 (as measured by reductions in photosynthetic rate) than the C_4-plant. Whether this generalization will extend to other C_3 versus C_4 plants remains to be tested.

Still a third type of photosynthetic pathway, characterizing succulents particularly of the family Crassulaceae, is known as Crassulacean Acid Metabolism (CAM). Such plants open their stomates at night to absorb CO_2. In the day time, when light is available to complete photosynthesis, the stomates are closed. CAM plants should therefore be more resistant to sources of air pollution which are more or less diurnal, such as photochemical oxidants or pollutants released by factories which reduce or close down operations at night.

Behavior

Many behavioral patterns of animals could be used to predict their likely response to stress. The increased resistance of tree-climbing small mammals to impact from track vehicles was noted in Chapter 8. Similarly organisms that can swim will be better able to escape oil spills or heated plumes than organisms that can only float. Organisms that build deep burrows will be more resistant to sonic booms, or fires, than those in shallow burrows or surface dwellers. Species with broad food preferences will be better able to survive partial habitat destruction than those with narrow, specialized food requirements (though see Chapter 8 on food webs and modules). Territorial species with fixed territory and home range sizes (many birds, small mammals) will be less able to survive partial habitat destruction without population decline than migratory, herding species whose territorial requirements are more flexible. The relationship of behavior to impact susceptibility is case specific; the use of behavior-based criteria for defining guilds (Chapter 8) is one means to incorporate such characteristics into a broader framework for predicting response to impact.

ORGANISMS AS INDICATORS OF ENVIRONMENTAL CONDITIONS

The type, abundance and health of organisms found at a site may be used as a guide to the favorableness of the habitat for the support of native biotic communities. Organisms or their attributes are thus being used as biological indicators of site conditions. Unlike the methods in the previous section, no predictions are being made here about response at a future time. Rather, organisms whose health is particularly sensitive to the environmental stressor of interest are being used as a guide to the health of all species on the site. We will consider use of biological indicators at the suborganismal, species, and guild levels of organization.

Suborganismal Level

A range of physiognomic or physiological attributes of species have been used as indicators of various stresses. In plants, foliar symptoms of damage to particular air pollutants (mottling, bronzing, discoloration, etc.) have long been used as ready visual guides to nature and intensity of stress (Lacasse and Moroz 1969, Treshow and Laccasse 1976; see Figure 9.8). Indeed, damaged or infected foliage can often be sensed as a distinct discoloration on infrared-sensitive film; this attribute has been used to track disease infection or air pollution damage to crops by remote sensing. Davies (1963) found that hairs on the stamens of the day flower (*Tradescantia paludosa*) are stunted in growth by ionizing gamma radiation. Alvarez and Sparrow (1965) have used this attribute as a quantitative index of radiation exposure. In addition individual cells on the stamen hairs in a particular strain of *Tradescantia* (Clone 4430) will change color from blue to pink as a result of mutation of a single gene from dominant (blue) to recessive (pink). This strain is heterozygous for cell color at a single gene locus. *Tradescantia* has been planted near nuclear power plants in Czechoslovakia, Japan, Mexico, Poland, and the United States to monitor mutagenic gamma radiation (Bedell 1983).

In animals, Valentine et al. (1973) found that fishes exposed to sewage pollution stress lost their bilateral symmetry in scale formation, and he used the degree of asymmetry as an index of pollution stress. Often the increased incidence of particular diseases (fin rot in fishes: Cross 1982; bark beetle infestation in pines: Miller and Elderman 1977) are indicators of an underlying environmental stress. Behavioral abnormalities have also been used, the classical example being the cessation of singing in canaries exposed to toxic levels of underground mine gases.

Figure 9.8. Morphological aberrations in plants exposed to air pollution. Here stems have widened ("fasciated"), and stem elongation has been inhibited in California sage (*Artemisia californica*) after 10 weeks of 40 hour per week exposure to 0.2 ppm SO_2 in a fumigation chamber. Some leaves have also become mottled (not apparent at this magnification). Photo courtesy K. P. Preston, L. B. Weeks, and W. E. Westman.

A number of biochemical monitoring methods have been developed. The respiration rate of bacteria on an organic substrate will be altered by the presence of toxic substances. Cairns and van der Schalie (1980) review several automated systems that monitor these respiratory rates using experimental bacterial cultures, as well as systems that monitor respiration, heart rate, or other activity (e.g., stone fly: Maki et al. 1973; crayfish: Maciorowski et al. 1977; fishes: Marvin and Burton 1973). Dillon and Lynch (1981) review physiological indicators of stress in marine and estuarine organisms.

Tingey et al. (1976) have used the evolution of ethylene by plants as an indicator of stress due to ozone. Bressan et al. (1979) also found ethane release to be a stress indicator in plants. Ivanovici and Wiebe (1981) suggest the level of energy-carrying compounds in organisms (ATP, ADP, and AMP, combined in an index termed the "adenylate energy charge" or AEC) as an indicator of stress in organisms. A wide range of organisms, from microorganisms to higher plants and animals, show a drop in adenylate energy charge under a variety of environmental stresses.

Omasa et al. (1980) used a thermal scanning device to map small changes in the surface temperature of a leaf that arose from differences in evapotranspiration induced by pollution effects on stomatal closure. Ellenson and Amundson (1982) used an image intensifier to detect differences in "delayed light emission" from a leaf following pollution exposure; leaves fluoresce slightly in the dark following light exposure, during a portion of the chemical reaction of photosynthesis. The SO_2 exposure causes characteristic changes in light emission over injured portions of the leaf before visible injury appears.

In addition tissue analysis of particular organs of indicator species are often used as indications of concentrations of chemicals in the environment. Thus Brooks (1972) discusses use of plant foliar analysis in mineral prospecting. The analyses of liver, brain, eggshell, or feather for heavy metal or pesticide monitoring was noted in Chapter 7.

Species Level

The presence or absence, or marked change in relative abundance, of certain species of known environmental tolerances and preferences have been used as indicators of habitat conditions. Species may be of indicator value either because they are very *intolerant* of degraded conditions, and therefore first to disappear following disturbance, or because they are unusually *tolerant* of degraded conditions, and survive where others won't. Species of greatest indicator value are those that are:

1. Sufficiently widespread to occur in most sites within the habitat type.
2. Sufficiently narrow in environmental tolerance to the particular stressor

so that changes in the species' relative abundance reliably indicate particular habitat conditions.

3. Sufficiently short in generation time so that population changes due to changing environmental conditions can occur rapidly.
4. Sufficiently abundant so that fluctuations in population numbers are substantial in magnitude with changing levels of environmental stress (see Figure 9.9).

Examples of the use of plants as indicators include the use of lichen abundances to indicate SO_2 pollution, the presence of rare species (e.g., certain orchids) as an indicator of lack of human disturbance, and the increased abundance of certain spiny or unpalatable shrubs on grazed lands as an indication of past grazing intensity (e.g., increased abundance of thistles). Steubing and Jäger (1982) review the use of higher plants as indicators of air pollution levels. Indications of changing environmental conditions are conveyed more subtly by changes in overall plant community composition; methods for such analyses are discussed in Chapter 10.

Among animals increases in the abundance of the polychaete marine worm, *Capitella capitata*, and the freshwater oligochaete worm, *Tubifex*, are used to indicate oxygen-poor conditions in water and sediments (Reish 1972). Karr (1981) finds that high abundance of the green sunfish (>20% of

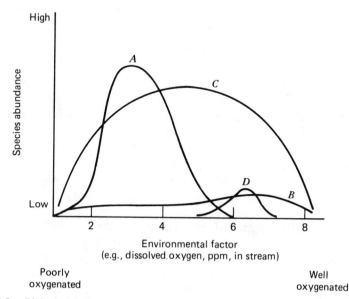

Figure 9.9. Biological indicators. The figure shows types of species distributions along an environmental factor axis. Species *A* would make the best biological indicator of poorly oxygenated conditions in water. Species *B* is too low in abundance to register dramatic change. Species *C* is too broad in tolerance to indicate particular environmental conditions. Species *D* is too narrow in tolerance range to be found in many sites.

individuals) in fresh water streams of the midwestern United States indicates degraded conditions; he also lists fish species that are intolerant, that is, disappearing soon after stream quality begins to decline. Table 9.4 summarizes the usefulness of different taxonomic groups as indicators of water quality, based on both their indicator qualities and ease of taxonomic identification.

The increased abundance of hybrids between two related species is often an indication of disturbed habitat conditions. Clifford (1954) and Pryor (1959) report this for *Eucalyptus* spp. in Australia, Anderson and Anderson (1955) for *Salvia* spp. (sage) in coastal California, and Greenfield et al. (1973) and Karr (1981) for cyprinid (minnow family) fish, sunfishes, carp, and goldfish. The phenomenon is very likely widespread and presumably occurs as a result of habitat disturbance which permits greater intermingling of species formerly separated due to substrate differences or other habitat patchiness.

Once tolerant or insensitive "indicator species" for a particular type of pollutant stress have been identified, it is possible to rate the condition of a site by the composition and abundance of such indicator species, in a "biotic index." Such biotic indexes have been developed particularly for water quality studies. In Europe the "saprobity index" is widely used, based on the abundance of detrivores (saprobes) with varying tolerance of anaerobic conditions (Herricks and Cairns 1982). Hellawell (1978) also reviews available biotic indexes of water quality.

Guild Level

An example of the use of guilds as indicators of habitat condition is provided by Karr (1981) in his analysis of freshwater streams in the midwestern United States. From an extensive study of such streams over a seven-year period, Karr found that omnivorous fish (which eat both plants and animals) increase in abundance in degraded streams. This is presumably because, as pollution kills off invertebrates, omnivorous fish can survive on plants. Streams with fewer than 20% of fish individuals as omnivores were of good water quality, whereas those with >45% omnivores were badly degraded. Second, he noted that insect-eating cyprinid fish decrease as omnivores increase; insectivorous cyprinids were therefore considered "intolerant" species. Additionally Karr noted that in degraded streams the top carnivore fish species that could normally occur in a stream of given size were typically reduced in numbers or absent. These fish species (smallmouth bass, walleye, grass pickerel, rock bass, and others) were therefore also useful as intolerant indicator species. Since the expected diversity of top carnivores varied with stream size, the abundance of the top carnivore guild, rather than the presence or absence of one particular fish species, was used as a criterion for stream quality.

In constructing a numerical index of stream quality, Karr used 12 criteria, including abundance of species in the three guilds. In addition he used total number of species (see Chapter 11), total number of individuals, numbers of species of indicator value (darters, sunfish, suckers, and others), the number of hybrid types, and abundance of disease or growth abnormalities. Each of these criteria were ranked on a three-point scale and summed to form a biotic index. The summing of ordinal ranks is of course mathematically inappropriate (Chapter 4), and the assumption of equal weighting for each criterion arbitrary. The idea of using more than one criterion for assessing habitat quality, however, is a good one. An alternative approach, rather than summing to a grand index, is to plot the values for each parameter separately. Bakelaar and Odum (1978) plotted nine ecological parameters as percentages of their highest value in comparing fertilized and unfertilized old fields and termed such graphs "ecosystem profiles." An analogous graphical representation of Karr's (1981) data for three streams is presented in Figure 9.10.

Such graphical approaches rapidly become cluttered and awkward when many streams are to be compared, or many parameters used. They are also difficult to interpret, since high levels of some parameters indicate the same quality as low levels of others (e.g., omnivores, insectivorous cyprinids). An alternative, mathematically valid approach is to rank ("ordinate") the

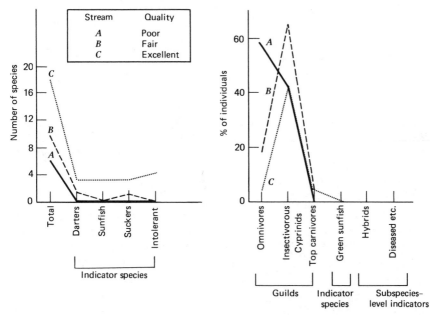

Figure 9.10. "Ecosystem profiles" showing 11 of 12 biological indicators of water quality used in the biotic index of Karr (1981) for headwaters of three freshwater streams. Stream *A* is Station 6, stream *B* is Wertz Woods station, both on Black Creek, Indiana. Stream *C* is Jordan Creek, Illinois. The additional criterion used in Karr's index (not plotted) is total number of individuals (63, 122, and 153 for streams *A, B,* and *C*).

Table 9.4. Usefulness of Various Taxonomic Groups

Group	Distribution (Species Patterns)	Taxonomy	Working Level	Ease	Sampling
Bacteria	Cosmopolitan	Incomplete	Genus, species	±	Descriptive, statistical
Fungi	Cosmopolitan	Incomplete	Genus	−	Descriptive
Protozoa	Cosmopolitan	Complete	Species	+	Descriptive
Algae					
Periphyton	Cosmopolitan	Approximately complete	Species	±	Descriptive, approximately statistical
Phytoplankton	Cosmopolitan	Approximately complete	Species	±	Descriptive, approximately statistical
Macrophytes	Geographic regions	Complete	Species	−SS	Sessile
Invertebrates					
Zooplankton	Cosmopolitan, geographic regions	Approximately complete	Genus	−	Descriptive, approximately statistical
Insects	Geographic regions limited	Incomplete	Genus family	−	Descriptive
Molluscs	Geographic regions limited	Approximately complete	Species	+	Descriptive, statistical
Worms	Cosmopolitan	Incomplete		−	Descriptive
Vertebrates					
Fish	Geographic regions limited	Incomplete (life stage limited)	Species	+	Descriptive
Mammals	Geographic regions limited	Complete	Species	+	Descriptive

Source: Reprinted with permission from *Water Research* **16**. E. E. Herricks and J. Cairns, Jr., Biological monitoring Part III—Receiving system methodology based on community structure. Copyright 1982, Pergamon Press Ltd.

| Movement | Indicator Strength | | | Interpretative Strength |
	Organics/ Nutrients	Metals	General Water Chemistry	
Generally stationary	√		√	Limited by sampling or analysis difficulties—high indicator strength
Generally stationary	√			Limited by sampling or analysis difficulties
Generally stationary	√	√	√	Ecological relationships poorly understood—saprobic importance
Generally stationary	~	√	√	Good indication of water chemistry and enrichment
Mobile	~	√	√	Good indication of water chemistry and enrichment
			√	Limited by sampling or analysis difficulties
Mobile		√	√	Limited by sampling or analysis difficulties = ecological relationships poorly understood—high value in lentic ecosystems, early trophic effects
Sessile		√	√	Ecological relationships
Sessile	~		√	Site-specific indicators
			√	
Mobile	√	√	√	Limited by sampling difficulties and early life stage identification
Mobile				

Key: (+) = relative ease of identification; (−) = difficult identification; (±) = group or life stage difficult identification; −SS = difficult identification is season specific; √ = taxa used to indicate pollution effects: ~ = mixed: group or life stage specific.

stream communities based on a similarity of species composition. In "weighted averages ordination" (see Chapter 10) one may separately weight indicator species if so desired. One may then correlate the resulting ordination axis with specific physical or chemical water quality parameters to determine the underlying source of variation in species composition along the axis. Streams can then be ranked, based on biotic composition, along an axis which represents one or more known water quality variables. The details of such methodologies are described in Chapter 10.

POLLUTION EFFECTS: ECOSYSTEM-LEVEL APPROACHES

A knowledge of the effects of toxic chemicals on individual species is extremely useful in predicting pollution effects. Nevertheless, to characterize the full range of interactions that can be generated by introducing a pollutant into an ecosystem, an understanding of ecosystem-level interactions is essential. Four approaches to this understanding that are commonly used are (1) microcosm studies, (2) experimental manipulation of natural ecosystems in the field, (3) synecological studies of natural ecosystems along disturbance gradients, and (4) computer models of natural ecosystems subjected to disturbance.

Microcosm Studies

Microcosms are whole samples of ecosystems placed in an artificial container and maintained in a laboratory environment. The terrarium made by shearing an intact slice of moss cover from the forest floor and keeping it watered in a glass-sided tank is a familiar, but imperfect, example. Such terraria are not very representative slices of the ecosystem from which they were taken, since they exclude all of the larger organisms (trees, shrubs, vertebrates) in their environment, and mimic only poorly the natural inputs and drainage of the hydrologic cycle. Nevertheless, such systems, containing intact assemblages of mosses, lichens, microorganisms, and invertebrates, are more realistic ecosystem samples than would be obtained by placing together some soil, one or two moss species, and some earthworms. The latter system, which contains a few species artificially assembled, is termed a *gnotobiotic* system.

Microcosms have been more widely assembled for aquatic than terrestrial ecosystems probably since the main primary producers (phytoplankton) are more quickly and easily reared in small spaces than are rooted trees and shrubs. In the aquatic microcosms of Cooke (1967), for example, a sample of algal mass from a farm pond had been added to one liter of water and allowed to stand for several months. Small samples of this mixture (0.07 mg dry weight) were then added to 300 mliter of nutrient-enriched water in 18 replicate beakers kept at 21°C with 12 hours of artificial light per day. The changes in organismal composition, abundance, biomass accumulation, and net photosynthesis were observed over a three-month period. To such systems other investigators have added known concentrations of a toxic substance and observed changes relative to the control (see, e.g., Fisher et al. 1974, Mosser et al. 1972).

The small size of most aquatic microcosms results in a number of problems. Shallow depths of tanks relative to real ponds or oceans result in unrealistically high influences of the sediment and associated benthos (sediment-dwelling organisms) on overall nutrient fluxes and decomposition. Special chambers to minimize contact between sediment and water can improve realism in this regard [National Research Council (NRC) 1981]. Shallow depths also distort vertical migration patterns of zooplankton and induce artificial settling by phytoplankton (NRC 1981). Also the high surface-to-volume ratio of tanks disproportionately favors the growth of surface-adhering organisms (periphyton). Most aquatic microcosms can only operate for a few months before they become overgrown by organisms reflective of the artificial conditions. Thermal stratification and mixing are hard to simulate realistically, and the introduction of larger animals (e.g., fish) quickly distorts the size balance of the scaled-down system.

Despite these problems microcosms have the advantage that the systems can be replicated and toxic substances readily introduced in a safe, controlled manner. Furthermore, as microcosms are made more complex and realistic, the changing nature of their dynamics can indicate the role of the added elements or compartments in ecosystem function.

An example of the latter is the study by Patten and Witkamp (1967) of nutrient flux in the floor of a temperature deciduous forest. Microcosms were established in 60 mliter glass funnels. Each microcosm had oak leaf litter and one or more of the following: microflora, five millipedes, and/or 1 cm of soil. The radioactive tracer cesium-137 was incorporated in the oak leaves, and water was allowed to leach through each system. The amount of Cs-137 to accumulate in each compartment, and to be leached through the funnel, was measured over time. The flux of the radionuclide was modeled by adjusting an analog computer model to fit the experimental data. In other words, by obtaining empirical constants for equations that model rates of transfer of cesium between compartments, a model of flux of cesium through the system was built. The use of the computer model expands the usefulness of the microcosm considerably, since new tests can be performed on the computer model alone. It is evident from Figure 9.11 that the addition of ecosystem compartments, particularly soil, drastically alters losses of cesium from the system in leachate. A somewhat more advanced apparatus (Marinucci 1982) that permits measurement of losses of CO_2 in respiration of the litter, as well as losses of nutrients in leachate, is illustrated in Figure 9.12.

Additional discussions of the use and limitations of microcosms in applied ecological research are those of Dudzik (1979), Giesy (1980), and Harte et al. (1981).

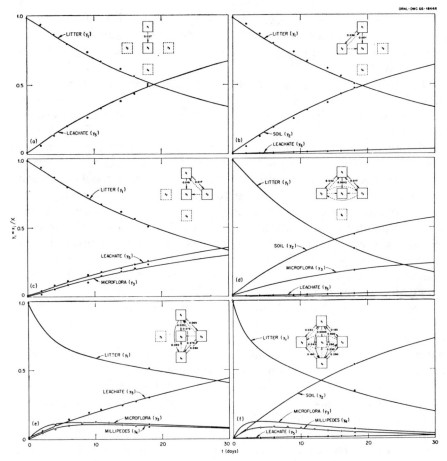

Figure 9.11. Terrestrial microcosms. Predicted (solid line) and observed (dots) changes in cesium[134]-content of ecosystem compartments when one or more compartments are present. From "Systems analysis of [134]cesium kinetics in terrestrial microcosms" by B. C. Patten and M. Witkamp, *Ecology* 1967, **48**, 813–824. Copyright (c) 1967 by the Ecological Society of America. Reprinted by permission.

Experimental Field Manipulations

In terrestrial ecosystems experimental manipulation may take the form of mowing (Harper 1969), clear-cutting (Bormann and Likens 1979), fertilizing (Bakelaar and Odum 1978, Shure and Hunt 1981), fumigating with air pollutants (Lauenroth 1985), applying pesticide (Likens et al. 1970, Giles 1970), excluding grazing animals by fencing (Ball 1974), exposing vegetation to ionizing radiation (Woodwell and Whittaker 1968), or other manipulations.

In aquatic ecosystems the problem of spatially containing the perturbation is more difficult. Schindler (1977) has fertilized entire lakes, while Goldman (1962) enclosed water columns in lakes with polyethylene cylinders

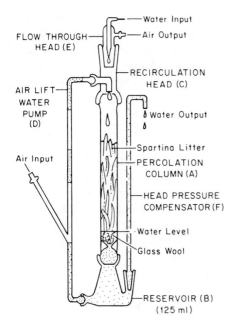

FLOW THROUGH HEAD (E)
— Water Input
— Air Output

— RECIRCULATION HEAD (C)

AIR LIFT WATER PUMP (D)
— Water Output

Air Input

— Spartina Litter
— PERCOLATION COLUMN (A)

— HEAD PRESSURE COMPENSATOR (F)

·Water Level
Glass Wool

— RESERVOIR (B)
(125 ml)

Figure 9.12. A flow-through microcosm for measuring decomposition of litter of the salt-marsh grass, *Spartina alterniflora*. Drops of seawater are lifted by the air stream and drip onto the litter. At the same time water is pumped through the percolator, and CO_2 is collected from the output air. From Marinucci (1982), reprinted with permission of the *Biological Bulletin*, Marine Biological Laboratory, Woods Hole, Mass.

before adding nutrients and Merks (1968) enclosed portions of marshland before adding DDT. Grice and Reeve (1982) provide examples from the marine environment. Such field manipulations typically occur over a moderate-sized area (100 m² or more) and are accompanied by adjacent control plots. It is not possible to obtain identical plots in the field for any experiment, due to inherent spatial variability in nature, but some measure of replication is nevertheless possible. A more serious problem for impact prediction is that the results of experimental perturbations should be observed for at least one full cycle of seasons, and preferably for several years.

The literature on experimental manipulation studies is large. Some recent volumes containing discussions of such experiments include Bormann and Likens (1979), Mooney and Godron (1983), and Ward (1978).

Synecological Field Studies

The principle involved in synecological field studies is to observe changes in a field site that has been subjected to prior human disturbance, relative to a control site (e.g., studying sites downwind of a factory relative to upwind controls). In many cases, however, the disturbance is so widespread (ozone dispersion, DDT application) that no "control" sites exist close enough to the area of disturbance to have the same habitat and biological composition. In such instances epidemiological approaches must be utilized.

As an example, Westman (1979) sampled the vegetation and 43 associated habitat variables on plots of coastal sage scrub at 67 locations in coastal California and northern Baja California. The habitat variables examined included community structure, topography, substrate, climate, fire and grazing history, and air pollution. In a preliminary analysis of data, correlation coefficients between each habitat variable and the percentage foliar cover by native species on each site revealed that mean annual levels of oxidants were the most strongly correlated ($r = -0.58$; $p < 0.0001$) with % cover of native species. The negative correlation indicates that higher oxidant levels are associated with lower levels of native cover. Correlations, however, cannot demonstrate causation. Oxidants might merely be associated with some other, causal variables.

One way to examine this possibility further is to test alternative hypotheses. The hypothesis that vegetation becomes less dense due to climate or other natural factors in the region of high pollution could not be tested directly by "control sites," since the entire basin inland of Los Angeles, where the polluted sites occurred, had high levels of ozone. The hypothesis had to be tested more indirectly, by ordinating samples from other regions of comparable climate and predicting what floristic cover and composition should occur in the polluted region based on trends from cleaner areas. When this was done, the hypothesis that climate or other natural variables tested could explain the decline in native cover in the polluted region was not supported (Westman 1983).

The partial correlation of oxidants with declining native cover remained high when covariations with other highly correlated variables (elevation, mean maximum temperature of the warmest month, distance from the coast) were extracted, indicating that an effect due to ozone was likely. Several path models were constructed; these postulate alternate routes of causation of the observed effect through interdependent variables. The likelihood that a particular path model represents a route of causation is enhanced if the size and sign of correlations between variables in the path model are consistent with the postulated model of their interrelations (see, e.g., Duncan 1975, Heise 1975). A highly significant path model (Figure 9.13) was found, in which the role of oxidants as the cause of the decline in native cover appeared central. Despite these rather complex statistical manipulations, no proof of causation was achieved, nor can statistical methods ever achieve such proof. Such methods merely indicate the likelihood that a particular hypothesis can correctly predict the observed effect.

The hypothesis that ozone was causing the decline in cover of native species would be further strengthened if it could be shown that coastal sage species suffered damage when exposed to levels of oxidant occurring in the region, when other growth factors were nonlimiting. Ten species of coastal sage scrub were therefore subjected to ozone levels of 0.1, 0.2 or 0.4 ppm for 40 hours per week for 10 weeks, in forced-draft, open-topped fumigation chambers (Figure 9.14). Such chambers introduce filtered air, with known amounts of ozone added, at the base of the chamber, and strong fans push the air out the top of the chamber, preventing mixing with ambient air. Results showed numerous evidences of damage to the species, even at the

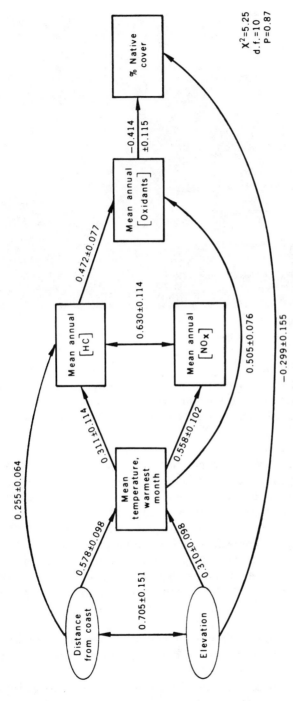

Figure 9.13. Path model relating oxidants and correlated factors to the observed reduction in foliar cover of native species of coastal sage scrub in southern California. Single-headed arrows are causal paths (numbers are path coefficients ± standard errors); double-headed arrows indicate correlations; HC indicates total hydrocarbons; NO$_x$ indicates oxides of nitrogen. From W. E. Westman, *Science* **205:**1001–1003. Copyright (1979) by the American Association for the Advancement of Science. Reprinted with permission.

Figure 9.14. Open-topped, forced-draft fumigation chambers used in testing effects of air pollutants on plant growth. Photo courtesy K. P. Preston.

lowest levels of exposure (Westman, W., K. Preston, and L. Weeks, unpublished). These results also do not "prove" that oxidants cause the death of coastal sage plants in nature; they only demonstrate that such is possible and increase the weight of available evidence for the hypothesis.

The foregoing example illustrates some of the difficulties of field-based epidemiological–synecological experiments. On the one hand, they are time-consuming and expensive; on the other hand, they will never prove a cause-and-effect relationship between pollutant exposure and species damage. Nevertheless, scientists have virtually no alternative if they are to demonstrate effects in the field. No amount of experimental, toxicological evidence of pollutant damage, combined with monitoring of pollutant levels in the field, will prove that the pollutant is causing damage in the field. Given this, a demonstration of correlations between pollutant levels and field-observed damage can lend weight to the hypothesis of causation.

Computer Models

Statistical models based on empirically derived regressions, and simulation models based on differential equations derived from established assumptions, are two approaches toward quantitative ecosystem modeling (NRC 1981). During the late 1960s and early 1970s the International Biological Program (IBP) encouraged ecologists to gather sufficient field data on samples of a few major biomes to be able to model the biomes, or ecosystem representatives of them, by computer. The IBP effort and other research has resulted in several major working models or partial models of ecosystems. Examples include the grassland biome (Innis 1978, Breymeyer and van Dyne 1980), the tundra biome (Bliss and Moore 1981), the chaparral ecosystem (Miller 1981), the temperate deciduous forest biome (Botkin et al. 1972), the arid-lands biome (Goodall and Perry 1981), and others. Most of these models have subsequently been used to test the effects of perturbations. The effect of clear-cutting (Botkin et al. 1972) or air pollution (Shugart et al. 1980) on deciduous forests and the effects of SO_2 (Lauenroth 1985) or grazing (van Dyne 1981) on grasslands are some examples. Major works on ecosystem modeling include Biswas and Biswas (1979), Hall and Day (1977), Innis and O'Neill (1979), Matis et al. (1979), Patten (1971–76), and articles in such journals as *Ecological Modelling* and *Ecology*.

Computer models are invaluable in predicting effects of impacts on ecosystem components much more quickly and inexpensively than can be achieved by field survey or experimental approach. The problem with such computer models is their generality. Lengthy and complex as they may be, they are never perfect models of natural ecosystems. To the extent that the models are oversimplified, their predictions can be in error. "Validation" of model predictions with new data not used in model building is as yet not very common, since many ecosystem models have only recently been built. As validation tests are performed, and experience with the models increases, the reliance that can be placed on them by impact analysts will increase. Complex computer models are often best understood by those who programmed them; thus mechanisms for impact analysts to communicate with the programmers are most likely to result in appropriate application of such models in impact assessment.

MONITORING: SOME TECHNOLOGICAL ADVANCES

Monitoring may take the form of characterizing the physical and chemical environment, or measuring the health of the biological community directly. In order to observe correlations of the type needed for epidemiological studies, it is important to have information on both aspects. Other aspects of monitoring are discussed in Chapters 7, 10 and 11; we review here a few recent advances in monitoring technology.

Remote-sensing techniques are being increasingly used to obtain data on air or water quality and biological health. Fiber-optic cables can now be used to send laser sources of light into remote places; a second fiber in the cable collects reradiated or fluoresced light for analysis by a spectrophotometer. By such means fiber-optic cables have been used to measure concentrations of uranium and plutonium in nuclear fuel reprocessing solutions; temperatures between 135°–700°C in nuclear reactors or other reaction vessels; and concentrations of pollutants in groundwater. The savings in drilling costs for monitoring leakage around a hazardous waste dump may be as much as $500,000 per site (Maugh 1982).

Lasers have also been used to detect the concentrations of gaseous air pollutants at a distance from the laser source. Lasers are "tuned" to emit specific wavelengths of light, which are absorbed by particular gaseous pollutants; the amount of absorption is an index of the pollutant concentration. Such devices have been designed to monitor nitric oxide (Kaldor et al. 1972), ammonia, ethylene (Hinkley and Kelley 1971), NO_2, CO, SO_2, and other gases. They can detect levels as low as 1 ppm at a distance exceeding 1.0 km (Hodgeson et al. 1973). Other remote-sensing approaches for air pollutants are discussed by Charlson et al. (1973) and Hodgeson et al. (1973). Infrared imagery was earlier mentioned as a remote-sensing tool for detecting pollution damage to plant tissue, as well as concentrations of heat, sediment, or dissolved oxygen in rivers (Chapter 7).

Laser holography has also been used in biological monitoring. John Cairns, Jr. and colleagues have for some years been working to perfect an automated system for the identification of species of diatoms that attach to glass slides in a stream. The diversity and abundance of diatom species that attach to a glass slide can be used as an indicator of stream quality, since polluted conditions can cause a reduction in richness and an increased concentration of dominance (Patrick et al. 1954; see Figure 11.5). Diatoms may reproduce or die in response to pollution stress within minutes or hours (Cairns et al. 1982b). Consequently the development of an automated system to identify diatom species could provide a means to use the glass-slide technique as a continuous in-stream biological monitoring device.

The device of Cairns et al. (1982b) produces three-dimensional holographs (images) of major diatom species using a laser beam and holographic or matched spatial filter. A laser beam is focused through a microscope onto a single diatom from the diatom array on a glass slide taken from the stream. The characteristic holographic pattern produced when the laser beam passes through the diatom and a set of fibers is compared automatically to a holograph of a known diatom species. When the patterns match, the result is recorded by minicomputer. In this way as many diatoms as there are reference holographs for can be identified automatically. Figure 9.15 shows how a particular diatom species can be identified from an array of several species.

The detection of diatoms by laser holography is at the sophisticated end of a wide range of biological monitoring techniques now in development or use. For further reading on the use of artificial substrates (glass slides, rock-filled baskets for benthic organisms, etc.) in biological monitoring, see Cairns (1982a). For more general references on biological monitoring for water quality, see, for example, Hellawell (1978), Cairns et al. (1977b, 1979), Cairns et al (1982a), and Word (1980). For references on biological monitoring for air quality, see, for example, Ashmore et al. (1978), Donagi and Goren (1979), Manning and Feder (1980), Martin and Coughtrey (1982), Nouchi and Aoki (1979), Posthumus (1982), and Steubing and Jäger (1982). For general reading on ecotoxicology, see Butler (1978).

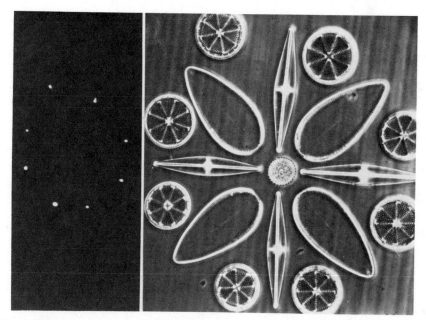

Figure 9.15. Laser-holographic identification of diatoms. At right, the circular diatom *Heliopelta mettii* arranged on a slide with other species. At left, the correlation dots after successful matching of holographic patterns of the test species with a reference pattern. Slide of arrayed diatoms courtesy of Drs. Ruth Patrick and Charles Reimer, Philadelphia Academy of Science. Figure used with permission of the author, Dr. J. Cairns, Jr., and the publisher, Dr. W. Junk, from "The ABC's of diatom identification using laser holography," *Hydrobiologia* 54(1):7–16, 1977.

REFERENCES

Abrahamson, W. G., and Gadgil, M. D. (1973). Growth form and reproductive effort in goldenrods (*Solidago*, Compositae). *Am. Nat.* **107**:651–661.

Alvarez, M. R., and Sparrow, A. H. (1965). Comparison of reproductive integrity in the stamen hair and root meristem of *Tradescantia paludosa* following acute gamma irradiation. *Radiation Bot.* **5**:423–430.

Anderson, E., and Anderson, B. R. (1955). Introgression of *Salvia apiana* and *Salvia mellifera*. *Ann. Mo. Bot. Gard.* **41**:329–338.

APHA (1971). *Standard Methods for the Examination of Water and Wastewater.* 13th ed. Amer. Public Health Assn., New York.

Ashmore, M. R., Bell, J. N. B., and Reily, C. L. (1978). A survey of ozone levels in the British Isles using indicator plants. *Nature* **276**:813–815.

Bakelaar, R. G., and Odum, E. P. (1978). Community and population-level responses to fertilization in an old-field ecosystem. *Ecology* **59**:660–665.

Ball, M. E. (1974). Floristic changes on grasslands and heaths on the Isle of Rhum after a reduction or exclusion of grazing. *J. Environ. Manage.* **2**:299–318.

Bedell, T. D. (1983). Nuclear reactions? *Horticulture* **61**(6):22–23.

Bell, J. N. B., Ayazloo, M., and Wilson, G. B. (1982). Selection for sulphur dioxide tolerance in grass populations in polluted areas. In R. Bornkamm, J. A. Lee, and M. R. D. Seaward, eds. *Urban Ecology*. Blackwell, Oxford, pp. 171–180.

Bennett, J. P. (1985). Regulatory uses of SO$_2$ effects data. In W. E. Winner, H. A. Mooney, and R. Goldstein, eds. *Sulfur Dioxide and Vegetation. Physiology, Ecology, and Policy Issues*. Stanford Univ. Press, Stanford.

Biswas, A. K., and Biswas, M. R., eds. (1979). *State-of-the-Art in Ecological Modelling. Environmental Science and Applications*. Vol 7. Pergamon, Oxford.

Bliss, L. C., and Moore, J. J. (1981). *Tundra Ecosystems: A Comparative Analysis*. Cambridge Univ. Press, Cambridge.

Bormann, F. H., and Likens, G. E. (1979). *Pattern and Process in a Forest Ecosystem*. Springer-Verlag, New York.

Botkin, D. B., Janak, J. F., and Wallis, J. R. (1972). Some ecological consequences of a computer model of forest growth. *J. Ecol.* **60**:849–872.

Bressan, R. A., Le Cureux, L., Wilson, L. G., and Filner, P. (1979). Release of ethane in response to various environmental stresses. *Plant Physiol. Suppl.* **63**:59.

Breymeyer, A. I., and van Dyne, G. M., eds. (1980). *Grasslands, Systems Analysis, and Man*. Cambridge Univ. Press, Cambridge.

Brooks, R. R. (1972). *Geobotany and Biogeochemistry in Mineral Exploration*. Harper & Row, New York.

Buikema, A. L., Jr., and Cairns, J., Jr., eds. (1980). *Aquatic Invertebrate Bioassays*. STP 715. Amer. Soc. Testing and Materials, Philadelphia, Pa.

Butler, G. C. (1978). *Principles of Ecotoxicology*. SCOPE 12. Wiley, New York.

Bystrom, B. G., Glater, R. B., Scott, F. M., and Bowler, E. S. C. (1968). Leaf surface of *Beta vulgaris* electron microscope study. *Bot. Gaz.* **129**:133.

Cain, S. A. (1950). Life-forms and phytoclimate. *Bot. Rev.* **16**:1–32.

Cain, S. A., and de Oliveira Castro, G. M. (1959). *Manual of Vegetation Analysis*. Harper, New York.

Cairns, J., Jr. (1969). Fish bioassays—reproducibility and rating. *Revista de Biologia (Lisboa)* **7**:7–12.

Cairns, J., Jr. (1982) *Artificial Substrates*. Ann Arbor Sci., Ann Arbor, Mich.

Cairns, J., Jr., et al. (1982a). *Biological Monitoring in Water Pollution*. Pergamon, Oxford.

Cairns, J., Jr., Almeida, S. P., and Fujii, H. (1982b). Automated identification of diatoms. *BioScience* **32**:98–102.

Cairns, J., Jr., and Dickson, K. L., eds. (1973). *Biological Methods for the Assessment of Water Quality*. STP 528. Amer. Soc. Testing and Materials, Philadelphia, Pa.

Cairns, J., Jr., and Dickson, K. L. (1978). Field and laboratory protocols for evaluating the effects of chemical substances on aquatic life. *J. Test Eval.* **6**:81–90.

Cairns, J., Jr., Dickson, K. L., and Slocomb, J. (1977b). The ABC's of diatom identification using laser holography. *Hydrobiologia* **54**:7–16.

Cairns, J., Jr., Dickson, K. L., and Westlake, G. F., eds. (1977a). *Biological Monitoring of Water and Effluent Quality*. STP 607. Amer. Soc. Testing and Materials, Philadelphia, Pa.

Cairns, J., Jr., and Gruber, D. (1979). Coupling mini- and microcomputers to biological early warning systems. *BioScience* **29:**665–669.

Cairns, J., Jr., Hall, J. W., Morgan, E. L., Sparks, R. E., Waller, W. T., and Westlake, G. F. (1973). The development of an automated biological monitoring system for water quality. Bull. 59. Virginia Water Resources Res. Center, Blacksburg, Va.

Cairns, J., Jr., Patil, G. P., and Waters, W. E., eds. (1979). *Statistical Ecology. Vol 11. Environmental Biomonitoring, Assessment, Prediction and Management.* Intl. Coop. Publ. House, Fairland, Md.

Cairns, J., Jr., and van der Schalie, W. H. (1980). Biological monitoring, Part I. Early warning systems. *Water Res.* **14:**1179–1196.

Chamel, A., and Garrec, J. P. (1977). Penetration of fluorine through isolated pear leaf cuticles, *Environ. Poll.* **12:**307.

Charlson, R. J., Vanderpol, A. H., Covert, D. S., Waggoner, A. P., and Ahlquist, N. C. (1973). Sulfuric acid-ammonium sulfate aerosol: optical detection in the St. Louis region. *Science* **184:**156–158.

Clifford, H. T. (1954). Analysis of suspected hybrid swarms in the genus *Eucalyptus.* Heredity **8:**259–269.

Cody, M. L. (1971). Ecological aspects of reproduction. In D. Farner and C. King, eds. *Avian Biology.* Academic Press, New York, pp. 462–512.

Cohen, D. (1966). Optimizing reproduction in a randomly varying environment. *J. Theor. Biol.* **12:**119–129.

Cohen, D. (1968). A general model of optimal reproduction. *J. Ecol.* **56:**219–228.

Cohen, D. (1971). Maximizing final yield within growth is limited by time or by limiting resources. *J. Theor. Biol.* **33:**299–307.

Cooke, G. D. (1967). The pattern of autotrophic succession in laboratory microcosms. *BioScience* **17:**717–721.

Council on Environmental Quality (1977). *Environmental Quality.* Eighth Ann. Rep. Washington, D.C.

Council on Environmental Quality (1978). *Environmental Quality.* Ninth Ann. Rep. Washington, D. C.

Cross, J. N. (1982). Trends in fin erosion among fishes on the Palos Verde shelf. In W Bascom, ed. *Biennial Report, 1981–82.* Southern Calif. Coastal Water Res. Proj., Long Beach, Calif., pp. 99–110.

Davies, D. R. (1963). Radiation-induced chromosome aberrations and loss of reproductive integrity in *Tradescantia. Radiation Res.* **20:**726–740.

Denison, W. C. (1973). *A Guide to Air Quality Monitoring with Lichens.* Lichen Technology, Corvallis, Oreg.

de Wit, A. (1976). *Epiphytic Lichens and Air Pollution in the Netherlands.* Cramer, Lehre, W. Germany.

Dillon, T. M., and Lynch, M. P. (1981). Physiological responses as determinants of stress in marine and estuarine organisms. In G. W. Barrett and R. Rosenberg, eds. *Stress Effects on Natural Ecosystems.* Wiley, New York, pp. 227–241.

Donagi, A. E., and Goren, A. I. (1979). Use of indicator plants to evaluate atmospheric levels of nitrogen dioxide in the vicinity of a chemical plant. *Environ. Sci. Tech.* **13:**986–989.

Dudzik, M., Harte, J., Jassby, A., Lapan, E., Levy, D., and Rees, J. (1979). Some considerations in the design of aquatic microcosms for plankton studies. *Intern. J. Environ. Stud.* **13**:125–130.

Duncan, O. D. (1975). *Introduction to Structural Equation Models.* Academic Press, New York.

Elkiey, T., and Ormrod, D. P. (1980). Sorption of ozone and sulfur dioxide by petunia leaves. *J. Environ. Qual.* **9**:93–95.

Ellenson, J. L., and Amundson, R. G. (1982). Delayed light imaging for the early detection of plant stress. *Science* **215**:1104–1106.

Fennelly, P. F., et al. (1976). *Environmental Assessment Perspectives.* Rep. PB-257-911. GCA Corp, Bedford, Mass.

Ferry, B. W., et al. eds. (1973). *Air Pollution and Lichens.* Oxford Univ. Press, New York.

Fisher, N. S., Carpenter, E. J., Remsen, C. C., and Wurster, C. F. (1974). Effects of PCB on interspecific competition in natural and gnotobiotic phytoplankton communities in continuous and batch cultures. *Microb. Ecol.* **1**:39–50.

Gadgil, M., and Solbrig, O. T. (1972). The concept of "*r*" and "*K*" selection: evidence from wild flowers and some theoretical considerations. *Am. Nat.* **106**:14–31.

Garsed, S. D., and Rutter, A. J. (1982). The relative sensitivities of conifer populations to SO_2 in screening tests with different concentrations of sulphur dioxide. In M. Unsworth and D. P. Ormrod, eds. *Effects of Air Pollutants in Agriculture and Horticulture.* Butterworths, London, pp. 474–475.

Giesy, J. P., ed. (1980). *Microcosms in Ecological Research.* Tech. Info. Center, U.S. Dept. Energy, Springfield, Va.

Gilbert, O. L. (1968). Bryophytes as indicators of air pollution in the Tyne Valley. *New Phytol.* **67**:15–30.

Giles, R. H. (1970). The ecology of a small forested watershed treated with the insecticide Malathion-S[35]. Wildlife Monogr. 24. The Wildlife Soc., Washington, D.C.

Gleason, H. A. (1926). The individualistic concept of the plant association. *Bull. Torrey Bot. Club* **53**:7–26.

Goldman, C. R. (1962). A method of studying nutrient limiting factors *in situ* in water columns isolated by polyethylene film. *Limnol. Oceanogr.* **7**:99–101.

Goodall, D. W., and Perry, R. A., eds. (1981). *Arid-land Ecosystems: Structure, Functioning and Management.* Vol 2. Cambridge Univ. Press, Cambridge.

Gordon, A. G., and Gorham, E. (1963). Ecological aspects of air pollution from an iron-sintering plant at Wawa, Ontario. *Can. J. Bot.* **41**:1063–1078.

Greenfield, D. W. F., Abdel-Hameed, F., Deckert, G. D., and Flinn, R. R. (1973). Hybridization between *Chrosomus erythrogaster* and *Notropis cornutus* (Pisces:Cyprinidae). *Copeia* **11**:54–60.

Grice, G. D., and Reeve, M. R., eds. (1982). *Marine Mesocosms. Biological and Chemical Research in Experimental Ecosystems.* Springer-Verlag, New York.

Grime, J. P. (1974). Vegetation classification by reference to strategies. *Nature* **250**:26–31.

Grime, J. P. (1977). Evidence for the existence of three primary strategies in plants and its relevance to ecological and evolutionary theory. *Am. Nat.* **111:**1169–1194.

Grime, J. P. (1979). *Plant Strategies and Vegetation Processes.* Wiley, New York.

Hall, C. A. S., and Day, J. W., Jr., eds. (1977). *Ecosystems in Theory and Practice: An Introduction with Case Histories.* Wiley-Interscience, New York.

Hanes, T. L. (1971). Succession after fire in the chaparral of southern California. *Ecol. Monogr.* **41:**27–52.

Harkov, R. S., and Brennan, E. (1982). An ecophysiological analysis of the response of woody and herbaceous plants to oxidant injury. *J. Environ. Manage.* **15:**251–261.

Harper, J. L. (1969). The roles of predation in vegetational diversity. *Brookhaven Symp. Biol.* **22:**48–62.

Harte, J., Levy, D., Rees, J., and Saegebarth, E. (1981). Assessment of Optimum Aquatic Microcosm Design for Pollution Impact Studies. Rep. EA-1989. Electric Power Res. Inst., Palo Alto, Calif.

Hawksworth, D. L. (1976). *Lichens as Pollution Monitors.* Arnold, London.

Hawksworth, D. L., and Ferry, B. W., eds. (1973). *Air Pollution and Lichens.* Univ. Toronto, Toronto.

Heise, D. R. (1975). *Causal Analysis.* Wiley, New York.

Hellawell, J. M. (1978). *Biological Surveillance of Rivers: A Biological Monitoring Handbook.* Water Res. Centre, Stevenage, U.K.

Herricks, E. E., and Cairns, J., Jr. (1982). Biological monitoring. Part III—Receiving system methodology based on community structure. *Water Res.* **16:**141–153.

Hinkley, E. D., and Kelley, P. L. (1971). Detection of air pollutants with tunable diode lasers. *Science* **171:**635–639.

Hodgeson, J. A., McClenny, W. A., and Hanst, P. L. (1973). Air pollution monitoring by advanced spectroscopic techniques. *Science* **182:**248–258.

Innis, G. S., ed. (1978). *Grassland Simulation Model.* Springer-Verlag, New York.

Innis, G. S., and O'Neill, R. V., eds. (1979). *Systems Analysis of Ecosystems.* Intl. Coop. Publ. House, Fairland, Md.

Ivanovici, A. M., and Wiebe, W. J. (1981). Towards a working "definition" of "stress": a review and critique. In G. W. Barrett and R. Rosenberg, eds. *Stress Effects on Natural Ecosystems.* Wiley, New York, pp. 13–47.

Janzen, D. H. (1971). Seed predation by animals. *Ann. Rev. Ecol. Syst.* **2:**465–492.

Kaldor, A., Olson, W. B., and Maki, A. G. (1972). Pollution monitor for nitric oxide: a laser device based on the Zeeman modulation of absorption. *Science* **176:**508–510.

Karr, J. R. (1981). Assessment of biotic integrity using fish communities. *Fisheries* **6:**21–27.

Katz, M. (1971). Toxicity bioassay techniques using aquatic organisms. In L. L. Ciaccio, ed. *Water and Water Pollution Handbook.* Vol 2. Dekker, New York.

Lacasse, N. L., and Moroz, W. J., eds. (1969). *Handbook of Effects Assessment. Vegetation Damage.* Center for Air Environ. Stud., Pennsylvania State Univ., University Park.

Lauenroth, W. K. (1985). Effects of SO_2 on plant community function. In W. E. Winner, H. A. Mooney, and R. Goldstein, eds. *Sulfur Dioxide and Vegetation. Physiology, Ecology, and Policy Issues.* Stanford Univ. Press, Stanford.

Likens, G. E., Bormann, F. H., Johnson, N. M., Fisher, D. W., and Pierce, R. S. (1970). Effects of forest cutting and herbicide treatment on nutrient budgets in the Hubbard Brook ecosystem in New Hampshire. *Ecol. Monogr.* **40**:23–47.

Linzon, S. N. (1978). Effects of airborne sulfur pollutants on plants. In J. O. Nriagu, ed. *Sulfur in the Environment. Part II. Ecological Impacts.* Wiley, New York, pp. 109–162.

MacArthur, R. H., and Wilson, E. O. (1967). *The Theory of Island Biogeography.* Princeton Univ. Press, Princeton.

Maciorowski, A. F., Cairns, J., Jr., and Benfield, E. F. (1977). Laboratory simulation of an in-plant biomonitoring system using crayfish activity rhythms to detect cadmium induced toxic stress. Abstract. 25th Ann. Mtg. N. Amer. Benthological Soc., Roanoke, Va.

Maki, A. W., Stewart, K. W., and Silvey, J. K. G. (1973). The effects of Dibrom on respiratory activity of the stonefly, *Hydroperba crosbyi*, hellgrammite, *Corydalis cornutus*, and the golden shiner, *Notemigonus crysoleucas*. *Trans. Am. Fish Soc.* **102**:806–815.

Malanson, G. P., and O'Leary, J. F. (1982). Post-fire regeneration strategies of Californian coastal sage shrubs. *Oecologia* **53**:355–358.

Malanson, G. P., and Westman, W. E. (1984). Post-fire succession in Californian coastal sage scrub: the role of continual basal sprouting. *Amer. Midl. Nat.:* in press.

Manning, W. J., and Feder, W. A. (1980). *Biomonitoring Air Pollutants with Plants.* Applied Science, Essex, U.K.

Marinucci, A. C. (1982). Carbon and nitrogen fluxes during decomposition of *Spartina alterniflora* in a flow-through percolator. *Biol. Bull.* **162**:53–69.

Marshall, D. R., and Jain, S. K. (1968). Phenotypic plasticity of *Avena fatua* and *A. barbata*. *Am. Nat.* **102**:457–467.

Marshall, D. R., and Jain, S. K. (1970). Seed predation and dormancy in the population dynamics of *Avena fatua* and *A. barbata*. *Ecology* **51**:886–891.

Martin, M. H., and Coughtrey, P. J. (1982). *Biological Monitoring of Heavy Metal Pollution. Land and Air.* Applied Science, Essex, U.K.

Marvin, D., Jr., and Burton, D. T. (1973). Cardiac and respiratory responses of rainbow trout, bluegills, and brown bullhead catfish during rapid hypoxia and recovery under normoxic conditions. *Comp. Biochem. Physiol.* **46A**:755–765.

Matis, J. H., Patten, B. C., and White, G. C., eds. (1979). *Compartmental Analysis of Ecosystem Models.* Intl. Coop. Publ. House, Fairland, Md.

Maugh, T. H., II. (1982). Remote spectrometry with fiber optics. *Science* **218**:875–876.

Menge, B. A. (1974). Effect of wave action on brooding and reproductive effort in a seastar, *Leptasterias hexactis*. *Ecology* **55**:84–93.

Merks, R. L. (1968). The accumulation of ^{36}CL ring-labelled DDT in a freshwater marsh. *J. Wildlife Manage.* **32**:376–398.

Miller, P. C. ed. (1981). *Resource Use by Chaparral and Matorral: A Comparison of Vegetation Function in Two Mediterranean Type Ecosystems.* Springer-Verlag, New York.

Miller, P. R., and Elderman, M. J. eds. (1977). Photochemical oxidant air pollutant effects on a mixed conifer forest ecosystem. EPA-600/3-77-104. Environmental Protection Agency, Corvallis, Oreg.

Mooney, H. A. (1977). Frost sensitivity and resprouting behavior of analogous shrubs of California and Chile. *Madroño* **24**:74–78.

Mooney, H. A. and Gordon, M., eds. (1983). *Disturbance and Ecosystems–Components of Response.* Springer-Verlag, New York.

Mosser, J. L., Fisher, N. S., and Wurster, C. F. (1972). Polychlorinated biphenyls and DDT alter species composition in mixed culture of algae. *Science* **176**:533–535.

Mountford, M. D. (1971). Population survival in a variable environment. *J. Theor. Biol.* **32**:75–79.

Mueller-Dombois, D., and Ellenberg, H. (1974). *Aims and Methods of Vegetation Ecology.* Wiley, New York.

National Research Council (1981). *Testing for Effects of Chemicals on Ecosystems.* Comm. to Review Methods for Ecotoxicology. Natl. Acad. Press, Washington, D.C.

Nouchi, I., and Aoki, K. (1979). Morning glory as a photochemical oxidant indicator. *Environ. Poll.* **18**:289–303.

Omasa, K., Fumiaki, A., Hshimoto, Y., and Aiga, I. (1980). Res. Rep. Natl. Inst. Environ. Stud. (Japan) **11**:239.

Orshan, G. (1983). Approaches to the definition of Mediterranean growth forms. In F. J. Kruger, D. T. Mitchell, and J. U. M. Jarvis, eds. *Mediterranean-Type Ecosystems: The Role of Nutrients.* Springer-Verlag, Berlin, pp. 86–100.

Palmbald, I. G. (1969). Population variation in germination of weedy species. *Ecology* **50**:746–748.

Patten, B. C., ed. (1971–76). *Systems Analysis and Simulation in Ecology.* 4 vols. Academic Press, New York.

Patten, B. C., and Witkamp, M. (1967). Systems analysis of ^{134}Cesium kinetics in terrestrial microcosms. *Ecology* **48**:813–824.

Posthumus, A. C. (1982). Biological indicators of air pollution. In M. H. Unsworth and D. P. Ormrod, eds. *Effects of Gaseous Air Pollution in Agriculture and Horticulture.* Butterworth, London, pp. 27–42.

Patrick, R., Hohn, M. H., and Wallace, J. H. (1954). A new method for determining the pattern of the diatom flora. Acad. Nat. Sci. Phila., *Notulae Naturae* **259**:1–12.

Pianka, E. R. (1970). On r- and K-selection. *Am. Nat.* **104**:592–597.

Pryor, L. D. (1959). Evolution in *Eucalyptus. Aust. J. Sci.* **22**:45–59.

Ramensky, L. G. (1924). The basic lawfulness in the structure of the vegetation cover [in Russian]. In Vestnik opytnogo dela Sredne-Chernoz Obl, Voronezh, pp. 37–73.

Raunkiaer, C. (1934). *The Life Forms of Plants and Statistical Plant Geography.* Clarendon, Oxford.

Reish, D. J. (1972). The use of marine invertebrates as indicators of varying degrees of marine pollution. In *Marine Pollution and Sea Life.* Fishing News (Books), London, pp. 203–297.

Schindler, D. W. (1977). Evolution of phosphorus limitation in lakes. *Science* **195**:260–262.

Sharma, G. K. (1975). Leaf surface effects of environmental pollution on sugar maple (*Acer saccharum*) in Montreal. *Can. J. Bot.* **53**:2312–2314.

Sharma, G. K., and Butler, J. (1975). Environmental pollution: leaf cuticular patterns in *Trifolium pratense* L. *Ann. Bot.* **39**:1087–1090.

Short, R. V. (1972). Reproduction in mammals. Species differences. In C. R. Austin and R. V. Short, eds. *Reproductive Patterns.* Cambridge Univ. Press, Cambridge, pp. 1–33.

Shugart, H. H., McLaughlin, S. B., and West, D. C. (1980). Forest models: their development and potential applications for air pollution effects research. In P. R. Miller, ed. *Symp. Effects of Air Pollutants on Mediterranean and Temperate Forest Systems.* U.S. Forest Service, Gen. Tech. Rep. PSW-43. Berkeley, Calif., pp. 203–214.

Shure, D. J., and Hunt, E. J. (1981). Ecological responses to enrichment perturbation in a pine forest. In G. W. Barrett and R. Rosenberg, eds. *Stress Effects on Natural Ecosystems.* Wiley, New York, pp. 103–114.

Skye, E. (1968). Lichens and air pollution. A study of cryptogamic epiphytes and environment in the Stockholm region. *Acta Phytogeographica Suecica* **52**:1–123.

Smith, C. C. (1970). The coevolution of pine squirrels (*Tamia sciurus*) and conifers. *Ecol. Monogr.* **40**:349–371.

Sprague, J. B. (1969). Measurement of pollutant toxicity to fish, I. Bioassay methods for acute toxicity. *Water Res.* **3**:793–821.

Sprague, J. B. (1971). Measurement of pollutant toxicity to fish, III. Sublethal effects and "safe" concentrations. *Water Res.* **5**:245–266.

Stearns, S. C. (1976). Life-history tactics: a review of the ideas. *Quart. Rev. Biol.* **51**:3–47.

Stearns, S. C. (1977). The evolution of life-history traits: a critique of the theory and a review of the data. *Ann. Rev. Ecol. Syst.* **8**:145–171.

Steubing, L., and Jäger, H.-J., eds. (1982). *Monitoring of Air Pollutants by Plants. Methods and Problems.* Tasks for Vegetation Science 7. Junk, The Hague.

Taoda, H. (1972). Mapping of atmospheric pollution in Tokyo based on epiphytic bryophytes. *Jap. J. Ecol.* **22**:125–133.

Taylor, G. E., Jr. (1978). Plant and leaf resistance to gaseous air pollution stress. *New Phytol.* **80**:523–534.

Tingey, D. T., Standley, C., and Field, R. W. (1976). Stress ethylene evolution: a measure of ozone effects on plants. *Atmos. Environ.* **10**:969–974.

Trama, F. B. (1954), The acute toxicity of copper to the common bluegill (*Lepomis macrochirus* Rafinesque). Acad. Nat. Sci. Phila. *Notulae Naturae* **257**:1–13.

Treshow, M., and Lacasse, N. L., eds. (1976). *Diagnosing Vegetation Injury Caused by Air Pollution*. Environmental Protection Agency, Washington, D.C.

Valentine, D. W., Soulé, M. E., and Samollow, P. (1973). Asymmetry: a possible indicator of environmental stress. U.S. Dept. Commerce Fishery Bull., Washington, D. C.

van Dyne, G. M. (1981). Response of shortgrass prairie to man-induced stresses as determined from modelling experiments. In G. W. Barrett and R. Rosenberg, eds. *Stress Effects on Natural Ecosystems*. Wiley, New York, pp. 57–70.

Ward, D. V. (1978). *Biological Environmental Impact Studies: Theory and Methods*. Academic Press, New York.

Warren, C. E. (1971). *Biology and Water Pollution Control*. Saunders, Philadelphia, Pa.

Weitzel, R. L., ed. (1979). *Methods and Measurements of Attached Microcommunities*. Amer. Soc. Testing and Materials, Philadelphia, Pa.

Westman, W. E. (1979). Oxidant effects on Californian coastal sage scrub. *Science* **205**:1001–1003.

Westman, W. E. (1981). Diversity relations and succession in Californian coastal sage scrub. *Ecology* **62**:170–184.

Westman, W. E. (1983). Xeric Mediterranean-type shrubland associations of Alta and Baja California and the community/continuum debate. *Vegetatio* **52**:3–19.

Westman, W. E., O'Leary, J. F., and Malanson, G. P. (1981). The effects of fire intensity, aspect and substrate on post-fire growth of Californian coastal sage scrub. In N. S. Margaris and H. A. Mooney, eds. *Components of Productivity of Mediterranean Regions—Basic and Applied Aspects*. Junk, The Hague, pp. 151–179.

Westman, W. E., Preston, K. P., and Weeks, L. B. (1985). Sulfur dioxide effects on the growth of native plants. In W. E. Winner, H. A. Mooney, and R. Goldstein, eds. *Sulfur Dioxide and Vegetation: Physiology, Ecology, and Policy Issues*. Stanford Univ. Press, Stanford.

Whittaker, R. H. (1960). Vegetation of the Siskiyou Mountains, Oregon and California. *Ecol. Monogr.* **30**:279–338.

Whittaker, R. H. (1975). *Communities and Ecosystems*, 2nd ed. Macmillan, New York.

Whittaker, R. H., and Niering, W. A. (1965). Vegetation of the Santa Catalina Mountains, Arizona: a gradient analysis of the south slope. *Ecology* **46**:429–452.

Winner, W. E., and Bewley, J. D. (1978). Contrasts between bryophyte and vascular plant synecological responses in an SO_2-stressed white spruce association in central Alberta. *Oecologia* **33**:311–325.

Winner, W. E., Koch, G. W., and Mooney, H. A. (1982). Ecology of SO_2 resistance. IV: Predicting metabolic responses of fumigated shrubs and trees. *Oecologia* **52**:16–21.

Winner, W. E., and Mooney, H. A. (1980). Ecology of SO_2 resistance. III: Metabolic changes of C_3 and C_4 *Atriplex* species due to SO_2 fumigations. *Oecologia* **46**:49–54.

Woodwell, G. M. (1970). Effects of pollution on the structure and physiology of ecosystems. *Science* **168**:429–433.

Woodwell, G. M., and Whittaker, R. H. (1968). Effects of chronic gamma irradiation on plant communities. *Quart. Rev. Biol.* **43:**42–55.

Word, D. L., ed. (1980). *Biological Monitoring for Environmental Effects.* Lexington Books, Lexington, Mass.

Wourms, J. P. (1972). The developmental biology of annual fishes. III: Pre-embryonic and embryonic diapause of variable duration in the eggs of annual fishes. *J. Exp. Zool.* **182:**389–414.

10

SAMPLING AND ANALYSIS
OF ECOLOGICAL DATA

Methods developed for sampling and analysis of biological communities can be applied to a wide array of impact studies. Several excellent texts review aspects of these methods in detail. Green (1979) reviews statistical methods with emphasis on designing appropriate sampling procedures. Clifford and Stephenson (1975), Gauch (1982), Gillison and Anderson (1981), Orloci (1978), and Whittaker (1978a, 1978b) review methods of classification and ordination in depth. More general reviews of quantitative ecological methods include those of Greig-Smith (1983), Kershaw (1973), Legendre and Legendre (1983), Pielou (1977), and Poole (1974). In this chapter we introduce and exemplify some methods of sampling and analysis of ecological data available for use in pre- and post-impact studies.

MODELS OF COMMUNITY STRUCTURE

Methods of ecological sampling involve assumptions about the structure of the data and the underlying community they represent. It is thus helpful both to have a conceptual model of community structure before beginning to sample and to test whether the model fits the observed data structure following sampling. If the actual structure of the community differs from that assumed by use of a particular statistical method, the method of statistical analysis must be judged inappropriate for use with those data, and a different method of analysis must be employed. Let us consider two examples of conceptual models of community structure.

Spatial Pattern at a Site

The first example concerns the spatial pattern of distribution of individuals at a site of interest. In Figure 10.1 three patterns are illustrated. Pattern 10.1a has been generated by placing points at random within the rectangular site, using a random numbers table to select coordinates for each point. A spatial

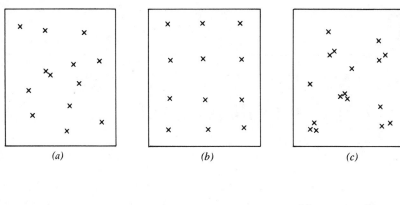

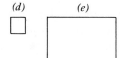

Figure 10.1. Types of spatial pattern of x's in relation to the rectangular area in which they are located. (a) Random; (b) regular; (c) aggregated; (d) and (e) sample plot sizes referred to in text.

pattern of points is random if every point on the ground within the study has an equal chance of being occupied by an individual (Pielou 1977). One may test for the randomness of spatial pattern by sampling for density (number of individuals per unit area) with an areal sample plot of fixed size. The proportion of sample plots that contains 0, 1, 2, up to n individuals is called a "frequency distribution." A frequency distribution generated in this way will typically have a mathematical form first described by Poisson:

$$P_r = \frac{\lambda^r e^{-\lambda}}{r!} \qquad r = 0, 1, 2, \ldots, n \tag{10.1}$$

where

P_r = the proportion of sample plots containing r individuals
r = numbers from 0 to n in discrete intervals of one
λ = mean number of individuals per sample unit
e = the Naperian log constant, approximately 2.732
$r!$ = r factorial, which is $(r)(r-1)(r-2)\ldots(1)$

A spatially random pattern will fit a Poisson distribution only when the number of individuals is small in relation to area (i.e., low density), so that the chances of finding nonmobile organisms at any one point are small; at high densities the number of individuals per sample unit, and the variance of such values, are equal when data fit a Poisson distribution. Whether the variance/mean ratio equals one has therefore been suggested as a simple test for the occurrence of data fitting a Poisson frequency distribution. The standard error of the difference between the variance/mean ratio and 1 for a

Poisson distribution is $(2/N-1)^{1/2}$, where N = number of sample units used. This may be compared to the observed standard error by means of a t-test.

Despite the simplicity of the variance/mean (v/m) test for fit to a Poisson distribution, the test has several drawbacks. A frequency distribution derived from a nonrandom spatial pattern may occasionally result in $v/m = 1$. Indeed, nonrandom spatial patterns will almost always have certain scales of sampling at which individuals occur in random spatial pattern. By "scale of sampling" we refer to the areal size of the sample unit. In Figure 10.1c it can be seen that individuals tend to be clumped into aggregates of a size roughly corresponding to a sample unit of size 10.1d. As a result such sample units placed at random in plot 10.1c will have a greater than average chance of being either very full (enclosing a clump) or empty. This will result in a frequency distribution with $v/m > 1$, indicating departure from randomness toward clumping. At a sampling scale of size 10.1e the number of individuals found from one sample unit to the next is likely to be nearly equal, resulting in a "regular" or even frequency distribution ($v/m < 1$), perhaps even tending toward random ($v/m = 1$). The physical spatial pattern has not changed: the scale of analysis has, and with it, the resulting frequency distribution.

Unfortunately the comparison of an observed frequency distribution with Poisson expectation by a chi-square test at a particular scale of sampling is also a fallible test of randomness (Table 10.1). The problem arises from the necessity of pooling frequency classes with low expectation (<5). Because this usually results in pooling data for sample units with large numbers of individuals in them, some of the most revealing evidence for spatial aggregation is obscured, and a truly nonrandom distribution may fail to depart significantly from a Poisson distribution (see Greig-Smith, 1964, pp. 68–70). A distribution that satisfies both the chi-square test for fit to a Poisson distribution, and the variance/mean test, is more likely to be truly random.

When the v/m was significantly less than unity, we referred to the spatial pattern as "regular." This is "regularity" in a statistical sense, departure from randomness toward a greater than random chance that each sample will have an equal number of individuals in it. The corresponding mathematical frequency distribution is called "uniform." Regularity occurs rarely in nature. Most often organisms are aggregated at certain scales of pattern and randomly distributed at the remainder. As a consequence it is not safe to assume if sedentary or sessile (attached) organisms are randomly placed at one scale, that they will be so at all scales. This conclusion has major significance for choice of sampling methods to be discussed here.

We have been discussing spatial pattern thus far with the tacet assumption that the pattern remains constant or homogeneous throughout the study area. There may, however, be a density gradient throughout the study site, which can be tested for by partitioning the site in half and comparing mean densities of samples from each subsection by a t-test. Alternatively one may detect heterogeneity in density by dividing the site into quadrants and calculating the coefficient of variation (standard deviation $\times$ 100/mean) between

Table 10.1.   A Test of Fit to a Poisson Distribution

Number of Plant Individuals per Quadrat (r)	Number of Quadrats with r Individuals in It		Difference	Chi-Square (Observed-Expected)^2/ Expected
	Observed	Expected		
0	165	132	33	8.25
1	63	107	44	18.0
2	34	33	10	2.27
3	38	12	26	56.3
>3	0	5	5	5.0
Total				89.8

Sample calculation[a]
For $r=3$, Poisson expectation is

$$300 \times \frac{0.8166^3 \times e^{-0.8166}}{(3)(2)(1)} = \frac{300 \times 0.5445 \times 0.4401}{6} = 12$$

Note: Three hundred 4 m^2 quadrats were sampled for density of black sage (*Salvia mellifera*), with the following results. Expected values for fit to Poisson distribution are P_r (from Eq. 10.1) × number of quadrats sampled (300 in this case); $\lambda = (63(1) + 34(2) + 38(3))/300 = 0.8166$.
[a] Degrees of freedom = $N - 2$, where N is the number of categories from which chi-square was calculated. In this example, $N = 4$; degrees of freedom = 2. The probability of a larger chi-square value by chance alone is <0.01. Therefore the deviation of the observed distribution from that expected if the plants were randomly (Poisson) distributed is highly significant. Correction for continuity was not made in this example because the number of quadrats sampled exceeded 200. For further discussion of chi-square tests, see, e.g., Grieg-Smith (1964, 1983), and Snedecor (1946).

quadrants. Severe clumping in one portion of the study site which can be recognized visually or by preliminary sampling should be segregated from the remainder of the study area for separate analysis of small-scale pattern.

Aggregation may take many forms, and numerous mathematical frequency distributions exhibiting "contagion" may be tested for fit to a particular pattern of aggregation (Figure 10.2; for additional examples, see Westman 1970, Table 1). There is no efficient means to determine which mathematical model of contagion describes a particular pattern of contagion except to test empirical frequency distributions for fit to a series of possible models (see, e.g., computer program Fit 5 De Mars, U.S. Forest Service, Southwest Forest and Range Experimental Station, Berkeley, Calif.).

Ecologists have also attempted to develop methods of determining the average scale of aggregation in spatial patterns. On the basis of experiments with simulated data, the original block size/variance technique of Greig-Smith (1952, 1961, 1964) has been shown to be rather complex to interpret

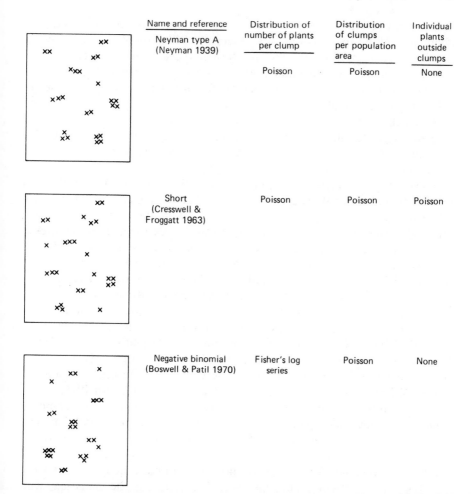

	Name and reference	Distribution of number of plants per clump	Distribution of clumps per population area	Individual plants outside clumps
	Neyman type A (Neyman 1939)	Poisson	Poisson	None
	Short (Cresswell & Froggatt 1963)	Poisson	Poisson	Poisson
	Negative binomial (Boswell & Patil 1970)	Fisher's log series	Poisson	None

Figure 10.2. Examples of aggregated spatial patterns and the mathematical functions that describe the resulting sample frequency distributions.

(Westman and Anderson 1970, Errington 1973, Usher 1975). Alternative methods have been proposed (Goodall 1974, Hill 1973a, Ludwig and Goodall 1978, Usher, 1975, Whittaker, Niering and Crisp 1979). Of these, the technique of Goodall (1974) appears most promising for detecting single-species pattern (Carpenter and Chaney 1983), and that of Whittaker et al. (1979) for multi-species pattern.

Species Distributions along Environmental Gradients

The second example of conceptual models of community structure concerns the form of species distributions along axes representing incremental change in some environmental factor or complex of factors strongly influencing the abundance of the species of interest. Figure 10.3 illustrates four of numerous possible models. Whittaker (1967) has termed the sequence of species distributions along an environmental gradient a *coenocline*. The full suite of species (coenocline) and associated habitat features which comprise a set of ecosystems arrayed along a gradient of change in some controlling environmental factor or factors is called an *ecocline*. The term "importance value" is a generic term for any measure of a species that reflects its role in usurping community resources, such as density, foliar cover, frequency, net primary production, and biomass.

The ecoclines in Figure 10.3 represent species importance values along an environmental gradient. Because the physical landscape is very patchy, with sharp transitions from one habitat condition to another, an environmental gradient must be reconstructed from such data by rearranging sample sites in rank order from low to high in occurrence of some environmental factor, such as available moisture for plant growth. Thus the environmental gradient in Figure 10.3 is a synthetic axis, unlikely to be found anywhere as a continuous stretch of physical landscape. The linear species distributions of Figure 10.3a are unlikely to be found except along short segments of an environmental gradient. Most species distributions are curvilinear, but a small enough segment of any curve can approximate a straight line, especially at the tails of a normal distribution. Austin (1976) found that only 5% of 135 published species distribution curves were of linear form. The Gaussian curves of Figure 10.3b were found in 26% of the cases (Austin 1976). However, 95% of the species distribution curves were curvilinear and nonmonotonic, (changing slope sign at least once), and 73% were unimodal (having a single peak). In other words, most species distribution curves are crudely bell shaped. Of the curvilinear, nomonotonic curves observed by Austin (1976), 29% were skewed, 16% were shouldered or plateaued, and 29% were bimodal. The most common, repeatable form, then, is the Gaussian, with some skewing or flattening of this shape being common.

Whittaker (1967) reviewed studies of terrestrial vegetation along moisture gradients corresponding to gradual changes in elevation, aspect, and position along a mountain slope from xeric to mesic and found a vegetation model of type 10.3b to be common. His curves were heavily smoothed, however, by averaging data from several intervals along the environmental axis and by further hand-fitting of curves to points. When less smoothing is employed, more departures from the pure Gaussian curve are typically observed (Figure 10.3d). Model 10.3c corresponds to the notion of discrete associations of species in community types, which are replaced rapidly by new associations as the environment changes across certain key regions of an environmental gradient. This ecocline form is rarely observed in nature, though occasional, isolated examples appear to exist (e.g., Westman 1983, Daubenmire 1966).

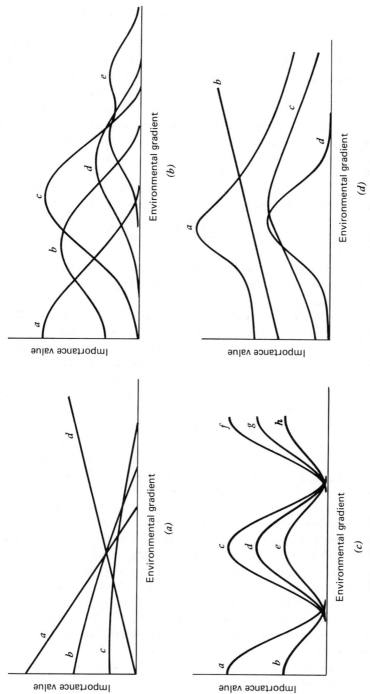

Figure 10.3. Four examples of possible species distribution curves along an environmental gradient.

RELATING SPECIES DISTRIBUTIONS TO ENVIRONMENTAL FACTORS

The models of species distributions just discussed have great relevance to the assumptions underlying statistical methods. For example, if an analyst wished to examine the correlation between levels of a pollutant and abundance of a species, he or she could only do so validly if the species distribution were linear in relation to the pollutant level, since correlation analysis assumes a linear relation between the dependent and independent variables. The same is true of linear or multiple linear regression analysis and of all methods based on linear regression analysis, such as principle components analysis, factor analysis, canonical correlation analysis (Greig-Smith 1974, Pielou 1977), binary discriminant analysis (Strahler 1978), and others. Since species distribution curves are only rarely linear, these statistical methods should not ordinarily be used with untransformed species importance data. One alternative that has been proposed for use with data of this type is "Gaussian analysis" (Westman 1980). Here one visually screens graphical data on species distribution curves initially to remove obviously linear or polymodal species distribution curves. One then tests the remaining approximately Gaussian curves more precisely against a series of environmental factor axes to determine which factor axis causes most species distributions to adopt significantly Gaussian forms. These factors are then considered to be those that most strongly influence species importance values. The method has been tested on coastal sage scrub (Westman 1980, 1981) with some success, using a Gaussian curve-fitting program of Gauch and Chase (1974).

Often a graphical plotting of species distributions along the environmental factor gradient of interest is revealing. These curves may then be fit to polynomial regression curves (see, e.g., Figure 7.7). More recently Austin et al. (1983, 1984) have fitted species distribution curves to general linearized models (Nelder and Wedderburn 1972) which permit the relative contributions of different environmental factors to the species curve to be assessed.

Another approach is to attempt to linearize the species distribution along the environmental factor axis (and hopefully to obtain independent and normal error distributions where replicate samples have been taken). For a single species such linearization can sometimes be accomplished by transformation of the data by arcsine, log, or other means. This is very much a trial-and-error statistical manipulation, however, and the biological meaning of the result will often be difficult to interpret. Alternatively sample sites may be ranked in an approximately linear fashion by ordering samples according to their similarity in species composition ("ordination") and the correlation of environmental factors to the ranked or "ordinated" sequence of stands then tested. The ordination of sample stands using species data is discussed later in the chapter and in Whittaker (1978a), Gauch (1982), and elsewhere. The ordination approach identifies key environmental factors influencing the distribution of whole suites of species rather than any one species alone.

DATA COLLECTION

The aim of a sampling program is normally to obtain an accurate representation of the average and range of conditions present at a site over space (and sometimes, time) within the constraints imposed by limited time, money, or labor. Since one aims to characterize an entire site by its average, the site should be as homogeneous as possible in the parameters of interest. Any initial heterogeneities that are observed or suspected (e.g., change in slope, aspect, position, soil type for terrestrial sites; change in flow rate, depth, or temperature of water for aquatic sites) can be used to stratify or divide the initial study area into homogeneous subsections.

Systematic Placement of Sample Units

The placement of sample units within each homogeneous study area can be either random or systematic (i.e., following a uniform pattern). Systematic samples do not satisfy the assumption of random sampling underlying parametric statistics. Therefore no error terms can be calculated validly from such data. Systematically placed samples have the further disadvantage that the pattern of sample placement may correspond to a pattern of regularity in the community being sampled (see Figure 10.4), so that certain features will be repeatedly over- or undersampled. Systematic samples should also sample the entire area evenly, which not all "regular" patterns do (Figure 10.5). Nevertheless, systematic sampling has the advantages over random sampling that (1) sample units can be located more quickly in the field, and (2) any undetected heterogeneities within the sample area are more likely to be sampled evenly. A random sample has the advantage that standard errors around the true mean value can be calculated. (A standard deviation is the true measure of error only if the entire population has been measured; for subsamples of the total population, a standard error is appropriate).

In a relatively homogeneous area that has been sampled at random, sample means should vary in a normally distributed fashion around the true mean. As the sample size approaches a complete survey or census (100% coverage of the sample area or tally of the population), the deviation of the sample mean from the true mean should decrease and finally equal the true mean (Figure 10.6). It is assumed that the organisms being randomly sampled are randomly distributed in space. As we noted earlier, organisms are

Figure 10.4. Systematic sampling. A site with a regular pattern of habitat heterogeneity (e.g., dune/trough) may be poorly sampled by systematically placed sample plots whose spacing corresponds to that of the environmental heterogeneity.

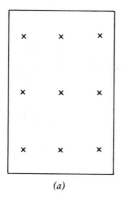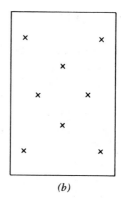

(a)   (b)

Figure 10.5. Regular versus systematic sampling designs. Samples x's are arranged systematically in both (a) and (b). In (a), however, the edge of the area is sampled more intensively than the center. Pattern (b), though appearing less regular, distributes samples more evenly between edge and center.

typically only randomly spaced at certain scales of sampling, so that to use the sample mean as a true estimate of the population mean, one must also establish that the organisms are randomly spaced at that scale of sampling. This can be done by testing the sample data for fit to a Poisson distribution with a chi-square test and/or a variance/mean test. If the random distribution of organisms in sample units is confirmed, one can proceed with the analysis. If the data do not fit a Poisson distribution, one must resample with sample plots of other sizes until a scale at which organisms are randomly distributed is found. Obviously this is a cumbersome procedure.

Although we have been discussing the problems of sampling organisms that are spatially aggregated, the same issues apply to aggregation in time. Thus in studying the water quality of a stream subject to periodic storm flood

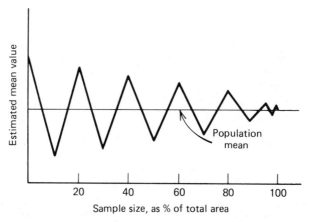

Figure 10.6. Effect of sample size on deviation from true (population) mean. This is a typical, though hypothetical, curve.

episodes, one is observing a stream in which major differences in water quality over time are clustered in the days following the storm event. Spacing one's samples at random or systematically over time is not valid for such a heterogeneous temporal pattern. An appropriate solution is to stratify the data, analyzing separately the water quality values in the storm episode periods and the between-episode periods. If one wishes an annual average of water quality, the water quality values from the two types of time periods (storm episode; nonstorm episode) can be weighted by frequency of occurrence and averaged.

Plotless Sampling

One- and Zero-Dimensional Sampling

In the case of sampling sessile organisms in space, the problem of requiring random spatial patterns can be avoided by using zero- or one- rather than two-dimensional sample units. Sample plots, be they circular, quadrangluar ("quadrats"), or some other shape, will inevitably establish a certain scale of analysis. A point (zero-dimensional) or line (one-dimensional) has no areal size and is therefore free of the problem of establishing a scale of sampling. A common one-dimensional technique is the line transect. Lines are placed at random across the study site. For foliar cover data the length of transect intercepted by the canopy of a plant is recorded. Percent foliar cover for a species can then be calculated as

$$\frac{\Sigma I}{L} \times 100 \qquad (10.2)$$

where

I = length of transect intercepted by the species of interest
L = total length of line transect observed

To obtain density measurements by line transect, the length of the maximum chord on the plant canopy parallel to the transect line is recorded (Figure 10.7). Density is then calculated as (McIntyre 1953)

$$\frac{\Sigma(1/D)}{L} \qquad (10.3)$$

where

D = maximum chord length parallel to transect of intercepted plant canopy
L = total transect length

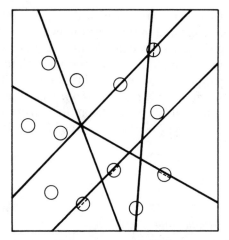

Figure 10.7. Sampling density with line transects. Solid lines are randomly placed line transects. Circles are plant canopies. Dotted lines show maximum chord lengths parallel to transects on intercepted canopies.

Maximum chord length values can be used to compute % foliar cover by the formula (McIntyre 1953)

$$\frac{0.785(\Sigma D)}{L} \times 100 \tag{10.4}$$

Methods involving maximum chord lengths assume that plant canopies are circular in outline.

Although line transect data can be calculated validly even for organisms that are spatially aggregated, the sample size required for accurate estimates when organisms are aggregated is much larger than when organisms are randomly spaced (Westman 1970). Also line transects are less accurate when organisms being observed are small in relation to the actual size of the line, since the line can no longer be assumed infinitesimal in width. Large numbers of transects must be observed to equal the area covered by areal sample plots. Line transects should not be confused with "belt transects" which consist typically of adjacent 1 m² quadrats in a line. Belt transects are a form of systematic quadrat sampling, subject to the problems of scale of pattern discussed earlier.

Point sampling comprises another scaleless approach. Recording the identity of plants occurring at random points in a study area, for example, permits computation of frequency (the percentage of all points at which an organism of the species is present). Frequency is itself a complex measure, however, since it will vary with both density and cover of a species.

Distance Methods

Another family of plotless techniques are the nearest neighbor or distance methods of sampling. In one example, the distance from a random sampling point to the nearest organism is measured (closest-individual method, Cottam 1947). The mean value $\bar{r}$, for a series of such measurements can be used to calculate the total number of individuals in the area (or the total per species, when $\bar{r}$ values are segregated by species), using the formula

$$\frac{\text{Total area of study plot}}{\pi \bar{r}^2} = \text{total number of individuals} \qquad (10.5)$$

This technique, and all related distance techniques, assume a random distribution of individuals at all scales of sampling, however. This is because the distance method, rather than being a true zero-dimensional technique, is actually a two-dimensional sampling technique in which the distance from point to organism can be thought of as the radius of an imaginary circular sample plot. Thus the "scale" of sampling is indeterminate, since it shifts with each distance measurement (Figure 10.8). As a result species must be assumed to be randomly spaced at all scales of sampling, a very unlikely

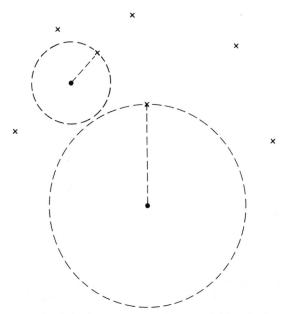

Figure 10.8. Closest-individual method. The x's are individuals in space. Dots represent random sample points. The distance from the random sample point to the nearest individual forms the radius of an imaginary circular sample plot, used to calculate number of individuals in Eq. 10.5. Note that the scale of sampling varies with the lengths of radii.

event. Therefore all such distance measurements, including the point-centered quarter method (Cottam and Curtis 1956), the pattern diversity measurement technique (Pielou 1966), and others, suffer from bias in populations which are aggregated at certain scales (see Lewin and Westman 1973 for an example of the problem applied to the field with data of known aggregation).

Sample Size and Precision of Estimates

How many random samples must be taken to achieve a desired precision in estimating the population mean value? Numerous approaches exist and are reviewed by Green (1979), Greig-Smith (1964), Poole (1974), and others. One very crude rule of thumb states that the number of samples required to achieve a particular level of precision, P, is approximately equal to $1/P^2$, where P is the standard error as a percent of the mean. Thus if mean foliar cover is to be estimated with a precision such that 2/3 of the sample estimates deviate from the mean value by less than $\pm 7.5\%$, then P, the standard error, is 7.5% of the mean, and the number of samples needed is $1/(0.075)^2 = 178$ (see Taylor 1961 and Green 1979 for refinements).

Another approach to estimating sample size involves conducting some preliminary sampling of the population. Suppose that s^2 is the variance of the preliminary sample. Then the sample size, n, required to obtain a sample mean within $\pm L\%$ of the true population mean 95% of the time is (Poole 1974)

$$ n = \frac{4s^2}{L^2} \tag{10.6} $$

For example, if the variance of % foliar cover for a species sampled with quadrats is 10%, and one wishes to estimate the true mean within 1% of its value 95% of the time, then $n = 4(10)^2/1^2 = 400$ quadrats. If the sample size, n, exceeds 10% of the total population, N, a corrected sample size can be calculated as $n' = n/[1 + (n/N)]$ (Poole 1974, p. 295). Poole (1974) also discusses optimizing sample size in view of constraints of both cost and precision.

Sampling Mobile Organisms

The discussion of sampling thus far has concerned organisms that do not move in space over time. For mobile animals various modifications of a basic technique known as mark-recapture are often used. In its simplest form (Lincoln or Peterson index), an investigator traps a random sample of individuals from the population, marks them, and releases them. After a period of time, a sample is retrapped. The proportion of marked individuals found in the second trapping is taken as the percentage of the total population size trapped in the first trapping. Thus suppose that, using 25 traps, one initially trapped 20 rabbits, marked, and released them. In the second trapping 15 rabbits were caught in the 25 traps, of which 5 were marked. Then 20 rabbits were 5/15 or one-third of the total population of 60 rabbits.

In general, if

a = the number of individuals initially trapped and marked
n = the number of individuals trapped in the second trapping
r = the number found marked in the second trapping

then the estimated total population size x is

$$x = \frac{an}{r} \tag{10.7}$$

and the variance is

$$\text{var}(x) = \frac{a^2 n(n-r)}{r^3} \tag{10.8}$$

These estimates require large sample sizes to be effective. Bailey (1951) and Poole (1974) discuss modifications for smaller sample sizes.

The mark-recapture method is subject to several assumptions (Poole 1974): that marked individuals mix randomly with the remainder of the population; that the population size does not change; that both samples are taken randomly, and all individuals have an equal chance of being trapped; that the time taken for sampling itself is negligible relative to the time between sampling periods; and that individuals who are trapped once do not become trap addicted or trap shy.

For additional discussions of sampling techniques for mobile organisms, see Poole (1974) and Southwood (1978), among others.

What to Sample

What species are of relevance to sample in impact studies? Listing all the species present, from microbes through top carnivores, is usually infeasible and unnecessary. Species of greatest interest are often those that are important to conservation, or to ecosystem function, or that are particularly vulnerable to impact. Thus species that are rare, threatened, or endangered should be listed, as should major primary producers, top carnivores, keystone, indicator, and critical-link species (Chapter 8).

In composing a species list of terrestrial plants, one may choose to limit the list to "dominants," due to time constraints. A plant is considered "dominant" if it appears to usurp a major portion of the resources of the habitat. Often this is judged by the biomass, or alternatively the extent of canopy cover ("foliar cover") exhibited by the species on the site. The less abundant species should be noted if time permits, however, since collectively they can play major roles in ecological function and individually they may play inconspicuous but vital roles in nutrient cycling, supplying food to herbivores, soil binding, or other functions. Rarer species (orchids, certain lichens, etc.) may also have great value as indicators of habitat conditions and in classification and ordination of sites. Nevertheless, the species list requires interpretation if it is to be meaningful to the nonecologist. Methods discussed later in this chapter suggest ways the data can be so interpreted.

Use of Qualitative versus Quantitative Data

Listing the presence of species gives only a qualitative indication of community composition. It tells the reader nothing about the relative importance of these species in the community. To convey the latter, any of several quantitative features can be compiled. For plants the common importance measures are listed in Table 10.2.

Table 10.2.   Common Measures of "Importance" of Terrestrial Plant Species in a Community

Importance Measure	Definition
Density	Number of individuals per unit area
Foliar cover	Percent of ground surface covered by leaves
Individual overlay foliar cover	Sum of cover percent counting each individual's canopy separately (can exceed 100%)
Projective foliar cover	Percent of ground surface covered one or more times by leaves of the species (cannot exceed 100%)
Leaf area index	Average number of leaves stacked above any point on the ground
Basal area	Cross-sectional area of tree trunks at breast height (1.37 m)
Frequency	Percent of small sample plots within a study site in which a species occurs
Constancy	Percent of large study sites of the same size, out of the total sampled, in which the species occurs
Presence	Percent of large study sites in which a species occurs when plots are not of the same size
Biomass	Total dry weight of living material of the species
Net primary production	Net increase in biomass per unit time before herbivory or decay

When sample plots contain very many species (more than about 80 per 0.1 ha), qualitative data are often sufficient to characterize differences between study sites. Thus Williams et al. (1973) found that qualitative data were as good as quantitative data in classifying large plots of tropical rain forest sites in Australia, mainly because the species richness of each site was so high, and the rate of change in species composition between sites (beta diversity; see Chapter 11) so great, that changes in composition alone could serve to characterize sites. The effects of chronic disturbance on such a community are at first quantitative (effects on growth) rather than qualitative (extinction of entire population) however, so that following temporal change even in species-rich sites will require use of quantitative data.

The establishment of precise boundaries for a study site (e.g., often 20 × 50 m in terrestrial woody vegetation), and subsequent sampling with line transects, belt transects (e.g., 50 1 m × 1 m quadrats in a row), or randomly placed sample plots, is a precise but time-consuming process. Such procedures are excellent for detecting spatial heterogeneity in community composition and in finding small, rare species.

When the site is considered highly homogeneous, some ecologists prefer the less precise but more rapid technique of recording a list of species by walking around the homogeneous site without establishing precise site boundaries. The boundary-free site is known as a *relevé*. Following completion of the species list, each species is rated visually based on both its foliar cover and abundance (number of individuals), using either a 5-point (Braun-Blanquet) or 10-point (domin) scale (Table 10.3). Its spatial pattern ("sociability") and stage of flowering ("phenophase") may also be noted (Table 10.3). Though being a more rapid procedure, this so-called "Braun-Blanquet" technique is much more subjective in its data collection process (e.g., plants in flower will appear more abundant than those of the same foliar cover that are not). The precision of information collected is less, as differences in cover are lumped into cover classes. Cover and abundance are confounded in the scales, although these are two quite distinct importance measures that will not necessarily vary equally with impact. Further it is easier to miss small, rare species. An approach to minimize the chances of missing inconspicuous species is to keep track of the number of new species being recorded per time interval as one meanders through the site. One can continue the search until the rate of discovery of new species on the site falls below a certain arbitrary level, such as, 1 species every 30 minutes (Goff et al. 1982). To compile a complete species list, it is important for a site to be revisited at least once during each major season of the year.

Since the number of species generally increases with the area sampled, the use of indefinite boundaries for survey results in noncomparable methods for obtaining measures of species richness at a site. This objection can be overcome by establishing site boundaries of constant size and shape (see also Chapter 11).

Table 10.3.   The Braun-Blanquet and Domin Scales for Cover Abundance

Braun-Blanquet	Domin

Braun-Blanquet

r one or a few individuals

+   occasional and less than 5% of total plot area

1   abundant and with very low cover, or less abundant but with high cover; in any case less than 5% cover of total plot area

2   very abundant and less than 5% cover, or 5–25% cover of total plot area

    2m very abundant

    2a 5–12.5% cover, irrespective of number of individuals

    2b 12.5–25% cover, irrespective of number of individuals

3   25–50% cover of total plot area, irrespective of number of individuals

4   50–75% cover of total plot area, irrespective of number of individuals

5   75–100% cover of total plot area, irrespective of number of individuals

Domin

+   one individual

1   rare

2   sparse

3   less than 5%, frequent

4   5–10%

5   11–25%

6   26–33%

7   34–50%

8   51–75%

9   76–90%

10   91–100%

Sociability

1   growing solitary, singly
2   growing in small groups of a few individuals, or in small tussocks
3   growing in small patches, cushions, or large tussocks; hummock builders
4   growing in extensive patches, carpets, or broken mats
5   growing in great crowds or extensive mats completely covering the whole plot area; mostly pure populations

Phenophase

v.   (vegetative)
fl.   (flowering)
fr.   (fruiting)
E.g., *Scirpus maritimus* 2.3 v. (cover.sociability phenophase)

DATA ANALYSIS

Comparing Species Composition in Space or Time

Having obtained a quantitative description of the relative importance of species on a set of sites, or on a single site at several time intervals, it becomes useful to summarize how the sites differ in composition and relative abundance of species in space or time. Indexes that compare the composition of pairs of sites are called *resemblance functions*. Resemblance functions that compare sites using only qualitative (presence/absence) data are called "coefficients of community" (CC). Those that compare sites using quantitative data (importance values) are called "indexes of percentage similarity" (PS).

There are many forms that each of these indexes can take. Two examples are the Jaccard (1912) and Czekanowski (1913) indexes. Examples are computed in Table 10.4. Although the two indexes give different values from the same data, there is little to choose between them. By the same token, their results should not be compared.

Two examples of indexes of percentage similarity are the Czekanowski (1909) and the Euclidean (e.g., Orloci 1967). All indexes vary between 0 and 100%, except the Euclidean, which varies between 0 and 144%. Because the Euclidean index involves differences between the squares of importance values, the index is particularly sensitive to differences in the importance value of dominant species on the site. A 10% difference in importance value for a species with 70% cover when squared ($7\%^2 = 49\%$) is much larger in absolute value than a 10% difference in a species with 10% cover ($1\%^2 = 1\%$). Hence the Euclidean index is of value when one wishes to explore changes in importance among dominant species. For example, if one's main concern is with the effect of an air pollutant on canopy cover, the Euclidean index would be appropriate to track changes in cover of dominants over time. A Euclidean index would similarly dramatize a grazing impact most effectively. The Czekanowski PS index, by contrast, does not give added weight to dominants relative to rarer species. This index would therefore be appropriate for following changes in the full suite of species, when one is equally concerned with relative changes in dominant and rare species (e.g., changes in composition of a nature preserve due to recreation). If one wishes to give special emphasis to changes in certain species, their importance

Table 10.4 Illustrative Computation of Coefficients of Community and Indexes of Percentage Simularity Using Hypothetical Data from Two Sites of Coastal Sage Scrub

Example Species	% foliar cover	
	Site A	Site B
Salvia mellifera	35	15
Bromus rubens	5	10
Artemisia californica	35	40
Yucca whipplei	25	20
Mirabilis californica	0	15

Table 10.4. *(Continued)*

Coefficients of Community

a = the total number of species in Site A
b = the total number of species in Site B
c = the number of species in common between the two sites, A and B

In the example $c = 4$, $a = 4$, $b = 5$
Jaccard's index: $[c/(a + b - c)] \times 100$; $[4/(4 + 5 - 4)] \times 100 = 80\%$
Czekanowski's index: $(2c/a + b) \times 100$; $(2 \times 4/4 + 5) \times 100 = 89\%$

Index of Percentage Similarity

a = importance value of species in Site A
b = importance value of species in Site B

Czekanowski's index

$$\frac{2 \sum\limits_{i} (\min a, b)}{\sum\limits_{i=1} (a + b)} \times 100$$

$(\min a, b)$ = the lesser importance value of a species present in both
sites, A and B

$$PS = \frac{2(15 + 5 + 35 + 20)}{100 + 100} \times 100 = 75\%$$

Euclidean Index

$$PS = \left[\left(\sum_{i=1}^{i} \frac{a}{\sqrt{\sum\limits_{i=1}^{i} a^2}} - \frac{b}{\sqrt{\sum\limits_{i=1}^{i} b^2}} \right)^2 \right]^{1/2} \times 100$$

$$= \left[\left(\frac{35}{\sqrt{35^2 + 5^2 + 35^2 + 25^2 + 0^2}} - \frac{15}{\sqrt{15^2 + 10^2 + 40^2 + 20^2 + 15^2}} \right)^2 \right.$$

$$+ \left(\frac{5}{\sqrt{35^2 + 5^2 + 35^2 + 25^2 + 0^2}} - \frac{10}{\sqrt{15^2 + 10^2 + 40^2 + 20^2 + 45^2}} \right)^2$$

$$+ \left(\frac{35}{\sqrt{35^2 + 5^2 + 35^2 + 25^2 + 0^2}} - \frac{40}{\sqrt{15^2 + 10^2 + 40^2 + 20^2 + 15^2}} \right)^2$$

$$+ \left(\frac{25}{\sqrt{35^2 + 5^2 + 35^2 + 25^2 + 0^2}} - \frac{20}{\sqrt{15^2 + 10^2 + 40^2 + 20^2 + 15^2}} \right)^2$$

$$\left. + \left(\frac{0}{\sqrt{35^2 + 5^2 + 35^2 + 25^2 + 0^2}} - \frac{15}{\sqrt{15^2 + 10^2 + 40^2 + 20^2 + 15^2}} \right)^2 \right]^{1/2}$$

$$= 52\%$$

values can be weighted (say by a factor of $2X$) before computing the PS index.

The coefficient of community weights changes in all species equally, regardless of changes in their relative abundance. This index is thus most sensitive to changes in composition due to species extinction. It might also be most useful in comparing sites for a survey of parcels for inclusion in park reserves, since it will emphasize differences in species composition that can be useful in seeking parcels that will maximize the number of different species preserved. In general, then, the PS indexes are preferred when one wishes to study subtle changes in composition and abundance due to chronic impact; CC indexes are preferred when gross differences in species composition (presence/absence) are of interest. Use of weighting on particular species can further refine the emphasis of each index. For a fuller discussion of resemblance functions, see, for example, Goodall (1978) and Whittaker and Gauch (1978).

Direct Gradient Analysis

Suppose one is interested in predicting how riparian (streamside) vegetation will change following the damming of a river upstream for irrigation of arid land. The riparian vegetation may have the appearance shown in Figure 10.9a. The streamside trees will obtain most of their water from roots reaching the groundwater table. Some annual herbaceous vegetation, however, will be dependent on surface soil moisture which remains available for a few months following major rain and will not have roots deep enough to reach the water table. Since the region is arid, it is reasonable to hypothesize that the major factor controlling plant distribution is moisture availability. An approach known as direct gradient analysis (Whittaker 1967) measures both the vegetation and its associated habitat features, as a basis for interpreting vegetative pattern.

In the present example, suppose a series of 0.1 ha plots are laid out perpendicular to the stream, and foliar cover of vascular plants in the plots estimated by line transect. The soil moisture tension at 15 cm depth (an average depth for shallow-rooted plants that have been excavated) is measured using a soil tensiometer probe, and the depth to groundwater with a soil corer. Recognizing that both of these soil water variables will change during the year with rainfall patterns, these measurements are repeated each month for a year, and average values are adjusted by the deviation of that year's annual precipitation and evaporation from a 20-year record of precipitation and evaporation available from a nearby weather station. Because such measurements are time-consuming, they are taken on only every fifth site. Suppose that the graph of foliar cover of five dominant species against groundwater depth looks like Figure 10.9b, and against average annual soil moisture tension like Figure 10.9c. Since groundwater depth did not change

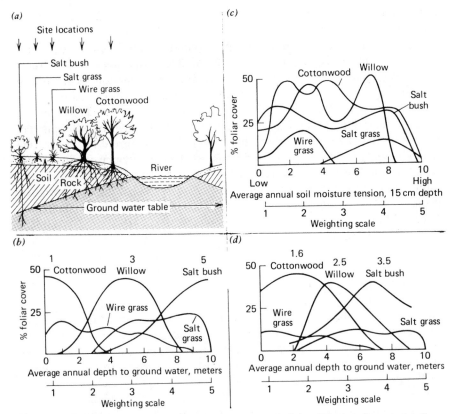

Figure 10.9. Examples of coenoclines of riparian vegetation referred to in text. (*a*) Cross section of stream and streamside vegetation; (*b*) distribution of foliar cover values of species in relation to groundwater depth; (*c*) distribution of foliar cover values of species in relation to soil moisture tension; (*d*) revised coenocline in relation to groundwater depth, using weighted stand scores to locate stands along the abscissa. The changed shape of species distribution curves (cf., 10.9*b*) reflects the shift in stand positions following weighted averaging (e.g., Table 10.5).

evenly with distance from the stream, the depth to groundwater axis does not correspond exactly with distance from the stream; sites have been located on the abscissa according to groundwater depth only. To smooth site-to-site variations, species data are averaged for the 4–6 sites sampled within each meter interval of increasing groundwater depth, having estimated groundwater depth by interpolation between sites of known depth.

Figures 10.9*b* and *c* show some species with nonmonotonic, unimodal curves and other species with highly polymodal curves. In Figure 10.9*b* the species showing the bell-shaped curves are the tree species with roots reaching the groundwater table; conversely, the species with shallow root systems show bell-shaped responses to surface soil moisture tension. Following the "leading dominants" or *weighted-averages technique* first developed by Curtis and McIntosh (1951) and later used extensively by Whittaker (1956, 1960, 1967), one can assign weights to the positions of the modal peaks on the graphs where dominant species rise to a clear unimodal maximum. In the original example a single environmental gradient was used; in our example two separate environmental gradients, appropriate to trees versus herbs, are used.

Because the location of four out of every five sites on the graphs was an estimation based on interpolation, one may now wish to refine the position of these sites using information on their vegetative composition. This procedure is known as iterative direct gradient analysis (Westman 1975). It consists of using the weights assigned to species (based on location of their modal peaks) to calculate weighted stand scores (Table 10.5). The rationale is that the modal peak position reflects the habitat preference of the species. Therefore stands with high importance in a particular species should occupy a position along the environmental gradient close to the modal peak of that species. Weighting stand composition by modal peak weights therefore uses

Table 10.5. Example of Computation of Weighted Average Stand Scores from Species Modal Peak Values

% Cover	Average Depth to Groundwater (mean of 4–6 measured sites each) ($\bar{X} \pm$ standard error)				
	1.1 ± 0.1	3.3 ± 0.2	5.1 ± 0.1	7.8 ± 0.4	9.2 ± 0.1
Cottonwood	48	20	0	0	0
Willow	0	20	45	20	0
Saltbush	0	0	8	25	45
Wire grass	15	15	10	5	1
Salt grass	0	2	20	19	25

Calculation of Weighted Stand Scores for
5 Stands within 1.1 ± 0.1 Groundwater Depth Interval

	Weight	Site 1	2	3	4	5
Cottonwood	1	45	40	48	52	55
Willow	3	1	0	0	0	1
Saltbush	5	0	1	0	1	0
Wire grass	2	12	10	20	18	15
Salt grass	4	1	1	0	0	0
Weighted stand score		1.29	1.33	1.29	1.31	1.24

$$\text{Weighting formula} = \frac{\sum_{i=1}^{i} (\% \text{ cover, species } i)(\text{weight, species } i)}{\sum_{i=1}^{i} \% \text{ cover, species } i}$$

information on vegetative composition to infer information about habitat. Of course some initial environmental information was used to position stands, and hence species curves, in the first place. In this sense the procedure is circular. But new information is also being used: namely the precise cover values for species in the stands. In this sense the relocation of stands based on vegetative information is an *iteration:* repetition of the cycle of computations, resulting in approximating the desired result ever more closely. In our case the desired result is to locate all stands most precisely in relation to the habitat gradient controlling species distributions.

We started with the assumption that groundwater depth was controlling the distribution of tree species, and surface soil moisture tension the distribution of herb species. However, by using foliar cover values to relocate stand positions (and shifting modal peaks accordingly; see Figure 10.9d), we are now introducing the influence of any other habitat features that may also have acted to determine foliar cover: soil salt content, wind action, past grazing, and so on. The result of iterative direct gradient analysis is to take us a step away from direct use of environmental information in locating stands and associated species distribution curves and a step toward the indirect ranking of stands by means of the cumulative information about habitat inherent in vegetative cover values. By assigning new weights to the modal peaks of species that have been plotted on the axis of weighted stand scores, we can repeat the calculation of weighted stand scores, find new modal peak scores, and continue our iterations. Eventually, stand scores will stabilize, and we will have reached a steady-state solution. This repeated iterative process of stand ranking is called "reciprocal averaging" ordination (Hill 1973b). It is a form of indirect gradient analysis in which vegetation information is used to rank stand positions along an axis whose correlation with environmental factors is now unknown. This technique of ordination works very well with species which have bell-shaped distribution curves, as has been shown with simulated data (Gauch et al. 1977).

We return to our problem of predicting changes in vegetative composition with damming. If, on the basis of hydrological studies, some prediction can be made of the extent of lowering of the groundwater table following damming, we can predict the new composition of trees at any given site by moving along the gradient in Figure 10.9b from the current groundwater depth to the predicted groundwater level (e.g., from 2 m to 4 m). Of course the vegetation will take some additional time to adjust to the new equilibrium, and the pace of change cannot be predicted from these graphs (see Chapter 12). Because precipitation and evaporation are not expected to change, the herbaceous cover can be expected to remain relatively unchanged, except to the extent that it is influenced by its changing overstory. Here it may be best to refer to Figure 10.9d as well as Figure 10.9c, since 10.9d reflects the actual floristic composition and abundance over a larger range of sites in which cover of dominants is changing. The actual change in herbaceous composition may be intermediate between its present composition and that in Figure 10.9*d*, since physical habitat is not changing, but biological coassociates are.

If historic groundwater levels have already been lowered by an unknown amount, a knowledge of past vegetative composition can be used to estimate water drawdown. Direct gradient analysis or the weighted averages techniques can be used to compare present to past vegetation. The shift in vegetation along the groundwater depth gradient can be used to estimate the extent of groundwater lowering during the time interval. Other techniques of gradient analysis are described by Whittaker (1967, 1975) and Gauch (1982).

An interesting application of direct gradient analysis techniques to the modeling of rate of fire spread in wilderness areas has been developed by Kessell (1979). Kessell uses direct gradient analysis to establish plots relating plant, animal, and fuel load quantities to four environmental gradients believed to be of major influence: elevation, date since last fire, moisture status, and position in relation to drainage. The wilderness area is characterized from aerial photographs. With a knowledge of weather conditions, rate of spread of fire through a vegetation type of particular live and dead fuel load and moisture status can be modeled (see Albini's model, Chapter 6). The gradient analyses are used to estimate fuel load at each site in the wilderness in considerable detail. Kessell (1979) has applied this approach to Glacier National Park, Montana, to chaparral in southern California (Kessell and Cattelino 1978), and more recently to Australian bushland (Kessell 1981; Kessell et al. 1984).

Indirect Gradient Analysis: Ordination

We saw in the previous section that one way to rank stands by biotic composition is to use position of species' modal peaks to weight stand positions along an axis and to recalculate species modal peak weights from new stand positions. Although in our example we started with a reasonable guess of stand positions based on measurement of a single environmental factor suspected to influence foliar cover strongly, in theory any initial ranking of stand scores will ultimately result in the same stable rank order of stands by the iteration procedure. Once the initial ordination axis is isolated, second- and higher-order perpendicular axes of biotic variation can be derived by an analogous procedure, while extracting variation due to previous axis directions.

Reciprocal averaging ordinates stands and species at the same time. Computer programs for reciprocal averaging, as well as weighted averages ordination, principal components analysis and polar ordination (Bray and Curtis 1957, Gauch 1982) are available as the ORDIFLEX package (Gauch 1977). An improved version of reciprocal averaging, in which higher-order axes are made more fully independent of previous axes and samples are spaced to achieve an even rate of turnover of species along the axes, has been developed. It is termed *detrended correspondence analysis* (DECORANA; Hill and Gauch 1980) and is also available as a packaged program (Hill 1979a). Readers interested in a more theoretical discussion of ordination methods can consult Whittaker (1978a), Gauch (1982), and the articles cited earlier.

Rebecca Sharitz and colleagues at the Savannah River Ecology Laboratory have successfully used DECORANA to elucidate the effect of thermal effluent from a nuclear power plant on southeastern U.S. swamp forests in the river downstream of discharge (see earlier accounts in Sharitz et al. 1974 and Sharitz and Gibbons 1979). Heated effluents were released into a stream (Four Mile Creek) in the Savannah River of South Carolina since the early 1950s. The vegetation is primarily a *Taxodium distichum* (bald cypress)–*Nyssa aquatica* (tupelo) swamp forest, interspersed with hardwoods on ridges in the bottomlands. Sharitz and colleagues were interested in predicting effects of thermal effluent on forest in an adjacent tributary (Steel Creek) which was expected to receive thermal effluent in the future. The effects of the thermal releases on the swamp forest vegetation are twofold: first, there is a direct impact from the high water temperature; second, dead and dying vegetation upstream releases increasing amounts of sediment, which buries and kills vegetation downstream.

Vegetation and habitat in the impacted region (the Four Mile delta) was first grossly stratified into six types using color infrared aerial photographs. Fifty-eight 100 m² plots were located at random within five of the six regions, (excluding the undisturbed cypress-tupelo swamp forest), and the number and basal area of trees and shrubs were recorded; biomass of herbs were obtained by harvesting 0.25 m² subsamples. Importance values based on the relative density and basal area of trees were used to quantify the tree and shrub component of each plot. The plots were then ordinated by DECORANA. Stand locations in relation to the first two axes are shown in Figure 10.10, based on tree and shrub data; a separate ordination of herbaceous data gave comparable results (Sharitz 1982). Four major vegetation types were recognized and segregated by the ordination, thus refining but supporting the initial photo-based classification. The four vegetation types were as follows:

1. An early successional "herbaceous marsh" type, dominated by willows (*Salix* spp.). Two subtypes were recognized: marshes exposed to the greatest heat stress (thermal) in the thermal delta and those exposed to lower temperatures ("post-thermal") occurring in the post-thermal recovery area (Figure 10.11). In the latter area sediment had accumulated, raising the sites above the predominant level of heated water.

2. The shrub-scrub zone, a permanently flooded region beyond the thermal delta, dominated by buttonbush (*Cephelanthus occidentalis*).

3. A "cypress-tupelo forest" zone of reduced (open) canopy due to thermal effluent and siltation.

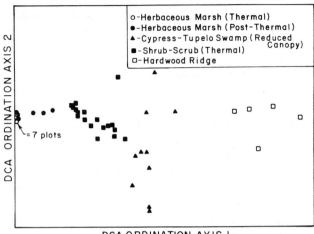

Figure 10.10. Detrended correspondence analysis ordination of importance values of woody species from 1982–83 quadrat sampling of Four Mile delta and swamp. Figure courtesy of R. R. Sharitz, Savannah River Ecology Laboratory.

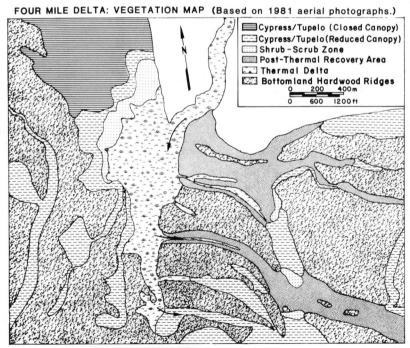

Figure 10.11. Vegetation map of Four Mile delta and swamp. Reprinted from Figure II.C-4 in Sharitz (1982), with permission.

4. A "bottomland hardwood ridge" zone of oaks (*Quercus nigra, Q. laurifolia*) and sweetgum (*Liquidambar styraciflua*) on elevated areas in the channel which were being reduced in size at the edges by sedimentation and thermal effluents.

This ordination allowed the investigators to confirm their initial vegetation-habitat classification, while observing the quantitative similarities and differences in vegetation between sample sites. The investigators proceeded to use the vegetation types to map vegetation in the Four Mile Creek delta area (Figure 10.11). By comparison of such vegetation maps made by aerial photos from 1966 and 1981, the successional relations between the community types became clear (Figure 10.12). The investigators then repeated the field sampling at nonpolluted Steel Creek delta, ordinated the samples, and prepared a vegetation map of it. By comparison of the Four Mile delta (polluted) and Steel Creek delta (unpolluted) vegetation maps, some broad predictions could then be made about the likely direction and extent of vegetation change in Steel Creek following introduction of thermal effluent. An additional way to quantify differences between sites on the two creeks would be to compute indexes of percentage similarity between comparable vegetation types on the two creeks, to see what changes in species composition and abundance were occurring within a given vegetation type under the influence of effluent.

Classification

The ordination in the previous example had the value that the investigator could portray visually the position of each stand in relation to every other, in order to compare similarities and differences. One of the uses to which the ordination was put was to classify the sites by visual clustering and to use these cluster types to map vegetation in the landscape. If classification is the sole aim of the data analysis, the investigator may choose to use a numerical classification technique directly, rather than to ordinate first. Note, however, that the boundaries of classes involve an arbitrary determination of what degree of difference between sites constitutes a significant difference, such that they belong in different clusters. In numerical classification techniques this decision rule is built into the classificatory program, so that degrees of clustering as portrayed by ordination can no longer be visualized in the outcome. For these reasons an ordination may be seen as a useful first step in classification, since it portrays the degree of diffuseness of clusters.

Many numerical classification techniques have been developed for use with biological data (see, e.g., Clifford and Stephenson 1975, Greig-Smith 1983, Pielou 1977, Whittaker 1978a). Of these only a few have been tested for robustness with simulated ecological data containing bell-shaped species distributions.

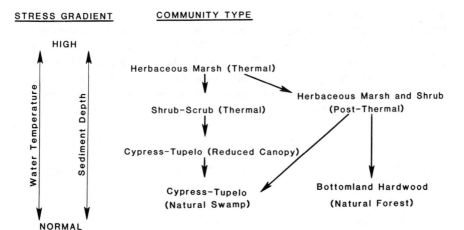

Figure 10.12. Generalized diagram of the successional (downward) or retrogressional (upward) relationships between plant communities subjected to thermal pollution stress (and associated erosional effects), based on studies of effects of nuclear reactor effluent in the Savannah River Swamp System. Figure courtesy of R. R. Sharitz, Savannah River Ecology Laboratory. Modified from Figure II.C-5 in Sharitz (1982).

The two studies that have tested classificatory methods with simulated data are those of Robertson (1979) and Gauch and Whittaker (1981). The classificatory methods that proved most robust with ecological data were MDISP and MINFO (Goldstein and Grigal 1972, Orloci 1978) and TWIN-SPAN (Hill 1979b). TWINSPAN has the advantage that it classifies species and sites simultaneously. It begins by ordinating sites by reciprocal averag-

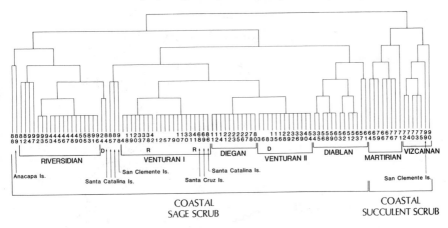

Figure 10.13. TWINSPAN classification of sites of coastal shrubland vegetation in California and Mexico, based on the samples shown in Figure 10.14. Site numbers corresponding to samples in Figure 10.14 are shown. Outlier sites are shown by letter beneath the site number: D = Diegan; R = Riversidian. Reprinted from Westman (1983), with permission of Junk, The Hague.

ing and then divides the set of sites on the basis of presence or absence of suites of species that occur more on one side or the other of the initial ordination axis. An example of a TWINSPAN classification of 99 xeric shrubland sites in Alta and Baja California is shown in Figure 10.13. This classification identified major floristic associations corresponding to geographic regions of gradually increasing evapotranspirative stress from northwest to southeast, shown in Figure 10.14. The classification also revealed

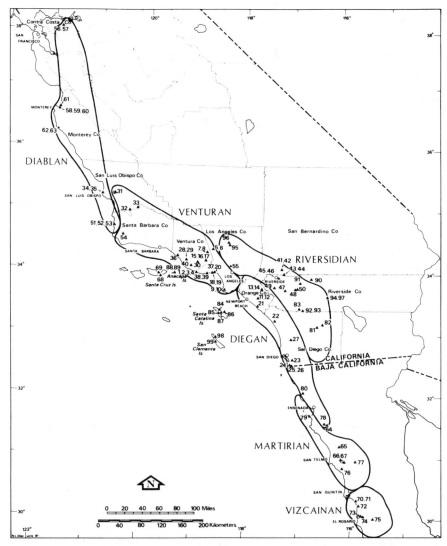

Figure 10.14. Location of major vegetation types, based on the TWINSPAN classification shown in Figure 10.13. Reprinted from Westman (1983), with permission of Junk, The Hague.

two distinct subassociations (Venturan I and II) within one of the floristic associations which had not emerged clearly in the initial examination of DECORANA ordination of the same data. Thus frequently ordination and classification of the same data can yield insights that neither approach alone will provide.

In this chapter we have touched extremely briefly on some major issues and approaches in ecological data sampling and analysis. Readers who wish a thorough grounding in these techniques and their theoretical underpinnings should refer to the citations provided in the text. In the last 10 years major strides have been made in the development of multivariate techniques for analysis of ecological data; these are now available in packaged programs for application to a wide range of ecological impact studies (see, e.g., Cornell Ecology Program series: Gauch 1976; Australian CSIRO statistical packages: Gillison and Anderson 1981; and CSIRO TAXON library: User's Manual, c/o G. N. Lance, Avon Universities Computer Centre, University of Bristol).

REFERENCES

Austin, M. P. (1976). On non-linear species response models in ordination. *Vegetatio* **33**:33–41.

Austin, M. P., Cunningham, R. B., and Good, R. B. (1983). Altitudinal distribution of several eucalypt species in relation to other environmental factors in southern New South Wales. *Aust. J. Ecol.* **8**:169–180.

Austin, M. P., Cunningham, R. B., and Fleming, P. M. (1984). New approaches to direct gradient analysis using environmental scalars and statistical curve-fitting procedures. *Vegetatio* **55**:11–27.

Bailey, N. T. J. (1951). On estimating the size of mobile populations from recapture data. *Biometrika* **38**:293–306.

Boswell, M. T., and Patil, G. P. (1970). Chance mechanisms generating negative binomial distributions. In G.P. Patil, ed. *Random Counts in Scientific Work*. Vol 1. Pennsylvania State Univ. Press, University Park.

Bray, J. R., and Curtis, J. T. (1957). An ordination of the upland forest communities of southern Wisconsin. *Ecol. Monogr.* **27**:325–349.

Carpenter, S. R., and Chaney, J. E. (1983). Scale of spatial pattern: four methods compared. *Vegetatio* **53**:153–160.

Clifford, H. T., and Stephenson, W. (1975). *An Introduction to Numerical Classification*. Academic Press, New York.

Cottam, G. (1947). A point method for making rapid surveys of woodlands. *Bull. Ecol. Soc. Amer.* **28**:60.

Cottam, G., and Curtis, J. T. (1956). The use of distance measures in phytosociological sampling. *Ecology* **37**:451–460.

Cresswell, W. L., and Froggatt, P. (1963). *The Causation of Bus Driver Accidents: an Epidemiological Study*. Oxford Univ. Press, New York.

Curtis, J. T., and McIntosh, R. P. (1951). An upland forest continuum in the prairie-forest border region of Wisconsin. *Ecology* **32**:476–496.

Czekanowski, J. (1909). Zur differential Diagnose der Neandertalgruppe. *Korrespbl. dt. Ges. Anthrop.* **40**:44–47.

Czekanowski, J. (1913). *Zarys Metod Statystycznyck.* Warsaw.

Daubenmire, R. (1966). Vegetation: identification of typal communities. *Science* **151**:291–298.

Errington, J. C. (1973). The effect of regular and random distributions on the analysis of pattern. *J. Ecol.* **61**:99–105.

Gauch, H. G., Jr. (1976). Catalog of the Cornell Ecology Programs Series. Ecology and Systematics, Cornell Univ., Ithaca, N.Y.

Gauch, H. G., Jr. (1977). *ORDIFLEX—A Flexible Computer Program for Four Ordination Techniques: Weighted Averages, Polar Ordination, Principal Components Analysis, and Reciprocal Averaging. Release B.* Cornell Univ., Ithaca, N.Y.

Gauch, H. G., Jr. (1982). *Multivariate Analysis in Community Ecology.* Cambridge Univ. Press, Cambridge.

Gauch, H. G., Jr., and Chase, G. B. (1974). Fitting the Gaussian curve to ecological data. *Ecology* **55**:1377–1381.

Gauch, H. G., Jr., and Whittaker, R. H. (1981). Hierarchical classification of community data. *J. Ecol.* **69**:135–152.

Gauch, H. G., Jr., Whittaker, R. H., and Wentworth, T. R. (1977). A comparative study of reciprocal averaging and other ordination techniques. *J. Ecol.* **65**:157–174.

Gillison, A. N., and Anderson, D. J., eds. (1981). *Vegetation Classification in Australia.* Austr. Natl. Univ. Press, Canberra.

Goff, F. G., Dawson, G. A., and Rochow, J. J. (1982). Site examination for threatened and endangered plant species. *Environ. Manage.* **6**:307–316.

Goldstein, R. A., and Grigal, D. F. (1972). Computer programs for the ordination and classification of ecosystems. Rep. ORNL-IBP-71-10. Oak Ridge Natl. Lab., Oak Ridge, Tenn.

Goodall, D. W. (1974). A new method for the analysis of spatial pattern by random pairing of quadrats. *Vegetatio* **29**:135–146.

Goodall, D. W. (1978). Sample similarity and species correlation. In R. H. Whittaker, ed. *Ordination of Plant Communities.* 2nd ed. Junk: The Hague, pp. 99–149.

Green, R. H. (1979). *Sampling Design and Statistical Methods for Environmental Biologists.* Wiley-Interscience, New York.

Greig-Smith, P. (1952). The use of random and contiguous quadrats in the study of the structure of plant communities. *Ann. Bot. Lond. NS* **16**:293–316.

Greig-Smith, P. (1961). Data on pattern within plant communities. I: The analysis of pattern. *J. Ecol.* **49**:695–702.

Greig-Smith, P. (1964). *Quantitative Plant Ecology.* 2nd ed. Butterworths, London.

Greig-Smith, P. (1983). *Quantitative Plant Ecology.* 3rd edn. University of California Press, Berkeley, Calif.

Hill, M. O. (1973a). The intensity of spatial pattern in plant communities. *J. Ecol.* **61**:225–235.

Hill, M. O. (1973b). Reciprocal averaging: an eigenvector method of ordination. *J. Ecol.* **61**:237–249.

Hill, M. O. (1979a). *DECORANA—A FORTRAN Program for Detrended Correspondence Analysis and Reciprocal Averaging.* Cornell Univ., Ithaca, N.Y.

Hill, M. O. (1979b). *TWINSPAN—A FORTRAN Program for Arranging Multivariate Data in an Ordered Two-Way Table by Classification of the Individuals and Attributes.* Cornell Univ., Ithaca, N.Y.

Hill, M. O., and Gauch, H. G., Jr. (1980). Detrended correspondence analysis, an improved ordination technique. *Vegetatio* **42**:47–58.

Jaccard, P. (1912). The distribution of the flora in the alpine zone. *New Phytol.* **11**:37–50.

Kershaw, K. A. (1973). *Quantitative and Dynamic Plant Ecology.* 2nd ed. Elsevier, New York.

Kessell, S. R. (1979). *Gradient Modeling.* Springer-Verlag, New York.

Kessell, S. R. (1981). Application of gradient analysis concepts to resource management modeling. *Proc. Ecol. Soc. Aust.* **11**:163–173.

Kessell, S. R., and Cattelino, P. J. (1978). Evaluation of a fire behavior information integration system for southern California chaparral wildlands. *Environ. Manage.* **2**:135–159.

Kessell, S. R., Good, R. B., and Hopkins, A. J. M. (1984). Implementation of two new resource management information systems in Australia. *Environ. Manage.* **8**:251–270.

Legendre, L., and Legendre, P. (1983). *Developments in Environmental Modeling. Vol 3: Numerical Ecology.* Elsevier, Amsterdam.

Lewin, D. C., and Westman, W. E. (1973). Pattern diversity in a *Eucalyptus* forest. *Aust. J. Bot.* **21**:247–251.

Ludwig, J. A. and Goodall, D. W. (1978). A comparison of paired- with blocked-quadrat variance methods for the analysis of spatial pattern. *Vegetatio* **38**:49–59.

McIntyre, G. A. (1953). Estimation of plant density using line transects. *J. Ecol.* **41**:319–330.

Nelder, J. A. and Wedderburn, R. W. M. (1972). Generalized linear models. *J. Roy. Statist. Soc. A* **135**:370–384.

Neyman, J. (1939). On a new class of "contagious" distributions, applicable in entomology and bacteriology. *Ann. Math. Statist.* **10**:35–57.

Orloci, L. (1967). An agglomerative method for classification of plant communities. *J. Ecol.* **55**:193–205.

Orloci, L. (1978). *Multivariate Analysis in Vegetation Research.* 2nd ed. Junk, The Hague.

Pielou, E. C. (1966). Species-diversity and pattern-diversity in the study of ecological succession. *J. Theor. Biol.* **10**:370–383.

Pielou, E. C. (1977). *Mathematical Ecology.* 2nd ed. Wiley-Interscience, New York.

Poole, R. W. (1974). *An Introduction to Quantitative Ecology*. McGraw-Hill, New York.

Robertson, P. A. (1979). Comparisons among three hierarchical classification techniques using simulated coenoplanes. *Vegetatio* **40**:175–183.

Sharitz, R. R. (1982). Plant community structure and processes. In M. H. Smith, R. R. Sharitz, and J. B. Gladden, eds. *An Evaluation of the Steel Creek Ecosystem in Relation to the Proposed Start of L-Reactor*. SREL-12,UC-66e. NTIS, Springfield, Va., pp. II-1–58.

Sharitz, R. R., and Gibbons, J. W. (1979). Impacts of thermal effluents from nuclear reactors on southeastern ecosystems. In R. A. Fazzolare and C. B. Smith, eds. *Changing Energy Use Futures*. Pergamon, New York, pp. 609–616.

Sharitz, R. R., Irwin, J. E., and Christy, E. J. (1974). Vegetation of swamps receiving reactor effluents. *Oikos* **25**:7–13.

Snedecor, G. W. (1946). *Statistical Methods Applied to Experiments in Agriculture and Biology*. 4th ed. Iowa State College Press, Ames.

Southwood, T. R. E. (1978). *Ecological Methods, with Particular Reference to the Study of Insect Populations*. Wiley, New York.

Strahler, A. H. (1978). Binary discriminant analysis: a new method for investigating species-environment relationships. *Ecology* **59**:108–116.

Taylor, L. R. (1961). Aggregation, variance and the mean. *Nature* **189**:732–735.

Usher, M. B. (1975). Analysis of pattern in real and artificial plant populations. *J. Ecol.* **63**:569–586.

Westman, W. E. (1970). Mathematical models of contagion and their relation to density and basal area sampling techniques. In G. P. Patil, E. C. Pielou, and W. E. Waters, eds. *Statistical Ecology Vol 1. Spatial Patterns and Statistical Distributions*. Pennsylvania State Univ. Press, University Park, pp. 515–536.

Westman, W. E. (1975). Edaphic climax pattern of the pygmy forest region of California. *Ecol. Monogr.* **45**:109–135.

Westman, W. E. (1980). Gaussian analysis: identifying environmental factors influencing bell-shaped species distributions. *Ecology* **61**:733–739.

Westman, W. E. (1981). Factors influencing the distribution of species of California coastal sage scrub. *Ecology* **62**:439–455.

Westman, W. E. (1983). Xeric Mediterranean-type shrubland associations of Alta and Baja California and the community/continuum debate. *Vegetatio* **52**:3–19.

Westman, W. E. and Anderson, D. J. (1970). Pattern analysis of sclerophyll trees aggregated to different degrees. *Aust. J. Bot.* **18**:237–249.

Whittaker, R. H. (1956). Vegetation of the Great Smoky Mountains. *Ecol. Monogr.* **26**:1–80.

Whittaker, R. H. (1960). Vegetation of the Siskiyou Mountains, Oregon and California. *Ecol. Monogr.* **30**:279–338.

Whittaker, R. H. (1967). Gradient analysis of vegetation. *Biol. Rev.* **42**:207–264.

Whittaker, R. H. (1975). *Communities and Ecosystems*. 2nd ed. Macmillan, New York.

Whittaker, R. H., ed. (1978a). *Ordination of Plant Communities*. 2nd ed. Junk, The Hague.

Whittaker, R. H., ed. (1978b). *Classification of Plant Communities*. 2nd ed. Junk, The Hague.

Whittaker, R. H. and Gauch, H. G., Jr. (1978). Evaluation of ordination techniques. In R. H. Whittaker, ed. *Ordination of Plant Communities*. 2nd ed. Junk, The Hague, pp. 227–336.

Whittaker, R. H., Niering, W. A., and Crisp, M. D. (1979). Structure, pattern, and diversity of a mallee community in New South Wales. *Vegetatio* **39**:65–76.

Williams, W. T., Lance, G. N., Webb, L. J., and Tracey, J. G. (1973). Studies in the numerical analysis of complex rain-forest communities. VI: Models for the classification of quantitative data. *J. Ecol.* **61**:47–70.

11

SPECIES AND
LANDSCAPE DIVERSITY

Consider two hillsides clad with fields of red poppies and white daisies. On hillside A, 99 poppies occur for every daisy; on hillside B, the proportions of the two species are equal (Table 11.1). Which hillside has the greater species diversity? Hillside A is less "diverse" in the sense that it is mostly poppies; yet the two hillsides are equally diverse in the sense that they both support two species. Thus there are two distinct properties subsumed within the term "diversity": the number of species at a site ("species richness" or "alpha diversity") and the relative abundance of species at a site ("equitability" or "evenness" of species, as measured by some importance value).

SPECIES RICHNESS AND SPECIES-AREA CURVES

The number of species encountered normally increases with the area surveyed. Thus in measuring species richness, it is important to standardize the species count by the extent of area sampled. One approach is to use a constant sample area of fixed size and shape. A 0.1 ha plot, 20 m × 50 m in dimensions, has been used widely for woody vegetation (see, e.g., Bond 1983, Peet 1978, Whittaker 1977). Additional 0.1 ha samples may therefore be compared in richness to a well-developed global data base. Smaller-sized plots, nested within the 0.1 ha (Figure 11.1), may be surveyed for species, so that the rate of increase in species with area can be plotted. The latter technique has the disadvantage from the point of view of sample error, that the number of quadrats of each area is unequal, but it is otherwise a rapid and efficient technique for obtaining information on the relation of richness to area.

The rate of increase in species richness with area varies with the size of area considered and with the patchiness of the habitat. Thus a variety of curve shapes may occur in vegetational samples. The curves of Figure 11.2a typically result as one plots species from 1 m² to 1000 m² in a "homogeneous" vegetation sample, using the quadrat method of Figure 11.1. The slightly curved, parabolic form often flattens to a straight line when area is

Table 11.1.   Richness and Equitability

Species	Relative Abundance of Species on a Hillside[a]	
	Hillside A	Hillside B
Poppy	99%	50%
Daisy	1%	50%

[a] Hillsides A and B are equal in species richness, but hillside B has a higher equitability

transformed logarithmically, so that the species-area curve fits the semilog, exponential equation:

$$S = \log b + d \log A \qquad (11.1)$$

A linear regression fitted to 1 m²–1000 m² data thus transformed results in two pieces of information. The y-intercept "log b" is a regression-derived number of species in one square meter; "d" is the slope or rate of increase of species richness with area. These two coefficients will vary independently in different vegetation types (e.g., Whittaker et al. 1979) and provide an alternative means to characterize species richness on a site. If the sample plot expands abruptly into a landscape patch of different habitat (change in substate, aspect, etc.), one can expect the arithmetic species-area curve to take a sudden rise before flattening out again (dotted line, Figure 11.2i).

A power-function curve (Figure 11.3) which fits an equation of the form

$$S = cA^z \quad \text{or} \quad \log S = \log c + z \log A \qquad (11.2)$$

is often encountered in sampling larger areas, from one hectare to islands or continents of many thousands or millions of hectares (MacArthur and

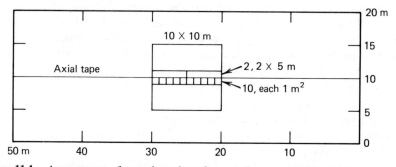

Figure 11.1. Arrangement of nested quadrats for sampling species richness in quadrats of increasing size (after Whittaker et al. 1979).

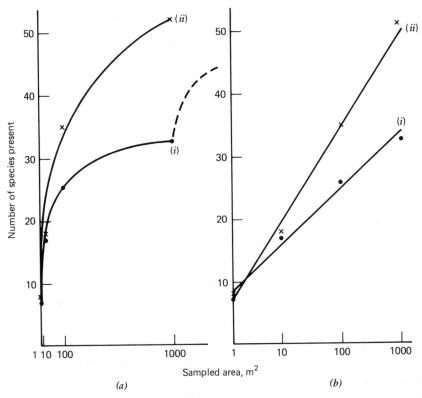

Figure 11.2   Species-area curves over the area range $1–10^3$ m², plotted by (*a*) arithmetic scale, (*b*) semilogarithmic scale. (*i*) Venturan I coastal sage scrub, Los Angeles County, four years after fire (A. Troeger, unpublished data). Dotted line shows hypothetical effect of expanding sample area into a different habitat. (*ii*) Mallee scrub, NSW, Australia, 13 years after fire (Whittaker et al. 1979).

Wilson 1963, 1967, May 1975, Whittaker 1972, Whittaker et al. 1979). The coefficient *c* typically will vary with taxon and biogeographic region. MacArthur and Wilson (1967) suggested that the exponent *z* typically falls in the range 0.20–0.35 for islands in an archipelago and 0.12–0.17 for samples within a larger land mass (e.g., sample on a continent). Considerably greater variation in these exponents has since been observed (Connor and McCoy 1979, Gilbert 1980a). Clearly there can be much scatter in such species-area curves (see, e.g., Figure 11.3, curve *ii*), due especially to habitat heterogeneities when sampling such large areas. Furthermore species-area curves do not necessarily fit either of these models best, and other models, including a straight line for untransformed data, may fit best in particular cases (Connor and McCoy 1979).

Because the shape of species-area curves differs in different community types and habitats, it is not possible to predict the total number of species for

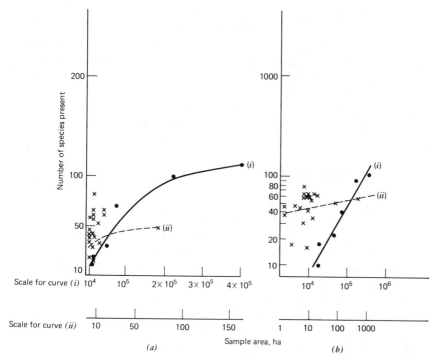

Figure 11.3. Species-area curves for islands (·) and mainland samples (x). (*i*) Land and freshwater bird species from some islands in the Sunda group of the South Pacific. Data from MacArthur and Wilson (1963). (*ii*) Total vascular plant species in natural forest woodlots in Kandiyohi County, Alexandria Moraine area of western Minnesota. Data from Scanlan (1981). (*a*) Arithmetic scales; (*b*) logarithmic scales. The large scatter of curve (*ii*) indicates that these mainland patches conform only poorly to the power function model.

an areal size different from that sampled, unless the shape of the underlying species-area curve is known. Even when a species-area curve has been determined for the $1-10^3$ m^3 area range, the extrapolation of results to larger areas by a semilog regression curve is unreliable, since the species-area curves might change from semilog to log-log form at the larger areal scales.

For reviews of measurement of species richness, see Whittaker (1972), Peet (1974), and Pielou (1975). We will return to a discussion of species-area curves in the section "Preservation of Diversity: Design of Nature Reserves."

HETEROGENEITY AND EQUITABILITY

The relative abundance of species at a site reflects the way in which species have partitioned the available habitat resources. Lloyd and Ghelardi (1964) suggest the term "evenness" to refer to the absolute distribution of relative abundances of species at a site, in which maximum "evenness" refers to a uniform distribution of relative abundances. Sometimes ecologists wish to compare the distribution of abundances relative to some other distribution such as a lognormal or geometric one. Lloyd and Ghelardi (1964) suggest that the relativized index in such an instance be termed "equitability." Although "evenness" is thus one type of equitability, it is common for ecologists to use the term "equitability" when referring to "evenness," and we shall use the term equitability in this latter sense.

Dominance-Diversity Curves

Richness and equitability vary somewhat independently of one another. A useful way to illustrate graphically both components of diversity is by means of the dominance-diversity curve (Whittaker 1965). By convention, the importance values of species ($\log_{10}$ transformed) are plotted on the ordinate in decreasing order of importance, along an abscissa in which species are equally spaced. In Figure 11.4a two of many possible shapes of dominance-diversity curve are shown. The straight line of curve i of Figure 11.4a indicates a relatively large concentration of resources in a few very abundant species, with comparable numbers of moderately abundant and rare species. Such a distribution of relative abundances can be described by a geometric distribution:

$$p_i = p_1 c^{i-1} \tag{11.3}$$

where

p_i = importance value of species i/Σ importance values of all species in sample
p_1 = relative importance of the first species
c = a constant between 0 and 1
i = position of the species in the sequence from first to last

The curve ii of Figure 11.4a shows a larger number of moderately abundant species, relative to very abundant or rare ones. The curve can be described by the lognormal distribution

$$S_r = S_0 e^{-(aR)^2} \tag{11.4}$$

where S_r = number of species in an octave R octaves distant from a modal octave containing S_0 species, and a is a constant (Whittaker 1972). An "octave" here refers to numbers in the series 2^n, $n = 1, 2, 3, \ldots$. Put more graphically, the curve of number of species in each of a series of abundance classes (increasing by $\log_2$) takes on the bell shape of a normal distribution,

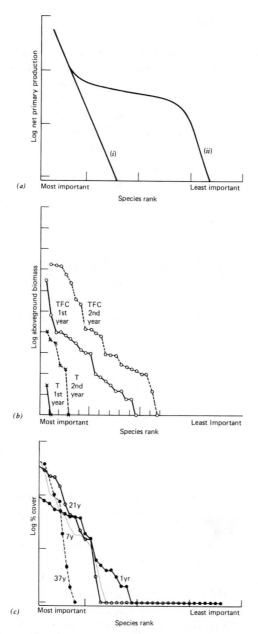

Figure 11.4. Dominance-diversity curves. (*a*) The shape of dominance-diversity curves that conform to (*i*) the geometric distribution; (*ii*) the lognormal distribution. (*b*) Changes in aboveground biomass one and two years after strip-mining of coastal sand dunes on North Stradbroke Island, Queensland. Rehabilitation treatment T had only topsoil replaced; treatment TFC involved adding topsoil, fertilizer, and a seeded cover of nonnative grass (*Melinis minutiflora*—molasses grass). From Figure 1 of Thatcher and Westman (1975). Reprinted with permission of the Ecological Society of Australia, Inc. (*c*) Changes in % foliar cover in the first growing season, and 7, 21, and 37 years after fire in four matched sites of coastal sage scrub in coastal Los Angeles. From "Diversity relations and succession in Californian costal sage scrub" by W. E. Westman, *Ecology*, 1981, **62**:170–184. Copyright (c) 1981 by the Ecological Society of America. Reprinted by permission.

so that most species are moderately abundant (peak of curve), and the number of more or less abundant species decreases from this mode.

The slope of the dominance-diversity curve expresses information on the rate of change in relative abundance across the species sequence, whereas the x-intercept records total species richness. Whittaker (1972) has suggested that some measure of the average slope of such curves could be used as a measure of equitability. One such measure is

$$E_c = \frac{S}{(\log_{10} n_1 - \log_{10} n_s)} \tag{11.5}$$

where

S = total species richness in the sample
n_1 = importance value (IV) of the species with highest IV
n_s = IV of the species with lowest IV

Since this is the inverse of the slope of the line connecting endpoint values ($\Delta X/\Delta Y$), it is not a very detailed reflection of curvilinear dominance-diversity curves.

Another measure is

$$E'_c = \frac{S}{4\sqrt{\Sigma \log_2 n_1 - \log_2 \bar{n})^2/S}} \tag{11.6}$$

where $\bar{n}$ = geometric mean of importance values (i.e., arithmetic average of $\log_2$ values). The term E'_c has the advantage that it incorporates the mean of all importance values and therefore is more reflective of changes in slope in the curve. Neither index is a wholly satisfactory reflection of the dominance-density curve, since (1) neither describe variations in slope along the curve, and (2) both are affected by S, the total richness of the sample. Indexes such as E_c and E'_c, which confound information on richness and equitability in a single number, are termed *heterogeneity indexes* (Peet 1974).

The measurement of equitability alone requires a knowledge of the total number of species in the community ("sample universe"). Such a number is virtually impossible to ascertain, since species-area curves never flatten out completely to a finite total number of species unless very arbitrary definitions of what constitutes the "community" are used. Using finite approximations to the total species number S results in indexes that vary widely with sample size (Peet 1974). As a result there are no truly practical ways to express equitability alone by use of an index; for this reason dominance-diversity curves, in which equitability can be shown visually and the curves fitted to a mathematical model, have particular value.

Whittaker (1965) observed that plant communities in harsh physical environments tended to exhibit geometric dominance-diversity curves (e.g., arctic tundra), whereas plant communities in more benign physical environments (e.g., tropical rain forest; temperate deciduous cove forest) or late successional stages tended to be curvilinear. Animal communities such as territorial temperate forest birds were also observed to exhibit curvilinear dominance-diversity curves (MacArthur 1957, Whittaker 1965). These generalizations must be viewed with caution, since they are based on only a few examples and are subject to exception.

The analysis of change in dominance-diversity curves of a particular community over time may nevertheless be of value in interpreting changes in diversity due to impact. In Figure 11.4b dominance-diversity curves have been plotted for 50 m $\times$ 50 m plots which had been rehabilitated following strip-mining of coastal dunes, as described in Chapter 8 (Figure 8.12). The treatment (TFC) in which topsoil, fertilizer, and a cover of nonnative grass was established showed much more rapid establishment and increase in species richness in the first two years following mining rehabilitation than did the plot in which only topsoil (T) was replaced (Thatcher and Westman 1975). The greater equitability of the topsoil-fertilizer-exotic cover treatment suggests that such treatment is more favorable to a wide range of species. Notice that the shape of the dominance-diversity curves remained relatively stable within treatments while richness increased.

In Figure 11.4c dominance-diversity curves are shown for a series of physically similar sites of different ages since last fire in coastal sage scrub. In this particular vegetation type richness and equitability are greatest in the first growing season following fire, due to the growth of many herbaceous animals and short-lived perennials. As shrub-cover increases, these herbs disappear, and by 37 years the sites are dominated by a few major shrub species (Westman 1981). In this case, then, succession proceeds from an equitable, species-rich site to a low equitability, species-poor site of almost geometric distribution, contrary to events in temperate deciduous forests (Whittaker 1975). Dominance-diversity curves can be used to assess successional status, once the expected sequence of changes for a given community type has been established. One technical problem often encountered is that the relative importance of rare species is difficult to measure precisely. In Figure 11.4b species of total biomass <4g ha^{-1} were excluded from the curves; in Figure 11.4c species of $<0.1\%$ foliar cover were plotted at the 0.1% level.

Species-Abundance Curves

A second graphical approach to examining effects of environmental impact on species richness and equitability is to plot the number of species found in each of several abundance classes (log scale). In the studies illustrated in Figure 11.5, the effect of chronic pollution stress was twofold. First, the stress differentially decreased the number of rare or moderately abundant

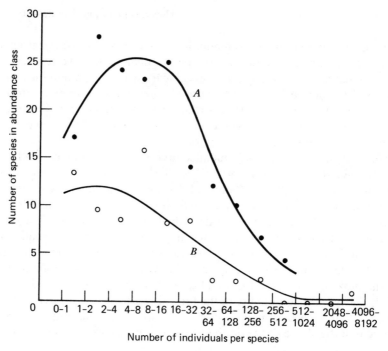

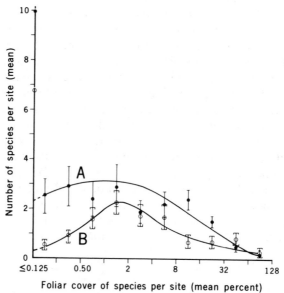

Figure 11.5. Species-abundance curves in unpolluted (*A*) and polluted (*B*) environments. (*i*) Diatoms colonizing a glass slide in a clean and a polluted stream in Pennsylvania. Redrawn from Figures 3 and 4 in R. Patrick, M. H. Hohn, and J. H. Wallace, 1954, A new method for determining the pattern of the diatom flora. *Notulae Naturae* **259**:1–12, with permission of the Philadelphia Academy of Natural Sciences. (*ii*) Cover of species of coastal sage scrub in 11 clean-air sites and 11 floristically comparable polluted-air sites. Reprinted, with permission, from W. E. Westman. 1979. Oxidant effects on California coastal sage scrub. *Science* **205**:1001–1003. Copyright (c) 1979 by the American Association for the Advancement of Science.

species. This had the effect of decreasing equitability. Second, pollution stress decreased the number of species in virtually all abundance classes, resulting in a drop in species richness. The mode of the lognormal curve shifted in one case (Figure 11.5 (i), but not the other.

Heterogeneity Indexes

Although graphical approaches allow richness and equitability to be illustrated separately, a numerical index that describes equitability is still desirable in order to examine statistically the changes in this feature with environmental conditions. Since no indexes of equitability alone are available, indexes of heterogeneity are often used for this purpose. This is unfortunate because indexes of heterogeneity, being expressions of both richness and equitability, cannot be clearly interpreted in terms of underlying changes in components of diversity. Changes in index value following environmental impact could be due to changes in species richness, equitability, or both. Sometimes opposing changes in these two components cancel each other out, so that the heterogeneity index shows little or no change.

Where heterogeneity indexes are used, it is at least useful to understand in what ways they are sensitive to changes in relative abundance and richness. A heterogeneity index that is particularly sensitive to changes in relative abundance of the one or two species of highest importance value is Simpson's index (1949). For an infinite sample the index is

$$C = \sum_{i=1}^{S} p_i^2 \tag{11.7}$$

where

p_i = the proportion of the total importance values
      of all species in species i
S = total number of species

For finite samples (i.e., in which only a portion of the community has been measured), the form is

$$C = \sum_{i=1}^{S} \frac{n_i(n_i - 1)}{N(N - 1)} \tag{11.8}$$

where

n_i = importance value of species i
N = sum of importance values for all species S

Simpson's index varies between values close to 0 (for a sample of high equitability) and 1 (for a sample completely dominated by one species). Since high values of the index indicate a high degree of concentration of resources in one or a few species, Simpson's index in this form is often called an index of *concentration of dominance*. To reverse the range of the index so that it increases to 1 as equitability increases, one may use the inverse $(1 - C)$ (Gini 1912) or the reciprocal $(1/C)$ (Williams 1964).

Peet (1974) has expressed the response of Simpson's index to changes in relative abundance by plotting a response curve along which paired opposite changes in importance of two species will result in an increase in the C index value (Figure 11.6a). The response curve is steepest near a relative importance of 100%, indicating that the index is most responsive to small percentage changes in importance value of dominant species and relatively insensitive to additions, or small changes in importance, of rare species.

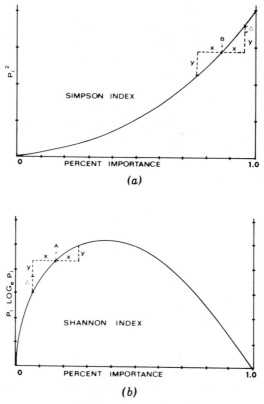

Figure 11.6. The response curves of two heterogeneity indexes. A divergence of two equally common species from any point results in a net loss (Δ) in heterogeneity in curve (*b*) and a net gain (Δ) in dominance in curve (*a*). (*a*) Simpson's index; (*b*) Shannon-Weaver index. From Peet (1974). Reproduced with permission from the *Annual Review of Ecology and Systematics*, Volume 5, (c) 1974 by Annual Reviews, Inc.

A second heterogeneity index that has been widely used is that of Shannon and Weaver (1949):

$$H' = - \sum_{i=1}^{s} p_i \log p_i \qquad (11.9)$$

where

p_i = percentage importance of the ith species

This formula also assumes an infinite sample, but as long as sample size is large, the bias is likely to be small if n_i/N is used to approximate p_i (Peet 1974). The response curve for the Shannon-Weaver index (Figure 11.6b) indicates that it is most sensitive to changes in species of moderate abundance (with p_i values close to $1/e$; Fager 1972, Whittaker 1972), but Peet (1974) notes that the index in fact responds most strongly to changes in importance of the rarest species, since small changes in abundance of a rare species could eliminate that species altogether, with a large effect on the index.

These response curves serve to emphasize that different heterogeneity indexes may indicate quite different magnitudes and directions of change in diversity using the same data set. As a result they cannot be recommended for use in impact studies for measurement of overall changes in diversity.

Two examples illustrate the problems in using heterogeneity indexes in pollution or other impact studies. In Figure 11.7 Shannon-Weaver diversity of aquatic invertebrate communities is shown above and below the discharge point of effluent outfalls. In both curves diversity declines immediately below the outfall, then rises to levels somewhat in excess of control values. We do not know if the final downstream values differ significantly from upstream controls, since no test of significance is available without replicate measurements (see Bowman et al. 1969, Hutcheson 1970, for tests of significance with this index). A more fundamental problem exists, however. We do not know the causes of the changes in the heterogeneity index values. It is possible that richness or equitability, or both, declined initially sufficiently to cause a decline in the index. Further downstream, the index may have risen as the new community became dominated by fewer but more pollution-tolerant species that shared the resources slightly more equally than before. As with any diversity index we do not know the composition of the final downstream community, relative to the control. Despite similar Shannon-Weaver values the downstream community might be composed of totally different and less desirable species (e.g., polychaete worms replacing oysters); the community may now contain nonnative opportunist species whose long-term survival is questionable.

The second example concerns the effect of dredging on benthic invertebrate communities (Figure 11.8). Again the heterogeneity indexes show an

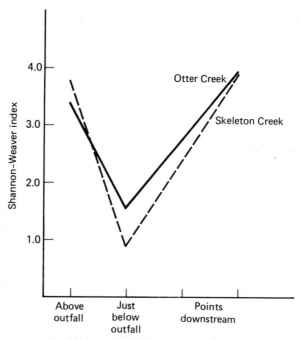

Figure 11.7.   Effect of effluent discharge on diversity of aquatic communities as measured by a heterogeneity index. (*a*) Otter Creek: oil field brines; (*b*) Skeleton Creek: domestic and oil refinery wastes. Data from Wilhm and Dorris (1968).

initial dip and a final rise above pre-impact levels. In this example, however, additional data on richness, density and relative abundance of dominant species allow us to interpret the heterogeneity index behavior more meaningfully. We see that what we hypothesized as a possibility in the last example has happened here. A large initial drop in richness and a concomitant dramatic rise in concentration of dominance by *Anisogammarus pugettensis* (an amphipod) caused a drop in both heterogeneity indexes. But despite these large declines in both richness and equitability, neither heterogeneity index changed very much. Nineteen days later richness was still depressed, but less so; the slight increase in relative abundance of *Anisogammarus* (1.7%) over pre-dredging levels (0.1%) caused the Shannon-Weaver index to rise dramatically, whereas the Gini index (1-Simpson's index) showed virtually no change from pre-dredge levels, despite the fact that the dominant species was now the amphipod *Photis brevipes* (which increased from 12.2% to 30.7%), and the pre-dredge dominant polychaete worm (*Owenia collaris*) was now rare (a change from 32.4% to 3.5%). This example illustrates that the collection of data on richness, density, and relative abundance of dominants is extremely helpful in following community changes. The heterogeneity indexes seem to have added little except the potential for confusion. If they are to be used at all, clearly they should not be the only index of

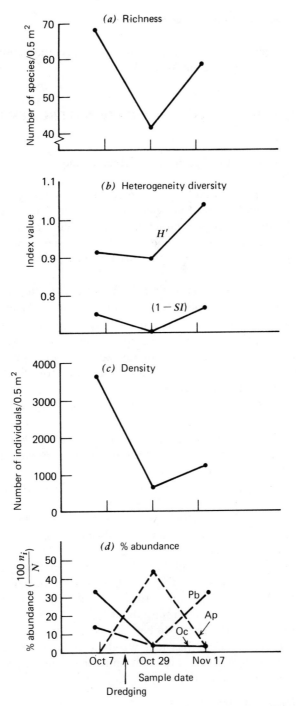

Figure 11.8. Effect of dredging on benthic invertebrates at Yaquina Bay, Oregon. In (*b*), *H'* is the Shannon-Weaver index; 1 − SI is the Gini index (inverse of Simpson's index). In (*d*), Oc = *Owenia collaris;* Ap = *Anisogammarus pugettensis;* Pb = *Photis brevipes.* Data from Swartz (1980).

community change to be observed, and their interpretation should be based on the larger set of information available.

PRESERVATION OF DIVERSITY: DESIGN OF NATURE RESERVES

Hypotheses to Explain the Species-Area Relationship

In discussing species-area curves, we noted the tendency for species richness to increase with area, following an exponential or power-function curve. We obtain such curves either for sample plots of ever larger size within a contiguous area or for islands of increasing size. There are several possible explanations for rising species-area curves:

1. *Habitat Heterogeneity Hypothesis.*   As area expands, new kinds of habitats and their associated species can occur (e.g., Williams 1964).
2. *Sample Hypothesis.*   As contiguous area expands, the statistical chance of encountering additional species increases, especially if the species are spatially aggregated or rare (e.g., Simberloff 1978, Connor and McCoy 1979). On islands, larger islands are more likely to capture randomly dispersing organisms by chance.
3. *Equilibrium Hypothesis.*   The number of species in an isolated area (e.g., island) at equilibrium results from the balance between the rate at which species immigrate to the area and the rate at which they go extinct in the area. Larger areas can support larger species populations, and the chances of extinction are thus reduced. As a result a higher number of species can survive at equilibrium (Figure 11.9; MacArthur and Wilson 1967).

Since the number of species in dynamic equilibrium results from a continued influx and extinction of species, one would expect to observe a turnover or change in species composition in an isolated area over time. Although the equilibrium hypothesis was initially formulated for biologically isolated areas such as islands, it could also apply to patches of habitat embedded within a large matrix of patches of different habitat type.

The empirical evidence from studies of species on islands or isolated patches on continental areas has lent weight to one or another of these hypotheses in particular cases, although more than one mechanism could also be operating simultaneously.

Habitat Heterogeneity

Westman (1983) sampled plant species within a single broad vegetation type on four of the California Channel Islands, using a constant plot size. He

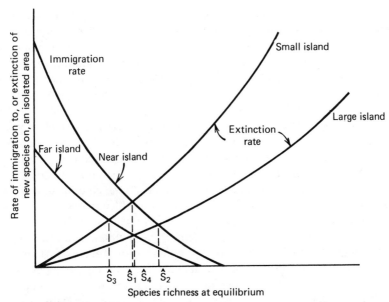

Figure 11.9. Equilibrium hypothesis. The number of species existing in dynamic equilibrium on a large island ($\hat{s}_2$, $\hat{s}_4$) is predicted to be greater than that on a small island ($\hat{s}_1$, $\hat{s}_3$) due to differing extinction rates on the two island sizes, according to the equilibrium model proposed by MacArthur and Wilson (1963, 1967). The larger immigration rates from islands near to mainland dispersal sources also lead to higher equilibrium levels of species ($\hat{s}_1$, vs. $\hat{s}_3$; $\hat{s}_2$ vs. $\hat{s}_4$). Adapted from MacArthur and Wilson (1963), with permission of *Evolution*.

found that species richness did not increase either with island size or the extent of island covered by the vegetation type. Thus the increase in plant species richness with island size earlier observed for these islands (Johnson et al. 1968) was likely due to habitat heterogeneity, the sample effect, or both. Buckley (1982) reached similar conclusions in his study of the Lowendal Islands off western Australia. Buckley divided each island in the archipelago into habitat types and measured the area and plant species richness of each. He developed regressions of species richness on area for each habitat type or combination of habitat types. He predicted total species richness on each island well ($R^2 = 0.81$) by regressions based on area of each habitat on the island. The richness predictions based on island area alone ($R^2 = 0.68$) or island area plus a Shannon-Weaver index of habitat heterogeneity (Juvik and Austring 1979) ($R^2 = 0.74$) were significantly less accurate ($P < 0.05$). Authors studying marine ecosystems who also suggested that habitat heterogeneity was a hypothesis favored by their data include Abele (1974), Dexter (1972), and Harman (1972).

Equilibrium Hypothesis

Many authors have studied species-area relations in order to test MacArthur's and Wilson's (1967) hypothesis; however, only a few such studies have met the rigorous conditions required for a clear-cut test (see Gilbert 1980a for a critical review). Such conditions include ensuring that islands are truly "at equilibrium," meaning that the total species richness is not changing over time; that species turnover is occurring, not counting species that are merely transients; and that alternative hypotheses have been ruled out. One set of studies that did exhibit a low, but indisputable, species turnover was that by Simberloff and Wilson (1969, 1970; Wilson and Simberloff 1969) and Simberloff (1969). They observed animal recolonization of tiny mangrove islets for two years following application of a broad-spectrum pesticide and found the species number returning to an equilibrium level in the presence of a species turnover rate of 1.5 species per year (Simberloff 1976).

Tallamy (1983) applied the equilibrium theory to the colonization of a new insect host species in North America (gypsy moth, introduced into North America in the late 1860s) by native and introduced insect parasites. Based on historical records, Tallamy documented the maintenance of an equilibrium number of parasite species on the gypsy moth. Tallamy further established an equilibrium number of parasite species (10.2) on the gypsy moth in North America predicted by the rate of "immigration" (introductions or colonization by native parasites) and extinction (based on date of last recorded appearance in North America) (Figure 11.10). The number of estab-

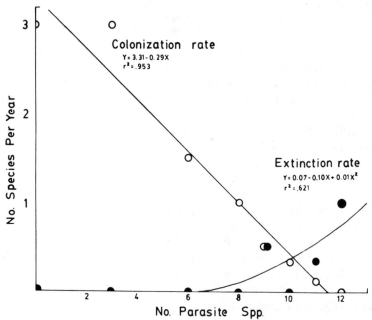

Figure 11.10. Colonization and extinction rate curves for parasites of the gypsy moth in North America. Data recorded from checklists of parasites reared from gypsy moth, 1906–1974. Reprinted with permission from D. W. Tallamy, *American Naturalist* **121**:244–254. Copyright (c) 1983 University of Chicago Press.

lished parasites is 9, although any one field population of moths averages 5.6 parasite species. These two studies give some support to the notion of species turnover under equilibrium conditions, but the situations observed were rather artificial, and the analogy to large, intact ecosystems is uncertain.

Sample Hypothesis

The number of studies that have examined the sample hypothesis directly are few. Coleman et al. (1982) established that the number of bird species on islands in a reservoir at the Pennsylvania–Ohio border conformed to the species-area curve predicted by assuming a random distribution of bird species, with number of species proportional to island area. Although this is evidence in favor of the sample hypothesis, it does not rule out the simultaneous operation of other mechanisms of assortment.

We may conclude from this much abbreviated review that there is some evidence favoring each of three alternative hypotheses or mechanisms to explain rising species-area curves. We cannot yet predict which mechanism or mechanisms will apply in particular cases.

Reserve Design and the Equilibrium Hypothesis

This last conclusion becomes relevant when we consider the recent literature on the design of nature reserves. A number of authors (e.g., Diamond 1975, Diamond and May 1976, Sullivan and Shaffer 1975, Terborgh 1974, Wilson and Willis 1975) have suggested park design principles for maximizing preservation of species richness that would follow if the equilibrium hypothesis held. These recommendations, which have been widely publicized (e.g., World Conservation Strategy, IUCN 1980) must be viewed with extreme caution for at least two reasons. First, there is not as yet an a priori basis for concluding that the equilibrium hypothesis, or the power function model for species-area curves, hold in a particular case; second, additional ecological factors, noted later in this section, must be taken into account in considering factors that influence immigration and extinction rates, the initial richness of landscape patches, and the long-term maintenance of diversity.

An example of the recommendations for reserve size, shape, and arrangement that have flowed from the equilibrium hypothesis is illustrated in Figure 11.11. Margules et al. (1982) have summarized the problems with a number of these recommendations; many are discussed here, along with additional points.

Design A flows from the species-area relation: a large reserve will encompass more species than a small one in the same area. As a reserve design recommendation it ignores the question of whether other park management considerations (cost of land, ease of management) are to take priority. Further Higgs (1981) and McCoy (1983) caution that the rate of increase in

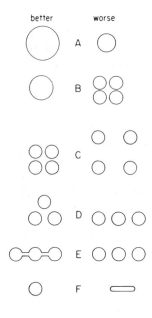

Figure 11.11. Implications for reserve design based on the equilibrium hypothesis. In each of the cases labeled *A* to *F*, species extinction rates would be lower for the reserve design on the left than that on the right. Reprinted from Figure 7 of Diamond (1975) with permission of Applied Science Publishers Ltd. and the author.

species with area will depend on the richness of the particular patches of landscape or islands. McCoy (1983) notes that judicious selection of several particularly species-rich patches may result in a richness equal or in excess of that in a single larger area. An additional consideration in choosing a reserve is to determine the minimum area needed to support sustainable, genetically diverse, breeding populations of the species. This will depend on the inherent genetic variability of the populations of concern, their territorial needs, breeding requirements, and so forth. Discussions of this point may be found in McCoy (1983), Shaffer (1981), Soulé (1980), and Chapter 8.

Design B asserts that one contiguous park is preferable to several separate ones of the same total area. This would follow from the equilibrium hypothesis, since a larger reserve would be expected to have a lower extinction rate. Although this is a plausible hypothesis, as noted earlier, the empirical evidence to support it is equivocal (Abele and Connor 1979, Gilbert 1980a, Miller 1978). The contiguous park might also be preferable based on species-area considerations, but whether this will be so depends on the slope of the species-area curve and the proportion of species in common between the separate reserves. Higgs and Margules (1980) use the index, *R*, to determine whether one contiguous reserve contains more species than two smaller reserves, of unequal size, whose combined area equals that of the single, contiguous reserve:

$$R = \frac{p^z + (1 - p)^z}{1 + P_v} \tag{11.10}$$

where

p = proportion of the total area in one small reserve (decimal fraction)
z = the slope of the power function for the species-area curve ($S = cA^z$)
P_v = Jaccard's coefficient of community between the two smaller reserves (Table 10.4), which can also be calculated as $[p^z + (1 - p)^z - 1]$ (Higgs and Usher 1980)

When $R > 1$, the two separate reserves contain more species than the single larger reserve. Game and Peterken (1980), Gilpin and Diamond (1980), Higgs and Margules (1980), Higgs and Usher (1980), and others have compared such situations for examples involving birds, mammals, and plants. The results vary; in a majority of cases two smaller patches contain more species than one large patch of the same total area, but a significant number of exceptions occur. A knowledge of the species composition and species-area relations is needed to resolve the question for a particular case. Equation 11.10 can be used, provided the species-area curve adequately fits a power function.

Designs C and D rely on the prediction of the equilibrium hypothesis that islands more distant from dispersal sources will have lower immigration rates, hence lower numbers of species at equilibrium (see Figure 11.9). When applied to reserves on the mainland, "isolation" is often less complete. A park surrounded by a cultivated lawn will still permit greater migration of terrestrial animals across it than would an ocean. In the case of plants some sun-loving plants may be able to grow both within open patches in the reserve and in such a buffer zone. Hence the degree to which the park is surrounded by a seminatural or open-space area will affect the extent to which immigration continues at a higher level than it would by a zone totally inhospitable to migrants. Further the degree of separation that serves as an effective boundary to dispersal between two habitat reserves differs between species. Some birds will migrate many kilometers between habitats; others may be stopped by a gap of only 10 m (MacArthur 1972). Some plants contain wind-dispersed or bird-dispersed seed that can travel many miles; others have heavy seeds that fall within a meter of the parent. Typically, r-selected species of frequently disturbed habitats can disperse over longer distances than K-selected species.

In a study of mammals, passerine birds, and lizards in native vegetation reserves surrounded by farmland in western Australia, Humphreys and Kitchener (1982) found that the species themselves could be classified into those that were limited to undisturbed habitats and those that also occupied cultivated fields, roadsides, and other disturbed areas. By examination of species-area curves, Humphreys and Kitchener found that as areal patch size became smaller, the habitats contained disproportionately more species characteristic of disturbed habitats. They concluded that, in considering minimum reserve size, it would be useful to distinguish between the specialist species requiring undisturbed habitat and the generalists able to tolerate a wider habitat range. To protect the specialist species, larger reserves would be required than for generalists, based on the difference in shape of species-area curve for the two species types.

Designs C and D also ignore habitat heterogeneity. To the extent that scales of habitat patchiness are large, reserves that are more distantly spaced are more likely to hold different species. Higgs and Margules (1980) found in Yorkshire a greater rate of turnover in grsss species composition in an east–west than a north–south direction. Westman (1981) found for coastal sage scrub in California that as one moved inland from the coast, there was a 50% change in herb species composition every 90 km, whereas 150 km of movement was required to achieve a half-change in shrub composition. The rate of change in species composition along an environmental gradient is termed *beta diversity* (Whittaker 1960, 1972).

A relatively simple way to measure beta diversity is as the slope tangent to the curve of change in species composition. Such a curve (Figure 11.12) can be constructed by plotting the percentage similarity (PS) or coefficient of community (CC) (Table 10.4) between successive sites along an environmental gradient and a starting site (Whittaker 1978). Alternatively, if an ordination such as reciprocal averaging or DECORANA (Chapter 10) has been performed, one may calculate the half-changes in species composition along the ordination axis by the formula (M. O. Hill in Whittaker et al. 1979):

$$HC = \frac{(12EV/(1 - EV))^{1/2}}{1.349} \tag{11.11}$$

where

EV = the eigenvalue of the ordination axis

An eigenvalue is a measure of the variance accounted for by the axis. For a discussion of eigenvalues, see, for example, Orloci (1978).

The greater the beta diversity along a landscape gradient, the greater the likelihood that more distant sites will be more different in species composition, the P_v (the coefficient of community) lower, the ratio R greater than 1 (Eq. 11.10), and two separated sites greater in total richness than one large site.

Design E reflects the effect of stepping stones or corridors in enhancing migration, hence immigration rates, between portions of the reserve. Clearly this consideration is important since, as noted earlier, species vary in their migratory abilities, some being inhibited (or endangered) by a road, pipeline, or other relatively narrow barrier or distance. The importance of such migration corridors will vary with the autecological requirements of the species involved. There is some evidence that wider corridors permit a greater num-

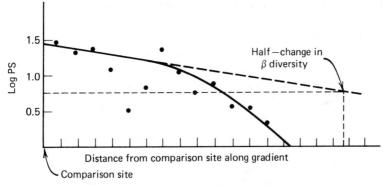

Figure 11.12. The graphical method for determining beta diversity. The declining percentage similarity of successive samples along an environmental gradient from the end point sample at left is here plotted using floristic data of California coastal sage scrub (W. E. Westman, unpublished). The slope of the tangent to the curve at its leftward end is taken as the rate of change in floristic composition, or beta diversity. This line can be extended to the point where a 50% change in PS, or floristic half-change, has occurred, and beta diversity is expressed as the corresponding distance along the abscissa.

ber of species to migrate in safety than narrow ones (Forman 1983). This is true because wider corridors permit both an "edge" and an internal or "core" habitat to develop. Larger animals migrating through wider corridors are less likely to be effectively spotted and hunted by predators (human or otherwise) outside the corridor. Sunny edge habitats provide suitable growing conditions for many plants that are shade intolerant. Certain animals are similarly limited to edge or core habitats, whereas others may require both for different parts of their behavior (e.g., feeding at edge, nesting in core). A disadvantage of corridors in reserves is that they also provide an efficient route for the transmission of pests and disease.

Design F is based on the notion that circular reserves minimize dispersal distances within the reserve, enhancing chances of reestablishing populations that are locally extinguished (extirpated) within the reserve. Game (1980) has noted that immigration rates may also be affected by the shape of a reserve. Furthermore much dissected, spread-out designs of the same total area may range into adjacent habitats of greater heterogeneity, from which immigration can occur; migratory birds with fixed flyways would be more likely to encounter a portion of a spread-out reserve than a compact circular one placed at random within the vicinity encompassed by the dissected design. Where extinction rates may be expected to dominate immigration rates (e.g., in isolating an existing habitat by developing a large urban zone around it), a circular shape is indeed optimal. In situations where immigration may be more important than extinction, (e.g., in rehabilitating strip-mined land with native species by natural dispersal), a more elongated or dissected shape is to be preferred (Game 1980).

Additional Considerations in Reserve Design

Many authors have noted that considerations other than strictly biogeographic ones are important in reserve design:

1. *Disease.* Soulé and Wilcox (1980) note that isolated reserves are desirable to prevent the spread of diseases (e.g., avian cholera, Dutch elm disease) between populations.

2. *Genetic conservation.* Large isolated populations are more likely to contain individual variation, enhancing the preservation of overall genetic diversity of the species. However, if isolated populations become too small, inbreeding can act to reduce genetic variability within the population (Chapter 8; Soulé 1980, Soulé and Wilcox 1980).

3. *Effects of catastrophic disturbance.* Fire, drought, or other catastrophic events are less likely to extinguish species if populations exist in well-separated reserves (Foster 1980, Gilbert 1980b). Higgs (1981) notes, however, that if a reserve is large enough, only a portion may be destroyed, and recolonization will be more rapid from remaining areas.

4. *Needs of species at different trophic levels.* A park suitable to the preservation of a small, sedentary plant can be much smaller than one needed for a top carnivore with large territorial and hunting range. Consideration of the preservation of any one species in isolation from its food web, however, is less than ideal. Gilbert (1980b), for example, discusses the case of a Costa Rican rain forest tree species—*Casearia corymbosa*—which acts as a kind of keystone species (Chapter 8), supporting large numbers of insects, birds, and fruit-eating mammals which in turn are interdependent with other plants. For example, the masked titrya is a bird that relies on *Casearia* trees for food during a 2–6 week period annually when other fruit sources are unavailable. In turn the masked titrya is an important seed dispersal agent for several other tree species in the forest. The disappearance of *Casearia* could lead to the extirpation not only of the masked titrya and other frugivores but other plant species that they help disperse. Gilbert (1980b) terms species (typically plants) that provide critical support to many animals and associated plants dependent on these, a *keystone mutualist.*

5. *Recreation, study, management, economics and other considerations.* Wilderness preserves often serve as places of recreation, scientific research, and education as well as repositories for the species themselves. Larger numbers of reserves, at least some of which are close to population centers, will enhance accessibility and use of such reserves by people, while increasing the level of potential disturbance and costs of management. Large numbers of separate reserves may aid in managing different reserve areas for different levels of use, although zoning of a single large reserve is also possible. At the same time the cost of maintaining staff and developing management plans is usually less for a single large reserve, or at least a

coordinated set of reserves of similar habitat type. Beyond all this the cost of acquiring parcels of land, the availability of suitable parcels, the cost of managing the park, the rarity of some of its components, and the political climate of the time will all influence the final selection of sites for nature reserves.

This discussion of factors influencing design of nature reserves is intended to highlight the fact that decisions regarding reserve design are dependent on many factors individual to the sites and circumstances. The early work of Preston (1948, 1960, 1962) and MacArthur and Wilson (1963, 1967) stimulated major advances in conceptual thinking about this problem through their formulation of models for the species-area curve and its origin. The fact that more recent work has lessened support for the universal application of the equilibrium hypothesis and its ramifications was presaged by MacArthur and Wilson (1967: Preface), who nevertheless emphasized the value of conceptual models in science. As the complexity of factors influencing diversity and its preservation come to be better understood, it is seen that spatial variation in the form of the species-area curve, in alpha (site) and beta diversity, and in the importance of immigration and extinction in particular situations, all contribute to variation in recommendations for reserve design that are to be made. The discussion presented here abbreviates the existing literature substantially, and further developments in the field are occurring rapidly.

Emphasis in this discussion has been on design of reserves in terrestrial ecosystems; application of the model to marine sanctuaries cannot be made without substantial modification. In ocean waters boundaries to migration are often vertical rather than horizontal, since organisms have narrow tolerances to hydrostatic and radiation changes. In addition patterns of long-distance migration are often influenced by currents and upwelling zones rather than by factors that can be modeled by random diffusion processes. Nevertheless, especially for attached marine organisms of rocky intertidal areas and coral reefs, the models developed for terrestrial systems have some applicability. Goeden (1979) discusses approaches to the design of coral reef reserves using some principles noted in this chapter.

LANDSCAPE ECOLOGY

Geographers have been concerned with human modifications of the landscape, and the constraints on its development imposed by nature, since the early nineteenth century. Planning for landscape development in light of natural features and constraints has been termed "landscape ecology" in Europe (Naveh 1982; Naveh and Lieberman 1984). In North America urban planners and landscape architects were particularly inspired by the application of these ideas to land planning through overlay mapping (see Chapter 6), eloquently described by Ian McHarg (1969).

More recently ecologists (e.g., Forman and Godron 1981, Forman 1982, Tjallingii and de Veer 1982) have sought to meld these geographic approaches to land planning with the ideas concerning immigration, emigration, and extinction of species between landscape patches that developed from consideration of the equilibrium hypothesis of island biogeography (MacArthur and Wilson 1963, 1967). The concept of the landscape as a mosaic of patches of different habitats is not new, but efforts to characterize the dynamics of flux of species, energy, and nutrients *between* such patches are relatively recent. Studies of species movement between habitat patches in temperate forests are reported, for example, in Burgess and Sharpe (1981). Movements of nutrients between habitats have been studied mainly in small watershed studies such as those at Hubbard Brook, New Hampshire (Likens et al. 1977, Bormann and Likens 1979), where nutrient losses from undisturbed and clear-cut temperate deciduous forests to stream drainages were quantified in detail.

The import or export of organic energy between ecosystems has been studied mainly in aquatic ecosystems, where the export of detritus from a stream or estuary to coastal waters can represent an important energy transfer. In terrestrial systems the changing balance of radiative energy flux through a vegetation canopy with change in the internal architecture of the stand has been studied (Ross 1981), but movements of organic energy across terrestrial boundaries have mostly been studied indirectly through movement of species, leaching, wind or water erosion of organic carbon, and dispersal of fruit, seeds, and pollen. Of course the larger movements of physical energy between ecosystems occur as climatological processes of air and water movement, with associated changes in heat content of the atmosphere and atmospheric moisture.

Thus fluxes of species, energy, and materials are occurring across the earth's surface continually, at a variety of scales; the study of these processes also occurs at a range of spatial scales. In the case of energy the global and regional scales of physical energy movement are the domain of climatology. The detailed flux of energy within an ecosystem is studied by autecologists or biophysical ecologists (see, e.g., Nobel 1974). The movement of both physical and organically bound energy between adjacent ecosystems falls within the domain of what is now being called landscape ecology. For example, Forman and Godron (1981, p. 733) propose the following definition of the scale of concern of landscape ecology:

A landscape is a kilometers-wide area where a cluster of interacting stands or ecosystems is repeated in similar form. The landscape is formed by two mechanisms operating together within its boundary—specific geomorphological processes and specific disturbances of the component stands. . . . Disturbances include both natural events and human activities such as fire, hurricanes, agricultural practices, or forest cutting.

Structural Features of the Landscape

One way to begin the examination of dynamics between the stands, ecosystems, or simply "patches" within such a landscape is to determine whether patches of different sizes and shapes have different characteristic patterns of flux of energy, species, and materials. Forman (1979) and Forman and Godron (1981) have suggested that a mosaic of landscape patches may be effectively characterized by reference to a few basic types of patches (Figure 11.13). A "spot patch" occurs following a localized disturbance (e.g., fire, clearing) within a matrix of the habitat type and normally is characterized by successional vegetation. A "remnant patch" is a small piece of the original habitat matrix that remains when disturbance occurs around it, such as a woodlot of original forest surrounded by farmland or an island of vegetation in a reservoir formed by flooding. Ecosystems here may remain in climax status, though they may suffer from reduced immigration and increased extinction rates. "An environmental resource patch" is Forman's term for a habitat patch in which some set of physical resources differ from those of their surroundings in ways that influence biotic composition (e.g., a particular soil type, a gully). An "introduced patch" is dominated by species introduced and maintained by people (e.g., a lawn, a garden). An "ephemeral patch" is a temporary aggregation of species due to natural causes, such as a flock of parrots nesting in a tree at night or a patch of desert annuals.

Forman (1982, Forman and Godron 1981, Forman 1983) has further defined different types of corridors (Figure 11.14). The "line corridor" (e.g., hedgerows, roads, trails, irrigation channels) is sufficiently narrow that its species composition is homogeneous, composed of species characteristic of open, sunny locations. "Strip corridors" are wider, containing a shaded interior portion inhabited by species different from those at the edges of the strip corridor. Corridors of either type may also form networks of intersecting corridors or may border streams (Forman and Godron 1981, Forman

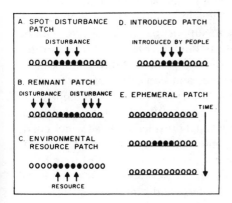

Figure 11.13. Types of landscape patches. OOO = matrix; ●●● = patch; disturbance = a sudden, severe environmental change. Reprinted with permission from Forman and Godron (1981), *BioScience* **31**:733–740. Copyright (c) 1981 by the American Institute of Biological Sciences.

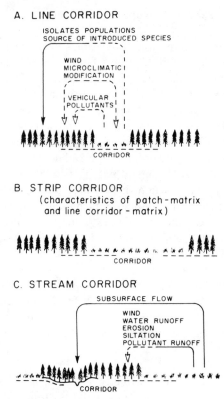

Figure 11.14.   Major fluxes between the matrix and three types of corridors. Reprinted with permission from R. T. T. Forman (1982). Copyright (c) 1982, PUDOC, Center for Agricultural Publishing and Documentation, Wageningen, The Netherlands.

1983). Figure 11.15 illustrates the occurrence of these patch types in an agricultural landscape.

This typology becomes of interest when it can be shown that patches or corridors of different sizes and shapes develop characteristic biotas, such that minimum habitat sizes and shapes to support particular species can be determined. Empirical evidence to date is limited. Figure 11.16 illustrates that, in transmission-line corridors of different widths cut through oak-hickory (*Quercus-Carya*) forests in Tennessee, and consisting of mixed grass and forb species and *Rubus* shrubs (blackberry), bird species composition changed markedly when corridor widths exceeded 30–60 m. Bird species not characteristic of the adjacent forest increasingly resided in these wider corridors, suggesting the development of an "internal" as well as "edge" habitat. In another example, Galli et al. (1976) and Forman et al. (1976) found that when bird species inhabiting old oak (*Quercus*) woodlots in an agricultural area of New Jersey were classified into "edge" and "interior" species, the interior species increased in number steadily, and more rapidly than edge

Figure 11.15. Portion of an agricultural landscape in New Jersey. *A*, spot disturbance patch (small opening in forest); *B*, strip corridor (powerline crossing stream corridor); *C*, narrow patch with no forest interior; *D*, strip corridor (wooded); *E*, tiny patches with no forest interior; *F*, peninsula; *G*, tiny remnant patch affected by proximity to larger patch; *H*, introduced patch (golf course); *I*, introduced line corridor (*Platanus* planted along road); *J*, large remnant patch (well-developed forest interior; patch edge about twice as wide to south as north); *K*, road network; *L*, dwellings clustered (village); *M*, introduced patch (cemetary conifers and grasses); *N*, environmental resource patch (lowland tree species on wet spot); *O*, temporal patch (area of shrubs and successional trees undergoing rapid change); *P*, wide stream corridor (containing both river and canal); *Q*, narrow stream corridor; *R*, matrix (corn and bean fields); *S*, line corridor (road); *T*, habitation (area of farm buildings); *U*, hedgerow network (connecting woods patches); *V*, small remnant patch (contains limited area of forest interior). Reprinted with permission from Forman and Godron (1981), *BioScience* **31**:733–740. Copyright (c) 1981 by the American Institute of Biological Science.

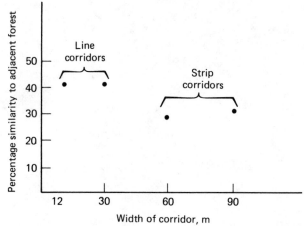

Figure 11.16.   Percentage similarity of bird species in transmission-line corridors of different widths to that in adjacent oak-hickory forest. Tennessee. Data of Anderson et al. (1977).

species, as the ratio of interior to edge area in the woodlots increased (Figure 11.17). Extensive studies of the effect of British hedgerow networks in preserving species are reported in Pollard et al. (1974).

Measuring Habitat Heterogeneity

As one enlarges patch areas to the point where an interior habitat can be differentiated from an edge habitat, one will encounter additional species characteristic of this second habitat (habitat heterogeneity hypothesis). The increase in species diversity due to this phenomenon is limited to those patches where the edge area/interior area ratio is close to one. When the ratio deviates significantly from one, the contribution of the less abundant habitat to the overall composition and relative abundance of species will decrease.

In the forgoing example the ratio of edge area to interior area for circular patches is

$$\frac{r_2^2 - r_1^2}{r_1^2} \tag{11.12}$$

where

r_1 = radius of the circle of ''interior'' habitat
r_2 = radius of the entire circular patch, including both interior and edge

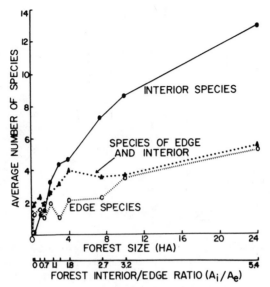

Figure 11.17. Diversity of interior and edge birds with increasing woodlot size. Each point is an average of eight samples. Data of Galli et al. (1976) and Forman et al. (1976). Reprinted with permission from R. T. T. Forman (1982). Copyright (c) 1982, PUDOC, Center for Agricultural Publishing and Documentation, Wageningen, The Netherlands.

More generally, a ratio of perimeter to area for patches of any shape can be expressed by using an index R which is 1 for a circle and increases to infinity as shapes become more oblong:

$$R = \frac{p}{2(\pi A)^{1/2}} \qquad (11.13)$$

where

p = perimeter
A = area

The index is often used by limnologists to express shoreline irregularity. Patton (1975) has suggested that if the total external perimeter of all contiguous habitat patches in a mosaic, plus edge length within the area, are measured and used for p in Eq. 11.13, the resulting edge area index could be used as a measure of "edge" habitat in a habitat mosaic (see Figure 11.18). Such an index could be useful to wildlife managers in seeking a guide to the amount of "edge" habitat in a forest mosaic. More generally, it is useful when the amount of contact between two habitats of different types is of interest. In addition it can be used as an index of edge/interior area if the width of "edge habitat" is constant for all habitats under consideration.

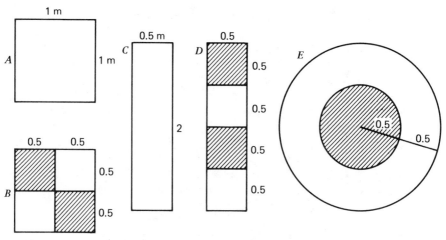

Figure 11.18. The perimeter/area ratio R (Eq. 11.13). The R value for each figure is $A = 1.13$, $B = 1.13$, $C = 1.41$, $D = 1.41$, $E = 1.00$. If the internal edges of contact between patches of different type are included in the numerator of R, the edge/area ratio (Patton 1975) for each figure is $A = 1.13$, $B = 1.69$, $C = 1.41$, $D = 1.83$, $E = 1.50$.

Measures of "habitat heterogeneity" may also be of interest in designing reserves or measuring ecological attributes of the landscape. A common approach is to use the Shannon-Weaver index (Eq. 11.9). If p_i is the proportion of the total area in a particular habitat patch, the index will measure the total "graininess" of the habitat patch mosaic; if p_i is the proportion in a habitat *type* (which may occur in several separate patches in the landscape), the index will measure the diversity of habitat types, regardless of how finely divided the landscape mosaic is (Figure 11.19). In using the index, one should be mindful of the response curve illustrated in Figure 11.6. Ecologists have devoted considerable study to the way species use resources in coarse-grained versus fine-grained habitats. For some examples of this literature,

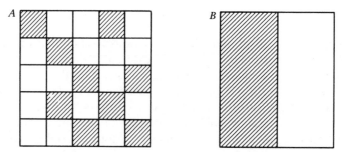

Figure 11.19. Two aspects of habitat heterogeneity. A and B are equal in the number of habitats they contain (two: shaded and white), but A has a finer grain to the habitat patch mosaic than B.

see Levin (1976), Pielou (1977), MacArthur (1972), and Whittaker and Levin (1977).

The study of diversity at several scales of spatial pattern in the landscape can reveal both structural and dynamic features of species distributions. When carefully chosen and interpreted, indexes of richness, dominance-diversity curves, and species-area curves can be of considerable use in impact assessment. The preservation of diversity in reserves is becoming more sophisticated as principles are developing for the use of knowledge of the species-area curve and of beta diversity. At the landscape level an enhanced appreciation of patch size, shape, and grain in relation to the habitat requirements of species, and their movements, can lead to better predictions of the effect of change in habitat patchiness on the associated biota. Work in this area is at an early stage.

REFERENCES

Abele, L. G. (1974). Species diversity of decapod crustaceans in marine habitats. *Ecology* **55**:156–161.

Abele, L. G., and Connor, E. F. (1979). Application of island biogeography theory to refuge design: making the right decision for the wrong reasons. In R. M. Linn, ed. Proc. First Conf. on Scientific Research in National Parks. U.S. Dept. Interior, National Park Service, Trans. & Proc. Vol. 5, pp. 89–94.

Anderson, S. H., Mann, K., and Shugart, H. H. (1977). The effect of transmission-line corridors on bird populations. *Amer. Midl. Nat.* **97**:216–221.

Bond, W. (1983). On alpha diversity and the richness of the Cape flora: a study in southern Cape fynbos. In F. J. Kruger, D. T. Mitchell, and J. U. M. Jarvis, eds. *Mediterranean-type Ecosystems. The Role of Nutrients.* Ecol. Stud. 43. Springer-Verlag, Berlin, pp. 337–356.

Bormann, E. H., and Likens, G. E. (1979). *Pattern and Process in a Forested Ecosystem.* Springer-Verlag, New York.

Bowman, K. O., Hutcheson, K., Odum, E. P., and Shenton, L. R. (1969). Comments on the distribution of indices of diversity. In G. P. Patil, E. C. Pielou, and W. E. Waters, eds. *Statistical Ecology.* Vol. 3. Pennsylvania State Univ. Press, University Park, pp. 315–359.

Burgess, R. L., and Sharpe, D. M., eds. (1981). *Forest Island Dynamics in Man-Dominated Landscapes.* Springer-Verlag, New York.

Buckley, R. (1982). The habitat-unit model of island biogeography. *J. Biogeogr.* **9**:339–344.

Coleman, B. D., Mares, M. A., Willig, M. R., and Hsieh, Y.-H. (1982). Randomness, area, and species richness. *Ecology* **63**:1121–1133.

Connor, E. F., and McCoy, E. D. (1979). The statistics and biology of the species-area relationship. *Am. Nat.* **113**:791–833.

Dexter, D. (1972). Comparison of the community structure in a Pacific and Atlantic Panamanian sandy beach. *Bull. Mar. Sci.* **22**:449–462.

Diamond, J. M. (1975). The island dilemma: lessons of modern biogeographic studies for the design of natural reserves. *Biol. Conserv.* **7**:129–146.

Diamond, J. M., and May, R. M. (1976). Island biogeography and the design of natural reserves. In R. M. May, ed. *Theoretical Ecology: Principles and Applications.* Blackwell, Oxford, pp. 163–186.

Fager, E. W. (1972). Diversity: a sampling study. *Am. Nat.* **106**:293–310.

Forman, R. T. T. (1979). The Pine Barrens of New Jersey: an ecological mosaic. In R. T. T. Forman, ed. *Pine Barrens: Ecosystem and Landscape.* Academic Press, New York, pp. 569–585.

Forman, R. T. T. (1982). Interactions among landscape elements: a core of landscape ecology. In S. P. Tjallingii and A. A. de Veer, eds. *Perspectives in Landscape Ecology.* PUDOC, Wageningen, The Netherlands, pp. 35–48.

Forman, R. T. T. (1983). Corridors in a landscape: their ecological structure and function. *Ekologia* (CSSR), in press.

Forman, R. T. T., Galli, A. E., and Leck, C. F. (1976). Forest size and avian diversity in New Jersey woodlots with some land use implications. *Oecologia* **26**:1–8.

Forman, R. T. T., and Godron, M. (1981). Patches and structural components for a landscape ecology. *BioScience* **31**:733–740.

Foster, R. B. (1980). Heterogeneity and disturbance in tropical vegetation. In M. E. Soulé and B. A. Wilcox, eds. *Conservation Biology: An Evolutionary-Ecological Perspective.* Sinauer, Sunderland, Mass., pp. 75–92.

Galli, A. E., Leck, C. F., and Forman, R. T. T. (1976). Avian distribution patterns within different-sized forest islands in central New Jersey. *Auk* **93**:356–364.

Game, M. (1980). Best shape for nature reserves. *Nature* **287**:630–632.

Game, M., and Peterken, G. F. (1980). Nature reserve selection in central Lincolnshire woodlands. In M. D. Hooper, ed. *Proc. Symp. on Area and Isolation.* Inst. Terrestrial Ecol., Cambridge, U.K.

Gilbert, F. S. (1980a). The equilibrium theory of island biogeography: fact or fiction? *J. Biogeogr.* **7**:209–236.

Gilbert, F. S. (1980b). Food web organization and conservation of tropical diversity. In M. E. Soulé and B. A. Wilcox, eds. *Conservation Biology. An Evolutionary-Ecological Perspective.* Sinauer, Sunderland, Mass., pp. 11–34.

Gilpin, M. E., and Diamond, J. M. (1980). Subdivision of nature reserves and the maintenance of species diversity. *Nature* **285**:567–568.

Gini, C. (1912). Variabilita e mutabilita. A-III, Part II. Studi Economico-Giuridici Fac. Giurisprudenza, Univ. Cagliari, Italy.

Goeden, G. B. (1979). Biogeographic theory as a management tool. *Environ. Conserv.* **6**:27–32.

Harman, W. N. (1972). Benthic substrates: their effect on fresh-water mollusca. *Ecology* **53**:271–277.

Higgs, A. J. (1981). Island biogeography theory and nature reserve design. *J. Biogeogr.* **8**:117–124.

Higgs, A. J., and Margules, C. (1980). Reserve area and strategies for nature conservation. In M. D. Hooper, ed. *Proc. Symp. on Area and Isolation.* Inst. Terrestrial Ecol., Cambridge, U.K.

Higgs, A. J., and Usher, M. B. (1980). Should nature reserves be large or small? *Nature* **285**:568–569.

Humphreys, W. F., and Kitchener, D. J. (1982). The effect of habitat utilization on species-area curves: implications for optimal reserve area. *J. Biogeogr.* **9**:391–396.

Hutcheson, K. (1970). A test for comparing diversities based on the Shannon formula. *J. Theor. Biol.* **29**:151–154.

IUCN (1980). World Conservation Strategy. Intl. Union for the Conservation of Nature and Natural Resources, Gland, Switzerland.

Johnson, M. P., Mason, L. G., and Raven, P. H. (1968). Ecological parameters and plant species diversity. *Am. Nat.* **102**:297–306.

Juvik, J. O., and Austring, A. P. (1979). The Hawaiian avifauna: biogeographic theory in evolutionary time. *J. Biogeogr.* **6**:205–224.

Levin, S. A. (1976). Population dynamic models in heterogeneous environments. *Ann. Rev. Ecol. Syst.* **7**:287–310.

Likens, G. E., Bormann, F. H., Pierce, R. S., Eaton, J. S., and Johnson, N. M. (1977). *Biogeochemistry of a Forested Ecosystem.* Springer-Verlag, New York.

Lloyd, M., and Ghelardi, R. J. (1964). A table for calculating the "equitability" component of species diversity. *J. Anim. Ecol.* **33**:217–225.

MacArthur, R. H. (1957). On the relative abundance of bird species. *Proc. Nat. Acad. Sci., Washington,* **43**:293–295.

MacArthur, R. H. (1972). *Geographical Ecology.* Harper & Row, New York.

MacArthur, R. H., and Wilson, E. O. (1963). An equilibrium theory of insular zoogeography. *Evolution* **17**:373–387.

MacArthur, R. H., and Wilson, E. O. (1967). *The Theory of Island Biogeography.* Princeton, Univ. Press, Princeton.

Margules, C., Higgs, A. J., and Rafe, R. W. (1982). Modern biogeographic theory: are there any lessons for nature reserve design? *Biol. Conserv.* **24**:115–128.

May, R. M. (1975). Island biogeography and the design of wildlife preserves. *Nature* **254**:177–178.

McCoy, E. D. (1983). The application of island-biogeographic theory to patches of habitat: how much land is enough? *Biol. Conserv.* **25**:53–61.

McHarg, I. (1969). *Design with Nature.* Natural History, New York.

Miller, R. I. (1978). Applying island biogeographic theory to an East African reserve. *Environ. Conserv.* **5**:191–195.

Naveh, Z. (1982). Landscape ecology as an emerging branch of human ecosystem science. *Adv. Ecol. Res.* **12**:189–237.

Naveh, Z., and Lieberman, A. S. (1984). *Landscape Ecology. Theory and Application.* Springer-Verlag, New York.

Nobel, P. S. (1974). *Introduction to Biophysical Plant Physiology.* Freeman, San Francisco.

Orloci, L. (1978). *Multivariate Analysis in Vegetation Research.* 2nd ed. Junk, The Hague.

Patrick, R., Hohn, M. H., and Wallace, J. H. (1954). A new method for determining the pattern of the diatom flora. Acad. Nat. Sci. Phila., *Notulae Naturae* **259**: 1–12.

Patton, D. R. (1975). A diversity index for quantifying habitat edge. *Wildl. Soc. Bull.* **394**:171–173.

Peet, R. K. (1974). The measurement of species diversity. *Ann. Rev. Ecol. Syst.* **5**:285–307.

Peet, R. K. (1978). Forest vegetation of the Colorado Front Range: patterns of species diversity. *Vegetatio* **37**:65–78.

Pielou, E. C. (1975). *Ecological Diversity.* Wiley-Interscience, New York.

Pielou, E. C. (1977). *Mathematical Ecology.* 2nd ed. Wiley-Interscience, New York.

Pollard, E., Hooper, M. D., and Moore, N. W. (1974). *Hedges.* Collins, London.

Preston, F. W. (1948). The commonness, and rarity, of species. *Ecology* **29**:254–283.

Preston, F. W. (1960). Time and space and the variation of species. *Ecology* **41**:611–627.

Preston, F. W. (1962). The canonical distribution of commonness and rarity. *Ecology* **43**:185–215.

Ross, J. (1981). *The Radiation Regime and Architecture of Plant Stands.* Tasks for Vegetation Science, Vol. 3. Junk, The Hague.

Scanlan, M. J. (1981). Biogeography of forest plants in the prairie-forest ecotone in western Minnesota. In R. L. Burgess and D. M. Sharpe, eds. *Forest Island Dynamics in Man-Dominated Landscapes.* Springer-Verlag, New York, pp. 97–124.

Shaffer, M. L. (1981). Minimum population size for species conservation. *BioScience* **31**:131–134.

Shannon, C. E., and Weaver, W. (1949). *The Mathematical Theory of Communication.* Univ. Illinois Press, Urbana.

Simberloff, D. S. (1969). Experimental zoogeography of islands: a model for insular colonization. *Ecology* **50**:296–314.

Simberloff, D. S. (1976). Species turnover and equilibrium island biogeography. *Science* **194**:572–578.

Simberloff, D. S. (1978). Colonisation of islands by insects: immigration, extinction, and diversity. In C. A. Mound and N. Waloff, eds. *Diversity of Insect Faunas.* Symp. Roy. Entomol. Soc., No. 9. London, pp. 139–153.

Simberloff, D. S., and Wilson, E. O. (1969). Experimental zoogeography of islands: the colonization of empty islands. *Ecology* **50**:278–296.

Simberloff, D. S., and Wilson, E. O. (1970). Experimental zoogeography of islands: a two-year record of colonization. *Ecology* **51**:934–937.

Simpson, E. H. (1949). Measurement of diversity. *Nature* **163**:688.

Soulé, M. E. (1980). Thresholds for survival: maintaining fitness and evolutionary potential. In M. E. Soulé and B. A. Wilcox, eds. *Conservation Biology: An Evolutionary-Ecological Perspective.* Sinauer, Sunderland, Mass., pp. 151–170.

Soulé, M. E., and Wilcox, B. A. (1980). Conservation biology: its scope and its challenge. In M. E. Soulé and B. A. Wilcox, eds. *Conservation Biology: An Evolutionary-Ecological Perspective.* Sinauer, Sunderland, Mass., pp. 1–8.

Sullivan, A. L., and Shaffer, M. L. (1975). Biogeography of the megazoo. *Science* **189**:13–17.

Swartz, R. C. (1980). Application of diversity indices in marine pollution investigations. In *Biological Evaluation of Environmental Impacts*. Rep. FWS/OBS-80/ 26. U.S. Dept. Interior, Fish and Wildlife Service, and Council on Environ. Quality, Washington, D.C., pp. 230–237.

Tallamy, D. W. (1983). Equilibrium biogeography and its application to insect host-parasite systems. *Am. Nat.* **121**:244–254.

Terborgh, J. (1974). Preservation of natural diversity: the problem of extinction-prone species. *BioScience* **24**:715–722.

Thatcher, A. C., and Westman, W. E. (1975). Succession following mining on high dunes of coastal southeast Queensland. In H. A. Nix and J. Kikkawa, eds. *Managing Terrestrial Ecosystems*. Proc. Ecol. Soc. Aust. 9, pp. 17–33.

Tjallingii, S. P., and de Veer, A. A., eds. (1982). *Perspectives in Landscape Ecology*. PUDOC, Wageningen, The Netherlands.

Westman, W. E. (1979). Oxidant effects on Californian coastal sage scrub. *Science* **205**:1001–1003.

Westman, W. E. (1981). Diversity relations and succession in Californian coastal sage scrub. *Ecology* **62**:170–184.

Westman, W. E. (1983). Island biogeography: studies on the xeric shrublands of the inner Channel Islands, California. *J. Biogeogr.* **10**:97–118.

Whittaker, R. H. (1960). Vegetation of the Siskiyou Mountains, Oregon and California. *Ecol. Monogr.* **30**:279–338.

Whittaker, R. H. (1965). Dominance and diversity in land plant communities. *Science* **147**:250–260.

Whittaker, R. H. (1972). Evolution and measurement of species diversity. *Taxon.* **21**:213–251.

Whittaker, R. H. (1975). *Communities and Ecosystems*. 2nd ed. Macmillan, New York.

Whittaker, R. H. (1977). Evolution of species diversity in land communities. In M. K. Hecht, W. C. Steere, and B. Wallace, eds. *Evolutionary Biology*, Vol. 10. Plenum, New York, pp. 1–67.

Whittaker, R. H. (1978). Direct gradient analysis. In R. H. Whittaker, ed. *Ordination of Plant Communities*. 2nd ed. Junk, The Hague, pp. 7–50.

Whittaker, R. H., and Levin, S. A. (1977). The role of mosaic phenomena in natural communities. *Theor. Popul. Biol.* **12**:117–139.

Whittaker, R. H., Niering, W. A., and Crisp, M. D. (1979). Structure, pattern, and diversity of a mallee community in New South Wales. *Vegetatio* **39**:65–76.

Wilhm, J. L., and Dorris, T. C. (1968). Biological parameters for water quality criteria. *BioScience* **18**:477–481.

Williams, C. B. (1964). *Patterns in the Balance of Nature*. Academic Press, London.

Wilson, E. O., and Simberloff, D. S. (1969). Experimental zoogeography of islands: defaunation and monitoring techniques. *Ecology* **50**:267–278.

Wilson, E. O., and Willis, E. O. (1975). Applied biogeography. In M. L. Cody and J. M. Diamond, eds. *Ecology and Evolution of Communities*. Harvard Univ. Press, Cambridge, Mass., pp. 522–536.

12

SUCCESSION AND
RESILIENCE OF
ECOSYSTEMS

In this chapter we examine several processes of temporal change in ecosystems. Development of ecosystem properties in the absence of major disturbance (or in the interval between major disturbances) is the process of *succession*. The gradual breakdown of ecosystem integrity under the influence of chronic stress is the process of *retrogression*. The resistance to change under stress is referred to as ecosystem *inertia*. Ecosystem properties governing the pace, manner, and extent of recovery following disturbance (acute or chronic) are components of ecosystem *resilience*.

SUCCESSIONAL CONCEPTS

Because organisms are adapted to particular environmental conditions, changes in habitat conditions due to natural or human disturbance will cause a change in species abundance or composition at the site. The biological changes initiated by habitat change will continue under several conditions: (1) The species themselves continue to modify the physical environment by additions of litter, increases in plant cover, enhancement of soil development or other changes (so-called "autogenic" changes). (2) External perturbations or changes to the environment continue (so-called "allogenic" changes); these may be natural (e.g., accumulation of sediment from eroding uplands, storms, fires, climatic shifts) or human induced (e.g., air pollution, ionizing radiation). (3) The species occupying the site are able to tolerate the changed conditions but are not able to reproduce there as effectively as species dispersing in. Consequently a gradual change in species composition will occur as species better adapted to the habitat gain dominance and perpetuate themselves on the site.

The temporal sequence of community change that occurs by any of these three mechanisms comprises a succession, and phases within it are often referred to as *seral stages,* following Clements (1916). If a community composition is reached in which the species existing on the site are able to recruit new individuals to the species populations in roughly the same proportions as they now exist, the composition of the site will have reached dynamic equilibrium with the physical environment. Such a community has been termed a *climax* community with respect to species composition. It must be recognized that stability in biomass, nutrient stocks, and other ecosystem features may still be changing, so that not all features of ecosystem structure and function necessarily reach equilibrium simultaneously.

Concepts of succession and climax have come under renewed scrutiny in the past decade (Connell and Slatyer 1977, Drury and Nisbet 1973, Horn 1976, McIntosh 1980, Noble and Slatyer 1980, Peet and Christensen 1980, West et al. 1981a, Whittaker and Levin 1977), following their early development by plant ecologists (Clements 1916, Gleason 1927, Whittaker 1953). Clements (1916) had early emphasized the importance of autogenic processes in development and hypothesized that this feature would result in a characteristic and predictable sequence of seral stages toward a single climax type determined by regional climate. Later authors acknowledged the diversity of processes, both autogenic and allogenic, that occur simultaneously. Superimposed on a landscape mosaic of different habitat types, these processes could lead to a regional climax pattern in which effects of soil, slope, aspect, and other relatively stable physical features impose their influence on the nature of the climax community, even though climate remains the predominant selective force (Whittaker 1953).

More recently authors have emphasized the importance of stochastic (chance or random) elements in site development. Because of this random element many different successional sequences on a site may occur, leading possibly to different climax communities, or to no true long-term stability at all. The major chance elements in succession include the sequence in which species arrive on the site to usurp resources, and the nature, intensity, and frequency of disturbances (storms, fires, etc.) that buffet the site over time. Because of these chance elements succession has more recently been seen as a process in which the individual attributes of the species that have dispersed to the site determine the outcome of competitive struggles to dominate the site. The degree to which the successional sequence is truly predictable will depend in part on the extent to which patterns of dispersal and competition between these species result in clear hierarchies of dominance.

The importance of the individual adaptations of species, and the role of spatial heterogeneity of physical factors other than climate, in determining successional processes are also concepts that can be found in some of the earliest literature on succession (Gleason 1917, 1926, 1927, Cooper 1913, 1923, 1926). The Clementsian hypothesis of autogenic succession having been found only occasionally applicable, ecologists now emphasize that any of several mechanisms of succession may predominate, resulting in sequences of vegetative change which are more difficult to predict.

As investigators have focused on different aspects of succession, various terms have been proposed to classify the processes and mechanisms involved:

1. Site Conditions at Start of Succession. If the site contains remnants of previous colonization by vegetation (e.g., soil, seeds, rootstocks), the succession that proceeds is termed *secondary* succession; if only bare rock or open water is present, the succession is *primary* succession (Clements 1916).

2. Composition of Initial Community following Disturbance. If the species that are ultimately to dominate the climax community may be found among the initial post-disturbance pioneers, the process of succession is one of gradual change in relative abundance of species among those already present. This process is called *initial floristics* by Egler (1954). If the pioneers are replaced by a different set of species later in succession and again in the climax stage, the process is termed *relay floristics* by Egler (1954) or "replacement succession" (Whittaker and Levin 1977). Replacement succession is the model primarily envisioned by Clements (1916) (Figure 12.1).

3. Patterns of Replacement of Species. The species that occur in the climax community may be completely, partially, or not at all dependent on preceding species to prepare the habitat for their colonization (Figure 12.2).

Direct Succession

In Figure 12.2*a* any of four species may colonize and dominate the site permanently. This pattern is termed "direct succession" by Whittaker and Levin (1977). When secondary succession is occurring, this pattern of immediate reestablishment of climax species may simply be a form of initial floristics in which remnants (e.g., seeds, rootstocks) of species formerly present on the site reassert dominance over time (e.g., in Mediterranean-climate shrublands). It may also occur, however, when gaps are formed in an otherwise completely filled habitat. The first species to arrive and grow in the gap can then retain the spot indefinitely. Horn (1976) refers to this mechanism as one characteristic of habitats in which disturbance is chronic and patchy. Gaps in rain forests may fill in this way (Webb, Tracey, and Williams 1972), as may patches of rock in the intertidal zone (Dayton 1971, Levin and Paine 1974). In the case of primary or secondary succession, such a pattern would imply that biotic competition is much less important than space as a limiting factor to survival.

Plateau Stages

In some cases species that colonize a site early in succession are able to occupy the site for a long time, inhibiting the growth of later arrivals, although the inhibition eventually breaks down (on the order of decades or centuries). Examples include the grass *Aristida oligantha* in Oklahoma,

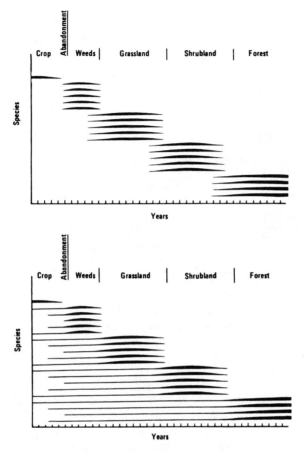

Figure 12.1. Hypothetical patterns of old-field succession. Upper: "relay floristics"; lower: "initial floristics." Reprinted with permission from Egler (1952–1954). Copyright (c) 1954, Junk, The Hague.

which releases a chemical inhibiting nitrifying bacteria (Rice and Pancholy 1972, 1973, 1974); monospecific stands of the shrub *Viburnum* which are able to resist invasion by forest tree species for 40 years or more (Niering and Egler 1955, Niering and Goodwin 1974); occupation of rocky intertidal areas by the gooseneck barnacle, *Pollicipes polymerus*, whose dominance is eventually broken down by the mussel *Mytilus californianus* (Sutherland 1974); occupation of floodplains in montane California for 500 years or more by redwoods (*Sequoiadendron giganteum*), even though, in the absence of fire or flood, the species will be replaced by the more shade-tolerant conifers of the area (e.g., Vankat 1977). These situations of unusually long-lasting seral stages have been termed "arrested succession" by Niering and Egler (1955) and "plateau stages" by Whittaker and Levin (1977). The process of inhibiting later arrivals has been termed the "inhibition pathway" by Connell and Slatyer 1977.

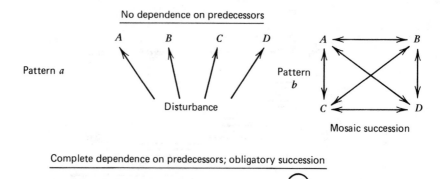

No dependence on predecessors

Pattern *a*

Disturbance

Pattern *b*

Mosaic succession

Complete dependence on predecessors; obligatory succession

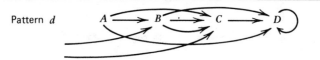

Pattern *c* $A \longrightarrow B \longrightarrow C \longrightarrow D$

Partial dependence on predecessors; tolerance pathway or competitive hierarchy

Pattern *d* $A \rightleftarrows B \cdots\!\!\rightarrow C \longrightarrow D$

Figure 12.2.   Four possible patterns of species replacement. Adapted, with permission, from Horn (1976). Copyright (c) 1976, Blackwell Scientific Publications, Oxford.

Mosaic Succession

In Figure 12.2*b* any species is shown capable of replacing any other. This would occur following the death of the predecessor, with the gap being colonized by the first species which is capable of growing in such a micro-habitat. Horn (1976, p. 193) notes that such a pattern of replacement would lead to "shifting successional mosaics of patches," a successional pattern that has been termed "mosaic succession" by Whittaker and Levin (1977). The more frequently disturbed the site is, the more likely it is that the potential colonists will reach a steady state in relative abundance, according to Horn (1976). Whether a steady state should even be expected depends on whether the sources and patterns of dispersal of colonists are uniform or variable.

Obligatory Succession

In Figure 12.2*c* each species is replaced by another in a sequence (i.e., relay floristics). This could occur through autogenic processes; for example, a plant could enrich the soil or shade the site, thus favoring colonization by new species. Connell and Slatyer refer to the latter as a "facilitation path-way," in which each species facilitates replacement by the next. Horn (1976) refers to it as "obligatory succession." The same sequence could also occur, however, if the habitat were being modified by an allogenic process such as incoming sediment, nutrients, or pollutants. Thus the observation of relay floristics as shown in Figure 12.2*c* does not allow one to state whether the cause of change is autogenic or allogenic. Often both processes occur simultaneously: the surface soil is enriched with nutrient both through nutrient input (from air or water) and by mineral "pumping": uptake of nutrients from deeper layers of soil or rock and deposition on the surface as litter.

Tolerance Pathway

In Figure 12.2*d* late successional or climax species can outcompete early arrivals to dominate a site; the climax species are also capable of colonizing the site in the absence of predecessors. Horn (1971) has studied such a case of "competitive hierarchy" for temperate forest trees. Considering light as a limiting factor in temperate forest succession, Horn finds that trees with leaves arranged in multiple layers in the canopy of the individual (e.g., conifers such as pine, *Pinus*) are capable of growing in full sunlight, but seedlings are intolerant of the shade. By contrast trees whose leaves do not overlap (monolayers; e.g., broad-leaved hardwoods such as maple, *Acer*) can grow in either sun or shade. Thus, if multilayered species colonize first, they will eventually be replaced by monolayered species; however, if monolayered species can occupy the site first, they and their offspring can occupy it. indefinitely. A successional pattern in which a species is successful whether or not preceded by another is termed the "tolerance pathway" by Connell and Slatyer (1977).

This brief review of successional concepts emphasizes several points: (1) processes driving change in community composition may be both internal and external to the community; (2) the sequence of species replacement is more predictable, the more a species is dependent on predecessors for habitat modification; and (3) the pace of succession is dependent on the extent to which remnants of former colonization were removed by disturbance, and the frequency, intensity and scale of subsequent disturbance. Having generated a larger number of models of possible successional patterns, ecologists are currently in the process of testing these in natural communities (e.g., Peet and Christensen 1980, West et al. 1981a) to determine which may be predominant in particular circumstances. A model of forest succession illustrating both classical and modern elements is shown in Figure 12.3.

ECOSYSTEM CHANGES IN STRUCTURE AND FUNCTION DURING SUCCESSION

Around 1970 several ecologists attempted to summarize understanding of successional trends in ecosystem structure and function (Odum 1969, Whittaker 1970, Woodwell 1970). Generalizations offered at the time were based largely on studies of temperate forests and abandoned fields. Large-scale ecosystem manipulation studies and computer modeling efforts in a variety of biomes in the years since have resulted in revised views of some of the

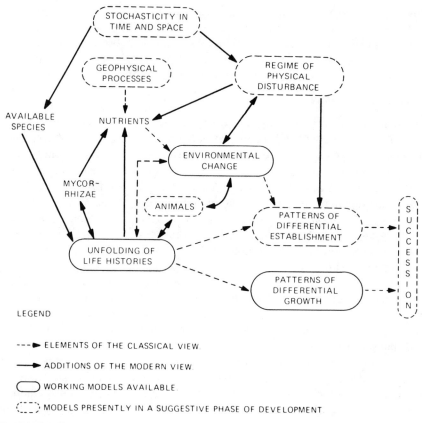

Figure 12.3.   Model of forest succession, showing classical and modern elements. Compiled by H. S. Horn and J. F. Franklin. Reprinted, with permission, from West et al. (1981b). Copyright (c) 1981, Springer-Verlag, New York.

early generalizations. We summarize some of these ideas to note how more recent research has modified our view of their generality.

Nutrient Cycling

Odum (1969) suggested that mineral cycles will tend to be open and "leaky" in early stages of succession and progress to a tightly recycling system in mature stages. Forests that have been recently clear-cut (e.g., Likens et al. 1970) or girdled (Johnson and Edwards 1979), and fertilized farm fields that lie fallow (Woodwell 1979) will indeed exhibit high losses of nitrate ion, due to microbial nitrification of organic matter in the early stages of secondary succession. In primary succession, however, where no preexisting organic pool exists, nitrate losses are low at first and only increase as the store of litter and soil organic matter increases (e.g., dune succession, Robertson and Vitousek 1981).

Growth

Generalizations about the gradual increase in net primary production, biomass, height, and stratal development of a community during succession hold reasonably well for temperate forest communities but much less well for nonforest communities subject to periodic disturbance. Mediterranean-climate evergreen shrublands, for example, respond to fire with a burst of growth; however, net annual additions to aboveground parts gradually decrease beyond the first five years (Specht 1969), until stimulated by the next fire. Stratal development, expressed as cover and richness of an herb as well as shrub layer, is greatest in the first few years following fire in these shrub lands, due to the stimulation of growth of herbs provided by the immediate post-fire environment. In the two or three fire-free decades that follow, herb growth is suppressed, and the community becomes virtually a single-stratum shrub community (Hanes 1971).

Diversity

Odum (1969) suggested that species richness and equitability increase during succession. This prediction has proved much too simple. As noted in Chapter 11, for wide application it does not hold for such communities as California coastal sage scrub (Figure 11.4) in which richness and equitability are highest in the first couple of years after fire. Whittaker (1975a) noted that richness and equitability were highest in Brookhaven temperate oak-pine forest in the middle stages of succession (6–15 years) and declined as the forest matured.

More recently several authors have suggested that the size and frequency of disturbance may influence diversity patterns during succession. Denslow (1980) notes that plant communities that frequently experience large-scale disturbance receive an influx of shade-intolerant pioneers when light gaps are created; as the canopy closes, species richness declines. Communities in which gap sizes are typically smaller contain species adapted to longer-term survival on the site, since the community normally replaces itself by recolonization of small gaps. In such communities Denslow (1980) postulates that species richness will gradually increase following large-scale disturbance.

Connell and Slatyer (1977) suggest that large areas of disturbance will be slower to colonize because internal sources of recolonization are absent. Connell (1978) and Huston (1979) suggest that habitats characterized by a moderate natural level of disturbance will provide the greatest opportunity for species turnover, colonization, and the persistence of high species richness.

All of these hypotheses require extensive testing with field data. They have in common a focus on the extent to which species capable of surviving in the habitat are adapted to disturbance of varying intensities. Those best adapted to responding to the natural level and type of fluctuation in that habitat, and to outcompeting other species vying for the site, are most likely to survive. If the community is repeatedly disturbed, species that would ultimately be outcompeted may be able to survive temporarily, thus enhancing species richness.

Can adaptations to natural stress pre-adapt a species to respond favorably to stresses of human origin? This is an area in which much research is needed. Westman et al. (1985) found that species of coastal sage scrub subjected to sulfur dioxide under controlled conditions are able to recover from death of foliage by the same basal resprouting mechanism that is used to recover from shoot death due to fire, herbivory, or frost. Some generalized adaptations to stress of species are discussed in Chapter 9.

PREDICTING SUCCESSIONAL TRENDS

The three most common methods of studying succession are by experimental manipulation, by observation of sites in a temporal sequence, and by numerical or conceptual modeling.

Experimental Manipulation

In the experimental approach vegetation recovery following a natural or human-induced disturbance is observed over a series of years. For example, the International Society for Vegetation Science has a European Working Group for Succession Research on Permanent Plots, whose investigators report results of changes in vegetation in areas earlier disturbed by grazing, fire, fertilizer, or some other agent (see, e.g., van der Maarel 1978, Trabaud and Lepart 1981). Techniques of measurement may involve permanent quadrats, transects, or mapping of individuals at a series of intervals. In North America the study of temperate forest recovery following clear-cutting in the Hubbard Brook watershed (Bormann and Likens 1979) is one of several long-term ecological research studies on succession. In Australia two of the longer-term studies of successional change are those of fertilizer application to heathlands in South Australia (Specht 1963, Heddle and Sprecht 1975) and recovery following exclosure from grazing in the arid zone of the Koonamore Reserve, South Australia (Carrodus et al. 1965, Hall et al. 1964). Such studies yield the most accurate and detailed picture of successional change, but they cannot be used to yield data on long-term trends very quickly.

Observation

The observational approach consists in studying sites of similar habitat, but different dates since disturbance, and predicting successional sequence by observing changes in biotic composition across the set of sites. Examples include the study of Cowles (1899) and Olson (1958) on vegetation of dunes of different ages surrounding Lake Michigan and that by Morrison and Yarranton (1973, 1974) on a similar dune sequence in Grand Bend, Ontario; the study of matched sites of coastal sage scrub in the Santa Monica Mountains of southern California at different ages since fire (Westman 1981); and the study comparing Chilean *Nothofagus* forests on recent landslides to older stands (Veblen and Ashton 1978).

Although this technique can provide successional sequences of a thousand years or more, the technique suffers several disadvantages: (1) the predisturbance vegetation may not have been the same on the different sites; (2) the intensity of disturbance may have differed between sites; (3) habitat conditions are never identical on different sites; and (4) the course of succession on different sites is assumed to be the same (i.e., determined by habitat), whereas stochastic elements are likely to cause differences both in the pathways of succession and the nature of the climax community. It is this last point that fuels the debate between those who hypothesize succession as largely a deterministic process and those who see it as a stochastic one. Normally elements of both are involved.

Ordination techniques (Chapter 10) may be used to rank plots of different ages, in order to plot the direction of succession (e.g., Austin 1977, for lawn plants; Allen and Koonce 1973, for algal seasonal dynamics; Figure 12.4). They may also be used to rank plots whose ages are not precisely known; the floristic axis extracted may be inferred to represent a temporal sequence, although obviously other habitat variables will influence the axis, so that this technique is less powerful (see, e.g., Goff 1968, Goff and Zedler 1972, and critique of this by Austin 1977). Austin (1977) and van der Maarel and Werger (1978) review the use of multivariate statistical techniques in analyzing successional data.

Modeling

The numerical approach involves computer simulation of successional patterns. Five such approaches to modeling growth of forest tree species are summarized in Table 12.1. The models vary in scale of analysis from single-species forest plantations through models that can simulate changes in composition alone, or composition and biomass, to models that predict changes in extent and type of forest on a regional level based on land use practices. The stand growth simulators have been used to predict effects on forest growth over several centuries due to chronic exposure to low levels of air pollutants (Kercher et al. 1980, Shugart et al. 1980, West et al. 1980).

Table 12.1. Five Types of Forest Models of Increasing Scale of Analysis

Model Type	Description	Application	Examples
Single-species stand simulators	Simulates yield of a plantation tree species based on species' growth over time	Can predict timber yield and economic gain from a single-species forest (even- or all-aged) over time	Dress (1970): general even-age stand simulator; Goulding (1972): Douglas fir stands
Population models	Simulates changes in stand composition by computing probabilities of replacement of one species by another (Markov process model).	Can predict changes in species composition of mixed species, all-aged stands	Hartshorn (1975): tropical forests; Horn (1975): temperate deciduous forests, New Jersey
Physiological stand models	Models physiological processes of water, nutrient uptake, photosynthesis and growth for dominants	Can predict physiologically based changes in forest growth due to toxic effects such as air pollutants	Luxmoore (1980): temperate deciduous forest
Stand growth stimulators	Models birth, death, and growth of each individual, including deterministic functions to represent inter specific competition	Can predict effects of altered habitat conditions of forest composition and yield	Botkin et al. (1972): northern hardwood forest, used to simulate effects of clearcutting. Mielke et al.; (1977): southern Arkansas upland forest simulator, used to test management strategies for endangered bird species dependent on forest insects

Table 12.1. (*Continued*)

Model Type	Description	Application	Examples
Large-scale simulators	Differential equations used to simulate changes in land area in different forest type categories for a large region	Can predict changes in diversity and extent of forest types at a regional level	Shugart et al. (1973): northern lower Michigan regional forests; Johnson and Sharpe (1976): Georgia Piedmont regional forests

Source: Adapted with modification from Shugart et al. (1981), with permission of Greenwood Press, Westport, Connecticut.

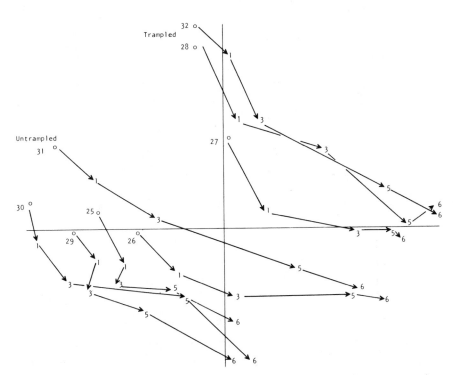

Figure 12.4. Use of ordination to plot trajectories of successional change. Permanent quadrats of trampled and untrampled lawn were measured at year 0, 1, 3, 5, and 6, and the plots ordinated by principal components analysis. Orthogonal axes represent major directions of floristic change in the sample set. Arrows connect the quadrat sample as its floristic composition changes over the years. Reprinted with permission from Austin (1977). Copyright (c) 1977, Junk, The Hague.

The JABOWA Model (Botkin et al. 1972) is a stand simulator in which annual birth, death, and growth of a set of species is simulated. The growth of each tree is increased yearly, depending on annual growing degree days, total canopy cover by taller trees, and size and number of other trees on the plot. The model has been used to predict recovery of biomass in the 500 years following clear-cutting in the Hubbard Brook temperate deciduous forest in New Hampshire (Bormann and Likens 1979) and the effects of elevated atmospheric CO_2 levels (Botkin 1973, 1977). It has also served as the basis for models of succession in other forests (e.g., FORET in southern Appalachian forests, Shugart and West 1977; Figure 12.5; SILVA in conifer forests of the California Sierra Nevada, Kercher et al. 1980). The SILVA model has been used to analyze effects of fire frequency and air pollution on succession (Figure 12.6).

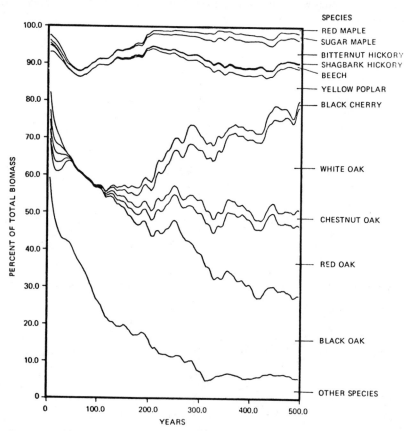

Figure 12.5. Percent changes in biomass of southern Appalachian deciduous forest species through 500 years of secondary succession, derived from the simulation model FORET (West et al. 1980). Reproduced from *Journal of Environmental Quality,* Volume 11, 1980, pp. 43–49, by permission of the American Society of Agronomy. Crop Science Society of America, and the Soil Science Society of America.

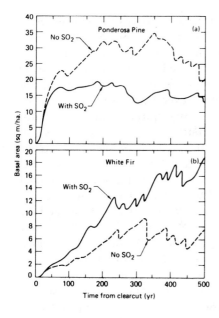

Figure 12.6. Basal area growth with and without SO₂ pollution for (*a*) *Pinus ponderosa* (Ponderosa pine) and (*b*) *Abies concolor* (white fir). Effects were simulated using a stand-growth model (SILVA) in which an SO₂ dose-response curve (Larson and Heck 1976) was incorporated. Reproduced from Kercher et al. (1980).

Not all modeling approaches to succession involve numerical computation. A conceptual approach to modeling is that proposed by Noble and Slatyer (1977), based on two vital attributes of key species in succession: mode of reproduction and of early establishment. Noble and Slatyer consider four types of reproduction or regeneration: seed dispersal, seed storage in soil or in fruits, and vegetative sprouting from above- or belowground parts. They also recognize three types of responses to competition with existing vegetation during early establishment, based on the tolerance, inhibition, or facilitation pathways outlined by Connell and Slatyer (1977) (Table 12.2).

Table 12.2.   Modes of Plant Reproduction and Early Establishment ("Vital Attributes") as Used in Successional Models of Noble and Slatyer (1977)

Mode of repro-duction	Seed dispersal (*D*)	Long-term viability in soil (*S*)	Protective cones or fruits (*C*)	Vegetative sprouting (*V*)	
Mode of early establishment	Tolerant of competition from established vegetation (*T*)		Intolerant of competition from established vegetation (*I*)		Requiring site preparation by previous vegetation (*R*)
Life history stages	Disturbance (*O*)	Replenishment of propagules (*p*)	Maturity (*m*)	Senescence or death (*l*)	Loss of propagules from site (*e*)

Different combinations of these vital attributes imply different life histories during the course of species persistence on a site. For example, a species that arrives by seed dispersal and is tolerant of competition from established vegetation (*T*) will establish itself by seed (*p*) on the site following disturbance (*O*), will grow to maturity from seed (*m*), will die (*l*), and disappear from the site (*e*) only at the time of the next disturbance (∞). By contrast, a species that regenerates by vegetative sprouting (*V*) and is intolerant of competition from established vegetation (*I*) will reestablish as both mature individuals (*m*) and seed (*p*), since the seed will be shed from flowering resprouts; the population will go extinct before the next disturbance since it is intolerant of competition by invaders and will not establish itself again until after the next disturbance. These and other life histories are illustrated in Figure 12.7. Noble and Slatyer (1977) originally developed this system for succession following fire but later (Noble and Slatyer 1980) generalized it to any type of acute disturbance. In Figure 12.8 the possible transitions between seed, juvenile, and adult form of species that will occur with and without fire are shown for each combination of vital attributes.

This system is of particular interest to environmental managers, since it has been used to model post-fire succession in a variety of fire-prone community types around the world (e.g., Australia, United States; Cattelino et al. 1979). Figure 12.9 shows possible successional sequences in an Australian *Eucalyptus* forest depending on the length of time between fires, based on vital attributes of forest dominants. The top line of the model shows that if a fire occurs when VI species only are present (*Eucalyptus, Daviesia*), these species will be able to reproduce. In the absence of fire SI species will be added (*Acacia, Dillwynia*), but after a century or so without fire the latter species and *Daviesia* will have died out. If a fire then occurs, only *Eucalyp-*

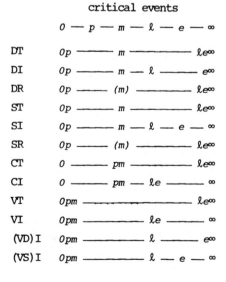

Figure 12.7. Life histories for species, with the modes of reproduction and establishment shown at left. Abbreviations from Table 12.2. Reproduced from Figure 3 of Noble and Slatyer (1977).

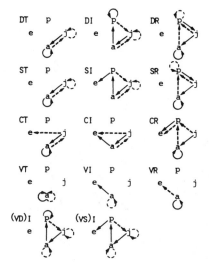

Figure 12.8. Transitions between states for species with combinations of vital attributes shown (abbreviations from Table 12.2). The states are p, present only as propagules; j, present as juveniles (and possibly propagules); a, present as adults (and possible propagules and juveniles); e, extinct from the community. Solid arrows show transitions in periods with no disturbance, and broken line arrows show the transitions due to fire. Reproduced from Figure 2 of Noble and Slatyer (1977).

tus will reappear, to which will be added SI species but not the short-lived *VI* (*Daviesia*). These features of succession can be predicted from Figure 12.9 without computation. Nevertheless, a detailed knowledge of the vital attributes of dominant species is necessary to build the model.

One feature revealed by Figure 12.9 is that a site may follow any of several successional pathways toward different predisturbance end points, depending on the attributes of individuals species and the timing of disturbance. The modeling approach, more than the experimental or observational

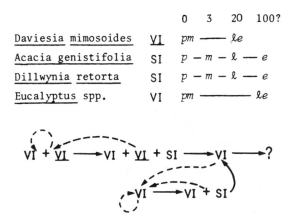

Figure 12.9. Successional sequences in an Australian eucalypt forest, predicted conceptually from vital attributes as shown in Figures 12.7–12.8. Broken line arrows show transitions due to fire. Solid arrows show successional changes over time, based on the maximum longevity of species in the absence of disturbance (indicated by numbers, in years, above life-stage events at top of figure). From Noble and Slatyer (1977).

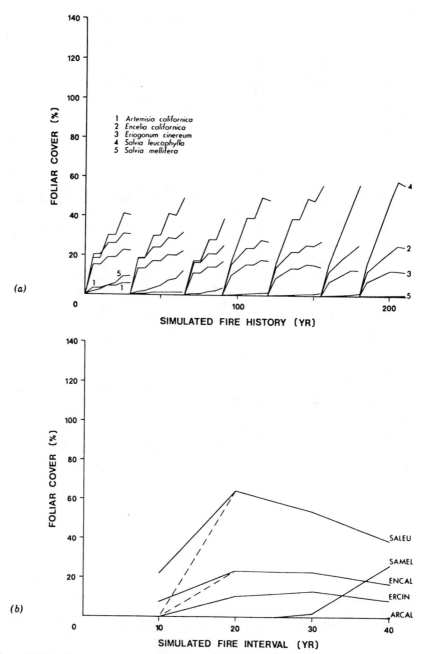

Figure 12.10. Computer simulation of post-fire succession in California coastal sage scrub, based on life-history attributes. (*a*) The change in foliar cover of five dominant species (1–5) on a site burned every 30 years ± 5 years, over a 210-year period. (*b*) The mean foliar cover of the same five dominant species after 200–210 years, with fires every 10, 20, 30, or 40 years. The dashed lines represent results under the assumption that resprouting vigor of 10-year-old shrubs is half that of 20-year-old shrubs. Part (a) reproduced from Malanson (1984), with permission of the author and of Elsevier Science Publishers, Amsterdam; part (b) reproduced from Malanson (1983), with author's permission.

approach, permits a clear understanding of the variety of possible successional pathways under different disturbance regimes over the long term. The
vital attributes approach of Noble and Slatyer permits predictions of the
change in species composition over time. The model predictions are qualitative, however, and do not incorporate quantitative changes in abundance
over time, nor the effects of competition.

Recently Malanson (1984) has developed a quantitative simulation model
of post-fire succession in California coastal sage scrub derived in part from
the Noble-Slatyer approach. Malanson simulates population processes of
birth, growth, and death in dominant species using a Leslie matrix, which
details recruitment from each age class to the next. He introduces the influence of competition by modeling growth of each species based on empirical
data and limiting further growth of the community, once foliar cover reaches
100%, to gaps in the canopy created by the death of shorter-lived individuals. The model successfully predicts changes in species composition and
abundance under various fire frequency regimes, involving different fire intensities. The model permits the testing of alternative strategies for prescribed burning in this vegetation type (Figure 12.10).

RETROGRESSION

The preceding discussion dealt with recovery of biotic communities following acute disturbance that removed all or most evidence of past colonization.
Such acute disturbances include fire, flood, landslide, volcanic eruption,
bulldozing, chaining, avalanche, tornado, and hurricane. When the disturbance is less severe but continual (i.e., chronic), the ecosystem changes
gradually, generally toward decline of biomass, soil thickness, and plant
cover, in a manner somewhat the reverse of succession. This so-called *retrogression* can be induced by such chronic disturbances as air pollution, low-
level ionizing radiation, toxic substance contamination, grazing, soil erosion, and trampling. In the case of aquatic systems such influences as
increased strength of currents or waves, introduction of toxic compounds in
low quantities, or increased sedimentation may be sources of chronic disturbance inducing retrogression. In the broadest sense both succession and
retrogression refer to changes in biotic composition over time. The terms are
intended to distinguish, however, between ecosystem development following a severe, short-lived stress (succession) and ecosystem deterioration
during a low-level, persistent stress (retrogression). Another way to view
retrogression is as the study of ecosystem change under chronic allogenic
influences of a stressful, rather than nurturing, nature.

Whittaker and Woodwell (1978) review cases of retrogression, including
such classics as the study of retrogressive changes in grassland communities
under grazing pressure (Dyksterhuis 1949), and of forest communities under
sulfur dioxide stress (Gordon and Gorham 1963) or chronic gamma irradia-

tion (Woodwell 1967, 1970, Woodwell and Whittaker 1968; see also Olsvig 1979). They conclude that to track retrogressive changes, using ecosystem-level attributes such as biomass, productivity, nutrient stocks, soil development, community height, stratal differentiation, or abundance of long-lived species, is fraught with difficulties. These attributes may be difficult to measure or insensitive as indicators of change, although in particular cases they may serve well. Whittaker and Woodwell (1978) favor the use of changes in species composition, either of indicator species or of whole communities, as indicators of retrogressive change. For this such techniques as weighted averages (Chapter 10) or changes in diversity may be used. As with the study of diversity often a knowledge of changes in species composition, combined with the study of other indicators of ecosystem structure noted here, may prove more revealing than any one index alone.

In his study of grazing effects Dyksterhuis (1949) introduced an approach that has since been widely adopted. After studying many rangelands of different "condition," that is, state of deterioration under grazing impact, he was able to classify the species found in the rangelands into three groups: (1) decreasers: sensitive "climax" species that were reduced in cover under even moderate grazing; (2) increasers: resistant species within the initial rangeland that increased in abundance under grazing influence; (3) invaders: species that were not present (or were restricted to local, disturbed areas) before grazing impact and increased in abundance following grazing. The percentage of each species type present in a rangeland can be used to "weight" composition along a gradient in condition from ungrazed to overgrazed. This is basically an application of the weighted averages technique (Chapter 10).

The U.S. Forest Service has classified a wide range of grassland species in the western United States by the decreaser-increaser-invader system (U.S. Forest Service 1970). This system may be used in successional, as well as retrogressional, studies. Kruse et al. (1979) used the system to evaluate successional changes in grassland species returning to two pinyon-juniper woodland sites that had been subject to acute disturbance during construction of a transmission-line corridor and a range-improvement project.

Winner and Bewley (1978), in a study of the effects of chronic SO_2 exposure on a white spruce (*Picea glauca*) association in central Alberta, observed an analogous grouping of species (Figure 12.11). Species like the dogwood (*Cornus canadensis*) and certain understory herbs (*Linnaea borealis, Mitella nuda, Pyrola asarifolia*) were "decreasers" (pattern *A*, Figure 12.11); grasses and horsetail (*Equisetum arvense*) were resistant to SO_2 and acted as "increasers" (pattern *B*, Figure 12.11); weedy "invader" species of lower (C_1) or higher (C) sensitivity to SO_2 included herbs like *Adenocaulon bicolor* and *Osmorhiza occidentalis*, and woody shrub-layer species, *Vaccinium vitis-idaea* and *Lonicera villosus*. An additional category of what may be called "opportunist" species increased in abundance as the cover of sensitive dominants declined, but these species were outcompeted by more

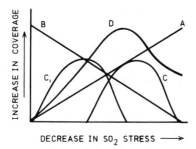

Figure 12.11. Patterns of response to chronic SO_2 stress observed in a white spruce forest in Alberta. *A*, decreasers; *B*, increasers; C_1, resistant weedy invaders; *C*, sensitive weedy invaders; *D*, native opportunists. Adapted from Winner and Bewley (1978) with permission of Springer-Verlag, New York.

SO_2-sensitive species at the lower-stress end of the SO_2 gradient; examples included two moss species, *Hylocomium splendens* and *Ptilium crista-castrensis*.

As with succession, patterns of retrogression vary with type of ecosystem and nature of stress. A knowledge of ecosystem behavior in the absence of stress, however, is an essential baseline for recognizing retrogressive changes following disturbance (see, e.g., study of changes following forest clear-cutting by Bormann and Likens 1979).

ECOSYSTEM INERTIA AND RESILIENCE

The dynamics of change in ecosystems can be viewed as generalized behavioral properties of complex systems. In the early ecological literature the dynamic properties of ecosystems in response to disturbance were referred to as stability properties (e.g., Holling 1973, May 1973, Orians 1975, Whittaker 1975b). During the 1970s a plethora of terms was proposed for a variety of such properties. We will use here three terms—inertia, amplitude, elasticity—in the sense proposed by Orians (1975) and two additional terms—hysteresis, malleability—defined by Westman (1978).

Inertia is the resistance to disturbance of a simple object (e.g., metal coil) or a complex homeostatic system (e.g., an ecosystem). Indeed, some authors have referred to this property simply as "resistance" (e.g., Harrison 1979, Harwell et al. 1977, Webster et al. 1975) or "resistance stability" (Smedes and Hurd 1981). Holling (1973) referred to this property as "resilience," but we will limit the term "resilience" to refer to the degree, manner, and pace of restoration of initial structure and function in an ecosystem *after* disturbance (Westman 1978).

Thus, whereas inertia refers to the resistance of a system in the face of a perturbing force, resilience properties refer to ways in which a disturbed system responds. We define four components of resilience:

1. *Elasticity.*   The time required to restore the system to its initial steady state.
2. *Amplitude.*   The zone of deformation from which the system will still return to its initial state.
3. *Hysteresis.*   The extent to which the path of degradation under chronic disturbance, and of recovery when disturbance ceases, are mirror images of each other.
4. *Malleability.*   The degree to which the new steady state established following recovery differs from the original steady state.

Examples of these properties are given in Table 12.3 and illustrated with a ball-and-cup model in Figure 12.12. We describe applications of these concepts to ecosystem studies that follow, emphasizing how they may be measured. In developing quantitative measures of these attributes, it must be recognized that the inertia or resilience of an ecosystem will differ depending on the nature of the disturbing force (fire, flood, etc.) and the attribute of the

Table 12.3.   Characteristics of Inertia and Resilience and Examples of Their Application

Characteristic	Definition	Example 1: A Metal Coil	Example 2: Ecosystem Subjected to Oil Spill
Inertia	Resistance to change	Force needed to stretch coil a given distance	Amount of oil that must accumulate over a given area in a given time period to cause a given level of ecosystem damage (such as local extinction of species X and Y)
Elasticity	Rapidity of restoration of a stable state following disturbance	Time required to spring back to initial size after stretching a given distance	Time required to recover initial structure or function following ecosystem damage (e.g., restoration of populations X and Y)

Table 12.3. (*Continued*)

Characteristic	Definition	Example 1: A Metal Coil	Example 2: Ecosystem Subjected to Oil Spill
Amplitude	Zone from which the system will return to a stable state	Distance beyond which coil cannot be stretched without being permanently deformed	Maximum amount of oil that can accumulate in an area such that damage sustained can be fully repaired (e.g., restoration of populations of X and Y)
Hysteresis	Degree to which path of restoration is an exact reversal of path of degradation	Degree to which region temporarily occupied by coil in springing back differs from region through which coil moved during stretching	Degree to which pattern of secondary succession is not an exact reversal of the pattern of retrogression experienced following impact (e.g., were the last species to die the first ones to return?)
Malleability	Degree to which stable state established after disturbance differs from the original steady state	Degree to which stretched coil remains stretched after deforming force is removed	Degree to which new climax ecosystem resembles the initial climax state (e.g., how closely do the species composition and equitability of new climax state resemble the old?)

Source: Reprinted with permission from Westman (1978), *BioScience* **28**:705–710. Copyright © 1978 by the American Institute of Biological Sciences.

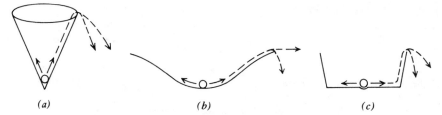

Figure 12.12.   Inertia and resilience analogized to a ball in a cup (cf. Hill 1975):

Inertia: $a > b > c$. How much disturbance is needed to cause the ball to move?

Elasticity: $a > b \gg c^*$. How fast will the ball return to its original position?

Amplitude: $c > a > b$ How much disturbance is needed to cause the ball to roll out of the cup?

Hysteresis: $a = b < c$ Will the ball roll back by the same route it took when initially displaced?

Malleability: $b > a > c$ How far away will the ball land when displaced from the cup with a given force of disturbance?

*c may stop somewhere on the flat surface (unstable equilibrium) rather than return to the central depression.

ecosystem being measured (cover, biomass, diversity, etc.). These features must be standardized when comparing inertia or resilience between ecosystems.

Inertia: Resistance to Change

One approach to the measurement of inertia is analogous to the acute toxicity test discussed in Chapter 9, in which the dose of toxicant required to kill 50% of a species population in 96 hours was recorded as an $LD_{50,96}$ index. By analogy, one can record the intensity of disturbance required to induce a 50% change in species composition or abundance (with a PS or CC index) from an ecosystem (Westman 1978).

One example is provided by Sousa (1980), who studied the algal communities growing on top of boulders in rocky intertidal areas. When these boulders are overturned so that the algae are under water, the algae gradually die from a combination of sea urchin grazing, inadequate oxygen, low light levels, and mechanical damage by crushing or abrasion. Sousa experimentally overturned 24 equal-sized boulders for a total of 17, 27, or 54 days, then righted them. The decline in percentage similarity of the algal community compared to the control (intact boulders) with these different levels of disturbance is shown in Figure 12.13. The inertia estimate at the PS = 50% level is 32 days.

Even with this carefully controlled experiment, several problems in arriving at an inertia estimate arise. First, the curve in Figure 12.13 is curvilinear,

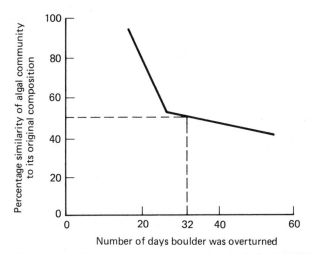

Figure 12.13. Measurement of inertia in an algal community attached to intertidal boulders in southern California. Data for eight boulders of each of three successional stages in algal community composition at onset of overturning are combined in the figure above. Number of days required to reduce PS to 50% is shown by dashed line. Data of Sousa (1980, Table 4).

so that more intervals of disturbance would ideally be needed to plot the curve accurately. Second, the arbitrary nature of selecting a PS level of 50% is emphasized. Beyond this Sousa found that other environmental variables, such as initial successional stage of the algal community before overturning and the artificial exclusion of urchin grazing, influenced the rate of decline in PS value during the period of disturbance. Thus to compare the inertia of algal communities in this boulder field in southern California to the effects of a similar perturbation elsewhere, boulders of comparable stages of succession should be used and other experimental conditions (e.g., inclusion or exclusion of grazers) standardized.

A second example concerns the resistance to change among vegetation strata in an oak-pine forest subjected to chronic gamma irradiation on Long Island, New York (Woodwell 1967). Woodwell found that species richness was a more linear indicator of compositional change than a PS index. Using the level of irradiation at which species richness is reduced 50%, the experiment indicated that forest vascular plants had a lower inertia (about 160 Roentgens per day for six months) than old-field herbs (1000 R/day) or forest lichens (2700 R/day) (Woodwell 1967, Woodwell and Gannutz 1967).

These two examples illustrate ways to measure ecosystem inertia experimentally. Frequently, however, impact analysts seek to predict the inertia of an ecosystem when experimental data are unavailable. Cairns and Dickson (1977) suggest a predictive index of inertia for aquatic ecosystems, based on six broad characteristics of the initial system. The index is the product of six parameters, each estimated on a three-point scale:

1. Degree of adaptation of the indigenous organisms to environmental fluctuation.
2. Degree of redundancy in function of the species within the ecosystem.
3. Cleansing capacity of the water body, measured by stream order, flow dependability, turbulent diffusivity, and flushing capacity of the water body.
4. Chemical buffering capacity of the water.
5. Proximity of parameters to a major ecological threshold such as shift from a cold to a warm water fishery.
6. Degree of effective management of water quality in the watershed.

Similar indexes could be developed for other ecosystem types; it is difficult to estimate any of these parameters uniformly and objectively, however (Westman 1978).

Harrison (1979) has suggested a way to express inertia mathematically, using rate of population growth as the ecosystem attribute of interest. He shows that the amount of negative feedback in the system (compensatory changes between two variables) and the sensitivity of the growth rate to environmental factors are the two attributes that most affect inertia in his model system. Most of the parameters in the index of Cairns and Dickson (1977) can be viewed as referring either to negative feedback properties within the biological system, sensitivity of organisms to the physical environment, or both. While there is no perfectly predictive index of inertia, clearly these two kinds of properties are of relevance to ecosystem inertia.

Harwell et al. (1977) discuss attempts to compare the inertia of eight major biome types simulated by a linear compartment model of nutrient flux (Webster et al. 1975). Harwell and coauthors found that the linear responses to stress exhibited by such models were unrealistic of the nonlinear behavior of ecosystems; nevertheless, their paper provides an interesting discussion of problems associated with modeling ecosystems for the measurement of inertia and resilience (particularly elasticity).

Components of Resilience

Elasticity: Restoration Time

Recovery to Steady State.   Elasticity can be measured as the time required to restore a particular ecosystem characteristic to within acceptably close limits of its pre-impact level. One would not expect a wholly recovered system to have a resemblance with the pre-impact site of PS = 100% in most cases, for at least two reasons: (1) natural spatial and sampling variation even within a single, undisturbed natural ecosystem will result in differences in species composition from plot to plot, so that an average PS among replicate samples may be 70–95% (see, e.g., Bray and Curtis 1957); (2) successional processes are a product of both stochastic and deterministic processes, as noted earlier in the chapter. Hence the probability of restoration of identical structure is small, even if the system is able to recover fully from artificial stress effects. One approach to this problem is to obtain an idea of the mean and range of values of composition in undisturbed steady-state systems of the type being disturbed, so that fiducial limits around the expected value can be set. This is similar to obtaining a mean ± standard deviation for the replicate PS value in point (1) but in reference to sites that have recovered from different disturbance episodes.

As with the measurement of inertia one could attempt to circumvent the problem of the time taken to return to the original state by measuring the time taken to return to, say, 50% PS with the original. One must be aware, however, that in such a measurement there is no guarantee that the system will eventually return fully to within acceptably close limits of its original state or that pace of recovery will be linear over time. Indeed, because longer-lived species tend to occupy later stages of succession in such biome types as forests, the pace of successional change tends to slow as climax is approached (Whittaker 1975).

Different ecosystem parameters will not change at the same rate through successional time. In the study of post-mining succession on Queensland sand dunes mentioned in Chapters 8 and 11, species richness on the fastest-recovering site (topsoil, fertilizer, exotic cover treatments) was within 53% of predisturbance levels (Westman and Rogers 1977), and heterogeneity (Simpson's index) was at 120% of predisturbance level within two years (Thatcher and Westman 1975). Floristic similarity to climax vegetation, however, was low (Sorenson's CC = 15%), and biomass was only 0.001% that of the climax forest (Westman 1978). Hence it is important to know which parameter one seeks to restore or to measure several ecosystem attributes simultaneously if overall ecosystem recovery is sought.

Recovery over short intervals.   Comparing the elasticity of recovering ecosystems over short, but comparable, periods may also be revealing. For example, in a similar study of post-mine rehabilitation in New South Wales to that in Queensland just cited, Clark (1975) found that the plots treated with fertilizer exhibited lower elasticity (PS = 0.17 with pre-mine composition) than those without fertilizer treatment (PS = 0.27). Such a measurement at least permits insight into factors controlling early post-disturbance colonization processes.

Harrison (1980–81) studied the relative elasticity of grasslands and heathlands in London's Green Belt to human trampling. Figure 12.14 plots elasticity on the ordinate, expressed as relative live cover of plants seven weeks after trampling disturbance (small dots), against inertia on the abscissa, expressed as relative live cover of plants still alive immediately after being subjected to 2000 passages of human footsteps during a summer period (400

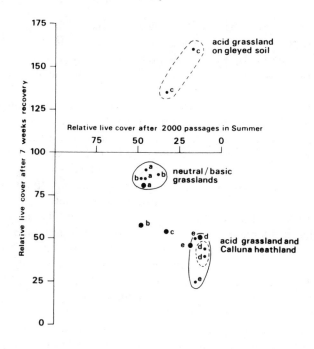

● = recovery after 8 months

Figure 12.14.   Inertia and elasticity of different British grasslands to trampling. Abscissa shows
live cover of vegetation at the end of summer trampling (a measurement of inertia); ordinate
shows relative live cover seven weeks after trampling ceased (a measurement of elasticity).
Cover values are expressed as percent of adjacent control plots. Reprinted, with permission,
from C. Harrison (1980–1981). Copyright Applied Science Publishers, Essex, England.

passages once a week for five weeks in July and August). The figure also
shows elasticity after eight months (large dots). The figure indicates that the
correlation between the inertia and elasticity of the different vegetation
types to trampling is weak, since sites of comparable inertia (*a, b, c*) show
quite different short-term elasticities (after seven weeks, site *c* has recovered much more than sites *a* or *b*). In addition the elasticity of a site can
change dramatically over time: elasticity of site *c*, and to a lesser extent site
b, has dropped by the eight-month sampling date, compared to the seven-
week sampling date, while elasticity of sites *a, d,* and *e* remain unchanged.
Also the fertility status of the habitat (acid vs. neutral-basic soil) seems to
influence the elasticity of the vegetation, although more replicates and a
longer sample period are needed to draw reliable conclusions.

Recovery over long intervals.   The study of elasticity in the mining and
human-trampling examples illustrates the need for much longer time intervals to judge the pace and path of vegetation recovery. One approach for
longer-term studies is the use of computer simulation models of the type

discussed earlier for forests. Indeed, Figure 12.5, which illustrates the recovery of forest composition in the first 500 years following clearing of a deciduous forest watershed, exemplifies the use of simulation models in estimating elasticity of an ecosystem in the longer term.

A second example of a modeling study of elasticity is that by Samuels and Lanfear (1982). These authors sought to model recovery time of two species of seabirds to damage from oil spills during oil production in the Gulf of Alaska. First, an estimate was made of the probability of oil spill occurrence, the likely movement of the spills, and the locations of seabird colonies along the coast, using the U.S. Geological Survey Oil Spill Trajectory Analysis model (Lanfear et al. 1979a, 1979b, Smith et al. 1980). This model combines data on the probability of an oil spill with a dispersion model based on current movements and data on the location of seabird colonies (U.S. Fish and Wildlife Service 1978).

To model effects on seabird populations, the authors considered the effect on population recovery of oil contact which affected only one-year-old birds, only adults, or all age classes. The extent of mortality of age classes was made sensitive to the season of contact with the oil spill. Population growth of the seabird colonies was modeled using a life-table approach, in which the proportion of each age class to survive to the next age class is estimated from empirical data. Results for different mortality scenarios for glaucous-winged gulls are shown in Figure 12.15. The authors found that if density-dependent mortality is assumed (i.e., greater mortality at larger population sizes, resulting in a logistic rather than exponential population growth curve; see Figure 6.22), recovery of the gull population would take 1.6 times longer than shown in Figure 12.15. Thus the model is sensitive to key assumptions about the nature of growth of the population. Further study of the bird populations in nature would be necessary to determine which assumptions are most realistic.

Amplitude: Brittleness

Foresters seeking the maximum sustainable yield of timber harvest from a site need to identify the threshold beyond which ecosystem repair to the initial state no longer occurs. Pastoralists similarly need to know the maximum stocking density of rangeland that will maintain a given composition of pasture sward indefinitely. The "amplitude" of an ecosystem is this threshold value, beyond which recovery approximately to the initial state will no longer occur. Not all ecosystems exhibit threshold behavior (Woodwell 1975). When dealing with complex systems, responses to perturbation from the system as a whole often appear to occur along a continuum. Nevertheless, components of the system, such as the population of a particular species, will often exhibit threshold behavior beyond which extinction of the population occurs and below which the population can recover.

An experimental study of amplitude is that of Baker (1971) on the effect of successive treatments of oil on the growth of a single species of British salt-

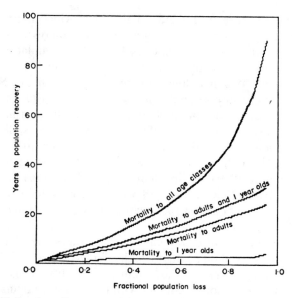

Figure 12.15. Estimation of long-term elasticity of glaucous-winged gull population to shore-line oil spill damage, using computer simulation. Time to recovery is shown as a function of the fraction of the gull population initially killed, using four alternative assumptions regarding age classes killed by oil. Reproduced with permission from W. B. Samuels and K. J. Lanfear, 1982. Simulations of seabird damage and recovery from oilspills in the northern Gulf of Alaska. *Journal of Environment Management* **15**:169–182. Copyright: Academic Press, Inc. (London) Ltd.

marsh grass, *Spartina anglica*. When the leaves of *S. anglica* are covered with oil, diffusion of oxygen to the roots of the plant is impaired, resulting in reduced growth or death. In a field experiment Baker measured the density of tillers (grass shoots) in treatment plots in which *S. anglica* shoots were painted with 4.5 liters of 90% crude oil residue 2, 4, 8, or 12 times over a 14-month period. Ten replicates of each 10 m² plot, plus a non-oiled control, were observed. Figure 12.16 shows the successive inhibition of recovery that resulted with increased duration of oil stress. The data suggest that 8 to 12 successive oilings may be the threshold beyond which species recovery will no longer occur; a longer post-disturbance monitoring period is necessary for a more certain conclusion.

Other cases where amplitude is of concern include tropical forests subjected to repeated fire by slash-and-burn agriculture (shifting cultivation; see Nye and Greenland 1960) or other causes (Fox 1976), aquatic communities subjected to pollutant discharges (e.g., Cairns and Dickson 1977, 1980), or other systems subjected to chronic, or repeated acute, disturbances. Threshold behavior may be studied with the use of simulation models (Shugart et al. 1980), experimental manipulations (Suter 1982), or observations of changes following disturbances of different intensities that may have occurred in an

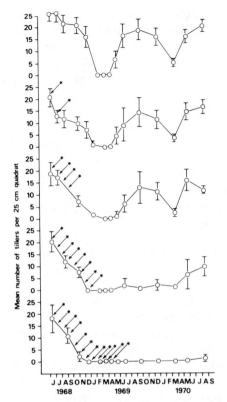

Figure 12.16. Amplitude. The results of successive oilings of plots of *Spartina anglica*. Oil was applied on the dates arrowed; 95% confidence limits (*t*-test) are indicated. Redrawn from Baker (1971) with permission of the Institute of Petroleum, London.

uncontrolled fashion (e.g., storms, floods, grazing of different intensities). Often the use of aerial photos or other forms of remote sensing over a time period may be used to follow changes after such disturbances (e.g., Hobbs 1980).

Hysteresis: Differences in Paths of Alteration and Recovery

The measurement of hysteresis involves comparing the ecosystem changes induced by chronic stress (retrogression) to those occurring during recovery (secondary succession). The measurement of hysteresis might be of value to impact analysts for a number of purposes. First, if only the retrogressive sequence, or only the successional one, is known, a knowledge of the degree of hysteresis can be useful in deciding whether the observed biological changes can be used to predict the sequence of changes in the reverse direction. For example, the Brookhaven oak-pine forest discussed under measures of inertia (Woodwell 1967) would appear to exhibit little hysteresis in relation to gamma irradiation. The plants most vulnerable to radiation damage were trees, shrubs, herbs, and lichens, in that order, whereas the pattern of secondary succession in such forests occurs normally in exactly the reverse order. Of course recovery in this forest would have to be studied further to establish the degree to which differences in degradation and recovery processes occurred in detail. If hysteresis appeared negligible, however, a detailed knowledge of succession or retrogression alone could be used to predict changes in the reverse process.

To the extent that hysteresis does exist, it implies that rehabilitation need not proceed by mimicking the reverse of retrogressive changes. Indeed, since, for example, retrogressive changes under grazing are likely to differ from those under air pollution, there is little a priori reason to expect secondary succession to be the reverse of the retrogression in both cases. Thus grazing often results in a retrogressive sequence ending with unpalatable or highly armed (spiny) species of plants; yet one would not plant such spiny or unpalatable species in starting to rehabilitate the rangeland. Enhanced understanding of hysteresis in the ecosystem could help an environmental manager to decide whether retrogressive changes are useful in guiding rehabilitation programs.

To exemplify the measurement of hysteresis, we may return to the experiment of Sousa (1980), in which algal-covered boulders in the intertidal zone were overturned for fixed periods, then righted and allowed to recover. Figure 12.17 plots retrogressive changes (under stress) and secondary successional changes (recovery) in two species of algae observed in this experiment. Hysteresis is represented by the gap between the curve for the stress period and that for recovery. Although hysteresis is observable, the standard error bars indicate that the hysteresis is not significant, at least with the limited sample size used. Thus the changes under stress may be a reasonable guide to the path of post-stress recovery in this situation.

Malleability: Ease of Permanent Alteration of Stable State

Malleability is a measure of the degree to which the steady state established after disturbance differs from the original. With a given level of disturbance, the more malleable ecosystem will be more drastically different when the new steady state is established. As with the measurement of elasticity and amplitude, one must establish a criterion for determining when a steady state has been achieved. Westman (1978) suggests as one possible criterion a stable state as one in which the mean difference in PS from each year to the next is no greater than 5% over x years and no greater than 10% between the first and the last year in the sequence. This definition could apply to a temporary equilibrium such as a late seral stage, as well as a more permanent one, depending on the number of years in the period of comparison. Even climax communities change over time, so that definitions of "steady state" should allow for some drift in ecosystem structure over time. As with elasticity and amplitude, historical records including remote sensing and simula-

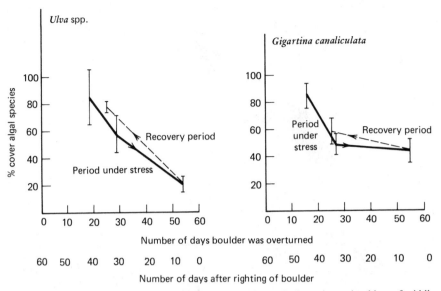

Figure 12.17. Hysteresis. Percent decrease in cover of two algal species on boulders of middle successional stage following overturning (stress) and average recovery 36 days later. Brackets indicate standard error. Hysteresis is shown by degree of difference between the two curves. Data from Sousa (1980, Tables 10, 11).

tion models can be used to estimate the degree of change in a system that has reached a new equilibrium following disturbance.

The Natural Environment Research Council (1976) developed an interesting predictive index of malleability of wading bird populations to the construction of freshwater reservoirs in the birds' feeding grounds. Reservoir construction was proposed for a tidal wetland area in England called the Wash. The diet and principal feeding grounds of the ten major wader species in the Wash were determined (e.g., Figure 12.18). The direct effects on the birds of loss of feeding habitat could be readily estimated from a knowledge of the location of proposed reservoir construction, and the densities of bird species feeding in each area were based on counts over a year (Table 12.4). The more difficult task, and the one for which a predictive index was needed, was the extent to which birds whose feeding grounds were destroyed would leave the Wash altogether, or be able to shift feeding location and food choice to survive in remaining Wash habitats.

The ecologists listed 10 considerations that would affect the risk that a bird species would leave the Wash permanently (Table 12.5). Based on these considerations, they rated each of the 10 wader species on a three-point scale of risk. By combining information on the proportion of each species whose feeding grounds would be destroyed by a particular reservoir scheme, with the risk score for the birds' ability to survive elsewhere in the Wash, an

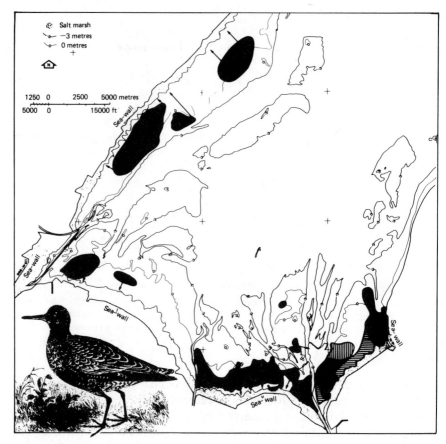

Figure 12.18.   Wash Study. Main feeding areas of the wading bird *Calidris canutus* ("knot") in the Wash. Black = main areas used by birds when exposed by tide. Hatching = different use on the eastern shore in the two years of study. Arrows indicate areas used as the tide ebbed and flowed. Reprinted from Figure 12 of NERC (1976), with permission of The Institute of Terrestrial Ecology, Cambridge, U.K.

ordinal scale of impact of each reservoir scheme on the bird populations was constructed (Table 12.4). This approach provides a prediction of the extent to which the impacted Wash will differ from the original in bird populations and hence constitutes a predictive index of malleability.

General Predictive Indexes of Resilience

In the previous example a predictive index of malleability was developed for the particular case at hand. There have been attempts, however, to isolate more general features of ecosystems which could serve as predictive indexes of resilience.

Table 12.4. Number (and % of Annual Peak) of Three Wader Species Displaced from Feeding Grounds by Each of Four Alternative Reservoir Schemes, Risk of Loss of Bird Species by Displacement, and Overall Index of Effect of Reservoir on Bird Species Loss

Alternative Reservoir Design Schemes	Number of Birds Lost	Species: % of Annual Peak[b]	Risk Category[c]	Overall Effect[d]
Oystercatcher				
Wingland	430	3	M	−1
Westmark	400	2		−1
Breast	370	2		−1
Bull Dog	2250	15		1
Curlew				
Wingland	1300	16	M	1
Westmark	1130	14		1
Breast	1060	13		1
Bull Dog	500	6		−1
Knot				
Wingland	10,560	18	H	2
Westmark	9350	16		2
Breast	12,680	22		3
Bull Dog	17,790	31		3

Source: From NERC (1976, Tables 5A, 7A), excerpted with permission of The Institute of Terrestrial Ecology, Cambridge, U.K.

[a] Number of birds whose feeding grounds would be removed by reservoir scheme, based on feeding as the tide ebbs and flows (rather than at low water).

[b] Percent of peak counts for the entire Wash for each species during the year feeding observations were made.

[c] L = low, M = medium, and H = high, based on attributes in Table 12.5.

[d] Categories of effect: −1 = small effect, less than 5% of the Wash population currently feeding in the proposed reservoir site; 1 = less than 10% of the Wash population of a high risk species currently feeding in the site, less than 20% of a medium risk species or less than 30% of a low risk species; 2 = 10–20% of a high risk species, 20–30% of a medium risk species, 30–40% of a low risk species, 3 = 20–40% of a high risk species, 30–50% of a medium risk species, 40–50% of a low risk species; 3+ = very large effect, more than 40% of a high risk species and more than 50% of all other species.

Table 12.5.   Characteristics of Wader Species Which Affect the Probability That the Numbers Permanently Lost to the Wash Will Equal the Numbers Directly Displaced by the Reservoirs

A. High Risk Characteristics	B. Low Risk Characteristics
1. Current immobility within the Wash or between estuaries and no indication of an ability to locate alternative feeding areas	1. High mobility within the Wash and between estuaries and demonstration of a current ability to locate new feeding grounds
2. A major part of the winter spent on the Wash	2. Passage migration or high population turnover
3. Specialist feeding on invertebrates restricted in distribution or size	3. Specialist feeding on widely distributed invertebrates and feeding on, or capable of adapting to, a wide range of prey species and sizes
4. Feeding on relatively sedentary prey items (cockles, *Macoma*, etc.)	4. Feeding on prey brought in by each tide (shrimps, crabs etc.)
5. A restricted range of feeding methods	5. A wide repertoire of feeding methods
6. Feeding by visual searching (would be more affected by an increase in bird density)	6. Touch feeding
7. Evidence of current depletion of a relatively high percentage of food resources by spring	7. Evidence of current depletion of a relatively low percentage of food resources
8. Signs of current difficulty in finding food (e.g., long time spent feeding, night feeding, fighting on feeding grounds)	8. No current signs of difficulty in finding food resources
9. Overt density-dependent fighting	9. No fighting
10. Apparent ceiling to numbers in preferred sites (as indicated by distribution as the birds arrive in autumn and spread out over the feeding grounds)	10. No evidence of an apparent ceiling to numbers in preferred sites

Source: Reprinted from NERC (1976), with permission of The Institute of Terrestrial Ecology, Cambridge, U.K.

One such is the recovery index of Cairns and Dickson (1977) which, like the "inertia index" noted earlier, is the product of six parameters, each estimated on a three-point scale:

1. Proximity of recolonization sources.
2. Mobility of propagules.
3. Physical suitability of habitat for recolonization.
4. Chemical suitability of habitat for recolonization.
5. Toxicity of disturbed habitat.
6. Effectiveness of human management structures to facilitate rehabilitation procedures.

This index is not a measure of any one component of resilience but is rather a measure of the probability of recolonization of a site. While such crude predictive indexes may be of value in approaching the study of an ecosystem about which little is known, in most cases one can gain a more detailed understanding of the likelihood of recolonization based on past efforts at rehabilitating particular ecosystems. Examples of books that discuss rehabilitation efforts in a variety of ecosystems are those of Bradshaw and Chadwick (1980), Cairns et al. (1977), Holdgate and Woodman (1976), Lenihan and Fletcher (1976), Lewis (1982), Schaller and Sutton (1979), Thorhaug (1979) and Wright (1978). Bibliographies include those of Czapowskyj (1976) and Goodman and Bray (1975).

Smith et al. (1975) and Suffling (1980) have suggested that the age of an ecosystem be taken as an index of its elasticity (which they call "sensitivity to disturbance"). Their reasoning is simple: older ecosystems will, by definition, take longer to replace with systems of the same age (Suffling 1980). Suffling suggests measuring the so-called "sensitivity to disturbance" of a landscape by summing the proportions of the landscape of each age, weighted by the log of the area of each age class. Such an index confounds several important ecological properties and can be very misleading. As we have seen, ecosystems vary in their rate of recovery of ecosystem properties; thus the age of a particular site tells us nothing about how long a particular property of interest (species richness, biomass, composition, etc.) has been in a steady state on the site. As a result the mere age of a site is a very poor indicator of the rate at which a particular attribute of the ecosystem will recover its original condition, if indeed the initial steady state is reachieved at all. Furthermore, since different ecosystems exhibit different elasticities, summing sites by age without regard to ecosystem type further confounds rates of recovery and is likely to result in an index of little or no predictive value.

Despite the desirability of finding generalized predictive indexes of resilience, it seems likely that the accumulated experience from field observations of inertia and resilience in particular ecosystems is likely to serve as a more accurate and efficient route towards developing a predictive theory of ecosystem response to disturbance at present. Post-impact monitoring of sites can be a rich source of information on which a future understanding of ecosystem inertia and resilience can be based. In this way projects for which environmental impact statements were prepared, and a post-impact monitoring program carried out, can provide useful information to basic research ecologists. The research scientists in turn can use these case studies to develop improved guidelines for impact prediction. It is through such a cooperative process of information feedback that a theoretically sound and predictive science of ecological impact assessment may develop most rapidly.

REFERENCES

Allen, T. F. H., and Koonce, J. F. (1973). Multivariate approaches to algal strategems and tactics in systems analysis of phytoplankton. *Ecology* **54**:1234–1246.

Austin, M. P. (1977). Use of ordination and other multivariate descriptive methods to study succession. *Vegetatio* **35**:165–175.

Baker, J. M. (1971). Studies on saltmarsh communities—successive spillages. In E. B. Cowell, ed. *The Ecological Effects of Oil Pollution on Littoral Communities.* Inst. Petroleum, London, pp. 21–32.

Bormann, F. H., and Likens, G. E. (1979). *Pattern and Process in a Forested Ecosystem.* Springer-Verlag, New York.

Botkin, D. B., Janak, J. F., and Wallis, J. R. (1972). Some ecological consequences of a computer model of forest growth. *J. Ecol.* **60**:948–972.

Botkin, D. B. (1973). Estimating the effects of carbon fertilization on forest composition by ecosystem simulation. In G. M. Woodwell and E. V. Pecan, eds. *Carbon and the Biota.* Brookhaven Symp. Biol. 24. NTIS, Springfield, Va., pp. 384–444.

Botkin, D. B. (1977). Forests, lakes, and the anthropogenic production of carbon dioxide. *BioScience* **27**:325–331.

Bradshaw, A. D., and Chadwick, M. J. (1980). *The Restoration of Land: the Ecology and Reclamation of Derelict and Degraded Land.* Univ. California Press, Berkeley.

Bray, J. R., and Curtis, J. T. (1957). An ordination of the upland forest communities of southern Wisconsin. *Ecol. Monogr.* **27**:325–349.

Cairns, J., Jr., and Dickson, K. L. (1977). Recovery of streams and spills of hazardous materials. In J. Cairns, Jr., K. L. Dickson, and E. E. Herricks, eds. *Recovery and Restoration of Damaged Ecosystems.* Univ. Virginia Press, Charlottesville, pp. 24–42.

Cairns, J., Jr., and Dickson, K. L. (1980). Risk analysis for aquatic ecosystems. In *Biological Evaluation of Environmental Impacts.* Rep. FWS/OBS-80/26. U.S. Dept. Interior, Fish and Wildlife Service, and Council Environ. Quality, Washington, D.C., pp. 73–83.

Cairns, Jr., Jr., Dickson, K. L., and Herricks, E. E., eds. (1977). *Recovery and Restoration of Damaged Ecosystems.* Univ. Virginia Press, Charlottesville.

Carrodus, B. B., Specht, R. L., and Jackman, M. E. (1965). The vegetation of Koonamore Station, S.A. *Trans. R. Soc. S. Aust.* **89:**41–57.

Cattelino, P. J., Noble, I. R., Slatyer, R. O., and Kessell, S. R. (1979). Predicting the multiple pathways of plant succession. *Environ. Manage.* **3:**41–50.

Clark, S. S. (1975). The effect of sand mining on coastal heath vegetation in New South Wales. In J. Kikkawa and H. A. Nix, eds. *Managing Terrestrial Ecosystems.* Proc. Ecol. Soc. Aust. 9, pp. 1–16.

Clements, F. E. (1916). *Plant Succession: An Analysis of the Development of Vegetation.* Carnegie Inst. Wash. Publ. **242:**1–512.

Connell, J. H. (1978). Diversity in tropical rainforests and coral reefs. *Science* **199:**1302–1310.

Connell, J. H., and Slatyer, R. O. (1977). Mechanisms of succession in natural communities and their role in community stability and organization. *Am. Nat.* **111:**1119–1144.

Cooper, W. S. (1913). The climax forest of Isle Royale, Lake Superior, and its development. *Bot. Gaz.* **55:**1–44, 115–140, 189–235.

Cooper, W. S. (1923). The recent ecological history of Glacier Bay, Alaska. II: The present vegetation cycle. *Ecology* **4:**223–246.

Cooper, W. S. (1926). The fundamentals of vegetation change. *Ecology* **7:**391–413.

Cowles, H. C. (1899). The ecological relations of the vegetation on the sand dunes of Lake Michigan. *Bot. Gaz.* **27:**95–117, 167–202, 281–308, 361–391.

Czapowskyj, M. M. (1976). Annotated bibliography on the ecology and reclamation of drastically disturbed areas. U.S. Forest Service, Gen. Tech. Rep. NE-21. Upper Darby, Pa.

Dayton, P. K. (1971). Competition, disturbance, and community organization: the provision and subsequent utilization of space in a rocky intertidal community. *Ecol. Monogr.* **41:**351–389.

Denslow, J. S. (1980). Patterns of plant species diversity during succession under different disturbance regimes. *Oecologia* **46:**18–21.

Dress, P. E. (1970). A system for the stochastic simulation of even-aged forest stands of pure species composition. Ph.D. dissertation. Purdue Univ. Lafayette, Ind.

Drury, W. H., and Nisbet, I. C. T. (1973). Succession. *J. Arnold Arb.,* Harvard Univ., **54:**331–368.

Dyksterhuis, E. J. (1949). Condition and management of range land based on quantitative ecology. *J. Range. Manage.* **2:**104–115.

Egler, F. E. (1954). Vegetation science concepts I. Initial floristic composition, a factor in old-field vegetation development. *Vegetatio* **4:**412–417.

Fox, J. E. D. (1976). Constraints on the natural regeneration of tropical moist forest. *For. Ecol. Manage.* **1:**37–65.

Gleason, H. A. (1917). The structure and development of the plant association. *Bull. Torrey Bot. Club* **44:**463–481.

Gleason, H. A. (1926). The individualistic concept of the plant association. *Bull. Torrey Bot. Club* **53:**1–20.

Gleason, H. A. (1927). Further views on the succession concept. *Ecology* **8**:299–326.

Goff, P. G. (1968). Use of size stratification and differential weighting to measure forest trends. *Am. Midl. Nat.* **79**:402–418.

Goff, P. G., and Zedler, P. H. (1972). Derivation of species succession vectors. *Am. Midl. Nat.* **87**:397–412.

Goodman, G. T., and Bray, S. A. (1975). Ecological aspects of the reclamation of derelict and disturbed land. GEO Abstracts, Norwich, U.K.

Gordon, A. G., and Gorham, E. (1963). Ecological aspects of air pollution from an iron-sintering plant at Wawa, Ontario. *Can. J. Bot.* **41**:1063–1078.

Goulding, C. J. (1972). Simulation techniques for a stochastic model of the growth of Douglas fir. Ph.D. dissertation. Univ. British Columbia, Vancouver.

Hall, E. A. A., Specht, R. L., and Eardley, C. M. (1964). Regeneration of the vegetation on Koonamore Vegetation Reserve, 1926–1962. *Aust. J. Bot.* **12**: 205–264.

Hanes, T. L. (1971). Succession after fire in the chaparral of southern California. *Ecol. Monogr.* **41**:27–52.

Harrison, C. (1980–81). Recovery of lowland grassland and heathland in southern England from disturbance by seasonal trampling. *Biol. Conserv.* **19**:119–130.

Harrison, G. W. (1979). Stability under environmental stress: resistance, resilience, persistence and variability. *Am. Nat.* **113**:659–669.

Hartshorn, G. S. (1975). A matrix model of tree population dynamics. In F. B. Golley and E. Medina, eds. *Tropical Ecological Systems*. Springer-Verlag, New York, pp. 41–51.

Harwell, M. A., Cropper, W. P., and Ragsdale, H. L. (1977). Nutrient recycling and stability: a reevaluation. *Ecology* **58**:660–666.

Heddle, E. M., and Specht, R. L. (1975). Dark Island Heath (Ninety-Mile Plain, South Australia). VIII: The effects of fertilizers on composition and growth, 1950–72. *Aust. J. Bot.* **23**:151–164.

Hill, A. R. (1975). Ecosystem stability in relation to stresses caused by human activities. *Can. Geogr.* **19**:206–220.

Hobbs, E. (1980). Effects of grazing on the northern populations of *Pinus muricata* on Santa Cruz Island, California. In D. M. Power, ed. *The California Islands: Proc. of a Multidisciplinary Symp.* Santa Barbara Museum Nat. Hist., Santa Barbara, Calif., pp. 159–165.

Holdgate, M. W., and Woodman, M. J., eds. (1976). *The Breakdown and Restoration of Ecosystems*. Plenum, New York.

Holling, C. S. (1973). Resilience and stability of ecological systems. *Ann. Rev. Ecol. Syst.* **4**:1–24.

Horn, H. S. (1971). *The Adaptive Geometry of Trees*. Princeton Univ. Press, Princeton.

Horn, H. S. (1975). Forest succession. *Sci. Amer.* **232**:90–98.

Horn, H. S. (1976). Succession. In R. M. May, ed. *Theoretical Ecology. Principles and Applications*. Saunders, Philadelphia, pp. 187–204.

Huston, M. (1979). A general hypothesis of species diversity. *Am. Nat.* **113**:81–101.

Johnson, D. W., and Edwards, N. T. (1979). The effects of stem girdling on biogeochemical cycles within a mixed deciduous forest in eastern Tennessee. II: Soil nitrogen mineralization and nitrification rates. *Oecologia* **40**:259–271.

Johnson, W. C., and Sharpe, D. M. (1976). An analysis of forest dynamics in the northern Georgia Piedmont. *For. Sci.* **22**:307–322.

Kercher, J. R., Axelrod, M. C., and Bingham, G. E. (1980). Forecasting effects of SO_2 pollution on growth and succession in a western conifer forest. In P. R. Miller, ed. *Effects of Air Pollutants on Mediterranean and Temperate Forest Ecosystems.* U.S. Forest Service, Gen. Tech. Rep. PSW-43. Berkeley, Calif., pp. 200–202.

Kruse, W. H. (1979). Community development in two adjacent pinyon-juniper eradication areas twenty-five years after treatment. *J. Environ. Manage.* **8**:237–247.

Lanfear, K. J., Nakassis, A., Samuels, W. B., and Shoen, C. (1979b). An oilspill risk analysis for the northern Gulf of Alaska (Proposed Sale 55) outer continental shelf. USGS Open File Rep. 79–284.

Lanfear, K. J., Smith, R. A., and Slack, J. R. (1979a). An introduction to the oilspill risk analysis model. In Proc. Offshore Oil Technology Conf., Houston, Tex., pp. 2173–2175.

Larson, R. I., and Heck, W. W. (1976). An air quality data analysis system for interrelating effects, standards, and needed source reductions. Part 3. Vegetation injury. *J. Air. Poll. Control Assn.* **26**:325–333.

Lenihan, J., and Fletcher, W. W., eds. (1976). *Environment and Man. Vol. 4: Reclamation.* Academic Press, New York.

Levin, S. A., and Paine, R. T. (1974). Disturbance, patch formation, and community structure. Proc. Natl. Acad. Sci., Washington, **71**:2744–2747.

Lewis, R. R., III, ed. (1982). *Creation and Restoration of Coastal Plant Communities.* CRC, Boca Raton, Fla.

Likens, G. E., Bormann, F. H., Johnson, N. M., Fisher, D. W., and Pierce, R. S. (1970). Effects of forest cutting and herbicide treatment on nutrient budgets in the Hubbard Brook watershed and ecosystem. *Ecol. Monogr.* **40**:23–47.

Luxmoore, R. J. (1980). Modeling pollutant uptake and effects on the soil-plant-litter system. In P. R. Miller, ed. *Effects of Air Pollutants on Mediterranean and Temperate Forest Ecosystems.* U.S. Forest Service, Gen. Tech. Rep. PSW-43. Berkeley, Calif., pp. 174–180.

Malanson, G. P. (1983). A model of post-fire succession in Californian coastal sage scrub. Ph.D dissertation. Univ. California, Los Angeles.

Malanson, G. P. (1984). Linked Leslie matrices for the simulation of succession. *Ecol. Modelling* **21**:13–20.

May, R. M. (1973). *Stability and Complexity in Model Ecosystems.* Princeton Univ. Press, Princeton.

McIntosh, R. P. (1980). The relationship between succession and the recovery process in ecosystems. In J. Cairns, Jr., ed. *The Recovery Process in Damaged Ecosystems.* Ann Arbor Sci., Ann Arbor, Mich., pp. 1–62.

Mielke, D. C., Shugart, H. H., and West, D. C. (1977). User's Manual for FORAR: a Stand Model for Composition and Growth of Upland Forests of southern Arkansas. Rep. ORNL/TM-5767. Oak Ridge Natl. Lab., Oak Ridge, Tenn.

Morrison, R. G., and Yarranton, G. A. (1973). Diversity, richness, and evenness during a primary sand dune succession at Grand Bend, Ontario. *Can. J. Bot.* **51**:2401–2411.

Morrison, R. G., and Yarranton, G. A. (1974). Vegetational heterogeneity during a primary sand dune succession. *Can. J. Bot.* **52**:397–410.

Natural Environment Research Council (1976). The Wash Water Storage Scheme Feasibility Study. A Report on the Ecological Studies. NERC Publ. Ser. C, No. 15.

Niering, W. A., and Egler, F. E. (1955). A shrub community of *Viburnum lentago,* stable for twenty-five years. *Ecology* **36**:356–360.

Niering, W. A., and Goodwin, R. H. (1974). Creation of relatively stable shrublands with herbicides: arresting "succession" on rights-of-way and pastureland. *Ecology* **55**:784–795.

Noble, I. R., and Slatyer, R. O. (1977). Post-fire succession of plants in Mediterranean ecosystems. In *Proc. Symp. Environmental Consequences of Fire and Fuel Management in Mediterranean Climate Ecosystems.* U.S. Forest Service, Gen. Tech. Rep. WO-3, Washington, D.C. pp. 27–36.

Noble, I. R., and Slatyer, R. O. (1980). The use of vital attributes to predict successional changes in plant communities subject to recurrent disturbances. *Vegetatio* **43**:5–21.

Nye, P. H., and Greenland, D. J. (1960). *The Soil under Shifting Cultivation.* Tech. Comm. 51, Commonwealth Bur. Soils, Harpenden. Commonwealth Agric. Bur., Farnham Royal, U.K.

Odum, E. P. (1969). The strategy of ecosystem development. *Science* **164**:262–270.

Olson, J. S. (1958). Rates of succession and soil changes on southern Lake Michigan sand dunes. *Bot. Gaz.* **119**:125–170.

Olsvig, L. S. (1979). Pattern and diversity of the irradiated oak-pine forest, Brookhaven, New York. *Vegetatio* **40**:65–78.

Orians, G. H. (1975). Diversity, stability and maturity in natural ecosystems. In W. H. van Dobben and R. H. Lowe-McConnell, eds. *Unifying Concepts in Ecology.* Junk: The Hague, pp. 64–65.

Peet, R. K., and Christensen, N. L. (1980). Succession: a population process. *Vegetatio* **43**:131–140.

Rice, E. L., and Pancholy, S. K. (1972). Inhibition of nitrification by climax vegetation. *Amer. J. Bot.* **59**:1033–1040.

Rice, E. L., and Pancholy, S. K. (1973). Inhibition of nitrification by climax ecosystems. II: Additional evidence and possible role of tannins. *Amer. J. Bot.* **60**:691–702.

Rice, E. L., and Pancholy, S. K. (1974). Inhibition of nitrification by climax ecosystems. III: Inhibitors other than tannins. *Amer. J. Bot.* **61**:1095–1103.

Robertson, G. P., and Vitousek, P. M. (1981). Nitrification potentials in primary and secondary succession. *Ecology* **62**:376–386.

Samuels, W. B., and Lanfear, K. J. (1982). Simulations of seabird damage and recovery from oilspills in the northern Gulf of Alaska. *J. Environ. Manage.* **15**:169–182.

Schaller, F. W., and Sutton, P., eds. (1979). *Reclamation of Drastically Disturbed Lands*. Amer. Soc. Agron., Madison, Wisc.

Shugart, H. H., Crow, T. R., and Hett, J. M. (1973). Forest succession models: a rationale and methodology for modeling forest succession over large regions. *For. Sci.* **19**:203–212.

Shugart, H. H., Hopkins, M. S., Burgess, I. P., and Mortlock, A. T. (1980). The development of a succession model for subtropical rain forest and its application to assess the effects of timber harvest at Wiangaree State Forest, New South Wales. *J. Environ. Manage.* **11**:243–265.

Shugart, H. H., Klopatek, J. M., and Emanuel, W. R. (1981). Ecosystems analysis and land-use planning. In E. J. Kormondy and J. F. McCormick, eds. *Handbook of Contemporary Development in World Ecology*. Greenwood, Westport, Conn., pp. 665–699.

Shugart, H. H., McLaughlin, S. B., and West, D. C. (1980). Forest models: their development and potential applications for air pollution effects research. In P. R. Miller, ed. *Effects of Air Pollutants on Mediterranean and Temperate Forest System*. U.S. Forest Service, Gen. Tech. Rep. PSW-43, Berkeley, Calif., pp. 203–214.

Shugart, H. H., and West, D. C. (1977). Development of an Appalachian deciduous forest succession model and its application to assessment of the impact of the chestnut blight. *J. Environ. Manage.* **5**:161–179.

Smedes, G. W., and Hurd, L. E. (1981). An empirical test of community stability: resistance of a fouling community to a biological patch-forming disturbance. *Ecology* **62**:1561–1572.

Smith, D. W., Suffling, R., Stevens, D., and Dai, T. S. (1975). Plant community age as a measure of sensitivity of ecosystems to disturbance. *J. Environ. Manage.* **3**:271–286.

Smith, R. A., Slack, J. R., Wyant, T., and Lanfear, K. J. (1980). The oil spill risk analysis model of the U.S. Geological Survey. USGS Open-File Rep. 80–687.

Sousa, W. P. (1980). The responses of a community to disturbance: the importance of successional age and species life histories. *Oecologia* **45**:72–81.

Specht, R. L. (1963). Dark Island Heath (Ninety-Mile Plain, South Australia). VII: The effect of fertilizers on composition and growth, 1950–1960. *Aust. J. Bot.* **11**:67–94.

Specht, R. L. (1969). A comparison of the sclerophyllous vegetation characteristic of Mediterranean type climates in France, California, and southern Australia. II: Dry matter, energy, and nutrient accumulation. *Aust. J. Bot.* **17**:293–308.

Suffling, R. (1980). An index of ecological sensitivity to disturbance, based on ecosystem age, and related to landscape diversity. *J. Environ. Manage.* **10:** 253–262.

Suter, G. W., II. (1982). Terrestrial perturbation experiments for environmental assessment. *Environ. Manage.* **6**:43–54.

Sutherland, J. P. (1974). Multiple stable points in natural communities. *Am. Nat.* **108**:859–873.

Thatcher, A. C., and Westman, W. E. (1975). Succession following mining on high dunes of coastal southeast Queensland. In J. Kikkawa and H. A. Nix, eds. *Managing Terrestrial Ecosystems*. Proc. Ecol. Soc. Aust. 9, pp. 17–33.

Thorhaug, A., ed. (1979). *Restoration of Major Plant Communities in the United States*. Elsevier, Amsterdam.

Trabaud, L., and Lepart, J. (1981). Changes in the floristic composition of a *Quercus coccifera* L. garrigue in relation to different fire regimes. *Vegetatio* **46:**105–116.

U.S. Fish and Wildlife Service (1978). Catalog of Alaskan Seabird Colonies. Rep. FWS/OBS 78-78. Washington, D.C.

U.S. Forest Service (1970). Range Environmental Analysis Handbook. Region 3, Albuquerque, New Mexico. U. S. Forest Service Handbook, 2209.21,R-3. Washington, D.C.

van der Maarel, E. (1978). Experimental succession research in a coastal dune grassland; a preliminary report. *Vegetatio* **38:**21–28.

van der Maarel, E., and Werger, M. J. A. (1978). On the treatment of succession data. *Phytocoenosis* **7:**257–278.

Vankat, J. L. (1977). Fire and man in Sequoia National Park. *Ann. Assn. Am. Geogr.* **67:**17–27.

Veblen, T. T., and Ashton, D. H. (1978). Catastrophic influences on the vegetation of the Valdivian Andes, Chile. *Vegetatio* **36:**149–167.

Webb, L. J., Tracey, J. G., and Williams, W. T. (1972). Regeneration and pattern in the subtropical rainforests. *J. Ecol.* **60:**675–695.

Webster, J. R., Waide, J. B., and Patten, B. C. (1975). Nutrient recycling and the stability of ecosystems. In F. G. Howell, J. B. Gentry, and M. H. Smith, eds. *Mineral Cycling in Southeastern Ecosystems*. ERDA Conf-740513, NTIS, Springfield, Va., pp. 1–27.

West, D. C., McLaughlin, S. B., and Shugart, H. H. (1980). Simulated forest response to chronic air pollution stress. *J. Environ. Qual.* **9:**43–49.

West, D. C., Shugart, H. H., and Botkin, D. B., eds. (1981a) *Forest Succession: Concepts and Application*. Springer-Verlag, New York.

West, D. C., Shugart, H. H., and Botkin, D. B. (1981b). Introduction. In D. C. West, H. H. Shugart, and D. B. Botkin, eds. *Forest Succession: Concepts and Application*. Springer-Verlag, New York, pp. 1–6.

Westman, W. E. (1978). Measuring the inertia and resilience of ecosystems. *BioScience* **28:**705–710.

Westman, W. E. (1981). Diversity relations and succession in Californian coastal sage scrub. *Ecology* **62:**170–184.

Westman, W. E., Preston, K. P., and Weeks, L. B. (1985). Sulfur dioxide effects on the growth of native plants. In W. E. Winner, H. A. Mooney, and R. Goldstein, eds. *Sulfur Dioxide and Vegetation. Physiology, Ecology, and Policy Issues*. Stanford Univ. Press, Stanford.

Westman, W. E., and Rogers, R. W. (1977). Biomass and structure of a subtropical eucalypt forest, North Stradbroke Island. *Aust. J. Bot.* **25:**171–191.

Whittaker, R. H. (1953). A consideration of climax theory: the climax as a population and pattern. *Ecol. Monogr.* **23:**41–78.

Whittaker, R. H. (1970). *Communities and Ecosystems*. Macmillan, New York.

Whittaker, R. H. (1975a). *Communities and Ecosystems.* 2nd. ed. Macmillan, New York.

Whittaker, R. H. (1975b). The design and stability of plant communities. In W. H. van Dobben and R. H. Lowe-McConnell, eds. *Unifying Concepts in Ecology.* Junk, The Hague, pp. 169–181.

Whittaker, R. H., and Levin, S. A. (1977). The role of mosaic phenomena in natural communities. *Theor. Popul. Biol.* **12:**117–139.

Whittaker, R. H., and Woodwell, G. M. (1978). Retrogression and coenocline distance. In R. H. Whittaker, ed. *Ordination of Plant communities.* 2nd. ed. Junk, The Hague, pp. 51–70.

Winner, W. E., and Bewley, J. D. (1978). Contrasts between bryophytes and vascular plant synecological responses in an SO_2-stressed white spruce association in central Alberta. *Oecologia* **33:**311–325.

Woodwell, G. M. (1967). Radiation and the patterns of nature. *Science* **156:**461–470.

Woodwell, G. M. (1970). Effects of pollution on the structure and physiology of ecosystems. *Science* **168:**429–433.

Woodwell, G. M. (1975). The threshold problem in ecosystems. In S. A. Levin, ed. *Ecosystems Analysis and Prediction.* Soc. Industr. Appl. Math., Philadelphia, Pa., pp. 9–21.

Woodwell, G. M. (1979). Leaky ecosystems: nutrient fluxes and succession in the Pine Barrens Vegetation. In R. T. T. Forman, ed. *Pine Barrens: Ecosystem and Landscape.* Academic Press, New York, pp. 333–342.

Woodwell, G. M., and Gannutz, T. P. (1967). Effect of chronic gamma irradiation on lichen communities of a forest. *Amer. J. Bot.* **54:**1210–1215.

Woodwell, G. M., and Whittaker, R. H. (1968). Effects of chronic gamma irradiation on plant communities. *Quart. Rev. Biol.* **43:**42–55.

Wright, R. A. ed. (1978). *The Reclamation of Disturbed Arid Lands.* SWARM/ AAAS, Colorado Mtn. College, Glenwood Springs, Colo.

INDEX